P9-CLI-712

W. R. Briggs, J. L. Spudich (Eds.)

Handbook of Photosensory Receptors

Further Titles of Interest:

M. Futai, Y. Wada, J. H. Kaplan (Eds.)

Handbook of ATPases

Biochemistry, Cell Biology, Pathophysiology

2004

ISBN 3-527-30689-7

S. Frings, J. Bradley (Eds.)

Transduction Channels in Sensory Cells

2004

ISBN 3-527-30836-9

M. S. McDonald

Photobiology of Higher Plants

2003

ISBN 0-470-85522-3

Winslow R. Briggs, John L. Spudich (Eds.)

Handbook of Photosensory Receptors

WILEY-VCH Verlag GmbH & Co. KGaA

Editors

Prof. Dr. Winslow R. Briggs
Department of Plant Biology
Carnegie Institution of Washington
260 Panama St.
Stanford, CA 94305
USA

Prof. Dr. John L. Spudich
Center for Membrane Biology
Department of Biochemistry & Molecular Biology
University of Texas Medical School
6431 Fannin Street, MSB 6.130
Houston, TX 77030
USA

Cover illustration
The blue and green figures nearest the sun belong to sensory rhodopsin II and its transducer. A dimer of two transducers (green) is flanked by two rhodopsin receptor (blue). These figures were prepared by Chii-Shen Yang and John Spudich.
The single image farthest from the sun is the LOV2 domain of oat phototropin 1, and was provided by Shannon M. Harper, Lori C. Neil, and Kevin H. Gardner.

Library of Congress Card No.: Applied for

British Library Cataloguing-in-Publication Data:
A catalogue record for this book is available from the British Library.

Bibliographic information published by Die Deutsche Bibliothek
Die Deutsche Bibliothek lists this publication in the Deutsche Nationalbibliografie; detailed bibliographic data are available in the Internet at <http://dnb.ddb.de>.

Printed in the Federal Republic of Germany

Printed on acid-free paper

Typesetting TypoDesign Hecker GmbH, Leimen
Printing betz-druck gmbh, Darmstadt
Bookbinding Litges & Dopf Buchbinderei GmbH, Heppenheim

ISBN-13 978-3-527-31019-7
ISBN-10 3-527-31019-3

Table of Contents

Handbook of Photosensory Receptors. Edited by W. R. Briggs, J. L. Spudich

ISBN 3-527-31019-3

Preface

Light is the foremost carrier of information in the biological world. Photosensory receptors, elegant molecular machines that transduce the electromagnetic energy of photons into the chemical language of cells, deliver this information to organisms that use light patterns, intensity, color, direction and duration to analyze their environment and adjust their behavior accordingly. Light provides images to animal eyes, and controls movement, growth, differentiation and development, circadian timing, and a host of gene-expression responses in diverse organisms from the simplest unicellular microorganisms through higher plants and animals.

During the past decade there has been extraordinary progress in identifying and characterizing the photosensory receptors in a broad range of organisms. The rapid growth in this field inspired the formation of a new ongoing Gordon Research Conference on Photosensory Receptors and Signal Transduction, the first of which was held at Il Ciocco in Italy in the spring of 2000. Many of these receptors, like our own visual pigments, are proteins that use the chromophore retinal as the prosthetic group for light absorption, while others consist of very different types of proteins using as chromophores, flavins, tetrapyrrole, coumaric acid, pterin, and hypericin. All carry out photosensory transduction, transmuting packets of electromagnetic energy into a chemical form recognizable by cellular biochemical pathways.

The literature on the various different photoreceptors is widely scattered and this is the first book to bring together information on all of them in a single volume. In this book leading experts on photosensory receptors from diverse organisms – animals, plants, and prokaryotic and eukaryotic microbes – take a close look at current research and its exciting possibilities. We assembled the volume to stimulate an integrated view and appreciation of Nature's elegant mechanisms of photosensory reception.

September, 2004
Houston, Texas, USA — *John Spudich*
Stanford, California, USA — *Winslow Briggs*

Handbook of Photosensory Receptors. Edited by W. R. Briggs, J. L. Spudich

ISBN 3-527-31019-3

List of Authors

Najmoutin G. Abdulaev
Center for Advanced Research in Biotechnology
University of Maryland Biotechnology Institute
9600 Gudelsky Drive
Rockville, MD 20850
USA

Alfred Batschauer
Philipps-Universität
Biology – Plant Physiology/Photobiology
Karl-von-Frisch-Str. 8
35032 Marburg
Germany

Carl Bauer
Interdisciplinary Biochemistry Program
Indiana University
Myers Hall, Rm 150
915 E. Third St.
Bloomington, IN 47405–7170
USA

Roberto A. Bogomolni
Department of Chemistry and Biochemistry
University of California, Santa Cruz
Sinsheimer Labs
1156 High Street
Santa Cruz, CA 95064
USA

Winslow R. Briggs
Carnegie Institution of Washington
Department of Plant Biology
260 Panama St.
Stanford, CA 94305
USA

Anthony R. Cashmore
Plant Science Institute
Department of Biology
University of Pennsylvania
Philadelphia, PA 19104
USA

John M. Christie
University of Glasgow
Plant Science Group
Bower Building
Division of Biochemistry and Molecular Biology
Institute of Biomedical and Life Sciences
University Avenue
Glasgow, G12 8QQ
Scotland, UK

Wim Crielaard
University of Amsterdam
BioCentrum Amsterdam
Swammerdam Institute for Lifesciences
Laboratory for Microbiology
Nieuwe Achtergracht 166
1018 WV Amsterdam
The Netherlands

Handbook of Photosensory Receptors. Edited by W. R. Briggs, J. L. Spudich

ISBN 3-527-31019-3

Sean Crosson
Stanford University School of Medicine
Department of Developmental Biology
Beckman Center – Room 351
279 Campus Drive
Stanford, CA 94305
USA

Jay Dunlap
Dartmouth Medical School
Department of Genetics
7400 Remsen
Hanover, NH 03755–3844
USA

Thomas G. Ebrey
University of Washington
Department of Botany
Box 351800
Seattle, WA 98195–1800
USA

Russell G. Foster
Department of Visual and
Neuroscience
Imperial College, Faculty of Medicine
Charing Cross Hospital
Fulham Palace Road
London W6 8RF
UK

Mark W. Hankins
Department of Integrative and
Molecular Neuroscience
Imperial College, Faculty of Medicine
Division of Neuroscience and
Psychological Medicine
Charing Cross Hospital
Fulham Palace Road
London W6 8RF
UK

Klaas Hellingwerf
University of Amsterdam
BioCentrum Amsterdam
Swammerdam Institute for
Lifesciences
Laboratory for Microbiology
Nieuwe Achtergracht 166
1018 WV Amsterdam
The Netherlands

Johnny Hendriks
University of Amsterdam
BioCentrum Amsterdam
Swammerdam Institute for
Lifesciences
Laboratory for Microbiology
Nieuwe Achtergracht 166
1018 WV Amsterdam
The Netherlands

Enamul Huq
University of Texas at Austin
Section of Molecular Cell and
Developmental Biology, and Institute
Molecular Biology
University Station, A6700
Austin, TX 78712
USA

Mineo Iseki
Japan Science and Technology Agency
Precursory Research for Embryonic
Science and Technology
4-1-8 Honcho Kawaguchi,
Saitama, 332–0012
Japan
and
National Institute for Basic Biology
38 Nishigonaka, Myodaijicho, Okazaki,
Aichi 444–8505
Japan

Kwang-Hwan Jung
Department of Life Science
Sogang University
Mapo-Gu Shinsu-Dong 1
Seoul 121–742
Korea

Baruch Karniol
Department of Genetics
University of Wisconsin-Madison
445 Henry Mall
Madison, WI 53706
USA

Masato Kumauchi
University of Washington
Box 351800
Department of Botany
Seattle, WA 98195–1800
USA

J. Clark Lagarias
University of California at Davis
Section of Molecular and Cell Biology
One Shields Avenue
Davis, CA 95616
USA

Jennifer J. Loros
Dartmouth Medical School
Department of Genetics
7400 Remsen
Hanover, NH 03755–3844
USA

Shinji Masuda
Laboratory for Photobiology
RIKEN Photodynamics Research
Center
The Institute of Physical and Chemical
Research
6519–1399 Aramaki, Aoba
Sendai 980–0845
Japan

Ferenc Nagy
Biological Research Centre
Temesvari krt. 62
Szeged H-6726
Hungary

Pill-Soon Song
Kumho Life & Environmental
Laboratory
1 Oryong-dong, Buk-gu
Kwangju 500–712
Korea
or:
Department of Chemistry
University of Nebraska
Lincoln NE 68588
USA

Peter H. Quail
University of California, Berkeley
Department of Plant and Microbial
Biology
and
USDA/UCB Plant Gene Expression
Center
800 Buchanan St
Albany, CA 94710
USA

Kevin D. Ridge
Center for Membrane Biology
Department of Biochemistry and
Molecular Biology
University of Texas Health Science
Center
6431 Fannin Street, MSB 6.210
Houston, TX 77030
USA

Aziz Sancar
University of North Carolina School of Medicine
Department of Biochemistry and Biophysics
Mary Ellen Jones Building CB 7260
Chapel Hill, North Carolina 27599
USA

Eberhard Schäfer
Biologisches Institut II
University of Freiburg
Schänzlestrasse 1
79104 Freiburg
Germany

Thomas F. Schultz
The Scripps Research Institute
Department of Cell Biology
10550 North Torrey Pines Road
La Jolla, CA 92037
USA

Oleg A. Sineshchekov
Center for Membrane Biology
Department of Biochemistry & Molecular Biology
University of Texas Medical School
6431 Fannin Street, MSB 6.130
Houston, TX 77030
USA

John L. Spudich
Center for Membrane Biology
Department of Biochemistry & Molecular Biology
University of Texas Medical School
6431 Fannin Street, MSB 6.130
Houston, TX 77030
USA

Noriyuki Suetsugu
National Institute for Basic Biology
Division of Photobiology
Nishigonaka 38
Myodaiji, Okazaki 444–8585
Japan

Trevor E. Swartz
Department of Chemistry and Biochemistry
University of California, Santa Cruz
Sinsheimer Labs
Santa Cruz, CA 95064
USA

Shih-Long Tu
University of California at Davis
Section of Molecular and Cell Biology
One Shields Avenue
Davis, CA 95616
USA

Michael van der Horst
University of Amsterdam
BioCentrum Amsterdam
Swammerdam Institute for Lifesciences
Laboratory for Microbiology
Nieuwe Achtergracht 166
1018 WV Amsterdam
The Netherlands

Russell N. Van Gelder
Washington University Medical School
Departments of Ophthalmology and Visual Sciences
660 South Euclid Ave. CB 8096
St. Louis, Missouri 63110
USA

Richard D. Vierstra
Department of Genetics
University of Wisconsin-Madison
425-6 Henry Mall
Madison, WI 53706
USA

Jocelyne Vreede
University of Amsterdam
BioCentrum Amsterdam
Swammerdam Institute for
Lifesciences
Laboratory for Microbiology
Nieuwe Achtergracht 166
1018 WV Amsterdam
The Netherlands

Masamitsu Wada
Tokyo Metropolitan University,
Graduate School of Science,
Department of Biological Sciences,
Minamiosawa 1–1,
Hashioji-shi,
Tokyo, 192–0397
Japan

Masakatsu Watanabe
Graduate University for Advanced
Studies
Department of Photoscience,
School of Advanced Sciences
Shonan Village, Hayama,
Kanagawa, 240–0193
Japan
and
National Institute for Basic Biology
38 Nishigonaka, Myodaijicho, Okazaki,
Aichi, 444–8585
Japan

Sergei Yeremenko
University of Amsterdam
BioCentrum Amsterdam
Swammerdam Institute for
Lifesciences
Laboratory for Microbiology
Nieuwe Achtergracht 166
1018 WV Amsterdam
The Netherlands

1
Microbial Rhodopsins: Phylogenetic and Functional Diversity

John L. Spudich and Kwang-Hwan Jung

1.1
Introduction

The first 30 years of research on microbial rhodopsins concerned exclusively four proteins that share the cytoplasmic membrane of the halophilic archaeon *Halobacterium salinarum*, and a few very close homologs found in related haloarchaea. These four haloarchaeal types were the only microbial retinylidene proteins known prior to 1999: the light-driven ion pumps bacteriorhodopsin [BR (Oesterhelt and Stoeckenius, 1973)] and halorhodopsin [HR (Matsuno-Yagi and Mukohata, 1977; Schobert and Lanyi, 1982)], and the phototaxis receptors sensory rhodopsin I [SRI (Bogomolni and Spudich, 1982)], and sensory rhodopsin II [SRII (Takahashi et al., 1985)]. Studies of the haloarchaeal rhodopsins by the most incisive biophysical and biochemical tools available produced a wealth of information making them some of the best understood membrane-embedded proteins in terms of their structure–function relationships. Crystal structures of three [BR (Essen et al., 1998; Grigorieff et al., 1996; Luecke et al., 1999), HR (Kolbe et al., 2000), and SRII (Gordeliy et al., 2002; Kunji et al., 2001; Luecke et al., 2001; Royant et al., 2001)] reveal a common seven-transmembrane α-helical structure with nearly identical helix positions in the membrane, despite their differing functions and identity in only ~25% of their residues. The positions differ from those of visual pigments, as shown by the crystal structure of bovine rod rhodopsin (Palczewski et al., 2000), but their overall topologies are similar, namely the seven helices form an interior binding pocket in the hydrophobic core of the membrane for the retinal chromophore. In both the microbial and visual pigments, the retinal is attached by a protonated Schiff base linkage to a lysine in the middle of the seventh helix and retinal photoisomerization initiates their photochemical reactions.

Starting in 1999, genome sequencing of cultivated microorganisms began to reveal the previously unsuspected presence of archaeal rhodopsin homologs in several organisms in the other two domains of life, namely Bacteria and Eukarya (Bieszke et al., 1999a; Jung et al., 2003; Sineshchekov et al., 2002). Further, in 2001, "environmental genomics" of populations of uncultivated microorganisms in ocean plankton

Handbook of Photosensory Receptors. Edited by W. R. Briggs, J. L. Spudich

ISBN 3-527-31019-3

showed the presence of a homolog in marine proteobacteria [hence given the name proteorhodopsin (Beja et al., 2000)], which has swiftly expanded so far to ~800 relatives identified in samples throughout the world's oceans (Beja et al., 2001; de la Torre et al., 2003; Man et al., 2003; Man-Aharonovich et al., 2004; Sabehi et al., 2003; Venter et al., 2004). Microorganisms containing rhodopsin genes inhabit diverse environments including salt flats, soil, fresh water, surface and deep sea water, glacial sea habitats, and human and plant tissues as fungal pathogens. They comprise a broad phylogenetic range of microbial life, including haloarchaea, proteobacteria, cyanobacteria, fungi, dinoflagellates, and green algae. The conservation of residues, especially in the retinal-binding pocket, define a large phylogenetic class, called *type 1* rhodopsins to distinguish them from the visual pigments and related retinylidene proteins in higher organisms (*type 2* rhodopsins).

Analysis of the sequences of the new type 1 rhodopsins, their heterologous expression and study, and in some cases study of the photosensory physiology of the organisms containing them, and spectroscopic analysis of environmental samples, have shown that the newfound pigments fulfill both ion-transport and sensory functions, the latter with a variety of signal-transduction mechanisms. The purpose of this chapter is to summarize what we have learned regarding the rapidly expanding group of retinylidene pigments comprising the microbial rhodopsin family.

1.2
Archaeal Rhodopsins

Many laboratories have characterized the four rhodopsins from *H. salinarum* with a battery of techniques because they provide model systems for the two fundamental functions of membranes: active transport and sensory signaling. Comprehensive reviews on mechanisms of BR (Lanyi and Luecke, 2001), HR (Varo, 2000), and the SRs (Hoff et al., 1997; Spudich et al., 2000) are available. Sixteen variants of BR, HR, SRI and SRII have been documented in related halophilic archaea, such as *Natronomonas pharaonis* and *Haloarcula vallismortis* (Table 1.1).

Identification of members of the type-1 family has been based primarily on the conservation of residues in the retinal-binding pocket, which is known from the structures of haloarchaeal members. Atomic resolution structures, which exist for only a small number of membrane proteins, have been obtained from electron microscopy and X-ray crystallography of three of the archaeal rhodopsins: BR and HR from *H. salinarum*, and SRII from *N. pharaonis* ("NpSRII"). These proteins share a nearly identical positioning of seven transmembrane helices forming an interior pocket for the chromophore, *all-trans* retinal. The retinal binding pocket is comprised of residues from each of the seven helices, and it is the conservation of these residues that provides the most definitive identification of archaeal rhodopsin homologs in other organisms. Conservation outside of the pocket is sparse (Figure 1.1). Even between members of the archaeal branch, conservation outside the pocket is limited. For example, the phototaxis receptor NpSRII is only 27% identical to BR in amino acid sequence, and exhibits typically ~40% identity with other archaeal sensory

Table 1.1 List of microbial rhodopsins with database accession numbers or other sources. We show all microbial opsin genes found in the NCBI database except proteorhodopsins, for which ~800 have been identified (see text). We selectively present a subset of proteorhodopsins for which absorption maxima have been published. BR, bacteriorhodopsin; HR, halorhodopsin; SRI & SRII, sensory rhodopsins I and II; PR, proteorhodopsin; NOPI, *Neurospora* opsin I; CSRA & CSRB, *Chlamydomonas* sensory rhodopsins A and B.

Species and Name	Accession Number	Comments
Archaea		
Haloarcula argentinensis BR	D31880	H^+ pump
Haloarcula japonica BR	AB029320	H^+ pump
Haloarcula sp. (Andes) BR	S76743	H^+ pump
Haloarcula vallismortis BR	D31882	H^+ pump
Haloarcula vallismortis HR	D31881	Cl^- pump
Haloarcula vallismortis SRI	D83748	phototaxis
Haloarcula vallismortis SRII	Z35308	phototaxis
Halobacterium marismotui BR	–	H^+ pump by homology with BR, Victor Ng, pers. com.
Halobacterium marismotui HR	–	Cl^- pump by homology with HR, Victor Ng, pers. com.
Halobacterium marismotui SRI	–	phototaxis,Victor Ng, pers. com.
Halobacterium marismotui SRII	–	phototaxis,Victor Ng, pers. com.
Halobacterium marismotui BR-2	–	unknown,Victor Ng, pers. com.
Halobacterium marismotui SRI-2	–	unknown,Victor Ng, pers. com.
Halobacterium salinarum BR	V00474	λ_{max} = 568 nm, H^+ pump
Halobacterium salinarum HR	D43765	λ_{max} = 576 nm, Cl^- pump
Halobacterium salinarum SRI	L05603	λ_{max} = 587 nm, phototaxis (attractant/repellent)
Halobacterium salinarum SRII	U62676	λ_{max} = 487 nm, phototaxis (repellent)
Halobacterium salinarum mex BR	D11056	H^+ pump
Halobacterium salinarum mex HR	P33970	Cl^- pump
Halobacterium salinarum port BR	D11057	H^+ pump
Halobacterium salinarum port HR	Q48315	Cl^- pump
Halobacterium salinarum shark BR	D11058	H^+ pump
Halobacterium salinarum shark HR	D43765	Cl^- pump
Halobacterium sp. AUS-1 BR	J05165	H^+ pump
Halobacterium sp. AUS-1 SRII	AB059748	phototaxis
Halobacterium sp. AUS-2 BR	S56354	H^+ pump
Halobacterium sp. NRC-1 BR	NP_280292	H^+ pump
Halobacterium sp. NRC-1 HR	NP_279315	Cl^- pump
Halobacterium sp. NRC-1 SRI	AAG19914	λ_{max} = 587 nm, phototaxis (attractant/repellent)
Halobacterium sp. NRC-1 SRII	AAG19988	λ_{max} = 487 nm, phototaxis (repellent)
Halobacterium sp. SG1 BR	X70291	H^+ pump
Halobacterium sp. SG1 HR	X70292	Cl^- pump

Table 1.1 Continued

Species and Name	Accession Number	Comments
Archaea		
Halobacterium sp. SG1 SRI	X70290	phototaxis
Halorubrum sodomense BR	D50848	H^+ pump
Halorubrum sodomense HR	AB009622	Cl^- pump
Halorubrum sodomense SRI	AB009623	phototaxis
Haloterrigena sp. Arg-4 BR	AB009620	H^+ pump
Haloterrigena sp. Arg-4 HR	AB009621	Cl^- pump
Natronomonas pharaonis HR	J05199	Cl^- pump
Natronomonas pharaonis SRII	Z35086	λ_{max} = 497 nm, phototaxis (repellent)
Eubacteria		
Anabaena sp. PCC7120	AP003592	Also known as *Nostoc*, λ_{max} 543 nm, photosensory
Gloeobacter violaceus PCC 7421	NP_923144	Unicellular cyanobacterium
Magnetospirillum magnetotacticum	–	genome.ornl.gov/microbial/mmag
γ-proteobacterium (BAC31A8)	AF279106	λ_{max} = 527 nm, H^+ pump; GPR
γ-proteobacterium (HOT75m4)	AF349981	λ_{max} = 490 nm, H^+ pump; BPR
γ-proteobacterium (HOT0m1)	AF349978	λ_{max} = 518 nm
γ-proteobacterium (PalE6)	AAK30200	λ_{max} = 490 nm
γ-proteobacterium (eBac64A5)	AAK30175	λ_{max} = 519 nm
γ-proteobacterium (eBac40E8)	AAK30174	λ_{max} = 519 nm
γ-proteobacterium (RSr6a5a6)	AAO21455	λ_{max} = 540 nm
γ-proteobacterium (RS23)	AAO21449	λ_{max} = 528 nm
γ-proteobacterium (RSr6a5a2)	–	λ_{max} = 505 nm, RSr6a5a6(V105E), from Oded Béjà
Fungi		
Botrytis cinerea	AL115930	
Botryotinia fuckeliana	–	cogeme.ex.ac.uk
Cryptococcus neoformans	CF192410	www.genome.ou.edu/cneo.html, Basidiomycetes
Fusarium sporotrichioides	BI187800	
Fusarium graminearum	BU067691	
Leptosphaeria maculans	AF290180	
Mycosphaerella graminicola	AW180117	two homologs are present
Neurospora crassa NR	AF135863	λ_{max} = 534 nm
Triticum aestivum	CA747087	
Ustilago maydis	CF642219	Basidiomycetes
Algae		
Chlamydomonas reinhardtii CSRA	AF508965	photomotility for high light intensity
Chlamydomonas reinhardtii CSRB	AF508966	photomotility for low light intensity
Pyrocystis lunula	AF508258	dinoflagellate
Guillardia theta	AW342219	cryptomonad
Acetabularia acetabulum	CF259014	green alga

rhodopsins; all 4 archaeal rhodopsins exhibit ~80% identity in the 22 residues that the crystal structures show form the retinal binding pockets in BR, HR, and NpSRII. 55–75% identity in these 22 residues is also found in the new rhodopsins (Figure 1.1).

The functions of the four archaeal rhodopsins have been well characterized. BR (λ_{max} = 568 nm) and HR (λ_{max} = 576 nm) are light-driven ion pumps for protons and chloride, respectively, absorbing maximally in the green–orange region of the spectrum (Oesterhelt, 1998; Varo, 2000). Their electrogenic transport cycles provide energy to the cell under conditions in which respiratory electron transport activity is low. Accordingly, their production in the cells is induced when oxygen is depleted in late exponential/early stationary phase cultures. Both BR and HR hyperpolarize the membrane to generate a positive outside membrane potential, thereby creating inwardly directed proton motive force. HR further contributes to pH homeostasis by hyperpolarizing the membrane by electrogenic chloride uptake rather than proton ejection, thereby providing an electrical potential for net proton uptake especially important in alkaline conditions.

SRI and SRII are phototaxis receptors controlling the cell's swimming behavior in response to changes in light intensity and color (Hoff et al., 1997). SRI (λ_{max} = 587 nm) is also induced in cells in late exponential/early stationary phase, and attracts the cells to orange light useful to the transport rhodopsins. SRI is unique among known photosensory receptors in that it produces opposite signals (attractant and repellent) depending on the wavelengths of stimulating light. To avoid guiding the cells into light containing harmful near-UV radiation, SRI uses its color-discriminating mechanism to ensure the cells will be attracted to orange light only if that light is not accompanied by near-UV wavelengths (Spudich and Bogomolni, 1984). The mechanism is based on photochromic reactions of the protein. If SRI absorbs a single photon (maximal absorption in the orange) it produces a photointermediate species called SRI-M or S_{373} (λ_{max} = 373 nm) that is interpreted by the cells' signal transduction machinery as an attractant signal. However, if S_{373} is photoexcited, it generates a strongly repellent-signaling photointermediate. Therefore single-photon excitation of SRI, such as occurs in orange light, attracts the cells, whereas two-photon excitation, as occurs in white light, repels the cells. SRII absorbs in the mid-visible range (λ_{max} = 487 nm in *H. salinarum* SRII) and appears to serve only a repellent function (Takahashi et al., 1990). It is the only rhodopsin in *H. salinarum* produced in cells during vigorous aerobic grow when light is not being used for energy and is therefore best avoided because of possible photooxidative damage.

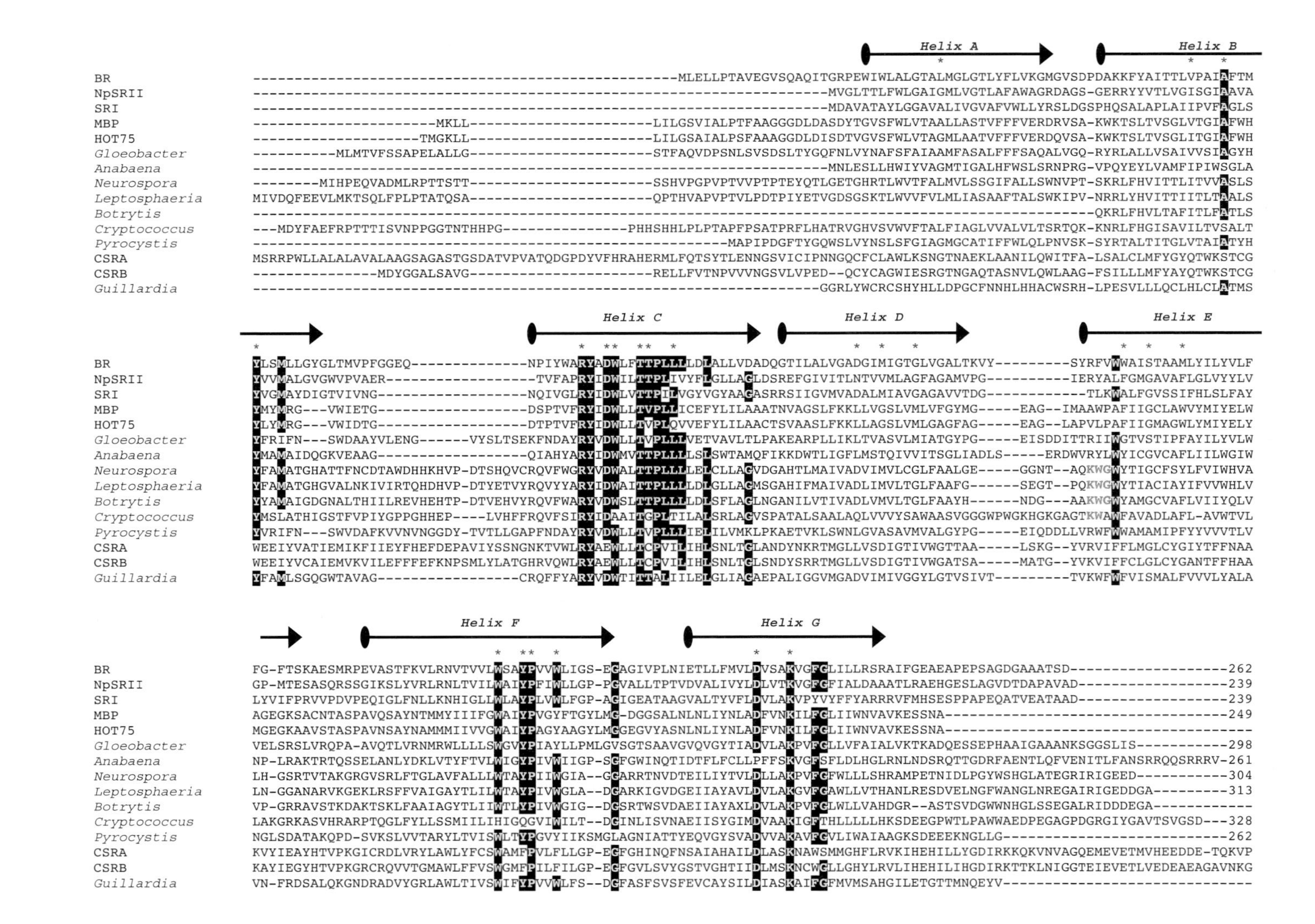
Helix A
Helix B
Helix C
Helix D
Helix E
Helix F
Helix G
BR
NpSRII
SRI
MBP
HOT75
Gloeobacter
Anabaena
Neurospora
Leptosphaeria
Botrytis
Cryptococcus
Pyrocystis
CSRA
CSRB
Guillardia

1.3 Clues to Newfound Microbial Rhodopsin Function from Primary Sequence Comparison to Archaeal Rhodopsins

A challenge posed by the newfound microbial rhodopsin genes is to identify the photochemical and physiological function of the proteins in the cells containing them, some of which are uncultivated microorganisms. Success has been obtained in several cases: as detailed below, the Monterey Bay surface water-proteorhodopsin functions as a light-driven proton pump for the γ-proteobacterium SAR86 in its native marine environment (Beja et al., 2001; de la Torre et al., 2003), and therefore its physiological function is similar to that of BR in *H. salinarum*. In contrast, the *Anabaena* (*Nostoc*) rhodopsin (Jung et al., 2003) and *Chlamydomonas reinhardtii* pigments CSRA and CSRB (Sineshchekov et al., 2002) have been demonstrated to serve photosensory rather than transport functions, and therefore are functionally more similar to the archaeal SRI and SRII. These known transport proteins and sensory proteins do not group separately in phylogenetic analyses (Figure 1.2); therefore the phylogenetic trees do not permit assignment of particular sequences as encoding transport or sensory proteins. Some individual residue differences, however, provide a clue, as do the photochemical reaction cycle kinetics of the proteins.

One difference in the primary sequence between BR, HR and the SRs stands out. Asp96 in BR functions as a proton donor, returning a proton to the Schiff base from the cytoplasmic side of the protein during the pumping cycle. This proton transfer improves the pumping efficiency of BR by accelerating the decay of its unprotonated Schiff base photocycle intermediate, M, and is present in all BR homologs in the haloarchaea. In the sensory rhodopsins, the corresponding M intermediates are signaling states of the receptor proteins (demonstrated unequivocally only for HsSRI), and longer M lifetimes increase the signaling efficiencies of the receptors. Accordingly, each of the five known haloarchaeal sensory rhodopsin sequences lacks a carboxylate residue at the position corresponding to Asp96 and contains Tyr or Phe instead. The residue corresponding to Asp85, which is the proton acceptor from the Schiff base, is a carboxylate residue in BR, SRI, and SRII and in each of the newly

◀ **Figure 1.1** Primary sequence comparison of 15 microbial rhodopsins. We selected several opsin genes from each domain of life. ***Archaea***- BR: *Halobacterium salinarum* bacteriorhodopsin, SRI: *H. salinarum* sensory rhodopsin I, NpSRII: *Natronomonas pharaonis* sensory rhodopsin II; ***Bacteria***- GPR: γ-proteobacterium (BAC31A8) proteorhodopsin, BPR: γ-proteobacterium (HOT75m4) proteorhodopsin, *Gloeobacter*: microbial rhodopsin from *Gloeobacter violaceus* PCC 7421, *Anabaena*: sensory rhodopsin from *Anabaena* (*Nostoc*) *sp.* PCC7120; ***Eukarya***- **Fungi**- rhodopsin from *Neurospora crassa*, *Leptosphaeria maculans*, *Botrytis cinerea*, and *Cryptococcus neoformans*, **Algae**- *Pyrocyctis*: rhodopsin from *Pyrocystis lunula*, CSRA & CSRB: *Chlamydomonas reinhardtii* sensory rhodopsins A and B (N-terminal portions), *Guillardia*: rhodopsin from *Guillardia theta*. Conserved residues are marked with black background and the 22 residues in the retinal-binding pocket are marked with asterisks. Bacteriorhodopsin Asp85 and Asp96 in helix C and corresponding residues in the other pigments are marked with blue background (see text). Red-colored KWG residues on helix E are nearly completely conserved in fungal rhodopsins.

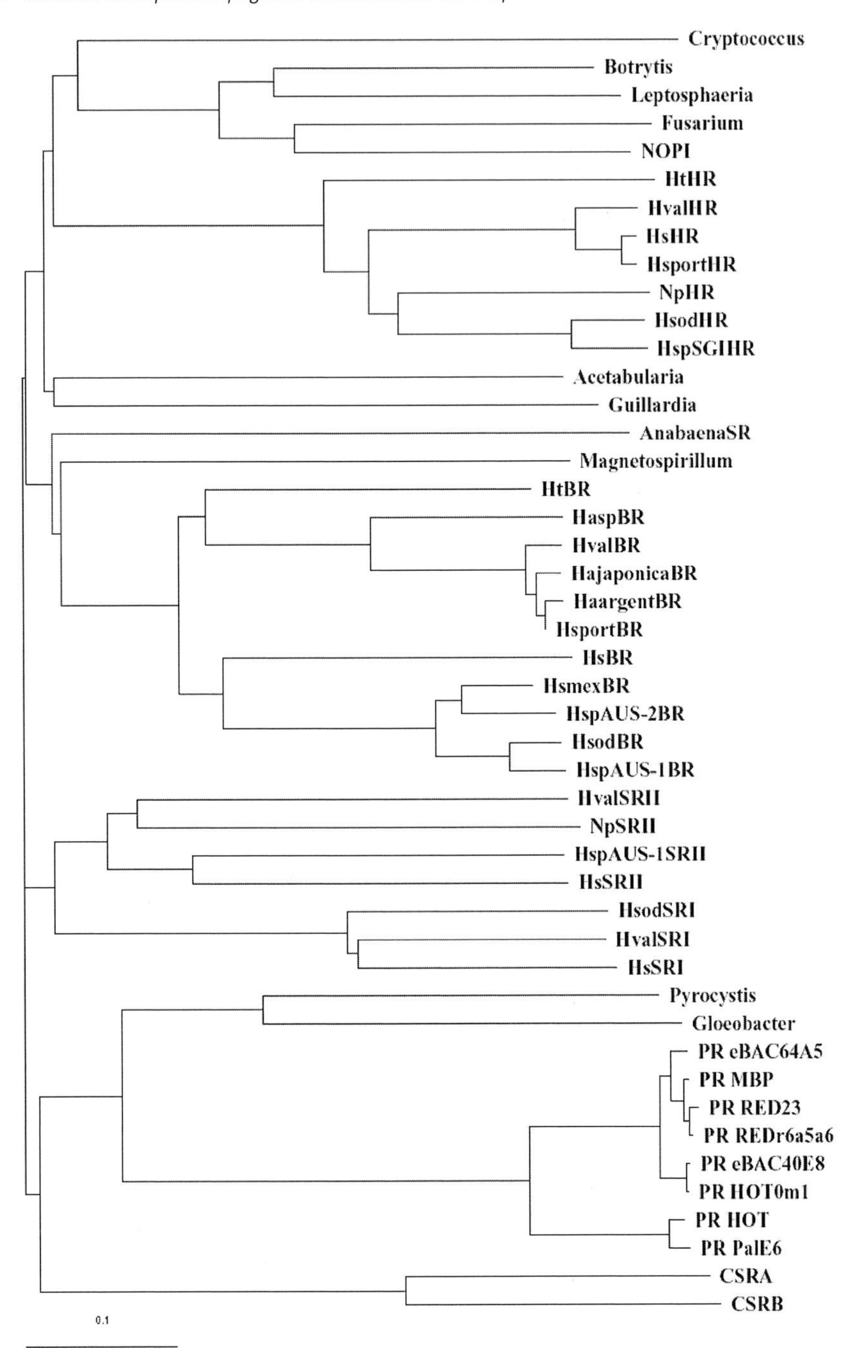
Cryptococcus
Botrytis
Leptosphaeria
Fusarium
NOPI
HtHR
HvalHR
HsHR
HsportHR
NpHR
HsodHR
HspSGIHR
Acetabularia
Guillardia
AnabaenaSR
Magnetospirillum
HtBR
HaspBR
HvalBR
HajaponicaBR
HaargentBR
HsportBR
HsBR
HsmexBR
HspAUS-2BR
HsodBR
HspAUS-1BR
HvalSRII
NpSRII
HspAUS-1SRII
HsSRII
HsodSRI
HvalSRI
HsSRI
Pyrocystis
Gloeobacter
PR eBAC64A5
PR MBP
PR RED23
PR REDr6a5a6
PR eBAC40E8
PR HOT0m1
PR HOT
PR PalE6
CSRA
CSRB
0.1

◀ **Figure 1.2** Phylogenetic tree of microbial rhodopsins. A neighbor-joining tree was constructed from CLUSTALX(1.81) alignment of 46 microbial rhodopsin apoproteins. The tree was constructed using the full-length sequences, except in the case of *Chlamydomonas* sensory opsins, in which only the rhodopsin domains were used (CsoA, 378 N-terminal residues; CsoB, 303 N-terminal residues). Scale represents number of substitutions per site (0.1 indicates 10 nucleotide substitutions per 100 nucleotides). 1000 bootstrap replicates were performed to determine the reliability of the tree topology. The tree was drawn using TreeView1.6.6. Abbreviations: NOPI (*Neurospora crassa* opsin I), HtHR (*Haloterrigena sp.* halorhodopsin), HvalHR (*Haloarcula vallismortis* halorhodopsin), HsHR (*Halobacterium salinarum* halorhodopsin), HsportHR (*Halobacterium salinarum port* halorhodopsin), NpHR (*Natronomonas pharaonis* halorhodopsin), HsodHR (*Halorubrum sodomense* halorhodopsin), HspSGIHR (*Halobacterium sp.* SG1 halorhodopsin), AnabaenaSR (*Anabaena (Nostoc) sp.* PCC7120 sensory rhodopsin), HtBR (*Haloterrigena sp.* bacteriorhodopsin), HaspBR (*Haloarcula sp.* bacteriorhodopsin), HvalBR (*Haloarcula vallismortis* bacteriorhodopsin), HajaponicaBR (Haloarcula japonica bacteriorhodopsin), HaargentBR (*Haloarcula argentinensis* bacteriorhodopsin), HsportBR (*Halobacterium salinarum port* bacteriorhodopsin), HsBR (*Halobacterium salinarum* bacteriorhodopsin), HsmexBR (*Halobacterium salinarum* mex bacteriorhodopsin), HspAUS-2BR (*Halobacterium sp. AUS-2* bacteriorhodopsin), HsodBR (*Halorubrum sodomense* bacteriorhodopsin), HspAUS-1BR (*Halobacterium sp. AUS-1* bacteriorhodopsin), HvalSRII (*Haloarcula vallismortis* sensory rhodopsin II), NpSRII (*Natronomonas pharaonis* sensory rhodopsin II), HspAUS-1SRII (*Halobacterium sp. AUS-1* sensory rhodopsin II), HsSRII (*Halobacterium salinarum* sensory rhodopsin II), HsodSRI (*Halorubrum sodomense* sensory rhodopsin I), HvalSRI (*Haloarcula vallismortis* sensory rhodopsin I), HsSRI (*Halobacterium salinarum* sensory rhodopsin I), PR eBAC64A5 (γ-proteobacterium proteorhodopsin), PR MBP (GPR; γ-proteobacterium proteorhodopsin BAC21A8), CSRA and CSRB (*Chlamydomonas reinhardtii* sensory rhodopsins A and B, respectively).

identified rhodopsins (Figure 1.1). HR does not produced an unprotonated Schiff base intermediate in its photocycle, and therefore does not contain a carboxylic acid residue in either of the positions corresponding to Asp96 and Asp85.

Notably the SAR86 proteorhodopsin demonstrated to be a light-driven proton pump does contain a carboxylate (Glu108) at the position corresponding to Asp96 in BR, and moreover Glu108 has been shown to participate in the reprotonation of the Schiff base in the latter half of the photocycle as does Asp96 in BR (Dioumaev et al., 2002; Wang et al, 2003). On the basis of available information, the presence of the carboxylate residue at this position appears to be a necessary, but not a sufficient condition for identification of a new rhodopsin as a proton pump. *Neurospora* rhodopsin also contains a glutamate at this position, but extensive analysis of the photoactivity of the protein expressed in *Pichia pastoris,* as well as purified and reconstituted into liposomes, reveals a non-transport photocycle with kinetics indicative of a sensory rhodopsin (Bieszke et al., 1999b). A caveat is that one cannot be certain that the protein is folded correctly when expressed heterologously, although the absorption spectrum in the visible range and photochemical reactivity of the expressed protein when reconstituted with *all-trans* retinal provides some assurance.

The demonstrated sensory rhodopsins, namely the archaeal SRI and SRII proteins, the *Anabaena* rhodopsin, and *Chlamydomonas* CSRA and CSRB, all lack a carboxylate residue at the homologous position of the BR Schiff base proton donor Asp96, while containing the carboxylate Schiff base proton-acceptor residue corresponding to

Asp85 in BR (Figure 1.1). Hence the absence of a carboxylate in the donor position in *Cryptococcus neoformans* (alanine in the corresponding position) and in a marine proteorhodopsin sequence recently deposited in GenBank (gi|42850614|gb|EAA92632.1, threonine in the corresponding position) strongly suggest sensory functions for these proteins.

More than 10-fold faster photocycling rates distinguish the archaeal transport from the sensory pigments; the first-found sensory rhodopsin, archaeal SRI, in fact was initially called "slow-cycling rhodopsin" for this reason (Bogomolni and Spudich, 1982; Spudich et al., 1995). The transport rhodopsins are characterized by photocycles typically <30 ms, whereas sensory rhodopsins are slow-cycling pigments with photocycle halftimes typically >300 ms (Spudich et al., 2000). This large kinetic difference is functionally important since a rapid photocycling rate is advantageous for efficient ion pumping, whereas a slower cycle provides more efficient light detection because signaling states persist for longer times.

The photocycle rate difference, which holds firm for the archaeal rhodopsins, may in some cases not be a definitive criterion for assigning a transport versus sensory function. The *Anabaena* rhodopsin which has been concluded to be a sensory protein based on other criteria has a photocycle half-time of 110 ms (Jung et al., 2003), intermediate between archaeal transport and sensory proteins. Furthermore, a deep sea proteorhodopsin from a Hawaiian Ocean-Time station plankton sample from 75 m depth exhibits a light-driven proton transport cycle that is ~10-fold slower (60 ms in cells) than that of the Monterey Bay proteorhodopsin (Wang et al., 2003). The slower photocycling rate of the deep sea pigment is explained as an adaptation to the ~10-fold decreased photon flux rate available to the BPR visible absorption band at 75 m.

1.4 Bacterial Rhodopsins

1.4.1 Green-absorbing Proteorhodopsin ("GPR") from Monterey Bay Surface Plankton

Among the most abundant and widely distributed of the type 1 rhodopsins are the proteorhodopsins, the first of which was identified by genomic analysis of marine proteobacteria in plankton from Pacific coastal surface waters. The proteorhodopsin gene was the first found to encode a eubacterial homolog of the archaeal rhodopsins and was revealed by BAC library construction and sequencing of naturally occurring marine bacterioplankton from Monterey Bay (Beja et al., 2000). The gene was functionally expressed in *Escherichia coli* and bound retinal to form an active, light-driven proton pump. The rRNA sequence on the same DNA fragment identified the organism as an uncultivated γ-proteobacterium (the SAR86 group), and the expressed protein was named proteorhodopsin. Phylogenetic comparison with archaeal rhodopsins placed proteorhodopsin on an independent long branch (Figure 1.2). The new pigment, designated GPR (λ_{max} = 525 nm), exhibited a photochemical reaction cycle with intermediates and kinetics characteristic of archaeal proton-pumping rhodopsins. Its transport, spectroscopic, and photochemical reactions have now been characterized by a number of laboratories in *Escherichia coli*-expressed forms (Beja et al., 2000; Beja et al., 2001; Dioumaev et al., 2002; Dioumaev et al., 2003; Friedrich et al., 2002; Krebs et al., 2002; Lakatos et al., 2003; Man et al., 2003). The efficient proton pumping and rapid photocycle (15 ms halftime) of the new pigment strongly suggested that proteorhodopsin functions as a proton pump in its natural environment. Asp97 and Glu108 in GPR function as Schiff base proton acceptor and donor carboxylate residues during the GPR pumping cycle, analogous to Asp85 and Asp96, respectively, at the corresponding positions in BR (Dioumaev *et al.*, 2002; Wang et al., 2003).

The next step was examination of the plankton samples directly for the newfound protein's activity. Retinylidene pigmentation with photocycle characteristics identical to that of the *E. coli*-expressed proteorhodopsin gene was demonstrated by flash spectroscopy in membranes prepared from Monterey Bay picoplankton (Beja et al., 2001). Estimated from laser flash-induced absorbance changes, a high density of proteorhodopsin in the SAR86 membrane is indicated, arguing for a significant role of the protein in the physiology of these bacteria. The flash photolysis results provided direct physical evidence for the existence of proteorhodopsin-like pigments and endogenous retinal molecules in the prokaryotic fraction of the Monterey Bay coastal surface waters, and provide compelling evidence that GPR functions as a light-driven proton pump photoenergizing SAR86 cells in their natural environment. Furthermore, the amplitude of the flash-photolysis signals permit a rough estimate of the total rate of solar energy conversion to proton motive force by marine proteorhodopsins; assuming for the calculation that the Monterey Bay sample has the average PR content, the conversion rate is on the order of 10^{13}–10^{14} W, a globally significant contribution to the biosphere.

Since the initial finding of GPR, a wide variety of similar genes has been identified in picoplankton from very different ocean environments: the Antarctic, Central North Pacific, Mediterranean Sea, Red Sea, and the Atlantic Ocean (Beja et al., 2001; de la Torre et al., 2003; Man et al., 2003; Man-Aharonovich et al., 2004; Sabehi et al., 2003). Genes have been isolated from both surface and deep-water samples, and both coastal and open-sea areas. New members from the PR family were recently reported to be found also in marine α-proteobacteria (de la Torre et al., 2003), and based on whole genome "shotgun sequencing" of microbial populations collected *en mass* on tangential flow and impact filters from sea water samples collected from the Sargasso Sea near Bermuda, a remarkable 782 different partial sequences homologous to proteorhodopsins were identified (Venter et al., 2004). Thus, microbial rhodopsin abundance and diversity within marine environments appears to be large.

1.4.2
Blue-absorbing Proteorhodopsin ("BPR") from Hawaiian Deep Sea Plankton

One of the variant groups (designated clade II) of proteorhodopsin genes, differing by ~22% in predicted primary structure from the clade I group defined by the GPR gene and its close relatives, was detected in both the Antarctic and in 75-m deep ocean plankton from Hawaiian waters (Beja et al., 2001). The Antarctic and Hawaiian PR genes when expressed in *E. coli* exhibit a blue-shifted absorption spectrum (λ_{max} = 490 nm; hence referred to as "BPR") with vibrational fine structure, unlike the unstructured spectrum of GPR (Beja et al., 2001). The stratification of the surface GPR and 75-m BPR with depth is in accordance with light spectral quality at these depths (Beja et al., 2001).

The different absorption spectra of GPR and BPR have provided an opportunity to examine "spectral tuning" in two rhodopsins with closely similar primary sequence. One of the most notable distinguishing properties of retinal among the various chromophores used in photosensory receptors is the large variation of its absorption spectrum depending on interaction with the apoprotein ("spectral tuning") (Birge, 1990; Ottolenghi and Sheves, 1989). In rhodopsins, retinal is covalently attached to the ε-amino group of a lysine residue forming a protonated retinylidene Schiff base. In methanol a protonated retinylidene Schiff base exhibits a λ_{max} = of 440 nm. The protein microenvironment shifts the λ_{max} [the "opsin shift" (Yan et al., 1995)] to longer wavelengths, e.g. to 527 nm in GPR and to 490 nm in BPR. With structural modelling comparisons and mutagenesis, a single residue difference in the retinal binding pockets at position 105 (Leu in GPR and Gln in BPR) was found to function as a spectral tuning switch and to account for most of the spectral difference between the two pigment families (Man et al., 2003). The mutations at position 105 almost completely interconverted the absorption spectra of BPR and GPR. GPR L105Q shifted to the blue and acquired vibrational fine structure like wild-type BPR, and BPR Q105L shifted to the red and lost the fine structure exhibiting spectra similar to those of GPR. Among both type 1 and type 2 rhodopsins the mechanisms of spectral tuning in general are still not well understood in physical chemical terms and the Q/L switch stands out as a simple spectral tuning model amenable to investigation. Spectral tuning is discussed in more detail in Section 1.6, below.

Another difference between GPR and BPR is their photocycle halftimes, 6.5 ms and 60 ms respectively in *E. coli* cells (Wang et al., 2003). The difference in photocycle rates and their different absorption maxima may be explained as an adaptation to the different light intensities in their respective marine environments, based on measured spectral distributions of intensities of solar illumination at the ocean surface and at various depths (Jerlov, 1976). Taking into account the blue shift of BPR, matching the lower photon fluence rate from solar radiation requires a 10-fold slower photocycle in BPR than in GPR. Therefore there is no selective pressure for a photocycle faster than that of BPR at that depth.

BPR may function to energize cells by light-driven electrogenic proton pumping, as does GPR. However, the contribution of solar energy capture from BPR is severely limited by the low light intensities in deep waters. This consideration raises the possibility of a regulatory rather than energy harvesting function of BPR, based either on its slow proton pumping or by yet unidentified protein–protein interaction with transducers in its native membrane.

1.4.3
Anabaena Sensory Rhodopsin

A rhodopsin pigment in a cyanobacterium established that sensory rhodopsins also exist in eubacteria. A gene encoding a homolog of the archaeal rhodopsins was found via a genome-sequencing project of *Anabaena (Nostoc) sp.* PCC7120 at Kazusa Institute (http://www.kazusa.or.jp/). The opsin gene was expressed in *E. coli,* and bound *all-trans* retinal to form a pink pigment (λ_{max} = 543 nm) with a photochemical reaction cycle containing an M-like photointermediate and 110 ms half-life at pH 6.8 (Jung et al., 2003).

The opsin gene was found in the genome to be adjacent to another open reading frame separated by 16 base pairs under the same promoter. This operon is predicted to encode a 261-residue protein (the opsin) and a 125-residue (14 kDa) protein. The rate of the photocycle is increased ~20% when the *Anabaena* rhodopsin and the soluble protein are co-expressed in *E. coli* (Jung et al., 2003), indicating physical interaction between the two proteins. Binding of the 14-kDa protein to *Anabaena* rhodopsin was confirmed by affinity-enrichment measurements and Biacore interaction analysis. The pigment did not exhibit detectable proton transport activity when expressed in *E. coli,* and Asp96, the proton donor of BR, is replaced with Ser86 in *Anabaena* rhodopsin. These observations are compelling that *Anabaena* opsin functions as a photosensory receptor in its natural environment, and strongly suggest that the 125-residue cytoplasmic soluble protein transduces a signal from the receptor, unlike the archaeal sensory rhodopsins which transmit signals by transmembrane helix–helix interactions with integral membrane transducers (Figure 1.3).

Chimeric constructs have established that the archaeal sensory rhodopsins SRI and SRII transmit signals to their cognate membrane-embedded taxis transducers by interaction with the transducers, transmembrane helices and a short membrane proximal domain (Jung et al., 2001; Zhang et al., 1999) and an extensive membrane-embedded transducer-binding region has been observed in the X-ray structure of *N. pharaonis* SRII (Luecke et al., 2001) co-crystallized with its taxis transducer fragment (Gordeliy et al., 2002; see also Spudich 2002). The interaction of the soluble 14-kDa protein, likely to be a signal transducer, with *Anabaena* rhodopsin, therefore would extend the range of signal transduction mechanisms used by microbial sensory rhodopsins.

Atomic resolution structures for the *Anabaena* pigment and its putative transducer have been obtained (Vogeley et al., 2004), but the physiological function of *Anabaena* SR has not been established. Several photophysiological responses of *Anabaena* with unidentified photosensory receptor(s) have been discussed (Jung et al., 2003; Mullineaux, 2001). One of these is light-modulation of the pigments contained in the light-harvesting complex of *Anabaena*, a photoresponse called chromatic adaptation. Green light such as absorbed by *Anabaena* rhodopsin has been found to modulate

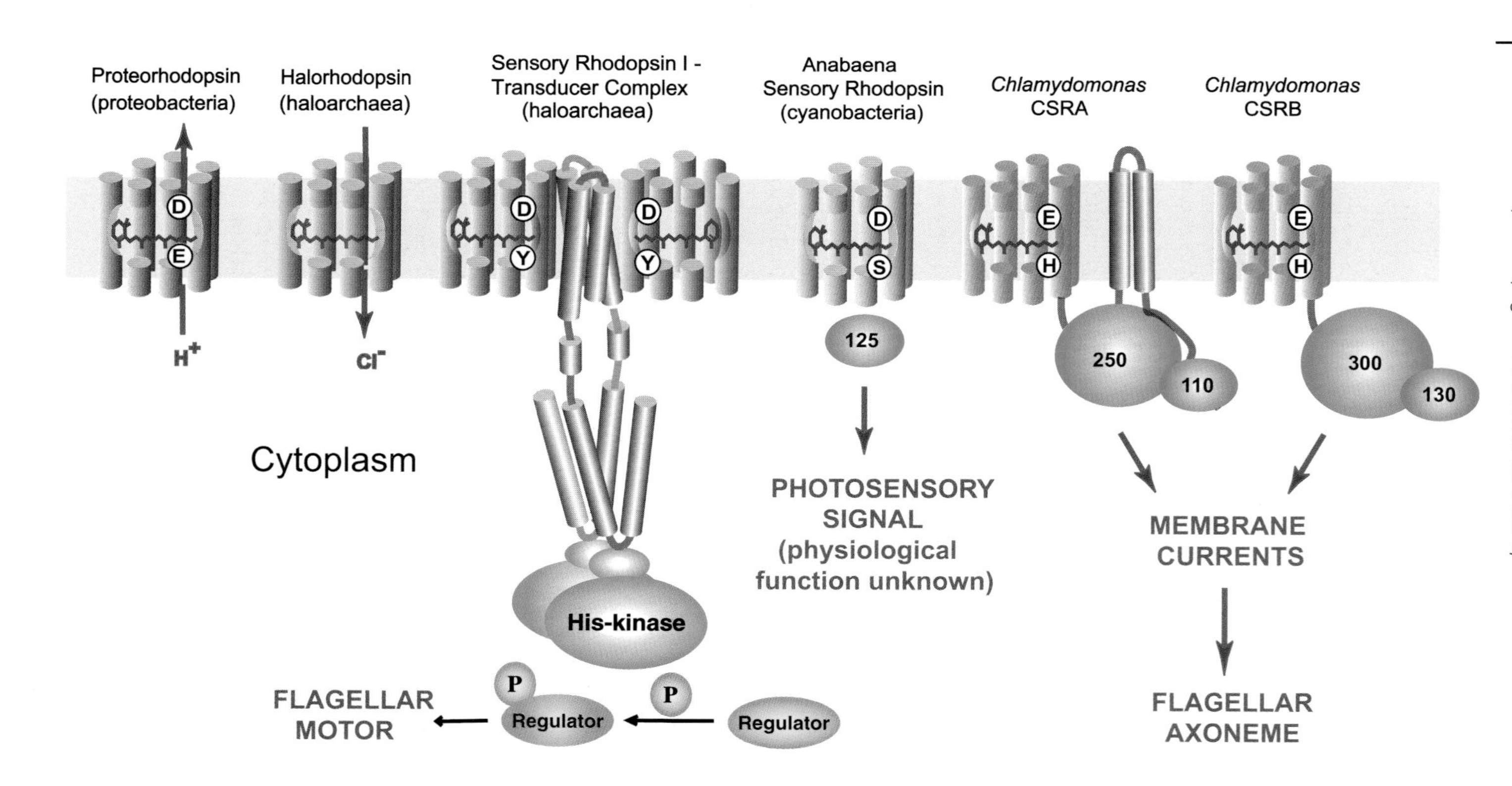
Proteorhodopsin
(proteobacteria)
H^+
Halorhodopsin
(haloarchaea)
Cl^-
Sensory Rhodopsin I -
Transducer Complex
(haloarchaea)
Anabaena
Sensory Rhodopsin
(cyanobacteria)
Chlamydomonas
CSRA
Chlamydomonas
CSRB
125
250
110
300
130
Cytoplasm
PHOTOSENSORY
SIGNAL
(physiological
function unknown)
MEMBRANE
CURRENTS
His-kinase
P
Regulator
FLAGELLAR
MOTOR
FLAGELLAR
AXONEME

complementary chromatic adaptation in the related cyanobacteria *Calothrix* and *Fremyella* (Grossman et al., 2001). The adaptation consists of differential biosynthesis of blue-absorbing phycoerythrins and red-absorbing phycocyanins depending on light quality. The genome of *Anabaena sp.* 7120 contains phycoerythrin subunits α and β (*pecA* and *B*), allophycocyanin subunits α and β (*apcA* and *B*), and phycocyanin subunits α and β (*cpcA* and *B*). Green light (optimally 540 nm) promotes phycoerythrin synthesis whereas red light (optimally 650 nm) promotes phycocyanin synthesis (Kehoe and Grossman, 1994). A phytochrome would be an attractive candidate for a red light sensor, and 3 phytochrome homologs are present in the *Anabaena sp.* 7120 genome. The *Anabaena* rhodopsin (λ_{max} = 543 nm) is a candidate for the green light sensor, and may function alone to discriminate color via its photochronic reactions (Vogeley et al., 2004). The two major biliproteins in *Synechocystis sp.* 6830 are phycocyanin (λ_{max} = 617 nm) and allophycocyanin (λ_{max} = 650 nm) (Toole et al., 1998). Interestingly, the phycoerythrin gene which is regulated by green light is missing in the genome of *Synechocystis* which does not contain an opsin-encoding gene.

1.4.4
Other Bacterial Rhodopsins

1.4.4.1 *Magnetospirillum*

Genome sequencing of the α-proteobacterium *Magnetospirillum magnetotacticum* revealed a microbial opsin gene (Table 1), proceeded by a homolog of *brp*, which encodes an enzyme for synthesis of retinal from β-carotene in *H. salinarum* (Peck et al., 2001). The two gene-operon contains a single promoter.

1.4.4.2 *Gloeobacter violaceus* PCC 7421

Gloeobacter is a unicellular cyanobacterium and the complete genome of this strain has been sequenced (Nakamura et al., 2003). There is only one type 1 rhodopsin gene, which unlike that of *Anabaena*, encodes a carboxylate (Glu) at the BR D96 position, suggesting a proton-pumping function.

◄ **Figure 1.3** Functional diversity among microbial rhodopsins. Domains of the two sensory rhodopsins from *Chlamydomonas reinhardtii*, CSRA and CSRB, based on secondary structure predictions, compared with those for proteorhodopsin, halorhodopsin and sensory rhodopsin I (in a dimeric complex with its cognate dimeric transducer) from haloarchaea and cyanobacterial sensory rhodopsin from *Anabaena*. The retinal chromophore (shown in red) is covalently linked to a conserved lysine residue in the seventh transmembrane helix in each protein. Colors shown are approximately the color of the pigments. Residues in helix C of proteorhodopsin that are important for proton translocation, Asp-97 and Glu-108, and the amino acid differences at their corresponding positions in the other rhodopsins are highlighted. The corresponding residues are not shown for halorhodopsin (see text) to avoid the impression that they are on the chloride translocation path. Transducer domains or proteins shown in green are presumably involved in the post-receptor signal transduction processes. For the cytoplasmic 14-kDa protein associated with ASR and for CSRA and CSRB, the numbers indicated correspond to the number of amino acid residues in each module.

1.5
Eukaryotic Microbial Rhodopsins

1.5.1
Fungal Rhodopsins

A genome sequencing project on the filamentous fungus *Neurospora crassa* revealed the first of the eukaryotic homologs, designated NOP-1 (Bieszke et al., 1999a), and search of genome databases currently in progress indicates the presence of archaeal rhodopsin homologs in various fungi including plant and human pathogens – Ascomycetes: *Botrytis cinerea, Botryotinia fuckeliana* (anamorph *Botrytis cinerea*), *Fusarium sporotrichioides, Gibberella zeae* (anamorph *Fusarium graminearum*), *Leptosphaeria maculans, Mycosphaerella graminicola* (two opsin homologs) and Basidiomycetes: *Cryptococcus neoformans* and *Ustilago maydis*. Each of these organisms contains genes predicted to encode proteins with the retinal-binding lysine in the seventh helix and high identity in the retinal-binding pocket. The Asp Schiff base counter ion and proton acceptor (Asp85 in BR) is conserved among all fungal opsin homologs and the carboxylate proton donor specific to proton pumps (Asp96 in BR) is also either Asp or Glu except in *Cryptococcus* which contains an Ala residue.

The *nop-1* gene was heterologously expressed in the yeast *Pichia pastoris*, and it encodes a membrane protein that forms with *all-trans* retinal a green light-absorbing pigment (λ_{max} = 534 nm) with a spectral shape and bandwidth typical of rhodopsins (Bieszke et al., 1999b). Laser-flash kinetic spectroscopy of the retinal-reconstituted NOP-1 pigment (i.e. *Neurospora* rhodopsin) in *Pichia* membranes reveals that it undergoes a seconds-long photocycle with long-lived intermediates spectrally similar to intermediates detected in BR and other members of the type 1 family.

The physiological function of *Neurospora* rhodopsin has not yet been identified. Based on the long lifetime of the intermediates in its photocycle and its apparent lack of ion-transport activities [at least when heterologously expressed (Brown et al., 2001)], it seems likely to serve as a sensory receptor for one or more of the several different light responses exhibited by the organism, such as photocarotenogenesis or light-enhanced conidiation. *Neurospora* is non-motile, but phototaxis by zoospores of the motile fungus *Allomyces reticulatus* has been shown to be retinal-dependent (Saranak and Foster, 1997) and therefore photomotility modulation is a likely photosensory function of rhodopsins in this particular fungal species.

A blue-green light-induced photocycle in *Cryptococcus neoformans* native membranes has been detected and confirmed as deriving from the rhodopsin pigment by its absence in an opsin gene-deletion mutant. The photocycle is typical of a microbial rhodopsin exhibiting a blue-shifted intermediate characteristic of a deprotonated Schiff base species and a 100–150 ms half-life (pH 7.0, 25°C) (authors, unpublished).

1.5.2
Algal Rhodopsins

The rhodopsins of the green alga *Chlamydomonas reinhardtii* are the only ones in eukaryotic microbes to have an identified physiological function, namely photoreception controlling motility behavior (Sineshchekov et al., 2002). Two type 1 opsin genes were identified in the *C. reinhardtii* genome. A microbial rhodopsin homolog gene is also present in *Guillardia theta,* which is a small biflagellate organism considered both a protozoan and an alga, and opsin genes are also found in the dinoflagellate *Pyrocystis lunula* and *Acetabularia acetabulum* which is a unicellular green alga of the order Dasycladales found in warm waters of sheltered lagoons. The *CSOA & CSOB* (**C***hlamydomonas* **s**ensory **o**psin A and B) genes encode 712 and 737 amino acid proteins (Figure 1.1). The N-terminal 300 residues have a significant homology to archaeal rhodopsins with seven transmembrane helices and the conserved retinal binding pocket (Figure 1.1). The *Chlamydomonas* rhodopsins provide the first examples of evolution fusing the microbial rhodopsin motif with other domains.

Early work established that *Chlamydomonas* uses retinylidene receptors for photomotility responses. Restoration of photomotility responses by retinal addition to a pigment-deficient mutant of *C. reinhardtii* first indicated a retinal-containing photoreceptor (Foster et al., 1984). Subsequent *in vivo* reconstitution studies with retinal analogs prevented from isomerizing around specific bonds ("isomer-locked retinals") in several laboratories further established that the *Chlamydomonas* rhodopsins governing phototaxis and the photophobic response have the same isomeric configuration (*all-trans*), photoisomerization across the C13–C14 double bond (*all-trans* to *13-cis*), and 6-*s-trans* ring-chain conformation (co-planar) as the archaeal rhodopsins (Hegemann et al., 1991; Lawson et al., 1991; Sakamoto et al., 1998; Sineshchekov et al., 1994; Takahashi et al., 1991).

The proteins encoded by *cso*A and *cso*B complexed with retinal (called CSRA and CSRB) are the first of the eukaryotic archaeal-type rhodopsins for which we can assign physiological roles (Sineshchekov et al., 2002). RNA*i* suppression of the genes established that CSRA and CSRB mediate both phototaxis (Sineshchekov et al., 2002) and photophobic reactions (Govorunova et al., 2004) to high- and low-intensity light, respectively. The functions of the two rhodopsins were demonstrated by analysis of electrical currents and motility responses in transformants with RNA*i* directed against each of rhodopsin genes. CSRA has an absorption maximum near 510 nm and mediates a fast photoreceptor current that saturates at high light intensity. In contrast, CSRB absorbs maximally at 470 nm and generates a slow current saturating at low light intensity (Sineshchekov et al., 2002). The rhodopsin domains of CSRA (Nagel et al., 2002) and CSRB (Nagel et al., 2003) have been shown to exhibit light-induced proton-channel activity in *Xenopus* oocytes. The relationship of this activity to their control of motility-regulating currents in *C. reinhardtii* is not clear. A more detailed review of the sensory rhodopsins in this organism, including 3 additional genome-predicted sensory rhodopsins, CSRC, CSRD, and CSRE, called also cop5, 6, and 7 (Kateriya et al., 2004), appears in this volume (Sineshchekov and Spudich, Chapter 2).

To clarify a possibly confusing series of reports in the literature, we mention here a protein that binds radiolabeled retinal, and named on this basis "chlamyrhodopsin," that had been isolated from *Chlamydomonas* eyespot preparations (Deininger et al., 1995). For several years this most abundant protein in the eyespot membranes was assumed and often cited by the authors of that work as the photoreceptor for photomotile responses. However, its gene-predicted primary sequence, as well as that of a similar *Volvox* protein (Ebnet et al., 1999), suggest 2–4 transmembrane helices and no homology to archaeal opsins, nor is a photoactive retinal binding site evident from the sequences. Moreover, recently the "chlamyrhodopsin" has been ruled out as the photoreceptor pigment for either phototaxis or photophobic responses in *C. reinhardtii* (Fuhrmann et al., 2001).

1.6 Spectral Tuning

Comparison of the primary sequence (Figure 1.1) alone gives hints to distinguishing properties of different microbial rhodopsins, but most properties cannot be deduced from primary structure alone. The Leu/Gln spectral-tuning switch at position 105 in GPR and BPR discussed above was revealed through structural modeling and mutagenesis. However, we were fortunate that a relatively simple single-residue switch is responsible for most of the color difference in that case, and, more generally, detailed knowledge of atomic structure will probably be required to elucidate spectral tuning mechanisms of most microbial rhodopsins. An exemplary case is that of NpSRII (from *Natronomonas pharaonis*) which is unusual in that its maximal absorption is shifted 70–90 nm to the blue of the other archaeal pigments (Tomioka and Sasabe, 1995). Mutagenic substitution of 10 residues, in or near the retinal-binding pocket with their corresponding BR residues, produced only a 28-nm red shift of the NpSRII absorption maximum (Kamo et al., 2001; Shimono et al., 2000). Structural differences responsible for the shift are evident in the 2.4-Angstrom resolution structure (Luecke et al., 2001). One notable change is a displacement of the guanidinium group of Arg^{72} by 1.1 Å coupled with a rotation away from the Schiff base in NpSRII. This increase in distance reduces the influence of Arg^{72} on the counterion, thus strengthening the Schiff base/counterion interaction, shifting the absorption to shorter wavelengths. In addition the position of the positive charge destabilizes the excited state contributing further blue shift (Ren et al., 2001). Arg^{72} is repositioned as a consequence of several factors, including movement of its helix backbone by 0.9 Å and the cavity created by changes from BR: $Phe^{208} \rightarrow Ile^{197}$, $Glu^{194} \rightarrow Pro^{183}$, and $Glu^{204} \rightarrow Asp^{192}$. Hence the spectral tuning results from precise positioning of retinal binding-pocket residues and the guanidinium of Arg^{72}, which could not be deduced from primary structure, but required atomic resolution tertiary structure information.

1.7
A Unified Mechanism for Molecular Function?

The idea that in microbial rhodopsins the sensory signaling mechanisms result from evolution "tweaking" the transport mechanism was suggested by the observation that SRI carries out light-driven proton pumping, but only when it is free of its transducer (Bogomolni et al., 1994; Olson and Spudich, 1993; Spudich and Spudich, 1993; Spudich, 1994). The HtrI protein was found to close or prevent the opening of a cytoplasmic proton-conducting channel in SRI during its photocycle. This finding led to the notion that a chemotaxis receptor progenitor of HtrI evolved an interaction with a proton-transporter progenitor of SRI, coupling to its pumping mechanism and thereby blocking the pump and converting the transport rhodopsin to a sensory receptor. NpSRII was also observed under some conditions to exhibit light-driven proton transport which was also prevented by its interaction with its transducer, HtrII (Schmies et al., 2001; Sudo et al., 2001).

That the transducer-inhibition of light-driven transport occurs in both haloarchaeal rhodopsins further supports that the interaction blocking the transport is a critical aspect of the signaling mechanism. A tilting of helices, primarily helix F, contributes to opening a cytoplasmic channel in the latter half of the photocycle in BR (Subramaniam and Henderson, 2000). A unifying mechanism is that ion transport and sensory signaling use the same retinal-driven protein structural changes, which is the conformational change that opens the cytoplasmic channel in the proton transport cycle. The key feature of the model is that consequences of retinal photoisomerization, including light-induced disruption of the salt bridge between the protonated Schiff base on helix G and aspartyl counterion on helix C (Spudich et al., 1997), triggers tilting of helix F to which the Htr transmembrane helices are coupled. Supporting this mechanism are (i) the proton-pumping by the sensory rhodopsins, (ii) its inhibition by transducer interaction discussed above, and (iii) light-induced tilting of helix F in *N. pharaonis* SRII, concluded from site-directed spin-labeling measurements (Wegener et al., 2000). Furthermore, (iv) genetic evidence supports helix-F involvement in signaling (Jung and Spudich, 1998).In addition, (v) substitution of the Schiff base counterion Asp85 with asparagine induces helix-F tilting in the dark in BR and the corresponding substitution in *H. salinarum* SRII partially activates the receptor in the dark (Spudich et al., 1997). Finally (vi), helix F interacts with the two transmembrane helices of the HtrII fragment co-crystallized with NpSRII (Gordeliy et al., 2002).

Another prediction is that, since in the model SR helix tilting is transmitted to the Htr protein by direct helix–helix contacts, alterations in structure must occur in the Htr transmembrane domains between the receptor interaction sites and the cytoplasmic domain of the transducer, where the activity of the bound histidine kinase is controlled. Such structural alterations have been detected as light-induced changes in interactions between spin labels introduced into the NpHtrII transmembrane helices (Wegener et al., 2000) and changes of disulfide bond formation rates between engineered cysteines (Yang and Spudich, 2001). In both studies the data indicate that the second transmembrane segment (TM2) of NpHtrII is more conformationally active than TM1. The authors of the site-directed spin-labeling study further suggest a signal-transfer mechanism in which TM2 undergoes a rotary motion in response to the helix F tilt in the photoactivated receptor (Wegener et al., 2000), and interaction of the receptor's E-F loop with the membrane proximal domain of the transducer has been implicated in signal transfer (Chen and Spudich 2004; Yang et al., 2004).

In summary, the evidence is compelling that the conformational changes in transport and sensory rhodopsins in haloarchaea share essential features despite their differing functions. For study of the newfound microbial rhodopsins, an important question is whether the light-induced conformational change observed in BR and strongly implicated in the haloarchaeal sensory rhodopsin photocycles is a key conserved feature of their functional mechanisms.

1.8
Opsin-related Proteins without the Retinal-binding Site

Several other genes in the fungi *N. crassa (YRO2), Aspergillus nidulans, Saccharomyces cerevisiae, Schizosaccharomyces pombe, Coccidioides immitis, Coriolus versicolor,* and in the plant *Sorghum bicolor* (sorghum) encode proteins that exhibit significant homology to type 1 rhodopsins, but are missing the critical lysine residue in the 7th helix that forms the covalent linkage with retinal. The microbial-opsin-related proteins are therefore not likely to form photoactive pigments with retinal. The most conserved region in these proteins is along helix C, E, and the middle of helix F. It is intriguing that one of the yeast opsin-related proteins, HSP30 (heat shock protein 30), is implicated as interacting with a proton transport protein, the H^+ATPase. HSP30 downregulates stress stimulation of H^+ATPase activity under heat shock conditions (Piper et al., 1997; Zhai et al., 2001). It may be that the conformational switching properties of the archaeal rhodopsins have been preserved in these opsin-related proteins, while the photoactive site has been lost and its function replaced by another input module such as a protein–protein interaction domain. It is striking that opsin-related proteins lacking the retinal-binding lysine have been observed so far only in fungi and not in any of the many other classes of organisms containing type 1 rhodopsins.

1.9
Perspective

Type 1 rhodopsins are present in all three domains of life, and therefore progenitors of these proteins may have existed in early evolution before the divergence of archaea, eubacteria, and eukaryotes. If so, light-driven ion transport as a means of obtaining cellular energy may well have predated the development of photosynthesis, and represent one of the earliest means by which organisms tapped solar radiation as an energy source. As more rhodopsins are identified, their evolution and dissemination into such a wide variety of organisms, whether by divergence from a common progenitor or horizontal gene transfer, should become clearer.

There is a much work to be done to understand the physiological roles and molecular mechanisms of the rhodopsins so far identified in the various microbial species. It seems likely that we will see even more members of this family as genomic sequencing becomes ever more rapid. The vast majority of microbial species have never been cultivated in a laboratory. Therefore the use of microbial rhodopsin probes in environmental genomics, which expands the search for homologous genes to uncultivated organisms, is likely to be especially fruitful.

Acknowledgements

We thank Elena Spudich for stimulating discussions. The work by the authors referred to in this review was supported by grants from the National Institutes of Health, National Science Foundation, Human Frontiers Science Program, and the Robert A. Welch Foundation.

References

Béjà, O., Aravind, L., Koonin, E.V., Suzuki, M.T., Hadd, A., Nguyen, L.P., Jovanovich, S.B., Gates, C.M., Feldman, R.A., Spudich, J.L., Spudich, E.N., and DeLong, E.F. (2000) *Science* 289, 1902–1906.

Béjà, O., Spudich, E.N., Spudich, J.L., Leclerc, M., and DeLong, E.F. (2001) *Nature* 411, 786–789.

Bieszke, J.A., Braun, E.L., Bean, L.E., Kang, S., Natvig, D.O., and Borkovich, K.A. (1999a) *Proc Natl Acad Sci USA* 96, 8034–8039.

Bieszke, J.A., Spudich, E.N., Scott, K.L., Borkovich, K.A., and Spudich, J.L. (1999b) *Biochemistry* 38, 14138–14145.

Birge, R.R. (1990) *Annu Rev Phys Chem* 41, 683–733.

Bogomolni, R.A., and Spudich, J.L. (1982) *Proc Natl Acad Sci USA* 79, 6250–6254.

Bogomolni, R.A., Stoeckenius, W., Szundi, I., Perozo, E., Olson, K.D., and Spudich, J.L. (1994) *Proc Natl Acad Sci USA* 91, 10188–10192.

Brown, L.S., Dioumaev, A.K., Lanyi, J.K., Spudich, E.N., and Spudich, J.L. (2001) *J Biol Chem* 276, 32495–32505.

Chen, X. and Spudich, J.L. (2004) *J Biol Chem* 279, 42964–42969.

de la Torre, J.R., Christianson, L.M., Beja, O., Suzuki, M.T., Karl, D.M., Heidelberg, J., and DeLong, E.F. (2003) *Proc Natl Acad Sci USA* 100, 12830–12835.

Deininger, W., Kroger, P., Hegemann, U., Lottspeich, F., and Hegemann, P. (1995) *EMBO J* 14, 5849–5858.

Dioumaev, A.K., Brown, L.S., Shih, J., Spudich, E.N., Spudich, J.L., and Lanyi, J.K. (2002) *Biochemistry* 41, 5348–5358.

Dioumaev, A.K., Wang, J.M., Balint, Z., Varo, G., and Lanyi, J.K. (2003) *Biochemistry* 42, 6582–6587.

Ebnet, E., Fischer, M., Deininger, W., and Hegemann, P. (1999) *Plant Cell* 11, 1473–1484.

Essen, L., Siegert, R., Lehmann, W.D., and Oesterhelt, D. (1998) *Proc Natl Acad Sci USA* 95, 11673–11678.

Foster, K.W., Saranak, J., Patel, N., Zarilli, G., Okabe, M., Kline, T., and Nakanishi, K. (1984) *Nature* 311, 756–759.

Friedrich, T., Geibel, S., Kalmbach, R., Chizhov, I., Ataka, K., Heberle, J., Engelhard, M., and Bamberg, E. (2002) *J Mol Biol* 321, 821–838.

Fuhrmann, M., Stahlberg, A., Govorunova, E., Rank, S., and Hegemann, P. (2001) *J Cell Sci* 114, 3857–3863.

Gordeliy, V.I., Labahn, J., Moukhametzianov, R., Efremov, R., Granzin, J., Schlesinger, R., Buldt, G., Savopol, T., Scheidig, A.J., Klare, J.P., and Engelhard, M. (2002) *Nature* 419, 484–487.

Govorunova, E.G., Jung, K.H., Sineshchekov, O.A., and Spudich, J.L. (2004) *Biophys J* 86, 2342–2349.

Grigorieff, N., Ceska, T.A., Downing, K.H., Baldwin, J.M., and Henderson, R. (1996) *J Mol Biol* 259, 393–421.

Grossman, A.R., Bhaya, D., and He, Q. (2001) *J Biol Chem* 276, 11449–11452.

Hegemann, P., Gartner, W., and Uhl, R. (1991) *Biophys J* 60, 1477–1489.

Hoff, W.D., Jung, K.H., and Spudich, J.L. (1997) *Annu Rev Biophys Biomol Struct* 26, 223–258.

Jerlov, N.G. (1976) *in Marin Optics*. Amsterdam, Oxford, New York, Elsevier.

Jung, K.H. and Spudich, J.L. (1998) *J. Bacteriology* 180, 2033–2042.

Jung, K.H., Spudich, E.N., Trivedi, V.D., and Spudich, J.L. (2001) *J Bacteriology* 183, 6365–6371.

Jung, K.H., Trivedi, V.D., and Spudich, J.L. (2003) *Mol Microbiol* 47, 1513–1522.

Kamo, N., Shimono, K., Iwamoto, M., and Sudo, Y. (2001) *Biochemistry (Mosc)* 66, 1277–1282.

Kateriya, S., Nagel, G., Bamberg, E., and Hegemann, P. (2004) *News Physiol. Sci.* 19, 133–137.

Kehoe, D.M., and Grossman, A.R. (1994) *Semin Cell Biol* 5, 303–313.

Kolbe, M., Besir, H., Essen, L.O., and Oesterhelt, D. (2000) *Science* 288, 1390–1396.

Krebs, R.A., Alexiev, U., Partha, R., DeVita, A.M., and Braiman, M.S. (2002) *BMC Physiol* 2, 5.

Kunji, E.R., Spudich, E.N., Grisshammer, R., Henderson, R., and Spudich, J.L. (2001) *J Mol Biol* 308, 279–293.

Lakatos, M., Lanyi, J.K., Szakacs, J., and Varo, G. (2003) *Biophys J* 84, 3252–3256.

Lanyi, J.K., and Luecke, H. (2001) *Curr Opin Struct Biol* 11, 415–419.

Lawson, M.A., Zacks, D.N., Derguini, F., Nakanishi, K., and Spudich, J.L. (1991) *Biophys J* 60, 1490–1498.

Luecke, H., Schobert, B., Richter, H.T., Cartailler, J.P., and Lanyi, J.K. (1999) *J Mol Biol* 291, 899–911.

Luecke, H., Schobert, B., Lanyi, J.K., Spudich, E.N., and Spudich, J.L. (2001) *Science* 293, 1499–1503.

Man, D., Wang, W., Sabehi, G., Aravind, L., Post, A.F., Massana, R., Spudich, E.N., Spudich, J.L., and Béjà, O. (2003) *EMBO J* 22, 1725–1731.

Man-Aharonovich, D., Sabehi, G., Sineshchekov, O.A., Spudich, E.N., Spudich, J.L., and Béjà, O. (2004) *Photochem Photobiol Sci* 3, 459–462.

Matsuno-Yagi, A., and Mukohata, Y. (1977) *Biochem Biophys Res Commun* 78, 237–243.

Mullineaux, C.W. (2001) *Mol Microbiol* 41, 965–971.

Nagel, G., Ollig, D., Fuhrmann, M., Kateriya, S., Musti, A.M., Bamberg, E., and Hegemann, P. (2002) *Science* 296, 2395–2398.

Nagel, G., Szellas, T., Huhn, W., Kateriya, S., Adeishvili, N., Berthold, P., Ollig, D., Hegemann, P., and Bamberg, E. (2003) *Proc Natl Acad Sci USA* 100, 13940–13945.

Nakamura, Y., Kaneko, T., Sato, S., Mimuro, M., Miyashita, H., Tsuchiya, T., Sasamoto, S., Watanabe, A., Kawashima, K., Kishida, Y., Kiyokawa, C., Kohara, M., Matsumoto, M., Matsuno, A., Nakazaki, N., Shimpo, S., Takeuchi, C., Yamada, M., and Tabata, S. (2003) *DNA Res* 10, 137–145.

Oesterhelt, D., and Stoeckenius, W. (1973) Functions of a new photoreceptor membrane. *Proc Natl Acad Sci USA* 70, 2853–2857.

Oesterhelt, D. (1998) *Curr Opin Struct Biol* 8, 489–500.

Olson, K.D., and Spudich, J.L. (1993) *Biophys J* 65, 2578–2585.

Ottolenghi, M., and Sheves, M. (1989) *J Membr Biol* 112, 193–212.

Palczewski, K., Kumasaka, T., Hori, T., Behnke, C.A., Motoshima, H., Fox, B.A., Le Trong, I., Teller, D.C., Okada, T., Stenkamp, R.E., Yamamoto, M., and Miyano, M. (2000) *Science* 289, 739–745.

Peck, R.F., Echavarri-Erasun, C., Johnson, E.A., Ng, W.V., Kennedy, S.P., Hood, L., DasSarma, S., and Krebs, M.P. (2001) *J Biol Chem* 276, 5739–5744.

Piper, P.W., Ortiz-Calderon, C., Holyoak, C., Coote, P., and Cole, M. (1997) *Cell Stress Chaperones* 2, 12–24.

Ren, L., Martin, C.H., Wise, K.J., Gillespie, N.B., Luecke, H., Lanyi, J.K., Spudich, J.L., and Birge, R.R. (2001) *Biochemistry* 40, 13906–13914.

Sabehi, G., Massana, R., Bielawski, J.P., Rosenberg, M., Delong, E.F., and Beja, O. (2003) *Environ Microbiol* 5, 842–849.

Sakamoto, M., Wada, A., Akai, A., Ito, M., Goshima, T., and Takahashi, T. (1998) *FEBS Lett* 434, 335–338.

Saranak, J., and Foster, K.W. (1997) *Nature* 387, 465–466.

Schmies, G., Engelhard, M., Wood, P.G., Nagel, G., and Bamberg, E. (2001) *Proc Natl Acad Sci USA* 98, 1555–1559.

Schobert, B., and Lanyi, J.K. (1982) *J Biol Chem* 257, 10306–10313.

Shimono, K., Iwamoto, M., Sumi, M., and Kamo, N. (2000) *Photochem Photobiol* 72, 141–145.

Sineshchekov, O.A., Govorunova, E.G., Der, A., Keszthelyi, L., and Nultsch, W. (1994) *Biophys J* 66, 2073–2084.

Sineshchekov, O.A., Jung, K.H., and Spudich, J.L. (2002) *Proc Natl Acad Sci USA* 99, 8689–8694.

Spudich, E.N., and Spudich, J.L. (1993) *J Biol Chem* 268, 16095–16097.

Spudich, E.N., Zhang, W., Alam, M., and Spudich, J.L. (1997) *Proc Natl Acad Sci U S A* 94, 4960–4965.

Spudich, J.L., and Bogomolni, R.A. (1984) *Nature* 312, 509–513.

Spudich, J.L. (1994) *Cell* 79, 747–750.

Spudich, J.L., Zacks, D.N., and Bogomolni, R.A. (1995) *Isr J Photochem* 35, 495–513.

Spudich, J.L., Yang, C.S., Jung, K.H., and Spudich, E.N. (2000) *Annu Rev Cell Dev Biol* 16, 365–392.

Spudich, J.L. (2002) *Nature Struct. Biol* 9, 797–799.

Subramaniam, S., and Henderson, R. (2000) *Nature* 406, 653–657.

Sudo, Y., Iwamoto, M., Shimono, K., and Kamo, N. (2001) *Photochem Photobiol* 74, 489–494.

Takahashi, T., Tomioka, H., Kamo, N., and Kobatake, Y. (1985) *FEMS Microbiol Lett* 28, 161–164.

Takahashi, T., Yan, B., Mazur, P., Derguini, F., Nakanishi, K., and Spudich, J.L. (1990) *Biochemistry* 29, 8467–8474.

Takahashi, T., Yoshihara, K., Watanabe, M., Kubota, M., Johnson, R., Derguini, F., and Nakanishi, K. (1991) *Biochem Biophys Res Commun* 178, 1273–1279.

Tomioka, H., and Sasabe, H. (1995) *Biochim Biophys Acta* 1234, 261–267.

Toole, C.M., Plank, T.L., Grossman, A.R., and Anderson, L.K. (1998) *Mol Microbiol* 30, 475–486.

Varo, G. (2000) *Biochim Biophys Acta* 1460, 220–229.

Venter, J.C., Remington, K., Heidelberg, J.F., Halpern, A.L., Rusch, D., Eisen, J.A., Wu, D., Paulsen, I., Nelson, K.E., Nelson, W., Fouts, D.E., Levy, S., Knap, A.H., Lomas, M.W., Nealson, K., White, O., Peterson, J., Hoffman, J., Parsons, R., Barden-Tillson, H., Pfannkoch, C., Rogers, Y.H., and Smith, H.O. (2004) *Science* 306, 66–74.

Vogeley, L., Sineshchekov, O.A., Trivedi, V.D., Sasaki, J., Spudich, J.L. and Luecke, H. (2004) Anabaena Sensory Rhodopsin: A Photochromic Color Sensor at 2.0 Å. *Science* 306, 1390–1393.

Wang, W.W., Sineshchekov, O.A., Spudich, E.N., and Spudich, J.L. (2003) *J Biol Chem* 278, 33985–33991.

Wegener, A.A., Chizhov, I., Engelhard, M., and Steinhoff, H.J. (2000) *J Mol Biol* 301, 881–891.

Yan, B., Spudich, J.L., Mazur, P., Vunnam, S., Derguini, F., and Nakanishi, K. (1995) *J Biol Chem* 270, 29668–29670.

Yang, C.S., and Spudich, J.L. (2001) *Biochemistry* 40, 14207–14214.

Yang, C.S., Sineshchekov, O.A., Spudich, E.N. and Spudich, J.L. (2004) *J Biol Chem* 279: 42970–42976.

Zhai, Y., Heijne, W.H., Smith, D.W., and Saier, M.H., Jr. (2001) *Biochim Biophys Acta* 1511, 206–223.

Zhang, X.N., Zhu, J., and Spudich, J.L. (1999) *Proc Natl Acad Sci USA* 96, 857–862.

2
Sensory Rhodopsin Signaling in Green Flagellate Algae

Oleg A. Sineshchekov and John L. Spudich

2.1
Introduction

2.1.1
Retinylidene Receptors

The sensory rhodopsins of the green flagellate alga *Chlamydomonas reinhardtii* are recent additions to the large and diverse family of microbial rhodopsins, proteins that are characterized by their visual pigment-like domain, consisting of 7-transmembrane helices forming an internal pocket for the chromophore retinal (see Spudich and Jung, Chapter 1). The first members of this family were observed in halophilic Archaea (reviewed in Hoff et al., 1997; Schäfer et al., 1999; Spudich et al., 2000). In haloarchaea, two are light-driven ion pumps [bacteriorhodopsin (BR) and halorhodopsin (HR)], and two are sensory receptors for phototaxis [sensory rhodopsins I and II (SRI and SRII)]. Over the past four years, cloning, heterologous expression, and functional analysis of homologous genes from cultivated-microorganism genome projects as well as environmental genomics of uncultivated microbes have revealed photoactive archaeal-rhodopsin homologs in the other two domains of life as well, i.e. Bacteria (Béjà et al., 2001; Jung et al., 2003) and Eucarya (Bieszke et al., 1999a; Sineshchekov et al., 2002). Both transport and sensory rhodopsins exist in Archaea and both functional classes are also found in Bacteria (e.g. proton-pumping marine proteorhodopsins and *Anabaena* sensory rhodopsin); so far only sensory members have been demonstrated in microbial Eucarya, namely the *C. reinhardtii* sensory rhodopsins reviewed here.

The prokaryotic rhodopsins rank among the best-understood membrane proteins in terms of structure and function at the atomic level. Atomic resolution structures, which exist for <60 membrane proteins, have been obtained from electron and X-ray crystallography of BR (Grigorieff et al., 1996; Pebay-Peyroula et al., 1997; Essen et al., 1998; Luecke et al., 1999), HR (Kolbe et al., 2000), SRII (Luecke et al., 2001; Royant et al., 2001), and Anabaena sensory rhodopsin (Vogeley et al., 2004). This knowledge and the fact that they can be activated by light, which permits precise temporal reso-

Handbook of Photosensory Receptors. Edited by W. R. Briggs, J. L. Spudich

ISBN 3-527-31019-3

lution, have made them foremost model systems for membrane-embedded transport and sensory-signaling proteins.

The microbial rhodopsins share with visual rhodopsins: (i) a seven-transmembrane helix structure, (ii) the attachment of the retinal as a protonated Schiff base to the ε-amino group of a lysine residue near the middle of the seventh helix, and (iii) their activation by photoisomerization of retinal followed by (iv) transfer of the proton from the retinylidene Schiff base to a carboxylate residue on the third helix of the protein. Despite these mechanistic similarities, there is no evident sequence homology between the microbial group (called type 1 rhodopsins based on phylogenic analysis) and visual pigments (type 2 rhodopsins). The isomeric configuration and conformation of the retinal is another difference between the two pigment families. Visual pigments contain *11-cis* retinal in the dark, which undergoes photoisomerization to the *all-trans* configuration to generate signaling states, whereas the retinal in published transport and sensory microbial rhodopsins photoisomerizes from *all-trans* to *13*-cis in their functional photoreactions. The retinal ring-polyene chain conformation in visual pigments is a nonplanar *6-s-cis* structure, whereas in microbial rhodopsins in the few well-studied cases (BR, HR, SRI, and SRII) the retinal is in a planar *6-s-trans* conformation (see Kumauchi and Ebrey, Chapter 3). However, recently we have found that *Anabaena* sensory rhodopsin contains predominantly *13-cis*-retinal in its light-adapted photoactive state (Vogeley et al., 2004).

2.1.2. Physiology of Algal Phototaxis and the Photophobic Response

Motile photosynthetic microorganisms, such as green flagellate algae, have developed fine-tuned mechanisms for light control of behavior. Two types of photomotility responses can be distinguished, phototaxis and the photophobic response (Figure 2.1) (Diehn et al., 1977). Phototaxis is the active adjustment of the swimming path with respect to the direction of light incidence. The photophobic, or photoshock, response is a brief stop followed by a short period of backward swimming, after which forward swimming is resumed in another direction. This response occurs upon an abrupt change in light intensity and does not depend on the light direction.

A single asymmetrically positioned photoreceptor apparatus is used for phototactic orientation (Foster and Smyth, 1980; Kreimer, 2001; Dieckmann, 2003). It consists of a layered eyespot, which serves as an accessory device, and the so-called "photoreceptor membrane", which is a part of the plasma membrane underlying the eyespot, where molecules of the receptor proteins reside. Illumination of the photoreceptor membrane during the helical swimming path is modulated by the eyespot and the rest of the cell. When the axis of a helical swimming path of the cell deviates from the light direction, changes in photoreceptor illumination during the rotation cycle give rise to unbalanced responses of the two flagella, which lead to a correction of the swimming path with respect to the direction of light. When the direction of the cell's movement becomes parallel with that of light, illumination of the photoreceptor is no longer modulated, and no corrective motor responses occur. A more detailed de-

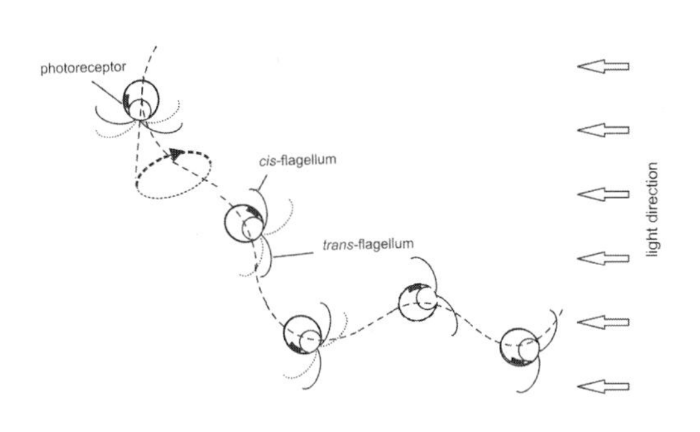

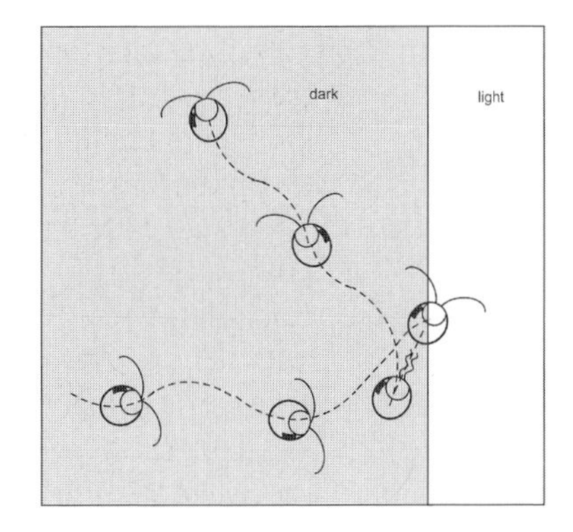

Figure 2.1 Schematic presentation of the two types of photomotility behavior in green flagellate algae.

scription of this mechanism can be found in several reviews (Foster and Smyth, 1980; Witman, 1993; Kreimer, 1994; Hegemann, 1997).

2.1.3
Photoelectrical Currents and their Relationship to Swimming Behavior

The photoreceptor site and the flagella are spatially separated in green flagellates, so that there is a question of how the light signal is transmitted within the cell. Photoexcitation of the receptor triggers a cascade of rapid electrical phenomena in the cell membrane, which plays a key role in the signal transduction chain for phototaxis and in the photophobic response. These currents are located asymmetrically and thus can be recorded extracellularly by means of the suction-pipette technique (Figure 2.2A) (Litvin et al., 1978; Sineshchekov et al., 1978), or with a population assay (Sineshchekov et al., 1992) (Figure 2.2C and D). In the first modification of the latter assay, microscopic electrical currents appear in suspensions of non-oriented cells in response to unilateral flash excitation due to the difference in the current amplitudes in the cells with illuminated and shaded photoreceptors (Figure 2.2C). In the second modification of the suspension assay, freely motile cells are pre-oriented by weak phototactically-active light (Figure 2.2D) or gravitaxis.

In early work, two major components of the photoelectric cascade in green flagellate algae *Haematococcus pluvialis* and *Chlamydomonas reinhardtii* were resolved in photocurrent transients (Figure 2.2B). A gradual inward photoreceptor current is generated in the eyespot region of the cell and is the earliest event detected so far in light regulation of behavior in green flagellate algae (Litvin et al., 1978; Sineshchekov et al., 1978; Harz and Hegemann; 1991). Under oscillating illumination (the temporal pattern of illumination of photoreceptors in freely motile rotating cell) periodic changes of the photoreceptor current occur in parallel with unbalanced changes in beating of the two flagella giving rise to phototactic orientation (Sineshchekov,

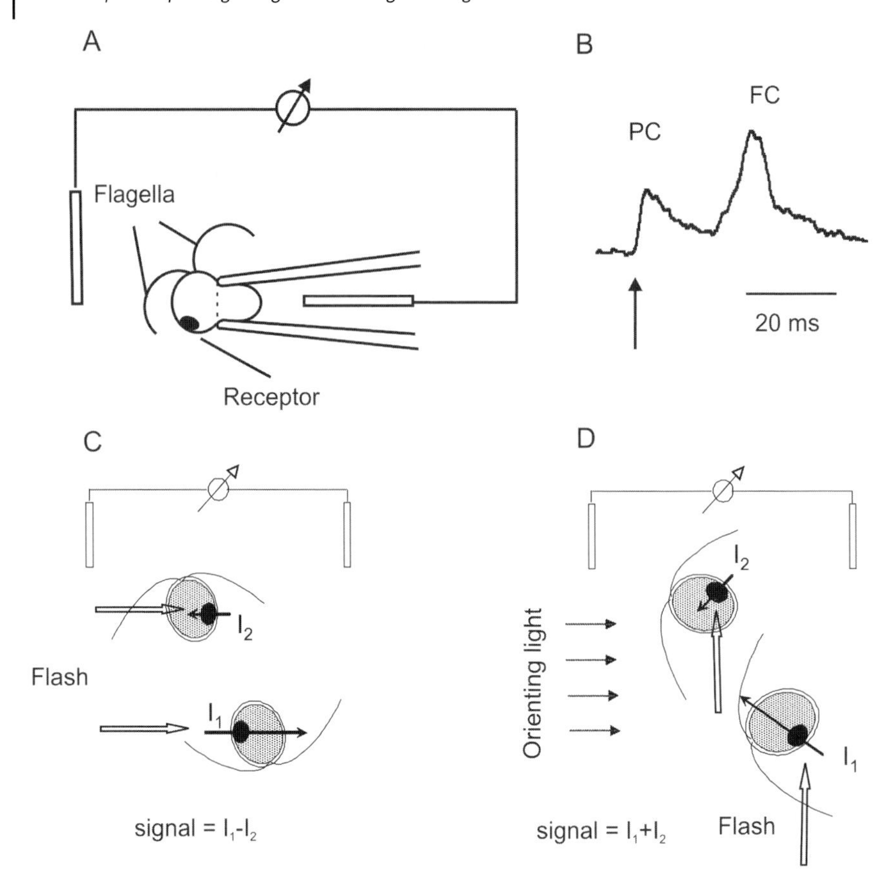

Figure 2.2 Measurement of photoelectric currents involved in photoreception in unicellular organisms. (A) Single cell recoding by a suction pipette. (B) Example of single cell recording of the photoreceptor current (PC) and flagella current (FC). (C) Recording of macroscopic currents in a suspension of non-oriented cells excited by a unilateral flash along the direction between measuring electrodes. (D) Recording of macroscopic current in a suspension of pre-oriented cells. I_1, photocurrent in cells with illuminated receptors; I_2, photocurrent in cells with shaded receptors.

1991a,b). When a change in light fluence exceeds a certain threshold, a transient regenerative response is superimposed on the photoreceptor current (Litvin et al., 1978). This current is a Ca^{2+} influx brought about by membrane depolarization induced by photoreceptor currents and is the basis for the photophobic response (Sineshchekov et al., 1991a,b; Holland et al., 1997). It is also called the flagellar current (Harz and Hegemann, 1991), because its most likely localization is the flagellar membrane (Beck and Uhl, 1994).

In recent years research has been focused on photoreceptor currents as the most specific part of the electrical cascade in phototactic algae. Kinetic analysis of laser flash-induced photoreceptor currents recorded with an improved time resolution showed that they are comprised of at least two components with different characteristics (Sineshchekov et al., 1990). The onset of the first component (the "early photoreceptor current") was observed to occur within the time resolution of the measuring system (< 30 μs). The second component (the "late photoreceptor current") appeared after a lag period of several hundreds of μs, the duration of which depended on the stimulus intensity. The late receptor current is sensitive to the physiological state of cell. Red background illumination, known to hyperpolarize the cell membrane (Sineshchekov et al., 1976), increases the amplitude mostly of the late photoreceptor current. Similar effects are observed after mechanical agitation of a cell suspension, or changes in the ionic composition of the medium, e.g. Ca^{2+} concentration (Figure 2.3). At least two exponential components with different time constants are also distinguishable in the current decay.

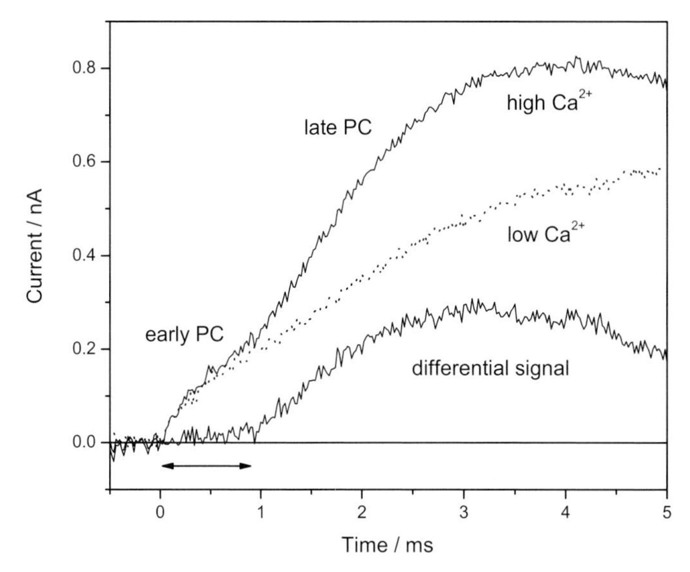

Figure 2.3 The influence of the external Ca^{2+} concentration on the photoreceptor current (PC) in *Chlamydomonas*. Excitation, white 3 μs flash at time zero. The two-headed arrow shows the duration of the delay of the late Ca^{2+}-dependent current component.

The complex nature of the photoreceptor current is also evident from the analysis of fluence-response curves of the current amplitude (Figure 2.4). The curves can be decomposed into two phases with different saturation levels (Sineshchekov, 1991a; Sineshchekov et al., 1992). The ratio between the amplitudes of the low- and high-saturating components varies from 1:10 to 1:5, and the ratio between their saturation levels from 1:300 to 1:50 in different species and culture states.

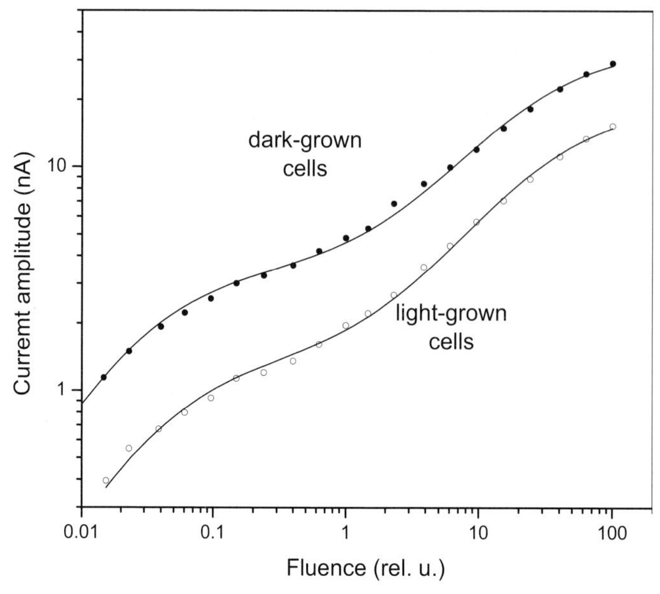

Figure 2.4 Fluence–response curves for photoreceptor current amplitudes in light- and dark-grown cells fit with sum of two hyperbolic functions.

2.2
The Photosensory Receptors: CSRA and CSRB

The presence of rhodopsin receptors for phototaxis and the photophobic response in green flagellate algae was initially proposed on the grounds of their action spectra (Foster and Smyth, 1980) and subsequently confirmed by the results of retinal and retinal-analog reconstitution studies in "blind" *Chlamydomonas* mutants (Foster et al., 1984; Lawson et al., 1991; Hegemann et al., 1991; Takahashi et al., 1991; Zacks et al., 1993; Sineshchekov et al., 1994). Both phototaxis and the photophobic response require chromophores with an *all-trans* polyene chain configuration in a planar ionone ring/polyene chain conformation, which is the same chromophore structure found in haloarchaeal rhodopsins. Different adaptation properties of the two types of behavior indicated a difference between their signal transduction mechanisms and/or their receptors (Zacks et al., 1993; Zacks and Spudich, 1994). Multiple bands resolved in the action spectra of photoelectric and behavioral responses in *Haematococcus pluvialis* suggested multiple receptor species in photomotility control (Sineshchekov and Litvin, 1988). The receptor proteins themselves were not identified until rhodopsin apoprotein-encoding genes were revealed in the *Chlamydomonas reinhardtii* genome.

2.2.1
Genomics, Sequence, and Predicted Structure

Search of a *Chlamydomonas* cDNA database revealed the presence of two sequences homologous to archaeal opsins. The encoded proteins contain 712 and 737 amino acid residues and mediate phototaxis and photophobic responses via two distinct photoreceptor currents (see below). According to their demonstrated sensory function, these proteins were named CSRA and CSRB (*Chlamydomonas* Sensory Rhodopsin A and B, respectively; Sineshchekov et al., 2002). The same proteins were independently reported by two other research groups under the names Chop1 and Chop2 (for **ch**annel**op**sin 1 and 2; Nagel et al., 2002; 2003) and Acop-1 and Acop-2 (for **A**rchaeal type ***C**hlamydomonas* **op**sin 1 and 2; Suzuki et a., 2003). In this review we will use CSRA and CSRB for the retinal-bound proteins and CSOA and CSOB for their apoprotein or "opsin" forms without retinal.

Both CSOA and CSOB are comprised of N-terminal domains of about 300 residues that form seven membrane-spanning helices (7-TM domain) followed by extensive primarily hydrophilic C-terminal domains (Figure 2.5). Sequence alignment of the N-terminal domains shows identity with the most conserved regions of archaeal rhodopsins, although different analyses resulted in different positions of helices A and B. According to one of the models, helix B has an exceptionally hydrophilic character (Suzuki et al., 2003). The residues known from crystal structures to form a chromophore-binding pocket in *Natronomonas pharaonis* sensory rhodopsin II (NpSRII) and in bacteriorhodopsin (BR) are conserved in CSRA and CSRB, including the lysine to which retinal is covalently bound. A glutamic acid residue is found in the position of the counterion to the protonated Schiff base (Asp-85 in BR). A histidine residue occupies the place of the Schiff base proton donor specific to proton pumps (Asp-96 in BR), which is substituted with a noncarboxylate residue also in archaeal and *Anabaena* sensory rhodopsins.

The C-terminal domains of both CSOA and CSOB appear to be membrane-associated when expressed in *E. coli*, and hydropathy plots predict two transmembrane he-

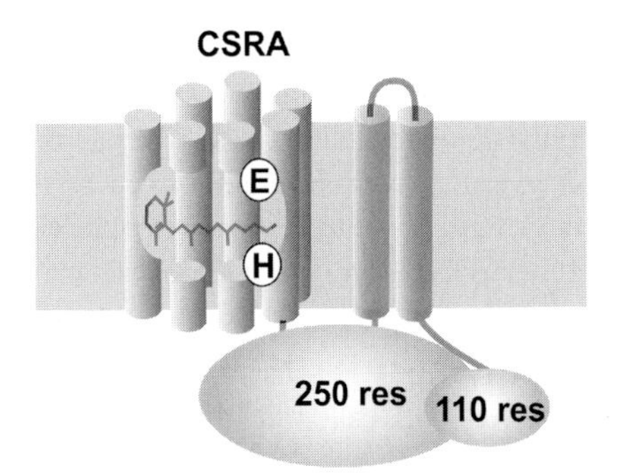

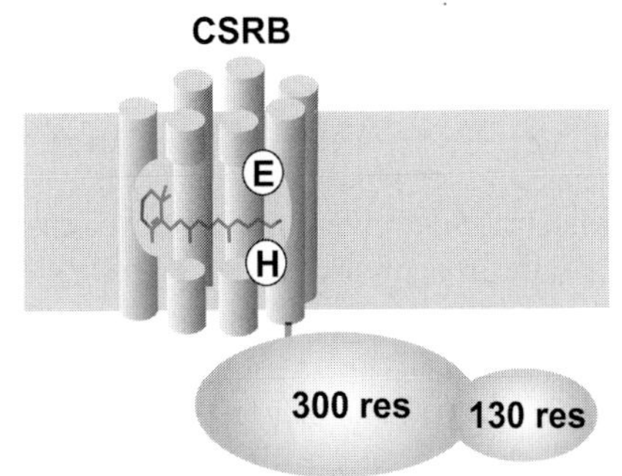

Figure 2.5 Model for the structures of CSRA and CSRB. E and H indicate Glu and His residues at the positions corresponding to the Schiff base proton acceptor and donor aspartates, respectively, in bacteriorhodopsin and proteorhodopsin (Spudich and Jung, Chapter 1).

lices in this domain in the CSOA. The C-terminal regions of about 110 residues show some homology to domain D of synapsin and contains several Ser/Thr residues predicted as phosphorylation sites.

Southern-blot hybridization indicated that the *Chlamydomonas* genome contains a single copy of each CSOA and CSOB gene, and no other sequences with high homology (Sineshchekov et a., 2002; Suzuki et al., 2003), a finding that was confirmed in the complete genome sequence (http://genome.jgi-psf.org/chlre2/chlre2.home.html). The complete genes encoding CSOA and CSOB were cloned from *Chlamydomonas* genomic DNA and were found to contain 14 and 19 introns, respectively (Sineshchekov et al., 2002). The number and location of introns in opsin genes are different among different species, e.g. *Neurospora* (2 introns), *Cryptococcus* (4 introns). Twelve introns of the CSOA and CSOB genes are found at the same loci.

2.2.2
Cellular Content and Roles in Phototaxis and Photophobic Behavior

The absolute amounts of CSRA and CSRB were measured by quantitative immunoblot analysis using *E. coli*-expressed C-terminal polypeptides of each protein to generate antibodies. Vegetative cells of wild-type *Chlamydomonas* contain 9×10^4 CSRA and 1.5×10^4 CSRB apoprotein molecules per cell (Govorunova et al., 2004). This value is somewhat higher than that for receptor cellular content estimated previously from photosensitivity measurements and retinal extraction yields (Foster and Smyth, 1980; Beckmann and Hegemann, 1991; Hegemann et al., 1991). The amounts of both pigments increase in dark-grown cells, leading to higher amplitudes and sensitivities of photoreceptor currents, especially the low-saturating one (Figure 2.4). The CSOA:CSOB ratio shifts in favor of CSOB in dark-grown cells (Nagel et al., 2003) and upon conversion of vegetative cells into gametes (Govorunova et al., 2004). Immunofluorescent analysis indicated localization of CSOA in the eyespot region of the cell (Suzuki et al., 2003), where the photoreceptor molecules are assumed to reside.

The functional roles of CSRA and CSRB in *Chlamydomonas* were established by measurement of photoreceptor currents in cells with reduced amounts of each of the proteins in cells transformed with respective RNA*i* constructs (Sineshchekov et al., 2002). The photoreceptor currents are the earliest events detected so far in the signal transduction pathways for both phototaxis and the photophobic response. Therefore, recording the photoreceptor currents provides the most suitable approach to examining the photoreceptor function. The measurement of photocurrents was especially important in understanding the roles of CSRA and CSRB because the two photoreceptors have overlapping signaling pathways, difficult to distinguish with motility assays.

RNA*i* constructs were generated using the first 6 and 7 exons of CSOA and CSOB, respectively, incorporated in forward and reversed orientations. Western-blot analysis revealed a decrease in the amount of the protein against which the RNA*i* construct was directed (Sineshchekov et al., 2002). Quantitative analysis has shown that the content of CSOA in the A-RNA*i* transformant A22 was ~9% of that in the wild type (Govorunova et al., 2004). Furthermore, the amount of the other protein (CSOB in the case of A-RNA*i* transformants and CSOA in the case of B-RNA*i* transformants) was increased, indicating a cooperative regulation of expression of the two proteins (Sineshchekov et al., 2002; Govorunova et al., 2004). As a result, the CSRA:CSRB ratio was significantly shifted toward CSRB in the A-RNA*i* transformants and toward CSRA in the B-RNA*i* transformants. In the A22 transformant this shift was ~30-fold (Govorunova et al., 2004). Thus, comparative analysis of the currents generated by the two transformants was carried out to test for the functions of CSRA and CSRB.

The kinetics of the photoreceptor currents recorded in A- and B-RNA*i* transformants was clearly different. Both rise and decay of the currents were much slower in CSRB- than in CSRA-enriched cells (Figure 2.6A). This result indicated that the fast (early) and the slow (delayed) components of the photoreceptor current observed earlier in the wild type green algae (Sineshchekov et al., 1990) were in fact mediated by the two separate photoreceptors. Superposition of the CSRA- and CSRB-mediated currents gives rise to the complex current kinetics observed in the wild type. Assuming that the small non-delayed current in CSRB-enriched cells (see dashed line in Figure 2.6A) and the slow current with decay ~30 ms in CSRA-enriched cells are generated by the residual amount of the opposite pigment, the kinetics of each current can be evaluated by mutual subtraction of the experimentally determined curves with corresponding coefficients (Figure 2.6B). Fitting the fluence-response dependence of the peak current amplitude with two saturation functions revealed that the relative contributions of the two components of the curve favored the low-saturating current in CSRB-enriched cells and the high-saturating current in CSRA-enriched cells. As a first approximation it appears that CSRA is responsible for generation of the fast high-saturating photoreceptor current and CSRB for the slow delayed low-saturating photoreceptor current.

Action spectra of the photoreceptor currents measured in RNA*i* transformants enriched in one or the other of the rhodopsins are clearly different, which indicated a

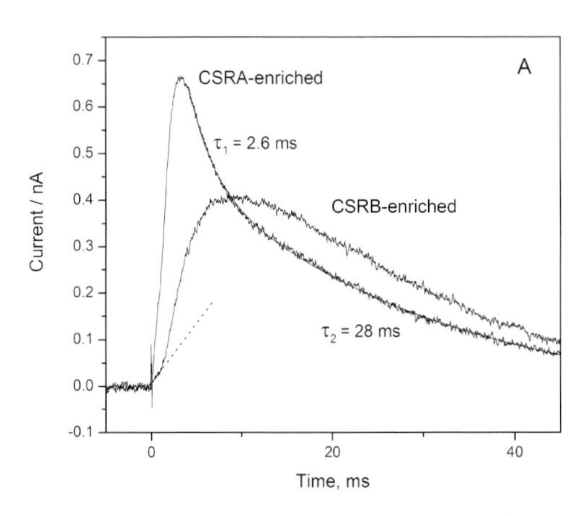

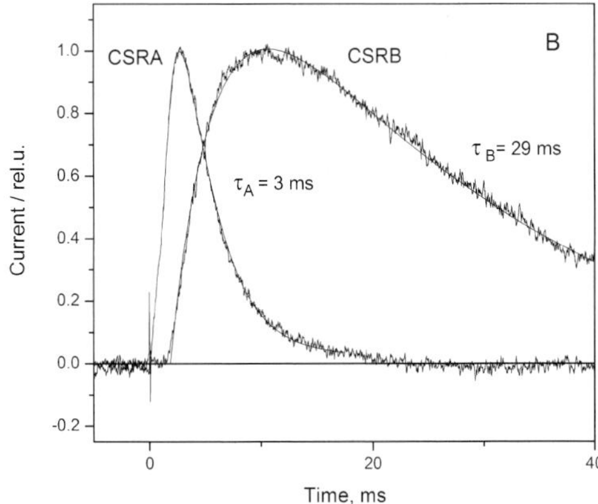

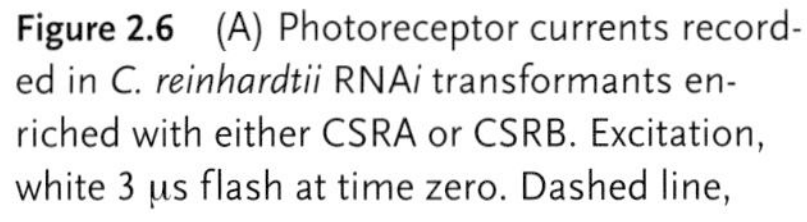

Figure 2.6 (A) Photoreceptor currents recorded in *C. reinhardtii* RNA*i* transformants enriched with either CSRA or CSRB. Excitation, white 3 μs flash at time zero. Dashed line, extended linear fitting of the initial non-delayed current. (B) Deconvolution of the CSRA- and CSRB-mediated currents (see text).

difference in the absorption spectra of CSRA and CSRB (Figure 2.7A). The spectrum in CSRA-enriched cells has a maximum between 500 and 510 nm with a shoulder at 475 nm, whereas in CSRB-enriched cells the maximal sensitivity is at 470 nm with a secondary band at 495–500 nm. In both cases a close correlation between the position of the minor maximum and that of the major maximum of the spectrum in the other transformant argues that the minor maxima likely reflect incomplete suppression of the respective rhodopsin by transformation with the RNA*i* constructs.

This difference in spectral sensitivity of transformants was observed with both modifications of the suspension method: upon unilateral excitation of a non-oriented suspension, when the absorption by the stigma and chloroplast would increase the signal (Figure 2.2C), and in pre-oriented cells, in which any screening decreases the signal (Figure 2.2D). This rules out the possibility that the difference in the action spectra observed in RNA*i* transformants is due to the difference in absorption/reflection of the eyespot (Sineshchekov et al., 2002).

The role of CSRA and CSRB in phototaxis was established by measuring relative efficiencies of the spectral bands predominantly absorbed by each of the two proteins. Photoorientation was measured independently by the photoelectric assay and the traditional light-scattering assay. In the photoelectric measurement, the amplitude of photoreceptor current in response to a standard test flash served as a measure of cell orientation. The test flash was applied in the direction perpendicular to the line between the measuring electrodes (Figure 2.2D) and thus did not elicit current components in the perpendicular direction, which are detected by electrodes. The higher is the degree of phototactic orientation to continuous light of different spectral composition, the larger is the amplitude of the test photocurrent. Orientation of cells also leads to changes in scattering, which can be monitored by infrared light (Uhl and Hegemann, 1990). The relative efficiencies of the two bands in phototactic orientation were different in the A- and B-RNA*i* transformants and correlated with the relative efficiencies of the two bands in generation of the photoelectric currents (Sineshchekov et al., 2002) (Figure 2.7B). Therefore, we concluded that both CSRA and CSRB mediate phototaxis via generation of their respective photoreceptor currents. Measurements of the spectral sensitivities of the photophobic response (Figure 2.7C) in cells enriched with CSRA or CSRB with a computerized motion analysis system indicated that both rhodopsins mediate photophobic responses as well as phototaxis (Govorunova et al., 2004).

Integration of the signaling pathways activated by photoexcitation of CSRA and CSRB occurs at the level of membrane depolarization (Figure 2.8). Although both CSRA and CSRB contribute to both types of photomotility responses in *Chlamydomonas*, the result of their different light-saturation levels is that CSRA dominates in the photophobic response, which appears under intense light stimulation, whereas CSRB dominates in the highly sensitive phototaxis response to low light. Such a preference explains the significant spectral shift between the two photoreceptor pigments. The modulation of photoreceptor illumination is the essential basis of phototaxis. Indeed, the combined absorption spectra of the eyespot and chloroplast (Schaller and Uhl, 1997) correspond approximately to the absorption spectrum of CSRB, which is responsible for low light-intensity phototaxis. On the other hand, any

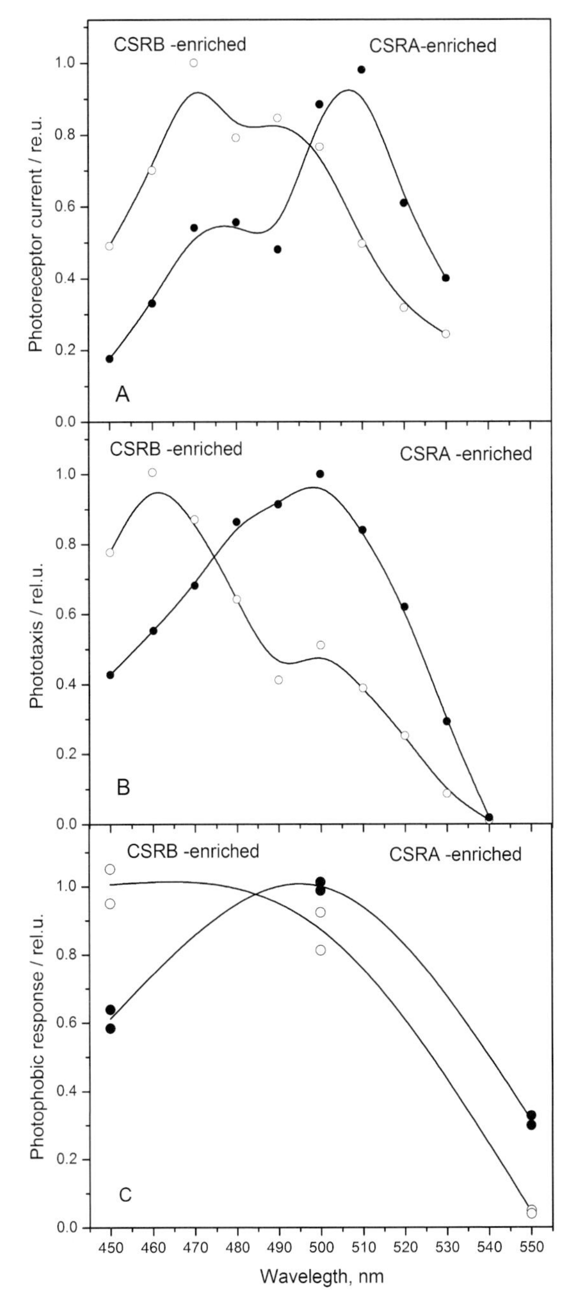

Figure 2.7 Spectral sensitivity of the photoreceptor currents, phototactic orientation and the photophobic response in *C. reinhardtii* cells enriched with either CSRA or CSRB. The normalized quantum requirement for equal response, calculated from fluence–response curves, is plotted in each panel.

screening of the photoreceptor pigment responsible for high light avoidance (photophobic) responses, would lead to loss of sensitivity. Accordingly, the maximum absorption of CSRA is shifted to longer wavelengths, outside the maximal absorption range of the eyespot and chloroplast.

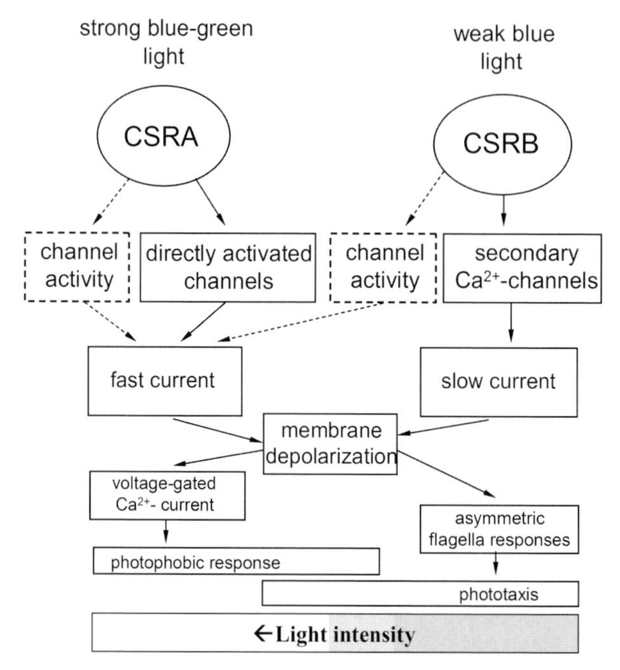

Figure 2.8 A scheme for photosensory transduction and control of motility in *Chlamydomonas*. Solid arrows indicate processes deduced from measurements in *Chlamydomonas* cells. Dashed arrows indicate processes demonstrated in *Xenopus* oocytes, but which have not been proven to play a role under physiological conditions in living *Chlamydomonas* cells.

2.2.3
Molecular Mechanism of Action

Early studies indicated that photocurrent generation likely involves a combination of different mechanisms (Sineshchekov, 1988; Sineshchekov et al., 1990; Sineshchekov 1991a,b). One is generation of the current by the receptor itself, or by a closely associated ion channel (Sineshchekov and Govorunova, 1999). The following evidence from the early work support this mechanism: (i) The fast component of the current appears without a measurable delay after the excitation flash. (ii) Saturation of the current amplitude occurs at high light intensities, which shows that it is limited only by photon absorption by the receptor. (iii) The fast current is only weakly dependent on physiological conditions, such as culture state, ionic composition of the medium, and red background illumination. This mechanism is likely to play a major part in the CSRA-mediated fast photocurrent (Figure 2.8).

Fast electrical signals with similar properties have been recorded from archaeal and visual rhodopsins, in which they derive from intramolecular charge movements triggered by photoexcitation (Trissl, 1990; Oroszi et al., 2002; Hong, 2004). Calculations, however, show that the amplitude of such displacement signals in *Chlamydomonas* cells, which contain relatively low concentrations of receptors, would probably be below the detection limit. Therefore, the fast current likely reflects ion translocation across the membrane (Sineshchekov and Govorunova, 2001).

The second suggested mechanism involved in photocurrent generation in *Chlamydomonas* and related algae is initiation of a biochemical cascade that regulates a secondary ion channel via a diffusible messenger, analogous to the signal-transduction process in animal vision. This suggestion is based on the observation of (i) a light-dependent delay of the slow component of the current, (ii) its low light saturation level, and (iii) its sensitivity to the physiological state of the cell and ionic composition of the medium, especially the extracellular Ca^{2+} concentration (Figure 2.3; Litvin et al., 1978; Sineshchekov, 1991a,b). The CSRB-mediated current appears to be primarily generated by this second mechanism. Several enzymes characteristic of signal transduction pathways in animals, including heteromeric GTPases, have been detected in isolated eyespot preparations of green flagellate algae (Schlicher et al., 1995; Linden and Kreimer, 1995; Calenberg et al., 1998), although whether they play a role in photomotility signaling has yet to be determined.

At a time when only one rhodopsin was considered as the receptor for generation of the photocurrents, it was suggested that it combines both ion transport and enzymatic functions (Sineshchekov and Govorunova, 2001). Now, since it has been established that there are two rhodopsins, CSRA and CSRB, involved in current generation and photomotility control in *Chlamydomonas*, the same possibility can be raised for both proteins, especially taking into account that light-induced channel activities have been demonstrated in a model system for both CSRA and CSRB expressed in *Xenopus* oocytes (Nagel et al., 2002, 2003).

In *Xenopus* oocytes, the light-dependent currents and current–voltage relationships were the same in both full-length and 7-TM domains of CSRA and CSRB, showing that the extensive C-terminal domain is not involved in the photocurrent measured in those experiments (Nagel et al., 2002, 2003). The action spectra of the observed photocurrents correlated with those measured for CSRA and CSRB in *Chlamydomonas* cells (Sineshchekov et al., 2002). The currents generated by CSRA expressed in oocytes were sensitive only to protons and not other ions present in the medium (Nagel et al., 2002). The proton dependence of the CSRA-mediated current supported the conclusion that CSRA expressed in oocytes behaves as a light-regulated proton channel. This would be the first known light-gated proton channel and, as such, is of interest as a prospective molecular tool for manipulation of intracellular pH and membrane potential. It is however unclear whether the proton conductance measured upon expression of this protein in oocytes also takes place in the native system and if so whether it plays a functional role in sensory transduction in *Chlamydomonas*.

The proton conductance measured upon expression of CSRA in oocytes may be related to the fast CSRA-mediated current recorded in intact *Chlamydomonas*. However a light-induced H^+ current could only be recorded in *Chlamydomonas* cells at pH below 4, i.e., under non-physiological conditions (Ehlenbeck et al., 2002; Gradmann et al., 2002). Furthermore, the pH-dependences of both the amplitude and the initial slope of CSRA-generated photoreceptor currents measured in *Chlamydomonas* within the physiological pH range (Sineshchekov et al., in preparation) are opposite to the pH-dependence of the electrochemical gradient for protons across the *Chlamydomonas* cell membrane (Malhotra and Glass, 1995). So far, no experimental data have been obtained to support a possible physiological role of a CSRA-mediated proton conductance in *Chlamydomonas*.

In contrast to CSRA, CSRB expressed and illuminated in oocytes is not only permeable to protons, but also to several mono- and divalent cations (Nagel et al., 2003). Its pore size has been estimated using a graded-size series of organic cations and found to be larger than that of a voltage-activated Na^+ channel. The relative conductance of CSRB was inversely proportional to atomic radius (with the exception of Mg^{2+}) and followed the sequence $Li^+ > Na^+ > K^+ > Rb^+ > Cs^+$ for monovalent cations, and $Ca^{2+} > Sr^{2+} > Ba^{2+} > Zn^{2+} > Mg^{2+}$ for divalent cations. The delayed Ca-dependent current in *Chlamydomonas* generated by CSRB does not appear to derive from this light-dependent conductance, since it has a delayed onset (see above). In CSRB-enriched cells, Li^+, which is the most permeable ion for CSRB in the oocyte, does not activate the photocurrent, but rather strongly inhibits it (Sineshchekov et al., in preparation).

It is likely that in the native system interaction of the receptor proteins with downstream elements of a signaling cascade results in the Ca^{2+} currents. The absence of these components in the oocyte membrane could bring about channel activity of *Chlamydomonas* rhodopsins analogous to the conversion of haloarchaeal sensory rhodopsins into proton pumps by transducer absence (Olson and Spudich, 1993; Bogomolni et al., 1994; Spudich, 1994, 1995; Sudo et al., 2001).

Substitution of His-173 with Asp (H173D), which corresponds in CSRA to the proton donor Asp-96 in BR, completely abolished the light-gated conductance. This observation and the fact that blue light did not quench the stationary currents led the authors to suggest that in CSRA the Schiff base is not deprotonated during the photocycle (Nagel et al., 2002). A firm conclusion is not warranted, since there was no determination of the relative amounts of the wild type and mutant proteins expressed, and the absence of a blue light effect could be explained by a low level of accumulation of blue-light absorbing (M) intermediate of the photocycle due to a short lifetime. Direct optical measurements of photochemical conversions of CSRA and CSRB are needed to discern the chemical events in these processes. Unfortunately, *Chlamydomonas* cells contain only small amounts of phototaxis receptor proteins, preventing their optical measurements in native membranes and impeding biochemical purification from the cells. Expression in oocytes and in mammalian cell culture in which photoactive pigments have been detected electrophysiologically (Nagel et al., 2003) is also not suitable for this purpose due to low concentrations of the pigments. Therefore, it would be highly desirable to develop an overexpression system for CSRA and CSRB production.

The apoproteins of CSRA and CSRB have been expressed in *Halobacterium salinarum, E. coli, Pichia pastoris, Saccharomyces cerevisiae,* monkey-kidney COS-1 cells and HEK293S cells (Sineshchekov et al., 2002; Suzuki et al., 2003; Jung K.-H., unpublished observations; Ridge K., unpublished observations). The heterologously expressed apoproteins are recognized by antibodies raised against their respective synthetic C-terminal peptides. However, no flash-induced absorption changes have been observed in either cells or membrane preparations. This is probably due to a low yield of correctly folded proteins able to incorporate exogenous *all-trans* retinal and form functional pigments.

Three additional sensory rhodopsin genes (cop5, cop6, and cop7) have been identified in the Chlamydomonas genome sequence (Kateriya et al., 2004). Like CSRA and CSRB, each contains the 7-transmembrane rhodopsin domain fused to extensive signal transduction domains. They lack residues critical in transport rhodopsins, and they contain domains highly homologous to well-studied histidinyl-kinases and phospho-acceptor regulators. Therefore we conclude they are sensory in function and refer to the cop5, cop6, and cop7 gene products as CSRC, CSRD, and CSRE, respectively.

2.3 Other Algae

Chlamydomononas reinhardtii is the only chlorophyte for which the photosensory receptors have been identified. Nevertheless, photoelectric cascades similar to that in *C. reinhardtii* were found in representatives of several genera of chlorophycean flagellates, namely, *Haematococcus, Polytomella, Spermatozopsis, Hafniomonas* and *Volvox* (Sineshchekov et al., 1990; Sineshchekov and Nultsch, 1992; Kreimer, 1994; Braun and Hegemann, 1999). Therefore it appears that the same basic scheme for photosignaling holds for these algae. Genomic analysis of these microorganisms is not yet available, but it seems likely that rhodopsin receptors will be found responsible for generation of the detected electric signals. Moreover, the complex kinetics, fluence response curves, and action spectra suggest that the two-pigment mechanism found in *Chlamydomonas* also occurs in other green algae. On the other hand, genes homologous to archaeal opsins (i.e. without the extensive C-terminal domains of CSRA and CSRB) emerged from DNA sequence database searches for the sedentary unicellular green alga *Acetabularia acetabulum,* motile gametes of which are phototactic (Crawley, 1966), for the cryptomonad *Guillardia theta* (Jung and Spudich, 2004), and for the dinoflagellate *Pyrocystis lunula* (Okamoto and Hastings, 2003). The functional significance of these genes has not been established, although the involvement of rhodopsin-like receptors in phototaxis in dinoflagellates has been proposed from action spectroscopy data (Foster and Smyth, 1980) and photoreceptor currents typical for those described above have been found in the fresh-water cryptomonad *Cryptomonas sp.* (Sineshchekov et al., in preparation). Retinal has been shown to be the chromophore for phototaxis by zoospores of the fungus *Allomyces reticulatus,* which exhibit photomotile behavior similar to that of unicellular algae (Saranak and Foster, 1997).

2.4
Conclusion and Future Perspectives

Chlamydomonas photomotility receptors are so far the only eukaryotic microbial rhodopsins whose function in the cell is known (Ebrey, 2002; Ridge, 2002; Jung and Spudich, 2004). Moreover, their mode of operation as photocurrent generators and their multi-domain structures are unique among microbial sensory receptors. Therefore, further studies on CSRA and CSRB promise to expand our understanding of rhodopsin diversity of structure and mechanism.

The dual-receptor system in *Chlamydomonas* extends the dynamic range of photosensing reactions and provides specificity at low and high light intensity, similar to other receptor systems in higher organisms; e.g. animal visual pigments, phytochromes, phototropins, cryptochromes (Lin et al., 1998; Briggs et al., 2001; Quail, 2002). The different spectral sensitivities of the two receptors establish that functionally significant color sensing occurs in unicellular flagellates.

The molecular mechanism of action remains one of the most interesting questions. As discussed above the mode of action *in vivo* appears to be dramatically different from that observed in heterologous model systems, highlighting the importance of photoelectric measurements in living *Chlamydomonas* cells. These measurements in combination with isolation of knock-out strains and expression of mutated opsins are expected to provide a powerful approach to investigation of the primary steps of photosensory transduction in green flagellates. Another challenging future task is biochemical purification of CSRA and CSRB, which are present in native cells in low concentrations amidst a large background of photosynthesis pigments. For purification, development of systems for heterologous expression is highly desirable. Once in pure form, the proteins will become accessible to a variety of incisive spectroscopic and crystallographic techniques.

Electrophysiological data indicate the involvement of an enzymatic amplification cascade and several types of ion channels in phototaxis and photophobic response in *Chlamydomonas* and other green algae (Sineshchekov and Govorunova, 2001). Identification of molecular elements of the photosensory signaling pathways downstream of the photoreceptors and the mechanisms of their interaction is a fascinating challenge.

As noted at the beginning of this Chapter, the microbial rhodopsin family is large and diverse, and therefore an important future direction would be extension of photophysiological studies to a wider range of species, including those outside Chlorophyta. Reports of type 1 rhodopsins in fungi (Bieszke et al., 1999a, b) and eubacteria (Béjà et al., 2000, 2001) show that these proteins are ubiquitously present in all main evolutionary lineages. Comparative analysis of rhodopsin-mediated signaling systems in different systematic groups will provide a deeper insight into their fundamental mechanistic principles and evolution.

Acknowledgements

We thank Elena Govorunova and Kwang-Hwan Jung for stimulating discussions. The work by the authors referred to in this review was supported by National Science Foundation Grant 0091287 and the Robert A. Welch Foundation.

References

Beck, C. and R. Uhl (1994) *J. Cell Biol.* 125, 1119–1125.

Beckmann, M. and P. Hegemann (1991) *Biochemistry* 30, 3692–3697.

Béjà, O., L. Aravind, E. V. Koonin, M. T. Suzuki, A. Hadd, L. P. Nguyen, S. Jovanovich, C. M. Gates, R. A. Feldman, J. L. Spudich, E. N. Spudich and E. F. DeLong (2000) *Science* 289, 1902–1906.

Béjà, O., E. N. Spudich, J. L. Spudich, M. Leclerc and E. F. DeLong (2001) *Nature* 411, 786–789.

Bieszke, J. A., E. L. Braun, L. E. Bean, S. Kang, D. O. Natvig and K. A. Borkovich (1999a) *Proc. Natl. Acad. Sci. USA* 96, 8034–8039.

Bieszke, J. A., E. N. Spudich, K. L. Scott, K. A. Borkovich and J. L. Spudich (1999b) *Biochemistry* 38, 14138–14145.

Bogomolni, R. A., W. Stoeckenius, I. Szundi, E. Perozo, K. D. Olson and J. L. Spudich (1994) *Proc. Natl. Acad. Sci. USA* 91, 10188–10192.

Braun, F. J. and P. Hegemann P. (1999) *Biophys. J.* 76, 1668–1678.

Briggs, W. R., C. F. Beck, A. R. Cashmore, J. M. Christie, J. Hughes, J. A. Jarillo, T. Kagawa, H. Kanegae, E. Liscum, A. Nagatani, K. Okada, M. Salomon, W. Rudiger, T. Sakai, M. Takano, M. Wada and J. C. Watson (2001) *Plant Cell* 13, 993–997.

Calenberg, M., U. Brohnsonn, M. Zedlacher and G. Kreimer (1998) *Plant Cell* 10, 91–103.

Crawley, J. C. W. (1966) *Planta* 69, 365–376.

Dieckmann, C. L. (2003) *Bioessays* 25, 410–416.

Diehn, B., M. Feinleib, W. Haupt, E. Hildebrand, F. Lenci and W. Nultsch (1977) *Photochem. Photobiol.* 26, 559–560.

Ebrey, T. G. (2002) *Proc. Natl. Acad. Sci. USA* 99, 8463–8464.

Ehlenbeck, S., D. Gradmann, F.-J. Braun and P. Hegemann (2002) *Biophys. J.* 82, 740–751.

Essen, L., R. Siegert, W. D. Lehmann and D. Oesterhelt (1998) *Proc. Natl. Acad. Sci. USA* 95, 11673–11678.

Foster, K.-W., J. Saranak, N. Patel, G. Zarrilli, M. Okabe, T. Kline and K. Nakanishi (1984) *Nature* 311, 756–759.

Foster, K.-W. and R. D. Smyth (1980) *Microbiol. Rev.* 44, 572–630.

Govorunova, E. G., K.-H. Jung, O. A. Sineshchekov and J. L. Spudich (2004) *Biophys. J.* 86, 2342–2349.

Gradmann, D., S. Ehlenbeck and P. Hegemann (2002) *J. Membr. Biol.* 189, 93–104.

Grigorieff, N., T. A. Ceska, K. H. Downing, J. M. Baldwin and R. Henderson (1996) *J. Mol. Biol.* 259, 393–421.

Harz, H. and P. Hegemann (1991) *Nature* 351, 489–491.

Hegemann, P. (1997) *Planta* 203, 265–274.

Hegemann, P., W. Gärtner and R. Uhl (1991) *Biophys. J.* 60, 1477–1489.

Hoff, W. D., K.-H. Jung and J. L. Spudich (1997) *Annu. Rev. Biophys. Biomol. Struct.* 26, 223–258.

Holland, E.-M., H. Harz, R. Uhl and P. Hegemann (1997) *Biophys. J.* 73, 1395–1401.

Hong, F. T. (2004) *CRC Handbook of Organic Photochemistry and Photobiology*, CRC Press, Boca Raton.

Jung, K.-H. and J. L. Spudich (2004) *CRC Handbook of Organic Photochemistry and Photobiology*, CRC Press, Boca Raton.

Jung, K.-H., V. D. Trivedi and J. L. Spudich (2003) *Mol. Microbiol.* 47, 1513–1522.

Kateriya, S., Nagel, G., Bamberg, E., and Hegemann, P. (2004) *News Physiol. Sci.* 19, 133–137.

Kolbe, M., H. Besir, L. O. Essen and D. Oesterhelt (2000) *Science* 288, 1390–1396.

Kreimer, G. (1994) *Int. Rev. Cytol.* 148, 229–310.

Kreimer, G. (2001) *Comprehensive Series in Photosciences*, Vol. 1 (Photomovement), pp. 193–227, Elsevier, Amsterdam.

Lawson, M.A., D. N. Zacks, F. Derguini, K. Nakanishi and J. L. Spudich (1991) *Biophys. J.* 60, 1490–1498.

Lin, C., H. Yang, H. Guo, T. Mockler, J. Chen and A. Cashmore (1998) *Proc. Natl. Acad. Sci. USA* 95, 2686–2690.

Linden, L. and G. Kreimer (1995) *Planta* 197, 343–351.

Litvin, F. F., O. A. Sineshchekov and V. A. Sineshchekov (1978) *Nature* 271, 476–478.

Luecke, H., B. Schobert, J. K. Lanyi, E. N. Spudich and J. L. Spudich (2001) *Science* 293, 1499–1503.
Luecke, H., B. Schobert, H. T. Richter, J. P. Cartailler and J. K. Lanyi (1999) *Science* 286, 255–260.
Malhotra, B. and A. D. M. Glass (1995) *Plant Physiol.* 108, 1527–1536.
Nagel, G., D. Ollig, M. Fuhrmann, S. Kateriya, A. M. Musti, E. Bamberg and P. Hegemann (2002) *Science* 296, 2395–2398.
Nagel, G., T. Szellas, W. Huhn, S. Kateriya, N. Adeishvili, P. Berthold, D. Ollig, P. Hegemann and E. Bamberg (2003) *Proc. Natl. Acad. Sci. USA* 100, 13940–13945.
Okamoto, O. K. and J. W. Hastings (2003) *J. Phycol.* 39, 519–526.
Olson, K. D. and J. L. Spudich (1993) *Biophys. J.* 65, 2578–2585.
Oroszi, L., A. Der and P. Ormos (2002) *Eur.Biophys. J.* 31, 136–144.
Pebay-Peyroula, E., G. Rummel, J. P. Rosenbusch and E. M. Landau (1997) *Science* 277, 1676–1681.
Quail, P. (2002) *Nat. Rev. Mol. Cell. Biol.* 3, 855–93.
Ridge, K. D. (2002) *Curr. Biol.* 12, R588–R590.
Royant, A., P. Nollert, K. Edman, R. Neutze, E. M. Landau, E. Pebay-Peyroula and J. Navarro (2001) *Proc. Natl. Acad. Sci. USA* 98, 10131–10136.
Saranak, J. and K.-W. Foster (1997) *Nature* 387, 465–466.
Schaller, K. and R. Uhl (1997) *Biophys. J.* 73, 1573–1578.
Schäfer, G., M. Engelhard and V. Müller (1999) *Microbiol. Mol. Biol. Rev.* 63, 570–620.
Schlicher, U., L. Linden, M. Calenberg and G. Kreimer (1995) *Eur. J. Phycol.* 30, 319–330.
Sineshchekov, O. A. (1988) *Phototrophic Microorganisms*, pp. 11–18, Acad. Sci. USSR, Puschino.
Sineshchekov, O. A. (1991a) *Light in Biology and Medicine*, Vol. II, pp. 523–532, Plenum Press, New York.
Sineshchekov, O. A. (1991b) *Biophysics of Photoreceptors and Photomovements in Microorganisms*, pp. 191–202, Plenum Press, New York.
Sineshchekov, O. A. and E. G. Govorunova (1999) *Trends Plant Sci.* 4, 58–63.
Sineshchekov, O. A. and E. G. Govorunova (2001) *Comprehensive Series in Photosciences* Vol. 1, pp. 245–280, Elsevier, Amsterdam.
Sineshchekov, O. A. and F. F. Litvin (1988) *Molecular Mechanisms of Biological Action of Optic Radiation*, pp. 412–427, Nauka, Moscow.
Sineshchekov, O. A. and W. Nultsch (1992) *Proc. Vth Int. Conf. on Retinal Proteins* Dourdan, France.
Sineshchekov, O. A., V. K. Andrianov, G. A. Kurella and F. F. Litvin (1976) *Fiziologia Rastenii* 23, 229–237.
Sineshchekov O. A., V. A. Sineshchekov and F. F. Litvin (1978) Doklady AN SSSR 239, 471–474.
Sineshchekov, O. A., F. F. Litvin and L. Keszthelyi (1990) *Biophys. J.* 57, 33–39.
Sineshchekov, O. A., E. G. Govorunova, A. Der, L. Keszthelyi and W. Nultsch (1992) *J. Photochem. Photobiol. B, Biol.* 13, 119–134.
Sineshchekov, O. A., E. G. Govorunova, A. Der, L. Keszthelyi and W. Nultsch (1994) *Biophys. J.* 66, 2073–2084.
Sineshchekov, O. A., K.-H. Jung and J. L. Spudich (2002) *Proc. Natl. Acad. Sci. USA* 99, 8689–8694.
Spudich, J. L. (1994) *Cell* 79, 747–750.
Spudich, J. L. (1995) *Biophys. Chem.* 56, 165–169.
Spudich, J. L., C.-S. Yang, K.-H. Jung and E. N. Spudich (2000) *Annu. Rev. Cell Dev. Biol.* 16, 365–392.
Sudo, Y., M. Iwamoto, K. Shimono, M. Sumi and N. Kamo (2001) *Biophys J.* 80, 916–922.
Suzuki, T., K. Yamasaki, S. Fujita, K. Oda, M. Iseki, K. Yoshida, M. Watanabe, H. Daiyasu, H. Toh and E. Asamizu (2003) *Biochem. Biophys. Res. Commun.* 301, 711–717.
Takahashi, T., K. Yoshihara, M. Watanabe, M. Kubota, R. Johnson, F. Derguini and K. Nakanishi (1991) *Biochem. Biophys. Res. Commun.* 178, 1273–1279.
Trissl, H.-W. (1990) *Photochem. Photobiol.* 51, 793–818.
Uhl, R. and P. Hegemann (1990) *Biophys. J.* 58, 1295–1302.
Vogeley, L., Sineshchekov, O.A., Trivedi, V.D., Sasaki, J., Spudich, J.L. and Luecke, H. (2004) Anabaena Sensory Rhodopsin: A Photochromic Color Sensor at 2.0 Å. *Science* 306, 1390–1393.
Witman, G. B. (1993) *Trends Cell Biol.* 3, 403–408.
Zacks, D. N., F. Derguini, K. Nakanishi and J. L. Spudich (1993) *Biophys. J.* 65, 508–518.
Zacks, D. N. and J. L. Spudich (1994) *Cell Mot. Cytoskeleton* 29, 225–230.

3
Visual Pigments as Photoreceptors

Masato Kumauchi and Thomas Ebrey

3.1
Introduction

3.1.1
General Considerations

This review is concerned with what is probably the most extensively studied photoreceptor system: the visual pigments. There have been a number of excellent reviews on this subject over the past four years (DeGrip et al., 2000; Essen, 2001; Meng et al., 2001; Menon et al., 2001; Okada et al., 2001; Teller et al., 2001; Sakmar, 2002; Stenkamp et al., 2002a; Stenkamp et al., 2002b; Hubbell et al., 2003), fueled in part by the X-ray structure of the prototypical visual pigment, bovine rhodopsin, first at 2.8 Å resolution (Palczewski et al., 2000; refined in Teller et al., 2001) and more recently at 2.6 Å resolution (Okada et al., 2002). This review is divided into four parts. The first two are on the initial (dark, unphotolyzed) state of vertebrate visual pigments and their light-activated (photolyzed, active) state, respectively. The third and forth sections will review what is known about these initial and light-activated states in the related invertebrate visual pigments.

Two more general questions should be addressed briefly before we proceed to the heart of this review. First, what defines a visual pigment and what kinds of species have them? Unremarkably, all vertebrate species investigated have been found to have eyes, photoreceptors, and visual pigments, presumably because the evolutionary advantage for the proto-vertebrate species was overwhelming. Perhaps remarkably, all invertebrate animals that have been investigated carefully have also been found to have some sort of a photoresponse (Figure 3.1). However, only a small number of invertebrate phyla have anything like a specialized body that could be called an eye, as opposed to a set of cells that are suspected of being photoreceptor cells. This group with image-forming eyes is not as large as one might expect, and the phyla which were known or suspected to have photoreceptors and pigments with retinal-based vision 50 years ago [see the nice reviews by Land (1981, 1992), Goldsmith (1972), and Crescitelli (1972)]: arthropods (insects, spiders, crustaceans, onychopho-

Handbook of Photosensory Receptors. Edited by W. R. Briggs, J. L. Spudich

ISBN 3-527-31019-3

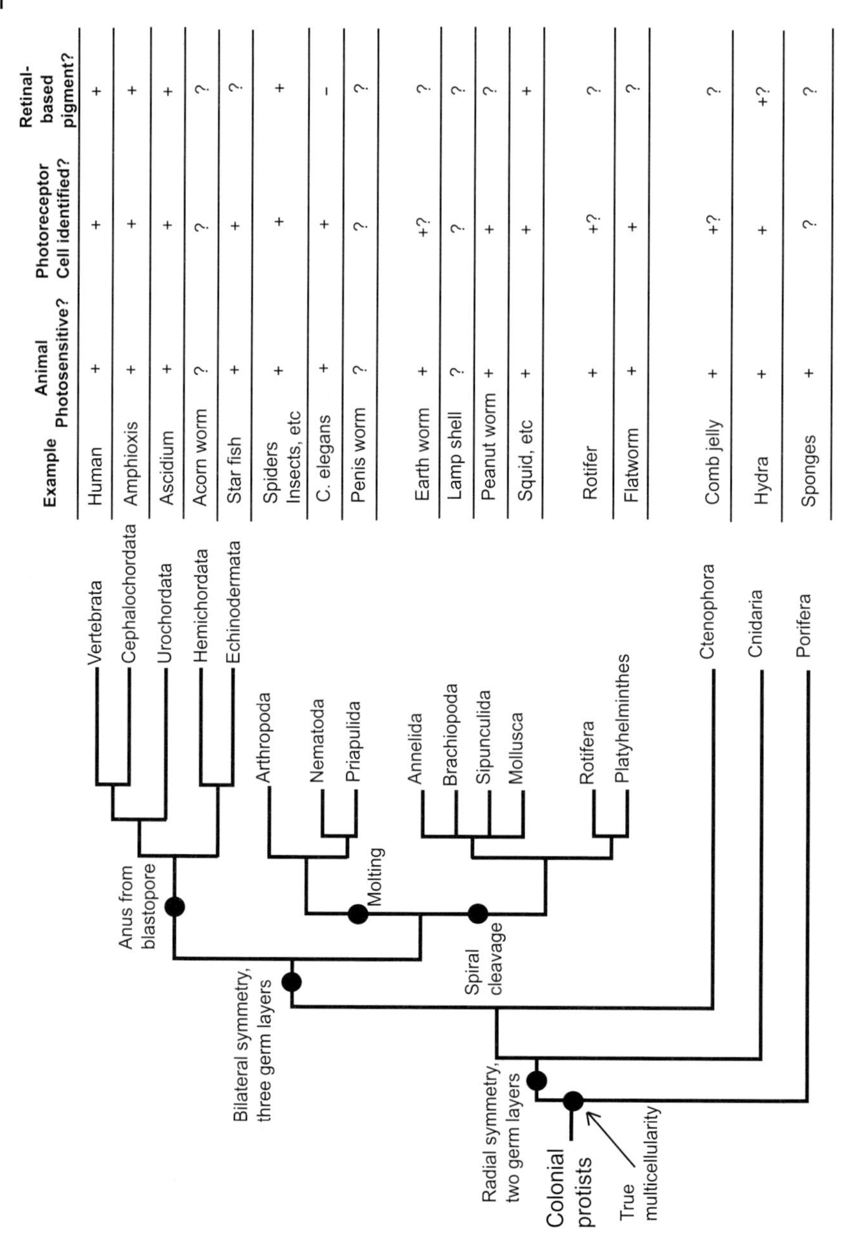

Example	Animal Photosensitive?	Photoreceptor Cell identified?	Retinal-based pigment?
Human	+	+	+
Amphioxis	+	+	+
Ascidium	+	+	+
Acorn worm	?	?	?
Star fish	+	+	?
Spiders Insects, etc	+	+	+
C. elegans	+	+	-
Penis worm	?	?	?
Earth worm	+	+?	?
Lamp shell	?	?	?
Peanut worm	+	+	?
Squid, etc	+	+	+
Rotifer	+	+?	?
Flatworm	+	+	?
Comb jelly	+	+?	?
Hydra	+	+	+?
Sponges	+	?	?

Figure 3.1 Identification of rhodopsins in animal phyla. There are about 30 multicellular animal phyla, one for vertebrates and the rest are for invertebrates. The scheme here, adopted from Gilbert, Developmental Biology, shows 15 of the invertebrate ones and the vertebrate one. Although there are examples in almost every invertebrate phylum of reports of photosensitivity (most easily found in Eakin, 1972), and many preliminary reports of photoreceptor cells (also Eakin, 1972), in only a few species have pigments been unambiguously identified (arthropods, mollusks, and most recently the protochordates, ascidian (Kusakabe et al., 2001) and amphioxus (Miyata, cited in Koyonagi et al., 2002).

ra), annelids (worms), and mollusks (e.g. octopus, squid), have had added to their numbers only two subphyla of protochordates, an ascidian (Kusakabe et al., 2001; Nakashima et al., 2003) and amphioxus (Miyata 2000, cited in Koyanagi et al., 2002).

Early authors, including Wald (1945) suggested that there may be a retinal protein mediating phototaxis in such motile unicellular organisms as the alga *Chlamydomonas*. Recently retinal proteins have been found in this green alga (Sineshchekov et al., 2002; see also Nagel et al., 2002), in fungi (Brown, 2004), and in marine eubacteria (Béjà et al., 2001). These retinal proteins are very different from visual pigments but are closely related to retinal-based pigments found in some halophilic members of the archaea kingdom, bacteriorhodopsin and its cousins, sensory rhodopsin I and sensory rhodopsin II. This microbial group is sometimes referred to as type 1 rhodopsins with visual pigments and their retinal-containing relatives in non-visual tissues referred to as type 2 rhodopsins (see Spudich and Jung, Chapter 1). All of these pigments seem to utilize *all-trans* retinal as their chromophore while visual pigments use 11-*cis* retinal.

Fairly early on it was recognized that visual pigments are proto-typical members of the G-protein Coupled Receptor (GPCR) super family (Strader et al., 1994). With the explosion of sequence data it is clear that the visual pigments have little sequence similarity to the microbial pigments, but there are still some possible ambiguities. These arise most cogently in two pigments, retinochrome found in mollusks (see Hara et al., 1991; Pepe et al., 1997), and RGR (for **R**etinal **G**-protein coupled **R**eceptor, (Shen et al., 1994)) which is found in the pigment epithelium of several vertebrates. Both retinochrome and RGR have significant sequence similarity to the visual pigments and other GPCRs; but each preferentially binds *all-trans* retinal and uses light to convert it to 11-*cis* retinal. That is, these pigments do not subsume a sensory function but rather act as photoisomerases, to produce 11-*cis* retinal, presumably for the visual system (see also Figure 3.2 and Table 3.1, below). Other related retinal-binding GPCRs will be discussed below.

Most GPCRs and all visual pigments seem to share several motifs. They all have seven transmembrane segments, with the amino terminus on the extracellular side and the C terminus on the cytoplasmic side of the cell membrane. They have glycosylation sites on their amino terminal region and phosphorylation sites on their C-terminal region. They have several conserved prolines and glycines in the transmembrane segments. They have two conserved cysteines which form a disulfide bond. And they have a triplet of residues on the third transmembrane helix near the cytoplasmic side (either ERY or DRY) that is a crucial part of the interaction of this surface with the G-protein, when the receptor is activated. Aspects of GPCRs which are more specific to visual pigments are discussed below.

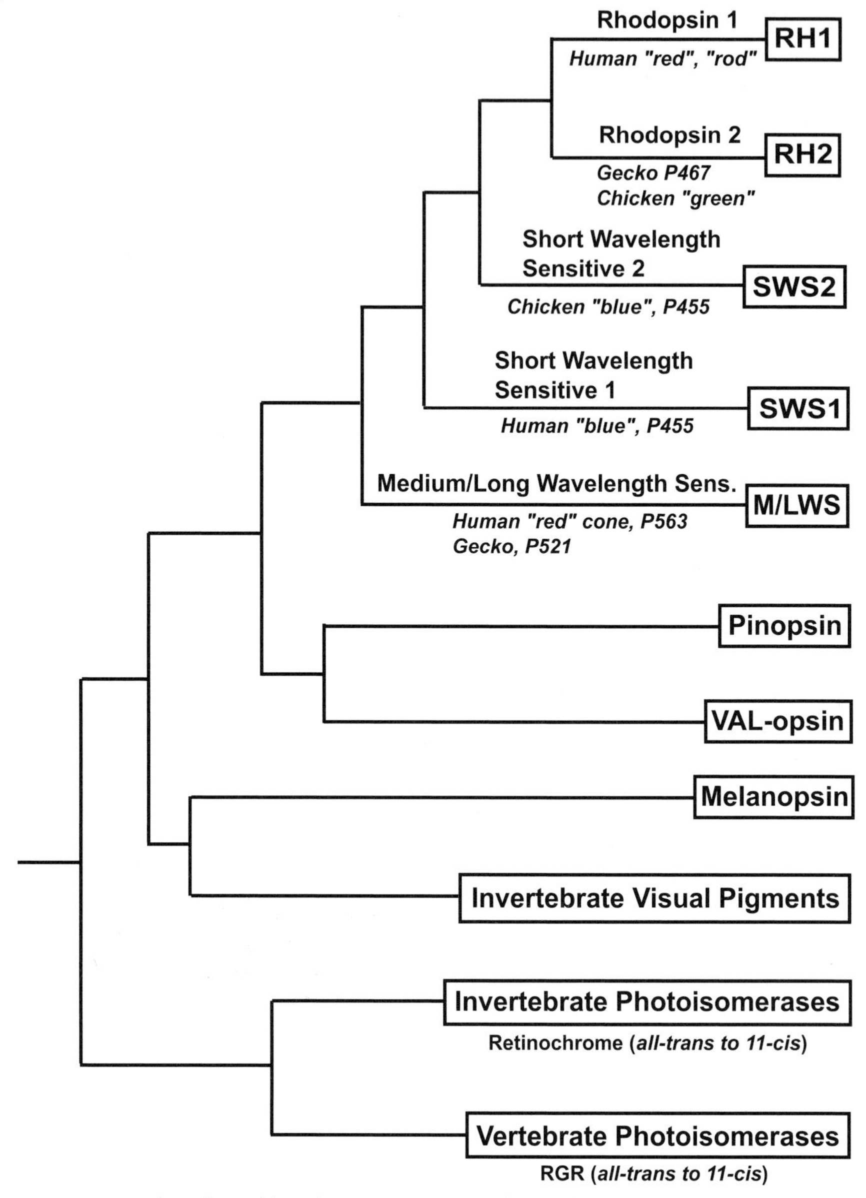

Figure 3.2 Families of retinal-based pigments in vertebrates and invertebrates. There are five visual pigment families found in the retinae of vertebrates as well as several non-rod, non-cone pigments in other cells of the retina as well as the photoisomerases.

Table 3.1 Amino acid residues in the binding-pocket of visual pigments and related retinal proteins

	Visual Pigments									Retinal Photoisomerases		Vert. Non-Rod Non-Cone
	Vertebrate					Uroch-ordate	Invertebrate					
Residue number	SWS1	SWS2	M/LWS	RH2	RH1	Ci-opsin	Mollusc	Insectia	Crusta-ceans	RGR	Retino-chrome	Melan-opsin
43	F	F	W,(F)	Y	Y	Y	F,Y	V,Y,F,L,T,I,(A)	F,(V,I)	G	G	I,(V)
44	M	M	M	M,(L,I)	M	M	I	Y,I,M,T,F	M,Y	T	S	L
47	V	L,I,(K,T)	V	L	L	V	C,V	L,I,F	T,L	L	L	G
83[1]	G	N	D,(E)	D,G,(N,S)	D,N	D	D	D,(N)	D	D	A	D
99[1]	S,C	G	A	G	G	G	G	M,A,N,T,Q	M,N	A	D	T,(S)
94	V	A,(S)	S	T	T	S	M,K,L	M,I,V,L	M,F	A	A	V,(I)
113	E,(D)	E	E	E	E	E	Y	Y,F	Y	H	M	Y
114	A,G	G	G	G	G,(A)	G	G,A	A,G,(S)	A,G	G	G	A
117	G	A,(S)	V	A	A	V	G	G	G	G	G	G
118	T,S,A	T	S,(A)	T	T	S	G	S,(A,G)	N,A	F	F	A
120	A,G,S,T	G	C	G	G	F	F	F,S,(T,Y,A)	C,T,F	T	G	F
121	G	G	G	G	G	G	G	G	G	A	G	G
122	L,(M)	M	I	Q,(E)	E,(Q,I)	V	L,F,V	C,I,M,(T,S)	C,V	L	M	I
164	G	G,A	A,S	A,(S)	A,(G)	S	S	S,A,T,C,V	S,(A)	S	G	S,A
167	V,(C)	A	W	C	C	W	W,L	W,F	W,(C)	W	W	W
178	Y,F	Y	Y,(F)	Y,(F)	Y	Y	Y	Y,F	Y	Y	Y	Y
181	E	E	H,(Y)	E	E	E	E	E	E	E	E	E
184	Q,(G)	Q,(H)	K	Q	Q,(K)	G	L,Q	L,M	L,(N)	Q	I,G	L,R,Q,M

Light grey: Residues in the binding site are conserved; so cannot regulate color.
Dark grey: There is good evidence these a.a residues do regulate color.
White, Plain = Part of the binding site that does vary from pigment to pigment but so far has not been shown to regulate color.

1) On the cytoplasmic side Asp83 H-bonds with Asn55 and carbonyl of residue 299 as well as Ala120 via a water (Palczewski et al., 2000); Nagata et al., 1998).

Table 3.1 Continued

	Visual Pigments									Retinal Photoisomerases		Vert. Non-Rod Non-Cone
	Vertebrate					Uroch-ordate	Invertebrate					
Residue number	SWS1	SWS2	M/LWS	RH2	RH1	Ci-opsin	Mollusc	Insectia	Crusta-ceans	RGR	Retino-chrome	Melan-opsin
186	S,A	S	S	S	S,(T)	S	N,S	S,A,T,(V)	G,S	C	S	S
187	C	C	C	C	C	C	C	C	C	C	C	C
188	G	G	G	G	G	A	S,T	G,S,T	G,S	T	T	S,T
189	P	P	P	P	V,I,(M,P)	P	F	F,T,I	T,Y	L	I	W
191	W	W	V	Y	Y,(H)	W	Y	Y,(F)	Y	Y	Y	Y
207	L,(I)	L,(I,S)	L	M,(L)	M	Y	I,F,L	I,F,V,I,T,A	L,(N)	M	L	C,(V)
211	C	C	C	H	H	C	M,C,F,L,V	Y,V,F,C,L,(P,A)	F,(Y)	N	S	I,L
212	F	F	C	F,(S)	F	F	F,C,G,I	I,L,T,C,V,F	L	F	F	P
261	F	F	Y,F	F	F,(Y)	F	F,S	W,F,(Y)	W	L	S	F,(Y)
265	Y	W	W	W	W,(C)[3]	W	W	W	W	W	W	W
268	Y	Y	Y	Y	Y,(N)[3]	Y	Y	Y	Y,C	Y	F	Y
269	A	A,T,S,(C)	T,A	A	A	A	A	L,G,A,(T,S)	L,A	A	G	S,A
292[2]	A,S	S	A,S	A	A,S	A	V,M	A,C,(S)	Y,A	A	P	A
293	F,(L)	C,V	Y,F	F	F	F	M,L	C,V,L,T	V,L	L	I	V
295	S	S,(C)	A	S	A	A	A	A,C	A	A	S	A
296	K	K	K	K	K	K	K	K	K	K	K	K
299	C,(S)	T	T	A,S,(C)	A,S	T	A,S	A,S	V,(C)	P	C	A

Light grey: Residues in the binding site are conserved; so cannot regulate color.
Dark grey: There is good evidence these a.a residues do regulate color.
White, Plain = Part of the binding site that does vary from pigment to pigment but so far has not been shown to regulate color.

2) Several authors have implicated the S292A change in regulating the absorption spectum of RH1, M/LWS visual pigments.

3) This residue should be reconfirmed.

3.1.2
Photoreceptors and Pigments

Vertebrate visual pigments are contained in two types of photoreceptors, rods and cones (reviewed in Ebrey and Koutalos, 2001). There are five distinct types of visual pigments found in these photoreceptors with several other types of retinal proteins found in other cells in the retina and in extraocular sites (Figure 3.2). Of the five types found in photoreceptors, the RH1 (Rhodopsin 1) type is usually found in rods while the other four, RH2, M/LWS (long and mid wavelength sensitive), SWS1 (short wavelength sensitive 1) and SWS2 are usually found in cones. But there are exceptions to each of these "rules". For example, a M/LWS sensitive pigment is found in a classic rod of the Tokay gecko (*Gecko gecko*) with a RH2 pigment found in another type of rod in the gecko eye. A third type of rod, whose ultrastructure has not been extensively studied, contains a SWS1 pigment. Another counter example is found in the chameleon retina (Kawamura et al., 1997) where a RH1 pigment, usually found in rods, is in cones.

It is reasonable but somewhat misleading to take the human photoreceptor situation as the standard for vertebrates. In our retinae, two closely related M/LWS pigments are found in our cones with a member of the SWS1 family in another type of cone (Nathans, 1987). The former are often called the "green" and "red" cone pigments while the latter is usually called the "blue" cone pigment. These color names are misleading and should be avoided unless used in the human context (see Ebrey and Takahashi 2001). Because human retinae contain two types of photoreceptors that are easily distinguished (rods and cones), humans and many other animals are usually called diurnal, with the reasonable assumption that rods are responsible for dim light vision while cones are responsible for bright light vision. However, from a visual pigment point of view this may be too simple. The SWS1 pigment found in human cones is about as distantly related to the M/LWS pigments as they are to the RH1 pigments found in human rods and so, perhaps one should say there are three kinds of photoreceptors in the human retina, those containing the RH1 pigment, those containing M/LWS sensitive pigments and those containing the SWS1 pigment. It has long been noted that human SWS1 cones differ in numerous ways compared to human M/LWS cones (see e.g. Wandel (1995); Mollon (1977)). It well may be that the human visual system should be considered not a diurnal but a triurnal one.

The photoreceptors of humans and other mammals contain just three of the five visual pigment families, RH1, SWS1, and M/LWS. As for the other vertebrate animal families, in the sparsely studied cyclostome (e.g. lamprey) and chondrichthyes (e.g. sturgeon) retinae, only RH1 pigments have been definitely identified but almost certainly other visual pigment families will be found. Much better data are available for teleosts (bony fish), amphibians, reptiles, and birds. While some retinae in these animal families do contain representatives of only two or three of visual pigment families, there are well documented cases in each of these animal families of retinae in which members of all five members of the visual pigment families are found (reviewed in Ebrey and Koutalos, 2001).

3.1.3
Non-photoreceptor or "Non-rod", "Non-cone" Retinal Pigments

Besides these "photoreceptor" visual pigments, a large number of other retinal-based pigments in animals have been found, whose functions have been difficult to assign with certainty. The entire group of these pigments, which include pinopsin, parapinopsin, peropsin, melanopsin, neuropsin, encephalopsin, vertebrate ancient (VA) opsin, and teleost multiple tissue (TMT) opsin, is reviewed by Hankins and Foster (Chapter 5). One or more of these pigments is probably involved in such light detection but in non-image forming (i.e. non-seeing) tasks as the photoentrainment of circadian rhythms (see this volume). Recently this photoentrainment function has been localized to non-photoreceptor cells in mammalian retinae. In mammals a pigment called melanopsin has been found in some of the retinal ganglion cells and it contributes to the photosystem responsible for entraining the light–dark cycle (see Lucas et al., 2003; Van Gelder et al., 2003). It is a protein which looks very much like an 11-*cis* retinal-binding, GPCR-like photopigment and is most similar to invertebrate pigments (Provencio et al., 1998). Melanopsin is found in a set of photosensitive retinal ganglion cells and its presence is correlated with the ability to maintain circadian rhythms (Panda et al., 2002; Ruby et al., 2002; Lucas et al., 2003). Most interestingly, the most important motif in GPCRs that couple them to the G-protein, the triplet of residues ERY (for vertebrate visual pigments, more generally E or D/R/Aromatic) is not found in melanopsin (see Provencio et al., 1998, for sequences), leaving open the door that either its signaling pathway is not a G-protein coupled one or that the neutral asparagine at position 135 in melanopsin is acting like an already protonated Asp, i.e. it is already in its signaling state (see below, Section 3.3.6) but the activation is inhibited by other factors.

An unsolved mystery is why melanopsin, found in vertebrates, has a sequence similarity closer to invertebrate visual pigments than vertebrate ones (see Figure 3.2). Recently Arendt (2003) has proposed that the common precursor to both vertebrates and invertebrates had two distinct types of opsin-producing cells. One of these evolved to the vertebrate pigment containing cells, the rods and cones. He suggests that the other opsin-containing cell type, which was the precursor to the rhabdomeric invertebrate photoreceptor pigments, remained with the vertebrates and evolved into ganglion and other cell types that contain melanopsin.

Perhaps related to melanopsin is pinopsin, which is found in the avian pineal organ (Okano et al., 1994). Mammals do not have such an organ and no similar pigment is found (melanopsin may be the corresponding pigment in mammals but both melanopsin and pinopsin are equally far in sequence from the visual pigments and from each other). Pigments similar to pinopsin have been found in amphibians and reptiles (e.g. Max et al., 1995). Teleosts seem to have their own version of a retinal binding GPCR-like protein which is a non-visual pigment, called VA (vertebrate ancient) opsin (Soni et al., 1997). Other pigments which are similar but distinct from these are parapinopsin (Blackshaw et al., 1997) and encephalopsin (Blackshaw et al., 1999).

3.1.4
Retinal Photoisomerases

As noted above, RGR is a distinct type of pigment found in the pigment epithelium (Shen et al., 1994), not in the retina, of several vertebrate species and appears to be quite similar to an invertebrate pigment, retinochrome (see Figure 3.2). Like retinochrome, this pigment binds *all-trans* retinal and light photoisomerizes it to the 11-*cis* isomer which is then released from the protein; it is thought that this pigment is not a photosensor but a pigment that uses the energy of the photon to regenerate 11-*cis* retinal for use by another, photosensory, pigment. Another pigment, peropsin (Sun et al., 1997) may have a similar function.

3.2
The Unphotolyzed State of Vertebrate Visual Pigments

3.2.1
Structure of Visual Pigments: the Chromophore

All known visual pigments consist of the same chromophore, 11-*cis* retinal (or its close analogues 11-*cis* 3, 4 dehydroretinal, 3-hydroretinal, and 4-hydroxretinal), covalently bound to a lysine of its apoprotein, opsin, via a Schiff base linkage. We will first discuss the chromophore, then the pigment in its dark, unphotolyzed state. The chromophore's shape has been investigated through a number of approaches. These have been used to both sketch out the shape of the 11-*cis* retinal binding pocket in the protein and to specify the chromophore's conformation *in vivo*. In particular the following questions have been addressed: What isomers of retinal can bind? Is the 6, 7 single bond, s-*cis* or s-*trans*? Is the 14, 15 single bond s-*cis* or s-*trans*? Is the chromophore chain between carbons C10 to C13 completely planar or is it twisted? What enantiomer of 11-*cis* retinal binds? The conformation and configuration of retinal in opsin have been extensively studied (Liu, 1990; Lou et al., 2000). While we would like to know its structure in the pigment, as determined by a complete solving of the structure at high resolution, so far the data (PDB 1L9H, 1F88, 1GZM and 1HZX) do not allow any strong conclusions about the configuration of the chromophore *in situ*. Therefore data has come from experiments designed to focus on the structure of the chromophore itself. A planar structure is energetically favorable for a unhindered conjugated polyene because the overlap of its π-orbitals is maximal; however, retinals in visual pigments are sterically hindered and so cannot be exactly planar. For all isomers, there is steric hindrance of the β-ionone ring and the chain due to the steric hindrance between 5-Me and 8-H, causing the ring to be twisted out of the plane. The ring-chain can be in an s-*cis* or s-*trans* conformation. Early studies of the retinal in rhodopsin employed solid-state ^{13}C-NMR spectra to show that the configuration of the C6–C7 bond is s-*cis* (Mollevanger et al., 1987; Smith et al., 1987). In contrast, in bacteriorhodopsin this bond is 6-s-*trans* (Harbison et al., 1985; Fujimoto et al., 2001). Nakanishi and coworkers (Fujimoto et al., 2001) confirmed that the configuration of the 6-*s*-bond of the chromophore in bovine rhodopsin was *cis* and determined the helicity between the β-ionone ring and polyene chain. Spooner et al., (2002) have estimated that the out-of-plane twist is about –28 degrees.

For the 11-*cis* isomer, there is also steric hindrance in the chain, between 13-Me and the 10-H, forcing the plane of the polyene chain between carbons 7 and 12 to be twisted away from the plane of the chain determined by C13 to the Schiff base. Indeed Pauling (1939) predicted that isomers like 11-*cis* would be too unstable to be found in nature. Nakanishi and coworkers (Fujimoto et al., 2001), using pigments that were regenerated with sterically locked chromophores which were enantiomeric, showed the absolute direction of this twist. Only one of the enantiomers formed a pigment (Lou et al., 2000; Fujimoto et al., 2001), indicating that the binding pocket of the visual pigment is constructed enantioselectively. Also with respect to the chain, solid-state NMR spectroscopic studies determined that the H-C10-C11-H torsional angle was $160\pm10°$ and the 10–11 bond was in the s-*trans* configuration (Feng et al., 1997). Creemers et al., (2002) made the complete assignment of ^{1}H and ^{13}C chemical shifts of the chromophore with 2D spectral measurements and their analyses.

In order to measure the NMR spectrum of the detergent-solubilized visual pigment, in most cases, solid-state spectra are taken. However, in a solution NMR study of rhodopsin, Klein-Seetharaman and coworkers (2002) were able to ^{15}N-label the eight Lys in the opsin, and a sharp ^{15}N signal from the Schiff base was obtained.

3.2.2
Overall Topology of the Pigment

The tremendous progress which has been made recently on the structure of rhodopsin has been reviewed in some detail in several papers mentioned in the first paragraph of this review. Here we will just discuss a few salient points.

One of the very first inferences made from the sequence is that rhodopsin has seven hydrophobic stretches of sequence which were reasonably incorporated into a model of seven transmembrane helices forming a circle and each connected to its adjacent helices (Ovchinnikov, 1982; Hargrave et al., 1983). Although a popular model, strong evidence for each of the seven helices, as well as their exact lengths was not quite as good as one would have liked (see review of Tang et al., 1995, about these concerns). The X-ray structure immediately confirmed the favored model, and, of course, added much more (Palczewski et al., 2000; Teller et al., 2001; Okada et al., 2002).

Besides establishing that the seven transmembrane helices (Hargrave et al., 1982; Ovchinnikov 1982) were arranged next to each other in a circular pattern, the structure confirmed the notion that several of the helices were not strictly perpendicular to the plane of the membrane (Baldwin et al., 1997; Unger et al., 1997). There were at least two major surprises from the structure. First, it was found that the loop between transmembrane helices 4 and 5 did not position itself in the cytoplasm as all previous models had supposed, but rather looped into the membrane, forming a beta-pleated sheet structure that is an important part of the chromophore binding pocket (see Figure 3.3). This explained many otherwise puzzling observations. The second surprise

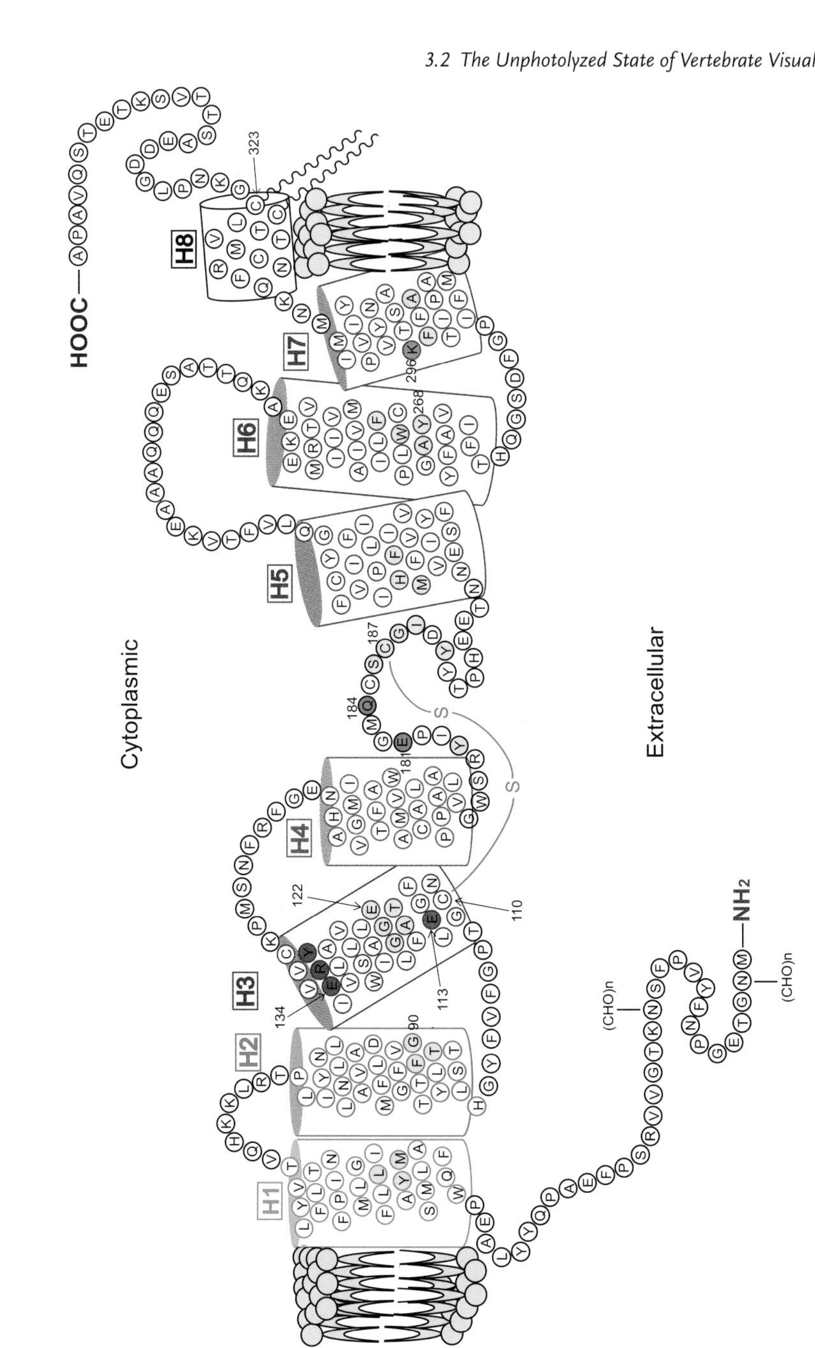

Figure 3.3 Two-dimensional rendering of the structure of bovine rhodopsin based on the X-ray structure. Note the beta-pleated sheet formed by the loop between Helix 4 and Helix 5 that penetrates into the center of the pigment, and Helix 8 which lies on the cytoplasmic surface of the pigment. Residues in the binding pocket (Palczewski *et al.*, 2000) are colored yellow, the chloride binding site in M/LWS pigments green, the Schiff base lysine K296 blue, the counterion E113 red, and the trio in helix3, crucial for activation, red.

was that part of the rather long C-terminal end of the molecule was partly folded into an alpha helix that lay approximately on the surface of the membrane. A very conserved motif in GPCRs is the NPxxY domain in the seventh transmembrane helix and at least part of the region of interaction of the eighth helix is with this peptide. Finally it should be noted that almost all of the seven transmembrane helices contained kinks associated with prolines and glycines, mostly near the center of the structure. The kinks aid in forming the binding pocket and they may also be involved in the activation process.

Besides the polypeptide itself there are several post-translational modifications which are important for the structure and function of visual pigments. Some are permanent, such as: the addition of one or two carbohydrate chains near the N-terminal end of the molecule; one or two presumed palmitoylated cysteines near the C-terminal end of the pigment (demonstrated so far just for the RH1 family); the chromophore itself, of course; and disulfide formation from two highly conserved cysteines at positions 110 and 187. Early reports of transient methylation of rhodopsin have not been confirmed but transient phosphorylation of the C-terminal region in the cytoplasm is a key reaction for shutting off the transduction cascade after light activation (Hurley et al., 1998; Kennedy et al., 2001; Arshavsky, 2002).

The visual pigment can conveniently be divided into three parts: the extracellular surface, the cytoplasmic surface, and the hydrophobic core. The extracellular surface is of interest not only for the beta-pleated sheet "plug" discussed above, but because, for many ligand-activated GPCRs, this is the side from which the activating ligand binds. The chromophore lies significantly closer to the extracellular side than to the cytoplasmic side, which is perhaps not too surprising since its binding site would be at a similar position as that of the small molecules like acetylcholine that activate such GPCRs as the muscarinic ACh receptor.

3.2.3
Cytoplasmic Domain

The cytoplasmic surface of rhodopsin interacts with at least four different kinds of proteins: a) the G-protein itself, which will become activated by the catalytic release of its GDP and the binding of GTP; b) a kinase, which will phosphorylate one or more residues on the C-terminal tail of rhodopsin (reviewed in Hurley et al., 1998); c) the small protein arrestin, which binds to the phosphorylated rhodopsin, inhibiting further binding of the G-protein, and finally d) a phosphatase, which must remove the phosphates from the light-activated rhodopsin, as part of the process of returning it back to its initial dark-adapted, regenerated, unphotolyzed state. There appear to be two or three key areas of interaction of the light-activated rhodopsin with the G-protein, the triad of amino acid residues, E/D//R//Aromatic, at the cytoplasmic end of Helix 3, and the NPxxYx(x) at the end of Helix 7 (Ernst et al., 2000; Marin et al., 2000); parts of the loop connecting helices 5 and 6 are also probably involved (Hofmann, 1999, 2000). Of course, it is necessary to remember that the G-protein-activating form of rhodopsin almost certainly has a different conformation than that of the unphotolyzed pigment. Nevertheless, the unphotolyzed structure is a plausible starting place to discuss the activation process (Section 3.3).

The location and structure of much of the cytoplasmic surface has also been examined with both site-directed spin labeling (Hubbell et al., 2003) and NMR (Yeagle et al., 2002) and both methods confirm, and in some cases, extend the X-ray results. These studies are particularly important since the cytoplasmic surface was the least well resolved in the X-ray studies.

3.2.4
The Hydrophobic Core of Rhodopsin and the Retinal Binding Pocket

Palczewski et al., (2000) noted that ca. 30 amino acid residues contribute to the binding pocket of 11-*cis* retinal. The king of the binding pocket is the residue to which the retinal is bound covalently, K296; in most cases the resulting Schiff base linkage is positively charged (see below, in Section 3.2 on color regulation); the queen is the counter-ion to the Schiff base, E113, whose presence was predicted 25 years ago (Honig et al., 1979b). In general the counter-ion residue is highly conserved in vertebrate visual pigments but interesting things happen in both invertebrate visual pigments and in the non-rod, non-cone pigments involved in photoentrainment (see below, Section 3.4). Table 3.1 shows residues present in the retinal binding pocket for each family of pigments as well as two residues that are not in the pocket but which can have large effects on the color of the pigment.

The two ends of the chromophore, the polyene chain attached to the lysine (from C15 to *ca.* C9), and the ring end of the retinal (from C1 to C8) have somewhat different types of pockets, probably to aid in solving different types of physiological problems. In particular the residues near the ring (See Figure 3.4) seem to hold the ring in place but are often not strictly conserved, but vary in such a way as to control the absorption spectrum (see below). The residues near the lysine and the retinal chain near the Schiff base tend to be conserved and we propose that they help set the high quantum efficiency of the phototransduction event. Note that many of these residues near the ring can be polar, probably helping to influence the absorption spectrum of the pigment even though they are in a generally non-polar environment. The role of some of the specific residues in the binding pocket will be discussed below.

Teller et al., (2001) noted that in several cases the residues on one of the helices formed hydrogen bonds with residues on other helices. Since the present structure does not resolve the position of many of the water molecules, which could bridge between hydrogen-bonding partners, the complete set of hydrogen bonds must await a higher resolution structure. In addition, some possible hydrogen bonds and salt bridges that might exist in bovine rhodopsin, cannot occur in many other visual pigments because the hydrogen-bonding or salt-bridging partners are not conserved through all the visual pigment families. (e.g. H211-E122 may form a salt bridge in RH1 pigments but not in the other families.)

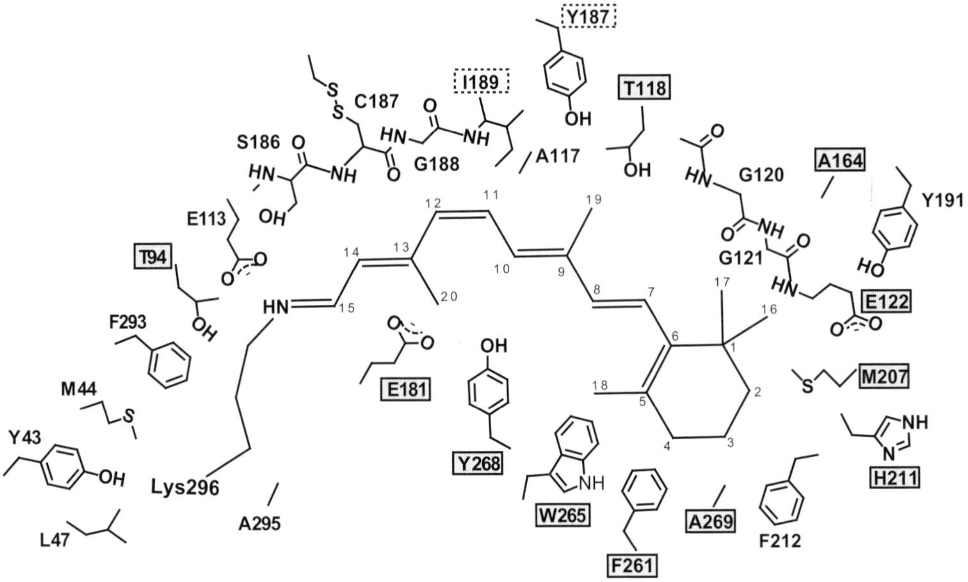

Figure 3.4 Two dimensional rendering of the residues in the binding pocket of the 11-*cis* retinal chromophore, following Palczewski et al. (2000). The residues that are tinted are known to be used to regulate the color of visual pigments. Note that they tend to be near the beta-ionone ring end of the pigment.

3.2.5
The Extracellular Domain of Rhodopsin

The extracellular domain is the site of several interesting properties of some rhodopsins and GPCRs. In the M/LWS visual pigments, the two residues usually conserved for the other four visual pigment families Glu181 and Gln184, on the link connecting helices 4 and 5, are altered to form a chloride-binding site (Wang et al., 1993). The effect of the chloride is both to shift the absorption spectrum to much longer wavelengths (Crescitelli et al., 1991) and to raise the pK of the Schiff base (Yuan et al., 1999).

3.2.6
Structure of Other Visual Pigments

Visual pigments from each of the five vertebrate families, the “non-rod, non-cone” retinal pigments, and invertebrate visual pigments have in some cases only 25% sequence identify with each other. Nevertheless they have many amino acid residues which are identical and these not only form the binding pocket and position the seven transmembrane helices, but also introduce the kinks and bends in the helices. The conservation of these residues can be seen in an alignment of the conserved residues in each of five vertebrate pigment families (Ebrey and Takahashi 2001). When representative members of the SWS1, SWS2 and M/LWS families (Stenkamp et al., 2002b; Atkinson, Parson, and Ebrey, cited in Ebrey and Takahashi (2001)) are threaded into the bovine rhodopsin structure and the resulting structure energy-minimized, no large changes compared to bovine rhodopsin are seen, suggesting that all these pigments have very similar structures.

3.2.7
Protonation State of Some of the Carboxylic Acids of Rhodopsin

Several of the carboxylic acids of visual pigments play key roles in the functional properties of the pigments. The protonation states of five conserved carboxyls in RH1 visual pigments have been determined: Glu113 is unprotonated and serves as a negative counter-ion to the positive protonated Schiff base. Asp83 and Glu122 are present in most RH1 pigments and may determine some of the differences between rod type and cone type pigments (Imai et al., 1997); both are protonated. Glu134 is unprotonated (Sakmar et al., 1995) and its possible protonation during the bleaching sequence of rhodopsin probably plays an important role in transducin activation (Fahmy, 1998). Yan et al., (2002) have argued that Glu181 is protonated, that is, neutral. More recently Yan et al., (2003) reported that the counter-ion to the protonated Schiff base may switch from Glu113 in the initial state to Glu181 in the photointermediate Meta I. The generality of the proposed mechanism is questionable since there is no Glu181 in the M/LWS visual pigment family. Birge and Knox (2003) proposed some alternative hypotheses. Of the carboxylic acids only Glu113, Glu134, and Glu181 (or 181 replaced by chloride) are found in every vertebrate visual pigment; in invertebrates Glu134 becomes Asp134. Note that two of these carboxyls are located in the hydrophobic interior of rhodopsin.

3.2.8
Internal Waters in Visual Pigments

Besides the apoprotein and the chromophore, there is a third component of every visual pigment: bound, internal water molecules. Maeda and co-workers (Maeda et al., 1992; reviewed in Maeda et al., 1997), using difference FTIR and isotopic substitutions, have shown that at least one water molecule is hydrogen bonded to the Schiff base/counter-ion; this finding is consistent with the hydrogen/deuteron exchange experiments in Deng et al., (1994) and theoretical considerations of the necessity of water at this location (Scheiner et al., 1991). Quite recently a water was located that is hydrogen bonded to Glu113 (Okada et al., 2002) (see Figure 3.5); this is almost certainly the water that Maeda and co-workers detected. An especially attractive aspect of the FTIR method is its ability to follow changes of the waters through each of the photointermediates.

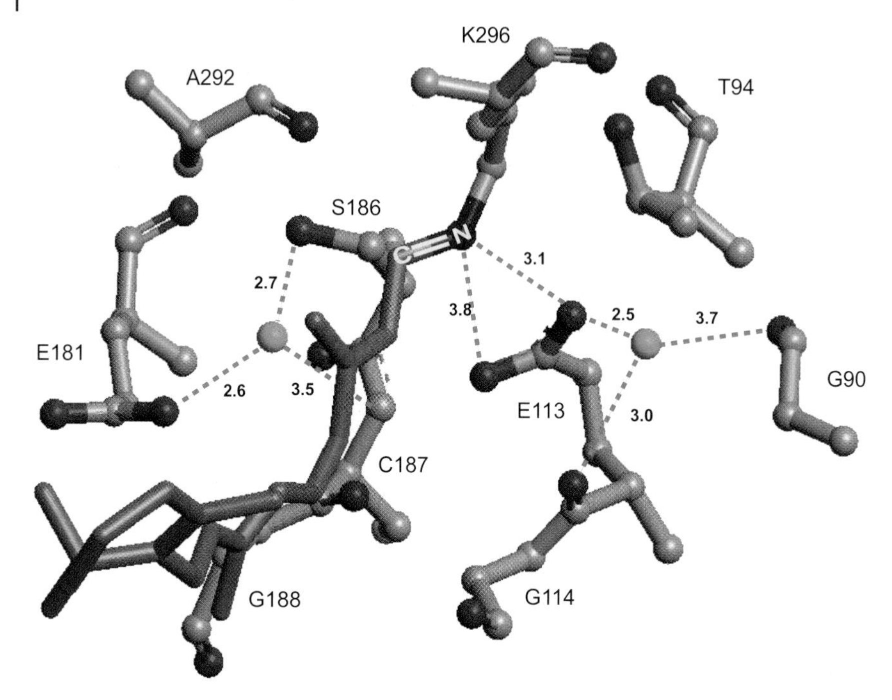

Figure 3.5 View of the region of the Schiff-base/counter-ion of bovine rhodopsin from the X-ray structure of Okada et al. (2002).

3.2.9 Is Rhodopsin a Dimer *in vivo*?

Evidence that many GPCRs are functional dimers [see e.g. Kunishima et al., (2000); and Bouvier, (2001)] have no doubt spurred new investigations on visual pigments. Recently there have been reports of dimerization of vertebrate visual pigments in situ based on Atomic Force Microscopy images (Fotiadis et al., 2003; Liang et al., 2003). This may be an artifact of the preparation, since there is convincing evidence that in their unphotolyzed state at room temperature in photoreceptor membranes that visual pigments are free to rotate and translate in the rod plasma membrane and do not form large aggregates (see summary in Chabre et al., 2003).

3.2.10 Functional Properties of the Unphotolyzed State of a "Good" Visual Pigment

There are at least three functional properties that a "good" visual pigment should have in its unphotolyzed state. These are a) control of its absorption spectrum, b) a low rate of thermally activation of the transduction cascade, and c) a high quantum efficiency for the photochemical process initiating activation of the visual pigment. Each of these properties is dealt with in this section.

3.2.10.1 The Absorption Spectrum of Visual Pigments

The light absorbing element of visual pigments is a retinal chromophore attached to a lysine of the apoprotein (opsin) as a Schiff base. There are two distinct but related factors which are involved in the control of the absorption spectrum of visual pigments: protonation state of the Schiff base and modulation of the default absorption spectrum. For a retinal based pigment to absorb much into the visible (beyond ca. 420 nm), the Schiff base must be protonated. (The longest-wavelength unprotonated Schiff base retinal pigment so far reported is the M412 intermediate of the bacteriorhodopsin photocycle (Doukas et al., 1978)). The known absorption maxima of visual pigments extends to *ca.* 575 nm for retinal$_1$-based pigments (or 620 nm for retinal$_2$ -based pigments). For RH1 retinal$_1$ pigments the range of absorption maxima is 465 to 525 nm, for RH2 it is 467 to 511 nm, for SWS2 it is 414 to 474 nm, for M/LWS it is 508 to 575 nm. For SWS1 pigments the range is 358 to 435 nm (see Tables in Ebrey and Takahashi, 2001). When a protonated Schiff base of retinal is placed in a bath of D_2O, the only exchangeable proton of the chromophore is that of the Schiff base. Accordingly, evidence of a significant D_2O-dependent shift of Schiff base vibrational frequency is taken as a demonstration of a protonated Schiff base. For RH1, RH2, SWS2 and M/LWS pigments there is good direct evidence from the vibrational spectrum of the chromophore of representative pigments (see discussion in Ebrey and Takahashi, 2001) that these pigments have protonated Schiff bases, as expected.

There are a number of visual pigments which do absorb below 420 nm and for these the options are either a protonated or an unprotonated Schiff base of retinal. Evidence recently presented by several groups of investigators suggests that some SWS1 pigments have protonated and some have unprotonated Schiff bases (Kusnetzow et al., 2001; Cowing et al., 2002; Dukkipati et al., 2002; Fasick et al., 2002). Which residues of the SWS1 pigments participate in determining the protonation state and absorption maximum are complex questions, depending very much on the vertebrate family, and have been well reviewed in Hunt et al., (2004).

The amino acid residues involved in setting the pK of the Schiff base of SWS1 pigments, and therefore its protonation state, seem to involve a correlated interaction of many residues in which no one residue has a dominating effect (Shi et al., 2003). A possible exception is residue 90 in bird SWS1 pigments, in which changing it from a Cys to a Ser does cause a *ca.* 30 nm shift in the absorption spectrum, probably due to switching from a protonated to an unprotonated Schiff base.

Since a typical model compound of a protonated Schiff base of retinal absorbs at *ca.* 440 nm (Kito et al., 1968), we take this as the default absorption maximum and ask how is it modulated. Since this is in the range of the SWS2 pigments, it is reasonable to consider the visual pigments from this family which are clustered around this wavelength as a starting point to investigate the mechanisms for controlling visual pigment spectra. The goal is not so much to provide a general theoretical explanation of the absorption maximum of the protonated Schiff base of retinal as to understand which residues are responsible for controlling the absorption maximum and what each residue's contribution is to the shifted spectrum. The later question assumes that each residue contributes independently of the others, which while surely a simplification, nevertheless seems to hold for many examples where there is wavelength modulation by alteration of several amino acids.

Because the protonated Schiff base carries a positive charge, and it was assumed that in the interior of a protein all positive charges will be compensated for by a negative charge, the need for a counter-anion of the protonated Schiff base was raised early on (Honig et al., 1979b). Although this argument held initially great sway and was a driving force for identifying the carboxylic acid residues in visual pigments which served this function, it is ironic that a negative counter-ion is now not believed to be required, especially in invertebrate pigments (Nakagawa et al., 1999).

3.2.10.2 **Mechanisms to Regulate Wavelength**

Given a protonated Schiff base for the chromophore, there are three possible ways of regulating its absorption spectrum: 1) changing the polarity of amino acid residues in the chromophore binding pocket, 2) changing those amino acid residues outside the binding pocket which would alter the distance between the chromophore and its counter-ion, or between the chromophore and polar residues in the binding pocket that interact with the chromophore, or 3) by introducing new charged species near the chromophore. While mechanism (2) appears to be involved in wavelength regulation in other retinal proteins like bacteriorhodopsin, so far for visual pigments only mechanisms 1 and 3 have been shown to operate. Table 3.2 lists those amino acid residues in or near the binding pocket who, when changed, lead to significant spectral shifts in one or more members of the SWS2, RH1, RH2, and M/LWS pigments families.

It is quite interesting to note that the amino acid residues in the binding pocket that have been shown to be varied in naturally occurring visual pigments in order to regulate wavelength are near the ring half of the chromophore, with only two exceptions known so far, Thr94 and Glu181. This is shown schematically in Figure 3.4, in which the distribution of amino acid residues in the binding pocket, following Palczewski et al. (2000), is shown with those residues known to regulate wavelength highlighted. Except for the residue at position 181, we suggest that all of the common residue changes that control the absorption spectrum do not exert much effect on the pK of the Schiff base. Nor do we suspect they have much effect on the quantum efficiency for the photochemistry (see below).

There is a special mechanism of wavelength control for most members of the M/LWS visual pigment family. Residue 181 (and 184) (see Figure 3.3) are conserved in all of the visual pigment families except the M/LWS one where these residues are changed to a His and a Lys respectively to form a chloride binding site (Wang et al., 1993). The binding or unbinding of chloride has a large effect not only on the wavelength (Crescitelli et al., 1991) but also on the pK of the Schiff base (Yuan et al., 1999). These effects were a mystery in that these residues were thought to be distant from the chromophore. The crystal structure of bovine rhodopsin solved that problem by showing that the extracellular loop between helices 4 and 5, which contained these residues, is folded into the transmembrane portion of the pigment and forms part of the retinal binding site (Palczewski et al., 2000).

Table 3.2 Mutations of amino acids in visual pigments which can cause significant spectral shifts

Mutation	Visual pigment Family	Spectra shift (nm)	Reference & Average (N)
D83G	RH1	+2	Nathans 1990a
D83N	RH1	−5	Average (7)
S90C	SWS1	−38	Average (3)
S90D	RH1	−17	Average (3)
T94I	RH1	−22	Ramon et al., 2003
T94S	RH1	−6	Ramon et al., 2003
S94A	SWS2	−14	Takahashi et al., 2003
T118S	RH1	−13	Nagata et al., 2002
E122D	RH1	−23	Average (3)
E122Q	RH1	−19	Average (7)
A164S	M/LWS, RH1	+4	Average (5)
M181Y	M/LWS	−28	Sun et al., 1997
L207M	RH2	+6	Yokoyama et al., 1999
H211C	RH1	−6	Nathans 1990b
H211F	RH1	−5	Nathans 1990b
Y261F	RH1	−8	Average (8)
W265Y	RH1	−15	Average (2)
Y265W	SWS1	+10, +12	Fasick et al., 1999
A269T	M/LWS, RH1	+14	Average (5)
S292A	RH1	+9	Average (4)
S292A	M/LWS	+23	Average (3)
S292A	SWS1	0	Fasick et al., 1998
A292C	RH 1	−1	Sun et al., 1997

For the four families of visual pigments for which there is a Glu at position 181, there is an intriguing question about the protonation state of this Glu (see Section 3.2.2). Its possible involvement as a counter-ion is discussed in the next paragraph.

The first five columns of Table 3.1 show which residues are conserved in the binding pocket for each family of the vertebrate rhodopsins. The variability of those residues involved in color regulation was noted above. The sixth column is for a urochordate, the ascidian *Ciona intestinalis.* The next to last column is for the vertebrate photoisomerase RGR. Any adjustment of the binding site that favors the *all–trans* isomer in RGR but the 11-*cis* isomer in the visual pigments is not immediately obvious. Some of the wavelength-regulating residues only (or mostly) vary between pigment families (e.g. 94, 181, 211, 265), while others seem to regulate wavelength within several different families (83, 122, 207, 261, 269, 292).

3.2.10.3 Nature of the Counter-ion

In all known vertebrate visual pigments the counter-ion to the Schiff base, whether protonated or unprotonated, is a carboxylic acid. In the Schiff base of bovine rhodopsin and probably all vertebrate visual pigments, with the possible exception of some of the SWS1 ones, this carboxylic acid, Glu113, is anionic, that is unprotonated. Exceptions are the invertebrate visual pigments (discussed in Section 3.4) and many of the non-rod, non-cone pigments such as RGR and melanopsin (see Table 3.1). In bacteriorhodopsin, NMR experiments indicated that the counter-ion to the protonated Schiff base of retinal in that pigment is distributed, and involves more than one residue, leading to the concept of a complex counter-ion (deGroot et al., 1989). Some of the same authors did similar NMR experiments for bovine rhodopsin and also found evidence for it having a complex counter-ion (Creemers et al., 1999). In bacteriorhodopsin it is known that two carboxylic acids, Asp85 and Asp212, are near the Schiff base although mutational studies suggest that Asp85 has the more important role. For rhodopsin, it seems from early mutational studies that only one carboxylic acid residue was involved, Glu113, but both the X-ray structure and studies of some invertebrate pigments indicate that Glu181 is also somewhat close to the Schiff base and, along with hydrogen bonded amino acid residues and waters, may comprise a complex counter-ion (see Figure 3.5). There is a controversy about the protonation state of Glu181 (also see above, Section 3.2.7). Terakita et al., (2000) showed that in retinochrome, which has a Met at the visual pigment counter-ion position of 113, Glu181 can serve as a counter-ion. This role requires that Glu181 be in its anionic form. In addition, no evidence for a protonated Glu181 was seen in FTIR difference spectra between bovine rhodopsin and its batho product; the presence of two other carboxylic acids already known to be protonated, Asp83 and Glu122 (Nagata et al., 2002), could be detected.

On the other hand, Yan et al., (2002) confirmed and extended an earlier study (Nathans, 1990a, b) that mutants of Glu181 did not cause significant spectral perturbations, suggesting that Glu181 was uncharged. Moreover, the model for counter-ion switching proposed by Yan et al., (2003) requires that Glu181 be protonated.

3.2.11
Quantum Efficiency of Visual Pigment Photochemistry

The second requirement for the initial state of a good visual pigment is to have a high quantum efficiency of photoisomerization. While this is obvious for rhodopsins, since they are the usual pigments in rods, and rods are very sensitive light detectors, it is not so obvious in cones, photoreceptors which usually work at high light levels. Nevertheless, the studies of Makino et al., (1996) and Okano et al., (1992) found the quantum efficiency for each member of five vertebrate visual pigment families that they examined had about the same high value, *ca.* 0.65 (Dartnall 1972). This is in contrast to the quantum efficiency of isomerization of a protonated Schiff base of retinal in solution, ca. 0.1. The reason for this high quantum efficiency of isomerization of visual pigments has just begun to be investigated. Nagata et al., (2002) proposed that the interaction of Thr118 with the 9-methyl group of retinal could help steer the photochemistry. And in a quite interesting paper by Liu and Colmenares (2003) a special role for Cys 187 was proposed.

3.2.12
Dark Noise Originating from the Photoreceptor Pigment

The third characteristic of a good visual pigment is that it have a low probability of spontaneous activation in the dark; that is, if one wants a high sensitivity photoreceptor, then the level of dark noise from the pigment itself should be very low. How this is accomplished is not known; however, Birge and Barlow (1995) put forth a very interesting hypothesis that the low dark noise of rods is related to the high pK of the Schiff base of rhodopsin. The pK of the Schiff base of the rod visual pigment (bovine rhodopsin) has been estimated to be very high, greater than 15 (Steinberg et al., 1993), while two other pigments whose pKs have been measured are much lower (Koutalos et al., 1990; Liang et al., 1994). Both of these are ca. 10.5, still much higher than the pK of the protonated Schiff base model compound, usually taken to be 6–7. One reason for a visual pigment to have a higher pK than the model compound would be to keep the Schiff base protonated so that its absorption spectrum will be at the desired long wavelength location. But this should at most require a pK of ca. 9.5, which would result in 99% of the Schiff bases protonated at any given time at the pH of a rod. Birge and Barlow (1995) noted that the very high pK of rhodopsin could be an explanation of the very low thermal noise of rods (for dark noise measurements see Rieke et al., 1996, 2000). Birge and Barlow's mechanism connecting these two was based on their assumption that a deprotonated Schiff base would isomerize more readily around a double bond and so more easily form an *all-trans* product thermally, which would be interpreted by the photoreceptor as an activated rhodopsin, i.e. noise. However, the formation of a deprotonated Schiff base should decrease the probability of thermal isomerization, not increase it (Ebrey, 2000); so that particular aspect of the hypothesis is unlikely. However, there still could well be a simple correspondence of noise with pK. This hypothesis (Ebrey, 2000) was based on experiments we had done with bacteriorhodopsin that showed that transient protonation of the counter-ion to the Schiff base catalyzes the thermal isomerization of its chromophore (see e.g. Balashov et al., 1993, 1996).

The linkage then of thermal noise with the pK of the Schiff base is through the pK of the counter-ion; the higher the pK of the Schiff base, the lower the pK of its counter-ion. So the counter-ion of a high Schiff base pK pigment will be very infrequently found in its protonated state and therefore very unlikely to catalyze the thermal isomerization of the chromophore. One appealing aspect of the hypothesis is that the only other vertebrate pigment whose Schiff base pK has so far been determined, gecko P521, has a pK of 10.4 (Liang et al., 1994). This is much lower than that of bovine rhodopsin and so one would predict that such a pigment would have a much higher rate of thermal isomerization than rhodopsin. Rieke and Baylor (2000) did find much higher noise in salamander cones containing a pigment very similar to the gecko pigment and so probably having a similar pK. So there is a perfect straight line correlation between pK and noise; a RH1 pigment/rod has a high pK and low noise while a M/LWS pigment/cone has a much lower pK and much higher noise; unfortunately there are only two points on the line. An appealing aspect of this hypothesis is that it pushes somewhat further the notion that an important difference between rod and cone pigments is their rate of producing dark noise.

Recently Firsov et al., (2002) tried to test the noise/pK hypothesis by measuring the pH dependence of the dark noise. Their experiments implicitly assume that the Schiff base will titrate in a simple Henderson–Hasselbalch manner. However, this assumption is wrong if the pK of the Schiff base is coupled to the pK of another group as we suspect (Kuwata et al., 2001). Then there can easily be a plateau (no change in concentration) in the fraction protonated of Schiff base or counter-ion when the pH is changed, even by more than 5 pH units (see Figure 2 in Balashov (2000) for such a long plateau for the titration of bacteriorhodopsin). So, unfortunately the Firsov et al., paper did not test the more plausible hypothesis.

There is an interesting alternative hypothesis linking pK with thermal noise. The cause of the thermal noise could still be related to the high pK of the Schiff base and not be due to thermal isomerization of the chromophore. This alternative hypothesis proposed here is based on some old observations of Matsumoto et al., (1975), initially reinforced by Crescitelli's work (1984) and recently tied to dark adaptation by Kefalov et al., (1999, 2001). The observation is that in some (probably all) M/LWS pigments (chicken iodopsin and the gecko P521 pigments were studied) added 9-*cis* retinal can displace in the dark the naturally occurring 11-*cis* chromophore, leading to a 9-*cis* pigment. For this displacement two things must occur: first the 11-*cis* chromophore must dissociate, albeit it may be very infrequently, from the opsin. Second, the 9-*cis* chromophore, when it combines with the opsin to form a pigment (probably by random choice assuming the opportunity for 11-*cis* and 9-*cis* being in the binding pocket is simply proportional to their concentrations, although it could be more complicated than this) must be more stable than the 11-*cis* native pigment. Kefalov et al., (2001) have provided a possible linkage of chromophore dissociation to dark adaptation. One interpretation of these experiments is that when the chromophore transiently dissociates from the opsin, the opsin plus retinal complex has a much greater possibility of acting like an activated rhodopsin than opsin by itself (which doesn't activate transducin very well (Melia et al., 1997)) or the regenerated pigment (which also can not activate transducin). That the addition of *all-trans* retinal to opsin makes the opsin much more effective in activating transducin even though a pigment is not formed, does suggest that a dissociated chromophore, non-covalently bound but sitting in the binding pocket, be it *all-trans* or 11-*cis,* would make this complex much more effective than the pigment itself in activating transducin. The correspondence with pK is that the probability of dissociation (hydrolysis) of the Schiff base should depend on the pK of the Schiff base. A pigment with a very high Schiff base pK will not dissociate very often.

3.3 Activation of Vertebrate Visual Pigments

3.3.1 Introduction

Soon after light absorption, the photolyzed rhodopsin reaches an activated state, in which it can interact with transducin to pass the activation event on to other members of the signaling cascade. At present, there are two somewhat parallel views one can take of the activation process. One view is to look at what has to be accomplished to activate rhodopsin, taken to mean allowing the cytoplasmic surface of the pigment to be altered so that it can interact with a transducin molecule, catalyzing the transducin's release of a normally bound GDP. The second way of looking at the changes is from the view of the changes in the absorption and vibrational spectra of rhodopsin that occur after light absorption and that can be correlated with the activation process. We start with what seems to be the sequence of required changes in rhodopsin that leads to its active state. There have been several excellent reviews of the activation process (Hofmann, 1999, 2000) as well as reviews that focus more on the photochemical changes undergone by rhodopsin after light absorption (Kliger and Lewis, 1995; Stuart et al., 1996).

3.3.2 The Primary Event, Photoisomerization

The initial action of light is to photoisomerize the chromophore from the 11-*cis* to the *all-trans* configuration, leading to a bathochromically shifted primary photoproduct called bathorhodopsin. The step has been characterized as the light changing a covalently attached antagonist, 11-*cis* retinal, to a covalently attached agonist, *all-trans* retinal. The key physical event is the storage of part of the photon's energy in bathorhodopsin (Honig et al., 1979a) which is then used to power other changes required to bring the pigment to its activated state. The amount of this energy has been estimated to be *ca.* 30 kcal mol^{-1} (Stuart et al., 1996). FTIR and Raman experiments have shown that the chromophore is twisted but in an *all-trans* conformation (see e.g. Bagley et al., 1985). Vibrational spectroscopy has also given the striking result that although isomerization has taken place when bathorhodopsin is formed, a strong hydrogen bond between the Schiff base and its acceptor is not changed and the nearby water(s) seems to undergo only small perturbations (see Maeda et al., 1997). In the next transition, from bathorhodopsin to the lumi intermediate, the Schiff base proton is much more weakly H-bonded in lumirhodopsin compared to rhodopsin or bathorhodopsin (Ganter et al., 1988).

3.3.3
The Meta I ↔ Meta II Transition

Meta II, a deprotonated Schiff base photo intermediate absorbing at 380 nm, is associated with the active state of rhodopsin (Hofmann, 1999, 2000). However, Meta II, when defined by its absorption spectrum, is not the same as the active state. The active state is created when Meta II is created, but the shutting off of the active state by phosphorylation of residues on the C-terminal end of the pigment and subsequent arrestin binding, does not lead to any changes in the absorption spectrum of Meta II. So the lifetime of the active state, the value of which is still under debate, is much shorter than the lifetime of Meta II.

It has been reported that illumination of crystals of rhodopsin under conditions that should lead to the stable formation of Meta I or Meta II causes the destruction of the crystals (Okada et al., 2001). The most pessimistic interpretation of this may be the correct one, that in attempting to form Meta I the pigment undergoes such large changes in its conformation that the crystal structure is destroyed. If so, then the best chance to obtain a structure for Meta I or Meta II is to crystallize directly these states of the pigment. One possible approach to this problem is discussed in Section 3.5.4.

Protonation and deprotonation of the Schiff base and of several other residues in rhodopsin have important roles in forming the active state. The initial indications of such protonation changes were the experiments of Matthews et al., (1963) which showed that there was a pH dependent equilibrium between Meta I and Meta II. The observed pH dependence was surprising in that it did not correlate with the protonation states of the Schiff base in Meta I and Meta II, but rather low pH favored Meta II, the deprotonated form of the Schiff base. The pH dependence suggests that the protonation state of another group shifts the equilibrium between Meta I and Meta II.

Kuwata et al., (2001) have reviewed the data on the protonation changes of rhodopsin and also evaluated the sometimes contradictory data on the kinetics and stoichiometry of proton uptake by rhodopsin following a bleaching flash. To summarize, several experiments monitoring light-induced protonation changes with pH sensitive dyes or with pH electrodes, pH effects on the Meta I/Meta II equilibrium, and photocalorimetric experiments find that just one proton is taken up in forming Meta II, although under some conditions a few workers have found more than one. Arnis and co-workers (1993) showed that the light–induced proton uptake by rhodopsin was completely abolished in a mutant in which Glu134 was changed to a neutral residue. In addition, using FTIR spectroscopy, Fahmy et al., (2000) have provided evidence that Glu134 becomes protonated when activated rhodopsin binds to transducin. The most coherent reading of all the data is that at neutral pH light activation leads to a single proton being taken up by rhodopsin, probably at a site on the cytoplasmic surface associated with Glu134, and with kinetics that follows the rate of transfer of a proton from the Schiff base to its counter-ion, Glu113.

Kuwata et al., (2001) presented evidence that the pK of the Schiff base (and its counter-ion, Glu113) depends on the protonation state of a distant group, Glu134. If so, then the reciprocal must also hold, that a change in the protonation state of Glu113 (as when Meta II is formed and Glu113 becomes protonated), could lead to a large change in the pK of Glu134, so that it could conceivably now pick up a proton from solution. This coupling of the pKs of two distant groups has been well-established in bacteriorhodopsin (Balashov et al., 1993, 1996). If there is such a coupling then it is quite possible that the pH-dependence of the percent of protonation of Glu113 and of the Schiff base do not follow a simple Henderson–Hasselbalch equation. Instead the titration curve could have a broad plateau in which the percent of extent of protonation of the Schiff base does not change with a large change of pH.

3.3.4 Molecular Changes upon the Formation of Meta I and Meta II

Many of the measures of conformational change show large changes upon the formation of Meta I and Meta II (Hofmann, 2000). Raman spectroscopy showed that the Schiff base is deprotonated when Meta I changes to Meta II (Doukas et al., 1978). Changes seen in FTIR difference spectra can often be assigned to specific modes of vibration or residues. Besides the Schiff base and the counter-ion, Glu122, Glu134 and Asp83 and internal waters undergo perturbation in forming Meta I and Meta II from bovine rhodopsin. Particularly important is the detection of the protonation of Glu113 upon the formation of Meta II, presumably from the Schiff base (reviewed in Kuwata et al., 2001).

3.3.5 Internal Water Molecules

Nagata et al., (1997) showed the involvement of at least one bound water molecule in the hydration of Glu113 of bovine rhodopsin. This water molecule (WAT-1) has its O–H stretching vibration at 3538 cm^{-1}. In the photochemical transition from the unphotolyzed state to bathorhodopsin, Nagata et al., (1997, 1998) observed changes in the H-bonding of a second water molecule (WAT-2), which has its O–H stretching vibration at 3564 cm^{-1} in the unphotolyzed state. The frequency of this second water molecule is slightly shifted to 3571 cm^{-1} in Glu113Gln, suggesting its connection also to Glu113. In forming bathorhodopsin, a few peptide carbonyl and amide bonds also undergo H-bonding changes. Rath et al., (1998) also nicely showed hydrogen bonding changes in internal waters upon forming Meta II.

3.3.6 Required Steps for Rhodopsin Activation

There are three changes that a vertebrate visual pigment must undergo before the pigment is transformed to its active state (see Figure 3.6). First, the chromophore must undergo *cis–trans* isomerization changing it from an inverse agonist, 11-*cis* retinal, to an agonist, *all-trans* retinal. If the chromophore cannot isomerize then activation cannot occur (Mao et al., 1981).

Second, Oprian and co-workers (Cohen et al., 1993) have developed the interesting concept that a required event for the activation of rhodopsin is the breaking of the

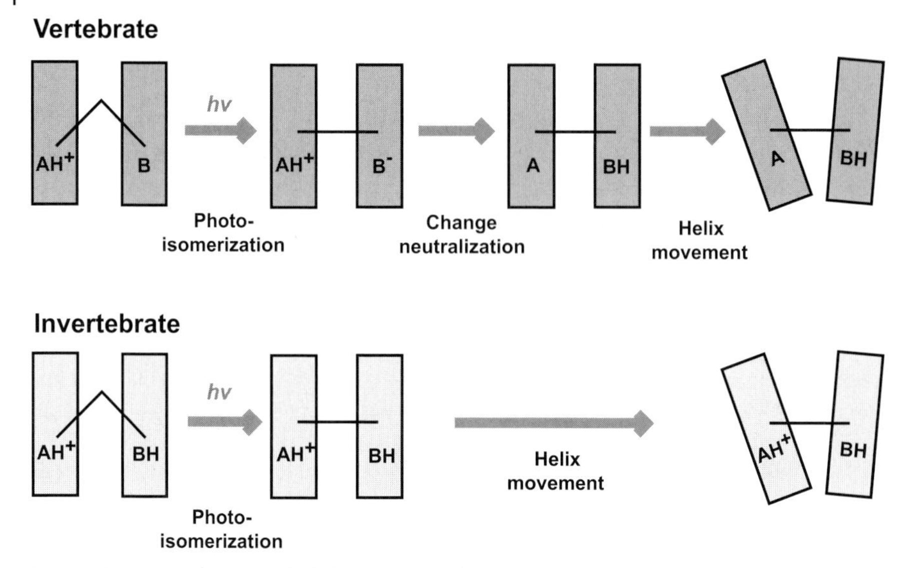

Figure 3.6 Required steps in the light activation of vertebrate and invertebrate visual pigments.

salt-bridge between the cationic Schiff base and the anionic counter-ion Glu113. This is usually accomplished by the transfer of a proton from the Schiff base to Glu113 when Meta II is formed, neutralizing both these residues. The deprotonated Schiff base and the protonated Glu113 are then "active" states of these residues and these changes are required before other changes could occur which would allow the photolyzed visual pigment to interact with and activate a G-protein.

The third change that photolyzed vertebrate rhodopsin must undergo before it can interact with the G-protein is for helix 3 to move away from helix 6. By engineering in covalent disulfide (Farrens et al., 1996) or zinc bridges (Sheikh at al. 1996) between helices, the activation of transducin could be inhibited by preventing this movement.

3.3.7
The Transmembrane Signaling Pathway

One idea for a transmembrane signaling pathway linking the Schiff base to the activating cytoplasmic surface of rhodopsin is for a hydrogen-bonded chain of groups including both amino acid residues and internal water molecules. For bacteriorhodopsin, Asp85, which is both the counter-ion and the acceptor of the Schiff base's proton, mediates proton release at the distant extracellular surface, while Asp96, the proton uptake site, is also located far away at the cytoplasmic surface. In the events leading to the L photointermediate, the interaction between Asp96 and Asp85 is mediated by the hydroxyl of Thr46 and the peptide carbonyl of Val49, along with internal water molecules which are present between these two aspartic acid residues (Yamazaki et al., 1996). Rhodopsin, transformed to its Meta II form so it can now activate the G-protein, seems to share at least a part of a common mechanism with bacteriorhodopsin, in that light causes a proton to be transferred from the Schiff base to the counter-ion, followed by (and probably causing) proton uptake at the cytoplasmic surface, at Glu134.

3.4 The Unphotolyzed State of Invertebrate Visual Pigments

3.4.1 Introduction

Invertebrate visual pigments are also G-protein coupled receptors, but the G-protein and the G-protein's effector targets are different from the vertebrate case (Zuker, 1996). So far there are no reports that a single activated invertebrate visual pigment molecule can activate many G proteins which has been shown for vertebrate rhodopsins (Liebman et al., 1987), so it is not known if this amplification feature is also present in invertebrates.

Although there are about 30 invertebrate phyla, sequence information is available only for pigments from two of them (see Figure 3.1), mollusks and arthropods, or maybe three if one includes the urochordate, ascidia, because the later also has part of its life cycle as a invertebrate. The recent review by Gartner (2000) listed over 60 invertebrate pigment sequences. Although there is only ca. 25–30% sequence identity, it is generally assumed that the invertebrate visual pigments will have a very similar structure to bovine rhodopsin because i) seven transmembrane helices can be identified in the hydropathy plots, and ii) a low resolution structure obtained by cryo-electron microscopy is similar to images of vertebrate pigments (Davis et al., 2001) and iii) many of the motifs that are conserved throughout the vertebrate visual pigment families are also present in invertebrates. These include the (again using the bovine rhodopsin number scheme) Cys110-Cys187 disulfide linkage, the Schiff base lysine in the seventh helix, the D/R/Aromatic triplet at positions 134/135/136, the Glu at position 181, the NPxxY motif near the Schiff base lysine, a conserved glycine at position 121, and conserved prolines at positions at or close to 170, 215, and 267. The corresponding residues in the bovine rhodopsin binding pocket (see Table 3.1) are not as conserved as in vertebrate pigments. However, there are many similarities. For instance Trp265 is conserved and position 268 always has an aromatic, either a Tyr or a Phe. One difference is the putative glycosylation site near the N-terminal end of the pigment that seems to be present only in some of the invertebrate pigments and is involved in a transient, not permanent, modification (see Gartner, 2000). In cephalopods the carbohydrate associated with the pigment has an unusual glycan structures (Zhang et al, 1997). Another difference is that at the counter-ion position, 113, instead of the carboxylic acid found in vertebrates there is either a tyrosine or a phenylalanine (the latter in the UV-absorbing arthropod pigments).

Until recently it has been possible to express invertebrate visual pigment mutants only via heterologous expression in Drosophila (Salcedo et al., 2003). Early reports of expressing octopus rhodopsin in cultured animal cells showed that although a protein could be made, it was denatured and so could not combine with 11-*cis* retinal to make a pigment. In 2000 Terakita et al., expressed retinochrome (see Table 3.1, Figure 3.2) and most recently Knox et al., (2003) showed that rhodopsin could be expressed in Xenopus oocytes. The Drosophila system has allowed some site-directed mutants to be made and tested (see, e.g. Salcedo et al., (2003)).

Resonance Raman spectroscopy has shown that, as expected, the Schiff base of octopus rhodopsin is protonated (Kitagawa et al., 1980; Pande et al., 1987). This probably holds for all the invertebrate pigments which absorb in the visible. However, as noted above the counter-ion position is occupied by either a tyrosine or a phenylalanine. The intriguing suggestion of Zhukovsky and Oprian (1992) that the tyrosine was anionic, a tyrosinate, was found to be not true (Nakagawa et al., 1999). A secondary candidate for the counter-ion is Glu181, which is conserved in all the invertebrate pigments (as well as the vertebrate ones except for the M/LWS family (see Table 3.1)). Although Glu181 is significantly further from the Schiff base than Glu113 in bovine rhodopsin (see Figure 3.5), it still may act to influence the Schiff base. It is interesting to point out that the pK of the Schiff base of octopus rhodopsin is 10.4 (Liang et al., 1994). It seems likely that one factor that lowers the pK from its very high value in bovine rhodopsin is the change in the counter-ion.

Finally, not unexpectedly, there is strong evidence for bound internal waters within invertebrate visual pigments (Nishimura et al., 1997a). These can be detected only in FTIR difference spectra and the lack of a way of making site directed mutants has slowed progress on exploring internal waters in these pigments.

3.4.2
Wavelength Regulation of Invertebrate Pigments

Because it is so difficult to do heterologous expression of the invertebrate rhodopsins, there has been very little experimental work. We note in Table 3.1 that many of the residues that vary to control the absorption spectrum also seem to vary from one pigment to another in the invertebrate pigments and so probably have similar roles. One striking exception is that in the UV-sensitive invertebrate pigments, the counterion of the vertebrate pigments, Glu113, is replaced by a phenylalanine. Salcedo and co-workers (2003) noted that in the invertebrate UV pigments, the residue at position 90 is usually a lysine and hypothesize that this change might lead to the lowering of the pK of the Schiff base so much that it deprotonates.

3.5
Mechanism of Activation of Invertebrate Visual Pigments

3.5.1
The Initial Photochemical Events

Everything known about the initial photochemical events in invertebrate eyes suggests that they are very similar to what is seen for vertebrate visual pigments. After light absorption, there is a rapidly formed, bathochromically shifted, photoproduct of the pigment, which is associated with an 11-*cis* to *all-trans* photoisomerization (reviewed in Tsuda et al., 1986). Besides the spectral shift, other small perturbations of the pigment, including perturbations of internal waters can be seen upon the formation of the bathoproduct (Nishimura et al., 1997b).

3.5.2
Formation of Acid Metarhodopsin

Besides the difference in the counter-ions, another difference between invertebrate and vertebrate visual pigments is that the Schiff base does not deprotonate in forming the active form of the photolyzed invertebrate pigment, Acid Metarhodopsin. A second difference is that the final photoproduct, Acid Metarhodopsin, is stable and does not decay over a period of many hours. The mechanism for the regeneration of photolyzed invertebrate pigments back to their initial states is unknown. Photoregeneration can lead to renewal of some of the pigment, but it cannot be used to get all of the pigment back to its unphotolyzed, dark-adapted state. Changes in the visual pigment after initial photolysis have been best studied in cephalopod rhodopsins (Nakashima et al., 2003). With octopus rhodopsin, not only is Acid Meta a protonated Schiff base, but also rather unusually it absorbs at longer wavelength than the initial pigment. In the series of events that occurs during the transition from the bathoproduct to Acid Meta, there is an isochromic species that precedes the stable Acid Meta. The life time of this species, Transient Acid Meta, is about 40 μs at room temperature and Transient Acid Meta probably is in equilibrium with Acid Meta (Nishioku et al., 2001). It is not clear if there are differences in the ability of these two species to activate the octopus G-protein.

3.5.3
Required Steps for Photolyzed Octopus Rhodopsin to Activate its G-protein

In the case of vertebrate rhodopsin at least three changes had to occur for the pigment to reach its active state after absorption of a photon. The chromophore must be isomerized from 11-*cis* to *all-trans*, the salt bridge between the protonated Schiff base and the counter-ion must be neutralized, and at least two of the transmembrane helices must move apart (see Section 3.4.1 and Figure 3.6). In this case the neutralization of the counter-ion results from the deprotonation of the Schiff base. Based on the available evidence, a reasonable hypothesis is that there need not be an unprotonated Schiff base species formed for octopus rhodopsin to get to the active state, because the salt bridge formed by the protonated Schiff base and the counter-ion that exists in the vertebrate pigments, is already neutral in the invertebrate pigments (Nakagawa et al., 1999). Presumably the third step in vertebrate activation, movement of the helices, also occurs in invertebrate pigments, but there is no evidence for this movement yet.

3.5.4
Purification of the Active Form of an Invertebrate Visual Pigment

Recently there has been a potentially quite exciting development. Tsuda and his colleagues (Ashida et al., 2004) showed that after octopus rhodopsin had a portion of its C-terminal "tail" proteolyzed, the resulting pigment had identical spectral properties compared to the unproteolyzed pigment. Upon irradiation of this species, the resulting mixture of (truncated) rhodopsin and (truncated) Acid Metarhodopsin could be separated on a DEAE column, resulting in a pure and stable preparation of the activating species, Acid Metarhodopsin. If this species can be crystallized then one would be able to have the structure of the active form of the receptor.

Acknowledgements

We would like to thank Akio Maeda and Ron Stenkamp for reading over our manuscript and John Spudich for close reading as well as editorial suggestions. This work was supported by NIH EYO1323.

References

Arendt, D. (2003) *Int J Dev Biol* 47, 563–571

Arnis, S., and Hofmann, K.P. (1993) *Proc Natl Acad Sci USA* 90, 7849–7853.

Arshavsky, V. (2002) *Trends Neuro Sci* 25, 124–126.

Ashida, A., Matsumoto, K., Ebrey, T.G., Tsuda, M. (2004) *Zoological Sci* 21, 245–250

Bagley. K., Balogh-Nair, V., Croteau. A, Dollinger, G, Ebrey, T.G., Eisenstein, L., Hong. M., Nakanishi, K., Vittitow, J. (1985) *Biochemistry* 24, 6055–6071.

Balashov, S.P., Govindjee, R., Kono, M., Imasheva, E.S., Lukashev, E., Ebrey, T.G., Crouch, R., Menick, D., Feng, Y. (1993) *Biochemistry* 32, 10331–10343.

Balashov, S.P. (2000) *Biochim Biophys Acta* 1460, 75–94.

Balashov, S.P., Imasheva, E.S., Govindjee, R., Ebrey, T.G. (1996) *Biophys J* 70, 473–481.

Baldwin, J.M., Schertler, G.F.X., Unger, V.M. (1997) *J Mol Biol* 272, 144–164.

Béjà, O., Spudich, E., Spudich, J., Leclerc, M., DeLong, E. (2001) *Nature* 411, 786–789.

Birge, R.R., and Knox, B. (2003) *Proc Natl Acad Sci U S A* 100, 9105–9107.

Birge, R.R., and Barlow, R.B. (1995) *Biophys Chem* 55, 115–126.

Blackshaw, S., and Snyder, S. (1997) *J Neuro Sci* 17, 8083–8092.

Blackshaw, S., and Snyder, S. (1999) *J Neuro Sci* 19, 3681–3690.

Bouvier, M. (2001) *Nat Rev Neurosci* 2, 274–286.

Brown, L.S. (2004) *Photochem Photobiol Sci*, submitted.

Chabre, M., Cone, R., Saibil, H. (2003) *Nature* 426, 30–31.

Cohen, G.B., Yang, T., Robinson, P.R., Oprian, D.D. (1993) *Biochemistry* 32, 6111–6113.

Cowing, J.A., Poopalasundaram, S., Wilkie, S.E., Bowmaker, J.K., Hunt, D.M. (2002) *Biochemistry* 41, 6019–6025.

Creemers, A.F.L., Kiihne, S., Bovee-Geurts, P.H.M., DeGrip, W.J., Lugtenburg, J., deGroot, H.J.M. (2002) *Proc Natl Acad Sci U S A* 99, 9101–9106.

Creemers, A.F.L., Klaassen, C.H.W., Bovee-Geurts, P.H.M., Kelle, R., Kragl, U.R.J., deGrip, W.J., Lugtenburg, J., deGroot, H.J.M. (1999) *Biochemistry* 38, 7195–7199.

Crescitelli, F. (1972) in *Handbook of Sensory Physiology, Vol. VII* (Dartnall, H.J.A., Ed) 245–363, Springer-Verlag.

Crescitelli, F. (1984) *Vision Res* 24, 1551–1553.

Crescitelli, F., and Karvaly, B. (1991) *Vision Res* 31, 945–950.

Dartnall, H.J.A. (1972) in *Handbook of Sensory Physiology, Vol. VII* (Dartnall, H.J.A. Ed). Springer-Verlag.

Davis, A., Gowen, B., Krebs, A., Schertler, G., Saibil, R. (2001) *J Mol Biol* 314, 445–463.

deGroot, H.J., Harbison, G.S., Herzfeld, J., Griffin, R.G. (1989) *Biochemistry* 28, 3346–3353.

deGrip, W.J., Rothschild, K.J. (2000) in *Handbook of Biological Physics*, (Stavenga, D.G., deGrip, W.J., Pugh, Jr. E.N. Eds) 3–54, Elsevier Science.

Deng, H., Huang, L., Callender, R.H., Ebrey, T.G. (1994) *Biophys J* 66, 1129–1136.

Doukas, A., Aton, B., Callender, R., Ebrey, T.G. (1978) *Biochemistry* 17, 2430–2435.

Dukkipati, A., Kusnetzow, A., Babu, K.R., Ramos, L., Singh, D., Knox, B.E., Birge, R.R. (2002) *Biochemistry* 41, 9842–9851.

Eakin, R.M. in *Handbook of Sensory Physiology, Vol. VII* (Dartnall, H.J.A. Ed). 625–684, Springer-Verlag.

Ebrey, T.G. (2000) *Methods Enzymol* 315, 196–207.

Ebrey, T.G., and Koutalos, Y. (2001) *Progress in Retinal and Eye Research* 20, 49–94.

Ebrey, T.G., and Takahashi, Y. (2001) in *Photobiology for the 21st Century* (Coohill, T.P., and Velenzeno, D.P. Eds) 101–133. Kansas, Valdenmar Publishing.

Ernst, O.P., Meyer, C.K., Marin, E.P., Henklein, P., Fu, W.Y., Sakmar, T.P, Hofmann, K.P. (2000) *J Biol Chem* 275, 1937–1943.

Essen, L-O. (2001) *Chembio Chem* 2, 513–516.

Fahmy, K. (1998) *Biophys J* 75, 1306–1318.

Fahmy, K., Sakmar, T.P., Siebert, F. (2000) *Biochemistry* 39, 10607–10612.

Farrens, D.L., Altenbach, C., Yang, K., Hubbell, W.L., Khorana, H.G. (1996) *Science* 274, 768–770.

Fasick, J.I., and Robsinson, P.R. (1998) *Biochemistry* 37, 433–438.

Fasick, J.I., Lee, N., Oprian, D.D. (1999) *Biochemistry* 38, 11593–11596.

Fasick, J.I., Applebury, M.L., Oprian, D.D. (2002) *Biochemistry* 41, 6860–6865.

Feng, X., Verdegem, P.J.E., Lee, Y.K., Sandström, M., Edén, M., Bovee-Geurts, P., deGrip, W.J., Lugtenburg, J., deGroot, H.J.M., Levitt, M.H. (1997) *J Amer Chem Soc* 119, 6853–6857.

Firsov, M.L., Donner, K., Govardovskii, V.I. (2002) *J Physiol* 539, 837–846.

Fotiadis, D., Liang, Y., Filipek, S., Saperstein, D.A., Engel, A., Palczewski, K. (2003) *Nature* 421, 127–128.

Fujimoto, Y., Ishihara, J., Maki, S., Fujioka, N., Wang, T., Furuta, T., Fishkin, N., Borhan, B., Berova, N., Nakanishi, K. (2001) *Chemistry* 7, 4198–4204.

Ganter, U.M., Gaertner, W., Siebert, F. (1988) *Biochemistry* 27, 7480–7488.

Gartner, W. (2000) in *Handbook of Biological Physics* (Stavenga, D.G., deGrip, W. J., Pugh, Jr.E.N. Eds) 297–388, Elsevier Science.

Goldsmith, T.H. (1972) in *Handbook of Sensory Physiology, Vol. VII* (Dartnall, H.J.A. Ed) 685–719. Springer-Verlag.

Hankins, M., Foster, R. (2004) in *Handbook of Photosensory Receptors*, (Briggs, W., Spudich, J.L., Eds.).

Hara, T., Hara, R. (1991) in *Progress in retinal research* (Osborne, N., Chader, J. Eds) 179–206, Pergamon Press.

Harbison, G.S., Smith, S.O., Pardoen, J.A., Courtin, J.M., Lugtenburg, J., Herzfeld, J., Mathies, R.A., Griffin, R.G. (1985) *Biochemistry* 24, 6955–6962.

Hargrave, P.A., Bownds, D., Wang, J.K., McDowell, J.H. (1982) *Methods Enzymol* 81, 211–214.

Hargrave, P.A., McDowell, J.H., Curtis, D.R., Wang, J.K., Juszczak, E., Fong, S.L., Rao, J.K., Argos, P. (1983) *Biophys Struct Mech* 9, 235–244.

Hofmann, K.P. (1999) in *Rhodopsins and Phototransduction* 58–180 *Novartis Foundation Symposium Vol. 224.*

Hofmann, K.P. (2000) in *Handbook of Biological Physics* (Stavenga, D.G., DeGrip, W.J., Pugh, Jr.E.N Eds) 91–142, Elsevier Science B.V.

Honig, B., Ebrey, T., Callender, R.H., Dinur, U., Ottolenghi, M. (1979a) *Proc Natl Acad Sci U S A* 76, 2503–2507.

Honig, B., Dinur, U., Nakanishi, K., Balogh-Nair, V., Gawinowicz, M.A., Arnaboldi, M., Motto, M.G. (1979b) *J Amer Chem Soc* 101, 7084–7086.

Hubbell, W.L., Altenbach, C., Hubbell, C.M., Khorana, H.G. (2003) *Adv Protein Chem* 63, 243–290.

Hunt, D.M., Cowing, J., Wilkie, S., Parry, J., Poopalasundaram, S., Bowmaker, J.K. (2004). *Photochem Photobiol Sci*, 8, 713–720.

Hurley, J.B., Spencer, M., Nieme, G.A. (1998) *Vision Res* 38, 1341–1352.

Imai, H., Terakita, A., Tachibanaki, S., Imamoto, Y., Yoshizawa, T., Shichida, Y. (1997) *Biochemistry* 36, 12773–12779.

Kawamura, S., and Yokoyama, S. (1997) *Vison Res* 37, 1867–1871.

Kefalov, V.J., Cornwall, M.C., Crouch, R.K. (1999) *J Gen Physiol* 113, 491–503.

Kefalov VJ, Crouch RK, Cornwall MC (2001) Neuron 29, 749–755.

Kennedy, M., Lee, K., Niemi, G., Craven, K., Garwin, G., Saari, J., Hurley, J. (2001) *Neuron* 31, 87–101.

Kitagawa, T., Tsuda, M. (1980) *Biochim Biophys Acta* 624, 211–217.

Kito, Y., Suzuki, T., Azuma, M., Sekoguti, Y, (1968) *Nature* 218, 955–957.

Klein-Seetharaman, J., Reeves, P.J., Loewen, M.C., Getmanova, E.V., Chung, J., Schwalbe, H., Wright, P.E., Khorana, H.G. (2002) *Proc Natl Acad Sci USA* 99, 3452–3457.

Kliger, D.S., Lewis, J.W. (1995) *Israel J Chemistry* 35, 289–307

Knox, B., Salcedo, E., Mathiesz, K., Schaefer, J., Chou, W., Chadwell, L., Smith, W., Britt, S., Barlow, R. (2003) *J Biol Chem* 278, 40493–40502.

Koutalos, Y., Ebrey, T.G., Gilson, H.R., Honig, B. (1990) *Biophys J* 58, 493–501.

Koyanagi, M., Terakita, A., Kubokawa, K., Shichida, Y. (2002) *FEBS Lett* 531, 525–528.

Kunishima, N., Shimada, Y., Tsuji, Y., Sato, T., Yamamoto, M., Kumasaka, T., Nakanishi, S., Jingami, H., Morikawa, K (2000) *Nature* 407, 971–977.

Kusakabe, T., Kusakabe, R., Kawakami, I., Satou, Y., Satoh, N., Tsuda, M. (2001) *FEBS Lett* 506, 69–72.

Kusnetzow, A., Dukkipati, A., Babu, K.R., Singh, D., Vought, B.W., Knox, B.E., Birge, R.R. (2001) *Biochemistry* 40, 7832–7844.

Kuwata, O., Yuan, C., Misra, S., Govindjee, R., Ebrey, T.G. (2001) *Biochemistry* (Moscow) 66, 1283–1299.

Land, M. (1992) *Annu Rev Neurosci* 15, 1–29.

Land, M., and Nilsson, D. (1981) *Animal Eyes*, Oxford University Press.

Liang, J., Steinberg, G., Livnah, N., Sheves, M., Ebrey, T.G., Tsuda, M. (1994) *Biophys J* 67, 848–854.

Liang, Y., Fotiadis, D., Filipek, S., Saperstein, D.A., Palczewski, K., Engel, A. (2003) *J Biol Chem* 278, 21655–21662.

Liebman, P.A., Parker, K.R., Dratz, E.A. (1987) *Ann Rev Physiol* 49, 965–991

Liu, R., Colmenares, L.U. (2003) *Proc Natl Acad Sci USA* 100, 14639–11444.

Liu, S.Y. (1990) *Biophys J* 57, 943–950.

Lou, J., Tan, Q., Karnaukhova, E., Berova, N., Nakanishi, K., Crouch, R.K. (2000) *Methods in Enzymology* 315, 219–237.

Lucas, R.J., Hattar, S., Takao, M., Berson, D.M., Foster, R.G., Yau, K.W. (2003) *Science* 299, 245–247.

Maeda, A., Sasaki, J., Shichida, Y., Yoshizawa, T. (1992) *Biochemistry* 31, 462–467.

Maeda, A., Kandori, H., Yamazaki, Y., Nishimura, S., Hatanaka, M., Chon, Y.S., Sasaki, J., Needleman, R., Lanyi, J.K. (1997) *J Biochem* 121, 399–406.

Makino, C.L., and Dodd, R.L. (1996) *J Gen Physiol* 108, 27–34.

Mao, B., Tsuda, M., Ebrey, T.G., Akita, H., Balogh-Nair, V., Nakanishi, K. (1981) *Biophys J* 35, 543–546.

Marin, E.P., Krishna, A.G., Zvyaga, T.A., Isele, J., Siebert, F., Sakmar, T.P. (2000) *J Biol Chem* 275, 1930–1936.

Matsumoto, H., Tokunaga, F., Yoshizawa, T. (1975) *Biochim Biophys Acta* 404, 300–308.

Matthews, R.G., Hubbard, R., Brown, P.K., Wald, G. (1963) *J Gen Physiol* 47, 215–240.

Max, M., McKinnon, P.J., Seidenman, K.J., Barrett, R.K., Applebury, M.L., Takahashi, J.S., Margolskee, R.F. (1995) *Science* 267, 1502–1506.

Melia, T.J., Cowan, C.W., Angleson, J.K., Wensel, T.G. (1997) *Biophys J* 73, 3182–3191.

Meng, E.C., and Bourne, H.R. (2001) *Trends Pharmaco Sci* 22, 587–593.

Menon, S.T., Han, M., Sakmar, T.P. (2001) *Physiol Rev* 81, 1659–1688.

Mollevanger, L.C., Kentgens, A.P., Pardoen, J.A., Courtin, J.M., Veeman, W.S., Lugtenburg, J., deGrip, W.J. (1987) *Eur J Biochem* 163, 9–14.

Mollon, J.D. (1977) *Nature* 268, 587–588

Nagata, T., Terakita, A., Kandori, H., Shichida, Y., Maeda, A. (1998) *Biochemistry* 37, 17216–17222.

Nagata, T., Oura, T., Terakita, A., Kandori, H., Shichida, Y. (2002) *J Phys Chem A* 106, 1969–1975.

Nagata, T., Terakita, A., Kandori, H., Kojima, D., Shichida, Y., Maeda, A. (1997) *Biochemistry* 36, 6164–6170.

Nagel, G., Ollig, D., Fuhrmann, M., Kateriya, S., Musti, A.M., Bamberg, E., Hegemann, P. (2002) *Science* 296, 2395–2398.

Nakagawa, M., Iwasa, T., Kikkawa, S., Tsuda, M., Ebrey, T.G. (1999) *Proc Natl Acad Sci USA* 96, 6189–6192.

Nakashima, Y., Kusakabe, T., Kusakabe, R., Terakita, A., Shichida, Y., Tsuda, M. (2003) *J Comp Neurol* 460, 180–190

Nathans, J. (1987) *Ann Rev Neurosci* 10, 163–194.

Nathans, J. (1990a) *Biochemistry* 29, 937–942.

Nathans, J. (1990b) *Biochemistry* 29, 9746–9752.

Nishimura, S., Kandori, H., Maeda, A. (1997a) *Photochem Photobiol* 66, 796–801.

Nishimura, S., Kandori, H., Nakagawa, M., Tsuda, M., Maeda, A. (1997b) *Biochemistry* 36, 864–870.

Nishioku Y, Nakagawa M, Tsuda M, Terazima M (2001) *Biophys J* 80, 2922–2927.

Okada, T., Palczewski, K. (2001) *Curr Opin Struct Biol* 11, 420–426.

Okada, T., Fujiyoshi, Y., Silow, M., Navarro, J., Landau, E.M., Shichida, Y. (2002) *Proc Natl Acad Sci USA* 99, 5982–5987.

Okano, T., Yoshizawa, T., Fukada, Y. (1994) *Nature* 372, 94–97.

Okano, T., Fukada, Y., Shichida, Y., Yoshizawa, T. (1992) *Photochem Photobiol* 56, 995–1001.

Ovchinnikov, Y.A. (1982) *FEBS Lett* 148, 179–191.

Palczewski, K., Kumasaka, T., Hori, T., Behnke, C.A., Motoshima, H., Fox, B.A., Le Trong, I., Teller, D.C., Okada, T, Stenkamp, R.E., Yamamoto, M., Miyano, M. (2000) *Science* 289, 739–745.

Panda, S., Sato, T.K., Castrucci, A.M., Rollag, M.D., DeGrip, W.J., Hogenesch, J.B., Provencio, I., Kay, S.A. (2002) *Science* 298, 2213–2216.

Pande, C., Pande, A., Yue, K.T., Callender, R. (1987) *Biochemistry* 26, 4941–4947.

Pauling, L. (1939) *Fortschr. Chem. Organ. Naturstoffe* 3, 203–235.

Pepe, I.M., and Cugnoli, C. (1997) *Visual Pigment Regeneration in Invertebrates. Retinal Photoisomerase, World Scientific.*

Provencio, I., Jiang, G., DeGrip, W.J., Hayes, W.P., Rollag, M.D. (1998) *Proc Natl Acad SciUSA* 95, 340–345.

Ramon, E., del Valle, L.J., Garriga, P. (2003) *J Biol Chem* 278, 6427–6432.

Rath, P., DeLange, F., DeGrip, W.J., Rothschild, K.J. (1998) *Biochem J* 329, 713–717.

Rieke, F., and Baylor, D.A. (1996) *Biophys J* 71, 2553–2572.

Rieke, F, and Baylor, D.A. (2000) *Neuron* 26, 181–186.

Ruby, N.F., Brennan, T.J., Xie, X., Cao, V., Franken, P., Heller, H.C., O'Hara, B.F. (2002) *Science* 298, 2211–2213.

Sakmar, T.P. (2002) *Curr Opin Cell Biol* 14, 189–195.

Sakmar, T.P., Fahmy, K. (1995) *Israel J Chem* 35, 325–337.

Salcedo, E., Zheng, L., Phistry, M., Bagg, E.E., Britt, S.G. (2003) *J Neurosci* 23, 10873–10878.

Scheiner, S., and Duan, X.F. (1991) *Biophys J* 60, 874–883.

Sheikh, S. P., Zvyaga, T.A., Lichtarge, O., Sakmer, T.P., Bourne, H.R. (1996) *Nature* 383, 347–350

Shen, D., Jiang, M., Hao, W., Tao, L., Salazar, M., Fong, H.K. (1994) *Biochemistry* 33, 13117–13125.

Shi, Y., Yokoyama, S. (2003) *Proc Natl Acad Sci USA* 100, 8308–8313.

Sineshchekov, O.A., Jung, K.H., Spudich, J.L. (2002) *Proc Natl Acad Sci USA* 99, 8689–8694.

Smith, S.O., Palings, I., Copie, V., Raleigh, D.P., Courtin, J., Pardoen, J.A., Lugtenburg, J., Mathies, R.A., Griffin, R.G. (1987) *Biochemistry* 26, 1606–1611.

Soni, B.G., and Foster, R.G. (1997) *FEBS Lett* 406, 279–283.

Spooner, P.J.R., Sharples, J.M., Verhoeven, M.A., Lugtenburg, J., Glaubitz, C., Anthony, W. (2002) *Biochemistry* 41, 7549–7555.

Sun, H., Macke, J.P., Nathans, J. (1997) *Proc Natl Acad Sci USA* 94, 8860–8865.

Steinberg, G., Ottolenghi, M., Sheves, M. (1993) *Biophys J* 64, 1499–1502.

Stenkamp, R.E., Teller, D.C., Palczewski, K. (2002a) *Chembiochem* 3, 963–967.

Stenkamp, R.E., Filipek, S., Driessen, C.A., Teller, D.C., Palczewski, K. (2002b) *Biochim Biophys Acta* 1565, 168–182.

Strader, C.D., Fong, T.M., Tota, M.R., Underwood, D., Dixon, R.A. (1994) *Ann Rev Biochem* 63, 101–132.

Stuart, J.A., and Birge, R.R. (1996) *Biomembranes 2A*, 33–139.

Sun, H., Macke, J.P., Nathans, J. (1997) *Proc Natl Acad Sci USA* 94, 8860–8865.

Takahashi, Y., and Ebrey, T.G. (2003) *Biochemistry* 42, 6025–6034.

Tang, L., Ebrey, T.G., Subramaniam, S. (1995) *Israel J Chem* 35, 193–209.

Teller, D.C., Okada, T., Behnke, C.A., Palczewski, K., Stenkamp, R.E. (2001) *Biochemistry* 40, 7761–7772.

Terakita, A., Yamashita, T., Shichida, Y. (2000) *Proc Natl Acad Sci USA* 97, 14263–14267.

Tsuda, M., Tsuda, T., Terayama, Y., Fukada, Y., Akino, T., Yamanaka, G., Stryer, L., Katada, T., Ui, M., Ebrey, T.G. (1986) *FEBS Lett* 198, 5–10.

Unger, V.M., Hargrave, P.A., Baldwin, J.M., Schertler, G.F.X. (1997) *Nature* 389, 203–206.

Van Gelder, R.N., Herzog, E.D., Schwartz, W.J., Taghert, P.H. (2003) *Science* 300, 1534–1535.

Wald, G. (1945) *Harvey Lectures XLI*. 117–160

Wandel, B.A. (1995) *Foundations of Vision*, Sinauer Associates, Inc.

Wang, Z., Asenjio, A.B., Oprian, D.D. (1993) *Biochemistry* 32, 2125–2130.

Yamazaki, Y., Tuzi, S., Saito, H., Kandori, H., Needleman, R., Lanyi, J.K., Maeda, A. (1996) *Biochemistry* 35, 4063–4068.

Yan, E.C.Y., Kazmi, M.A., Ganim, H.J.M., Pan, D., Chang, B.S.W., Sakmar, T.P., Mathies, R.A. (2003) *Proc Natl Acad Sci USA* 100, 9262–9267.

Yan, E.C.Y., Kazmi, M.A., De, S., Chang, B.S., Seibert, C., Marin, E.P., Mathies, R.A., Sakmar, T.P. (2002) *Biochemistry* 41, 3620–3627.

Yeagle, P.L., Albert, A.D. (2002) *Methods Enzymol* 343, 223–231.

Yokoyama, S., and Radlwimmer, F.B. (1999) *Genetics* 153, 919–932.

Yuan, C., Kuwata, O., Liang, J., Misra, S., Balashov, S.P., Ebrey, T.G. (1999) *Biochemistry* 38, 4649–4654.

Zhang, Y., Iwasa, T., Tsuda, M., Kobata, A., Takasaki, S. (1997) *Glycobiology* 7, 1153–1158

Zhukovsky, E.A., Robinson, P.R., Oprian, D.D. (1992) *Biochemistry* 31, 10400–10405.

Zuker, C.S. (1996) *Proc Natl Acad Sci USA* 93, 571–576.

4
Structural and Functional Aspects of the Mammalian Rod Cell Photoreceptor Rhodopsin

Najmoutin G. Abdulaev and Kevin D. Ridge

4.1
Introduction

Visual phototransduction is the process of transforming light energy absorbed by the specialized retinal cells (rods and cones) into an electrical response. The amplitude and duration of the response depends on light intensity and the type of cell. Rods (the dim-light photoreceptors) are more sensitive to light than cones (the bright-light photoreceptors), although the latter demonstrate a faster light response (Burns and Baylor, 2001). In rods (Figure 4.1A), a characteristic two-phase electrical response reflects a multistep, highly amplified series of biochemical reactions expressing activation and recovery phases of signaling. This elegant biochemical machinery is designed to exchange information between the rod outer segment (ROS) plasma membrane (PM) harboring cGMP-regulated ion channels and the light-sensitive pigment, rhodopsin, which is abundantly present in the ROS disk membranes (DM). Rhodopsin is the first link in the chain of biochemical reactions involved in visual phototransduction (Figure 4.1B). It is also a member of the large family of heptahelical G-protein coupled receptors (GPCRs). In this family, rhodopsin appears to be unique in that it displays very low noise levels and the rapid activation and inactivation kinetics necessary for high sensitivity and fast image processing. These requirements are realized by adopting 11-*cis* retinal as a chromophore sensitive to visible light. Of the ~1000 GPCRs identified, bovine rhodopsin is the only receptor for which a three-dimensional structure is available (Palczewski et al., 2000; Teller et al., 2001; Okada et al., 2002). The crystal structure has confirmed a wealth of biochemical and biophysical data on the dark-state of rhodopsin and provided a template for interpreting phenomenological data obtained by affinity labeling and site-directed mutagenesis. In addition, the availability of the rhodopsin structure has provided a useful model for understanding the structure and function of other members of the GPCR family. In this chapter, various structural and functional aspects of mammalian rod rhodopsin are highlighted. While this by no means represents a comprehensive mo-

Handbook of Photosensory Receptors. Edited by W. R. Briggs, J. L. Spudich

ISBN 3-527-31019-3

lecular description of rhodopsin, it is anticipated that the reader will gain a new sense of understanding and appreciation for this fascinating photosensory receptor.

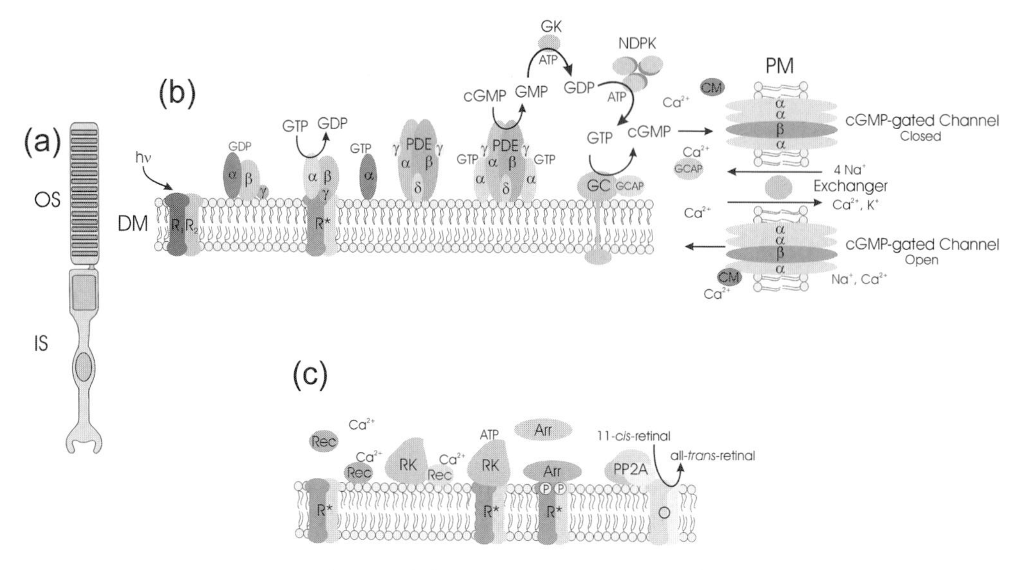

Figure 4.1 Mammalian visual phototransduction. a) Schematic of the highly differentiated rod cell with the outer segment (OS) and inner segment (IS) indicated. The OS is comprised of numerous stacked disks that contain the major phototransduction components. The IS contains the biosynthetic machinery of the cell. b) Signal propagation by light-activated rhodopsin. The major protein components of the disk membrane (DM) that lead to hyperpolarization of the rod cell include rhodopsin (R), the retinal G-protein transducin (G_t) $\alpha\beta\gamma$ heterotrimer, light-activated rhodopsin (R*), the cGMP phosphodiesterase (PDE6) α-, β-, γ-, and δ-subunits, guanylate cyclase (GC), and gunaylate cyclase activating protein (GCAP). The major protein components of the plasma membrane (PM) that lead to hyperpolarization of the rod cell include the cGMP-gated channel α- and β-subunits and the Na^+/Ca^{2+}, K^+ exchanger. Also indicated are guanylate kinase (GK) and nucleoside diphosphate kinase (NDPK), which are involved in guanine nucleotide metabolism, and calmodulin, a major Ca^{2+} binding protein in the ROS. c) Inactivation of light-activated rhodopsin. The major protein components of the DM that lead to the inactivation of R* include the Ca^{2+}-free and Ca^{2+}-bound forms of myristoylated recoverin (Rec), rhodopsin kinase (RK), arrestin (Arr), and protein phosphatase 2A (PP2A). In b) and c), R, R*, and opsin (O) are shown as distinctly shaded dimers in order to emphasize their proposed higher order organization. While atomic force microscopy evidence exists for the oligomeric state of murine rhodopsin (R_1R_2) and opsin (O_1O_2) in the DM (Fotiadis et al., 2003; Liang et al., 2003), the existence of R* dimers interacting with G_t, Arr, and RK remains an open question.

4.2
Rhodopsin and Mammalian Visual Phototransduction

The process of mammalian visual phototransduction can be divided into three distinct pathways. Only the first two pathways involve rhodopsin: signal propagation by light-activated rhodopsin (metarhodopsin II or R*) and inactivation of R*. The third pathway, inactivation of the catalytic subunit (α-subunit) of the heterotrimeric retinal G-protein, transducin (G_t), continues to attract much attention and those interested in a perspective on this process are referred to some recent reviews on the subject (Arshavsky et al., 2002; Ridge et al., 2003; Cabrera-Vera et al., 2003).

4.2.1
Signal Amplification by Light-activated Rhodopsin

Light activation of rhodopsin (R) in the DM to produce R* catalytically promotes the exchange of GDP for GTP in hundreds of G_t molecules to produce the GTP-bound form of $G_{t\alpha}$ (Figure 4.1B). $G_{t\alpha}$-GTP interacts with the γ-subunit of the cGMP phosphodiesterase (PDE6 in rods), removing an inhibitory constraint from the catalytic α or β subunits that results in cGMP hydrolysis. The fast depletion of cGMP results in the closure of cGMP-gated channels in the PM. A drastic reduction in the circulating "dark" current, due to a blockage of Na^+ and Ca^{2+} entry into the rod and Ca^{2+} depletion of the ROS as a result of continuous function of the PM Na^+, Ca^{2+}, K^+ exchanger, activates the Ca^{2+}-binding guanylate cyclase activating protein (GCAP). The low intracellular Ca^{2+} concentration promotes activation of guanylate cyclase (GC) by Ca^{2+}-free GCAP, catalyzing accelerated synthesis of cGMP from GTP supplied by the guanine nucleotide cycle. The latter is comprised of two distinct nucleotide-binding enzymes, guanylate kinase (GK) and nucleoside diphosphate kinase (NDPK). The release of bound Ca^{2+} from calmodulin (CaM) leads to its dissociation from the cGMP-gated channel conferring a lower affinity for cGMP. Thus, light-activation of rhodopsin to produce R* leads to changes in the net production of cGMP by perturbing two opposing catalytic activities: degradation of cGMP by PDE6 and synthesis of cGMP by GC. Since cGMP levels govern how many channels are opened, the signal is transformed from a physical stimulus (light) to a biochemical chain of reactions that culminates in the hyperpolarization of the rod.

4.2.2
Inactivation of Light-activated Rhodopsin

In the dark, when the concentration of intracellular Ca^{2+} is high, myristoylated recoverin (Rec) is in the Ca^{2+}-bound state and forms a complex with rhodopsin kinase (RK) at the DM, preventing phosphorylation of R* (Figure 4.1C). When the Ca^{2+}concentration in the ROS drops as a result of the light response, Rec releases its bound $Ca^{2+}$and dissociates from RK. The interaction between R* and RK leads to rapid phosphorylation of R*. The subsequent binding of arrestin (Arr) blocks R* signaling via G_t. The release of *all-trans* retinal, accompanied by enzymatic dephosphorylation by protein phosphatase 2A (PP2A), prepares the apoprotein, opsin, for 11-*cis* retinal binding and restoration of its inactive ground state.

4.3 Properties of Rhodopsin

4.3.1 Isolation of Rhodopsin

Bovine rhodopsin is the best characterized member of the large family of GPCRs. A virtual explosion of information on rhodopsin can be attributed to the high natural abundance of the pigment in the DM and the accessibility of bovine rhodopsin and its mutants in amounts amenable to extensive biochemical, biophysical, and structural studies. Polyacrylamide gel electrophoresis in the presence of sodium dodecyl sulfate (SDS-PAGE) shows that more than 90% of the protein complement of the DM is accounted for by rhodopsin. In fact, highly purified rhodopsin in the DM can be prepared by sucrose density gradient centrifugation of ROS obtained by simple shaking of bovine retinae in the appropriate buffer and low speed centrifugation (Schnetkamp and Daemen, 1982). ROS preparations in hypotonic buffers are then subjected to Ficoll flotation with subsequent high speed centrifugation (Smith et al., 1975). Further purification of rhodopsin is typically based on its solubilization in detergents in combination with conventional chromatographic techniques. Although many different detergents effectively solubilize rhodopsin (e.g., digitonin, cetyl trimethylammonium bromide, nonylglucoside), dodecylmaltoside remains the detergent of choice due to very high rhodopsin stability in this media and the retention of many functional properties (DeGrip, 1982; Ramon et al., 2003). Affinity chromatography based on separation with concanavalin A-Sepharose was an early purification method of choice (Litman, 1982). However, rhodopsin samples prepared by this method are often contaminated with concanavalin A requiring the introduction of additional purification steps. Affinity chromatography based on matrices with immobilized monoclonal antibodies, such as rho 1D4-Sepharose (Oprian et al., 1987), appears to provide an extraordinarily efficient method for the one-step purification of rhodopsin. Immunoaffinity chromatography is essential for purifying mutants of rhodopsin obtained by heterologous eukaryotic cell expression and also appears to be instrumental in separating unfolded or misfolded opsin from properly folded rhodopsin (Ridge et al., 1995). More recently, methods have been developed for the selective extraction of ROS rhodopsin from the DM using specific divalent cations (Zn^{2+} or Cd^{2+}) in conjunction with alkylglucoside detergents (Okada et al., 1998). This method allows for the effective purification of rhodopsin and was instrumental in obtaining diffraction-quality three dimensional crystals (Okada et al., 2000).

4.3.2
Biochemical and Physicochemical Properties of Rhodopsin

Bovine rhodopsin is a single polypeptide of ~40 kDa consisting of 348 amino acids. Its amino acid sequence is known from analyses of the opsin apoprotein and the corresponding DNA (Ovchinnikov et al., 1983; Hargrave et al., 1983, Nathans and Hogness, 1983). About 65% of the amino acid residues in bovine opsin are hydrophobic residues unevenly distributed along the polypeptide chain (Figure 4.2). This arrangement appears to be crucial for the membrane disposition of the protein. Rhodopsin purity is assessed by comparing the ratios of the absorption maxima at 280 nm (protein) and 500 nm (chromophore) in the dark. This ratio (A_{280}/A_{500}) for purified rhodopsin varies between 1.60 and 1.65. Purified preparations of rhodopsin appear on SDS-PAGE as a single but somewhat broadened and diffuse band. This property

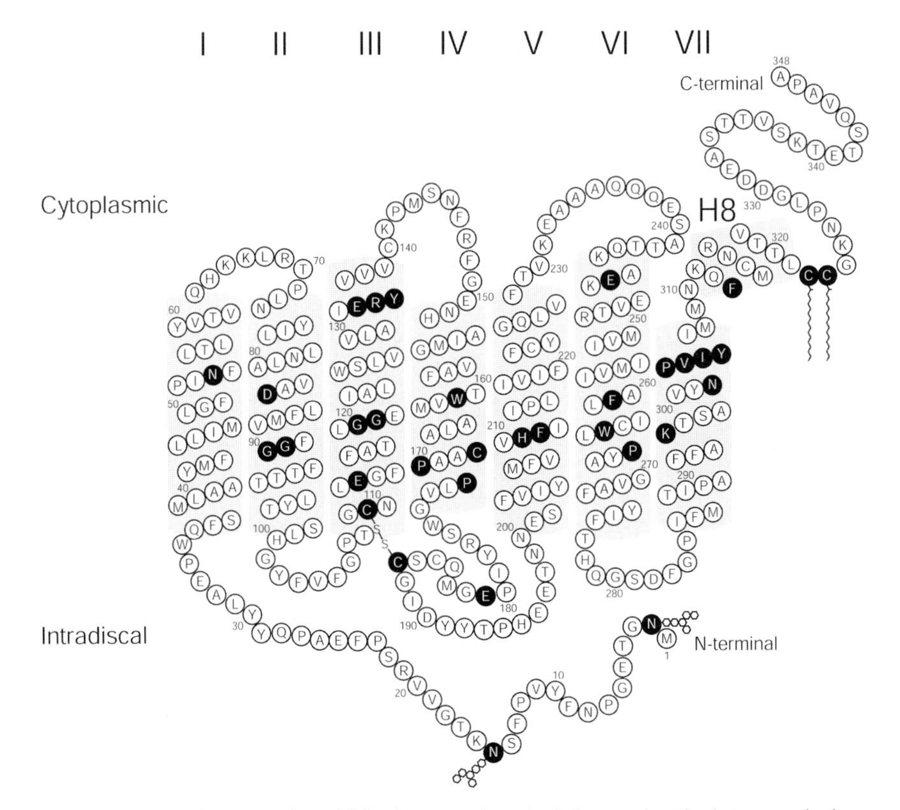

Figure 4.2 A two-dimensional model for the organization of bovine rhodopsin in the DM. The lengths of the transmembrane (TM) helices (I–VII) are based on the crystal structure of rhodopsin (Palczewski et al., 2000; Teller et al., 2001). The cytoplasmic helix, H8, extending from TM VII is also shown. The small hexagons in the N-terminal region represent branched oligosaccharide chains attached to Asn-2 and Asn-15 and the zigzag lines attached to Cys-322 and Cys-323 in the C-terminal region represent palmitoyl groups. The disulfide bridge between Cys-110 and Cys-187 is also shown. Key amino acid residues mentioned in the text are highlighted in black.

is characteristic for many membrane proteins and may represent possible heterogeneity in several post-translational modifications. In bovine rhodopsin, diffusion of the band appears to be primarily due to heterogeneity in asparagine-linked (*N*-linked) glycosylation (see Section 4.3.3). Additional factors contributing to the apparent heterogeneity of bovine rhodopsin may include phosphorylated rhodopsin species produced by exposure to light or lack of appropriate dark adaptation before or during retina processing. Finally, defects in fatty acylation (palmitoylation) may also contribute to rhodopsin heterogeneity (Qanbar and Bouvier, 2003).

4.3.3
Post-translational Modifications in Rhodopsin

As indicated in Section 4.3.2, the opsin polypeptide is subject to several post-translational modifications. First, the amino-terminal sequence, together with the acetylated amino-terminal amino acid, contains two consensus sequences allowing *N*-glycosylation at Asn-2 and Asn-15 (Fukuda et al., 1979). Carbohydrate analysis of bovine rhodopsin shows that ~70% of the oligosaccharides are accounted for by $Man_3GlcNAc_3$, 10% by $Man_4GlcNAc_3$, and 20% by $Man_5GlcNAc_3$ (Fukuda et al., 1979; Liang et al., 1979; Hargrave et al., 1984). What is not clear, however, is the distribution of these carbohydrate components between the two *N*-linked glycosylation sites. A second type of post-translational modification is palmitoylation of two vicinal cysteine residues at positions 322 and 323 via a thioester linkage (Ovchinnikov et al., 1988; Papac et al., 1992; Karnik et al., 1993). Third, two highly conserved cysteine residues, Cys-110 and Cys-187, are disulfide linked (Karnik and Khorana, 1990). This disulfide bridge is conserved not only in opsins, but also in the entire rhodopsin family of GPCRs and appears to play an important role in protein stability (Davidson et al., 1994; Garriga et al., 1996; Hwa et al., 2001). Another type of rhodopsin post-translational modification is the light-dependent phosphorylation of several threonines and serines at the extreme carboxyl terminus of the protein by RK (Hurley et al., 1998; Maeda et al., 2003). Despite 30 years of intensive effort, questions as to which of the serine and threonine residues become phosphorylated under physiological conditions remain unanswered. Finally, covalent binding of 11-*cis* retinal to the ε-amino group of Lys-296 via a protonated aldimine bond, a Schiff base, is a crucial and unique post-translational modification of rhodopsin.

4.3.4
Membrane Topology of Rhodopsin and Functional Domains

Seven hydrophobic stretches of about 25–30 amino acid residues interrupted by relatively short hydrophilic sequences is the signature feature of rhodopsin (Figure 4.2), other GPCRs, and numerous retinylidene microbial photoreceptors as well (see Spudich and Jung, Chapter 1). This odd number of transmembrane (TM) excursions positions the amino- and carboxyl-terminal ends of the protein on the opposing intradiscal (extracellular) and cytoplasmic surfaces, respectively. There is no obvious explanation for these so called "seven pillars of wisdom". The notion that seven hydrophobic TM helices are required to create a suitable binding pocket for low molecular weight agonists or antagonists (like 11-*cis* retinal) appears to be consistent with much of the experimental data, but this can not be extended to a vast number of GPCRs which have smaller ligands. Remarkably, a consensus on the heptahelical arrangement of rhodopsin was reached in the late 1970s (Albert and Litman, 1978; Michel-Villaz et al., 1979; Hubbell and Fung, 1979; Mas et al., 1980) with only limited available information on the primary structure, and subsequently confirmed by the crystal structure of bovine rhodopsin (Palczewski et al., 2000; Teller et al., 2001; Okada et al., 2002). The structure defines three major functional domains within rhodopsin: transmembrane, intradiscal, and cytoplasmic.

4.3.4.1 Transmembrane Domain of Rhodopsin

The TM domain of rhodopsin includes the seven predominantly α-helical segments, termed TM I –TM VII (Figure 4.2). TMI (residues 34–64) contains a highly conserved Asn-55 which forms hydrogen bonds with other residues (see Section 4.5.1). TM II, extending from residue 71 through 100, contains two adjacent glycine residues, Gly-89 and Gly-90. These two residues appear to be particularly susceptible to naturally occurring mutations in both mild and severe visual disorders (Rao, et al., 1994; Bosch et al., 2003; Abdulaev, 2003). TM II also contains a conserved Asp-83. TM III encompasses residues 106–139 and contains several important amino acid residues. First, the cytoplasmic end of TM III is capped by the highly conserved and functionally important triad: Glu-134, Arg-135, and Tyr-136 (the E(D)RY motif). The carboxyl group of Glu-113, which is located toward the center of this helix, serves as the dark-state counterion to the protonated retinylidene–Schiff base (Sakmar et al., 1989; Zhukovsky and Oprian, 1989; Nathans, 1990). Next, highly conserved cysteine residue Cys-110 forms a disulfide bond with Cys-187 from the second intradiscal loop (Karnik and Khorana, 1990). Interestingly, there are also two adjacent glycine residues, Gly-120 and Gly-121, which may contribute to the conformational flexibility of this helix (Ballesteros et al, 2001).

TM IV is the shortest of the TMs and extends from residues 150 to 172. Two proline residues, Pro-170 and Pro-171, are located at the cytoplasmic end of this helix. Cys-167, near the center of the helix, is an important component of the chromophore binding pocket (Palczewski et al., 2000; Teller et al., 2001). TM V incorporates residues 200–225, and with the exception of His-211 and Phe-212, appears not to be overburdened with functionally important amino acids. TM VI includes residues 244–276. Pro-267 of this helix appears to be essential for providing the necessary flexibility to support the dynamics of receptor activation (Ridge et al., 1999). Another important residue is Glu-247, which makes contact with Arg-135 of the E(D)RY motif of TM III (see Section 4.5.1).

TM VII extends from residues 286 to 309 and contains several important side chains. First, this helix contains Lys-296, the site of retinal attachment. The ε-amino group of this amino acid reacts with the aldehyde group of 11-*cis* retinal to form a Schiff base. Second, the cytoplasmic end of this helix contains the conserved NPXXY motif, considered to be a part of a larger functional unit known as the NPXXY$(X)_{5,6}$F motif (Fritze et al., 2003). It has been shown that a portion of this motif becomes accessible to a monoclonal antibody upon light activation of rhodopsin (Abdulaev and Ridge, 1998). Finally, Lys-296, Thr-297, Ser-298, and Ala-299 form a 3, 10 helix. All other helices in rhodopsin, both transmembrane and cytoplasmic, are regular α-helices (Riek et al., 2001).

4.3.4.2 **Intradiscal Domain of Rhodopsin**

The intradiscal (extracellular) domain of rhodopsin (Figure 4.2), long suspected in playing an important role in the structural stability of the pigment, incorporates a 33-residue *N*-glycosylated amino-terminal region and three connecting loops between TM helices II–III (amino acid residues 101–105), IV–V (residues 173–198), and VI–VII (residues 277–285). As indicated in Sections 4.3.3 and 4.3.4.1, Cys-187 of the second intradiscal loop forms a disulfide bridge with Cys-110 on the extracellular end of TM helix III (Karnik and Khorana, 1990). Only three of the seventeen positively charged surface residues in rhodopsin are found in the intradiscal domain, supporting the inside-positive rule characteristic for the arrangement of many polytopic membrane proteins (Sipos and von Heijne, 1993). The intradiscal domain appears to be highly structured and rarely tolerates even minor amino acid substitutions (Doi et al., 1990).

4.3.4.3 **Cytoplasmic Domain of Rhodopsin**

The cytoplasmic domain (Figure 4.2) includes the carboxyl-terminal tail of about 25 residues and loops connecting TM helices I–II (residues 65–70), III–IV (residues 140–149), and V–VI (residues 226–243). A fourth cytoplasmic loop (residues 311–323), now designated as H8 (Palczewski et al., 2001), extends from the cytoplasmic end of TM VII to the palmitoylated Cys-322 and Cys-323. In contrast to the highly structured and rigid intradiscal domain, the cytoplasmic domain appears to be designed to provide transient conformational changes that support activation of signaling and regulatory proteins (Palczewski et al., 2001; Sakmar, 2002). Valuable information on the cytoplasmic domain dark-state structure and the dynamics of its light-induced conformational changes became available through the painstaking work of the Khorana and Hubbell laboratories (Hubbell et al., 2003). The use of several experimental approaches, including site-directed spin labeling (SDSL) in combination with electron paramagnetic resonance spectroscopy (EPR) (Farahbakhsh et al., 1993), sulfhydryl reactivity (Ridge et al., 1995), and disulfide crosslinking (Yang et al., 1996), on more than 100 single cysteine mutants and 40 double cysteine mutants of rhodopsin, has allowed important information to be collected on the solution state conformation and the dynamics of this domain, as well as the boundaries of the loops. The resulting borders appear to be in good agreement with the published crystal structure of rhodopsin (Palczewski et al., 2000; Teller et al., 2001; Okada et al., 2002), although SDSL/EPR studies suggest that the third cytoplasmic loop might fold up further into TM V and TM VI (Altenbach et al., 1996). Quite interestingly, this latter finding is in agreement with a three-dimensional model generated from an alternative (trigonal) crystal form of rhodopsin (Li et al., 2004).

4.4 Chromophore Binding Pocket and Photolysis of Rhodopsin

The high quantum yield of retinal isomerization in rhodopsin is believed to be supported by the constrained conformation of the chromophore within the protein. Numerous efforts have been directed towards understanding both the ground- and excited-states of the rhodopsin chromophore. In the ground-state, the conjugated polyene chain is in the 11-*cis* configuration with the β-ionone ring primarily oriented to give a 6-s-*cis* conformer (Smith et al., 1987; Lugtenburg et al., 1988; Spooner et al., 2002). The chromophore binding pocket is predominantly comprised of hydrophobic amino acids with the β-ionone ring positioned between Phe-212 and Phe-261, whereas Trp-265 is closer to the center of the chromophore promoting the 11-*cis* form of the polyene chain (Palczewski et al., 2000; Teller et al., 2001). Although well protected from the aqueous environment, the chromophore binding pocket contains several hydrophilic side chains. Of keen interest is the carboxyl group of Glu-113, which serves as counterion to the protonated the retinylidene–Schiff base (Sakmar et al., 1989; Zhukovsky and Oprian, 1989; Nathans, 1990). This interaction appears to be of key importance in stabilizing ground-state chromophore–protein interactions. Altogether, around 30 amino acid side chains participate in forming the chromophore binding pocket (Palczewski et al., 2001).

Light-induced isomerization of 11-*cis* retinal to its *all-trans* configuration is the primary step in vision (Wald, 1968). This ultra-fast (femtosecond) photochemical process is followed by much slower events culminating in increased accessibility of the chromophore binding pocket to the aqueous environment with subsequent hydrolysis of the retinylidene–Schiff base (Schoenlein et al., 1991). These two extreme events in the photolysis of rhodopsin are mediated by several thermally stable dark processes resulting in photoproducts with defined spectral, kinetic, and functional properties. Following the light-induced *cis* → *trans* isomerization of retinal, the earliest photointermediate observed to date is photorhodopsin (Shichida et al., 1991). However, this intermediate is highly unstable so little is known about its structure. The first structurally characterized stable photointermediate detected is bathorhodopsin (Figure 4.3), which thermally decays to the blue-shifted intermediate (BSI), followed by lumirhodopsin, metarhodopsin I, and metarhodopsin II (Hug et al., 1990; Szundi et al., 1997). Of particular interest in this sequence is metarhodopsin II (or R*), the active form of rhodopsin with bound *all-trans* retinal and capable of G_t activation. Although the precise mechanism of decay for this signaling photointermediate is not fully understood (Lewis et al., 1997; Heck et al., 2003), the two products typically detected are metarhodopsin III and opsin plus free *all-trans* retinal.

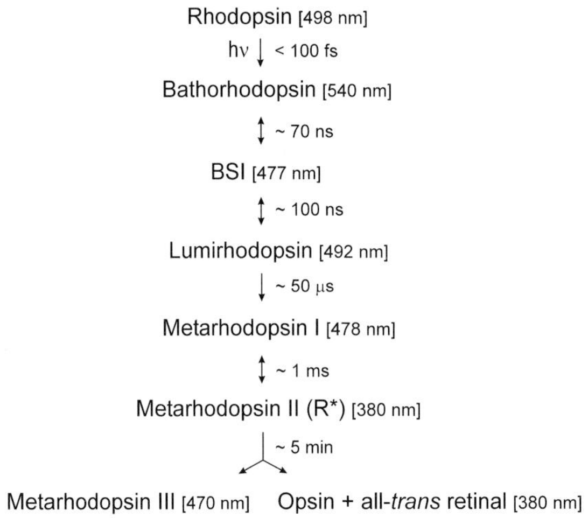

Rhodopsin [498 nm]
$h\nu$ ↓ < 100 fs
Bathorhodopsin [540 nm]
↕ ~ 70 ns
BSI [477 nm]
↕ ~ 100 ns
Lumirhodopsin [492 nm]
↓ ~ 50 μs
Metarhodopsin I [478 nm]
↕ ~ 1 ms
Metarhodopsin II (R*) [380 nm]
~ 5 min
Metarhodopsin III [470 nm]   Opsin + all-*trans* retinal [380 nm]

Figure 4.3 Photointermediates of rhodopsin photolysis. The photobleaching sequence of rhodopsin according to kinetic spectroscopy studies is shown. The absorbance maxima for the various photointermediates and their approximate half lives at 20°C are indicated.

4.5 Structure of Rhodopsin

4.5.1 Crystal Structure of Rhodopsin

Having outlined the three topologically distinct functional domains of rhodopsin in two dimensions in Section 4.3.4, it is now of interest to examine the three-dimensional structure of rhodopsin in order to visualize how these domains are arranged to support tertiary contacts that ensure ground-state stability. While several aspects of this issue, including orientation of the TM helices, structural irregularities (e.g., tilts, kinks), chromophore–protein interactions, and the structural consequences of naturally occurring rhodopsin mutations are of particular importance, many of these have already been covered in the original report (Palczewski et al., 2001) as well as several recent reviews (Ballesteros et al., 2001; Sakmar, 2002; Menon et al., 2001; Okada et al., 2001; Filipek et al., 2003). Here, we focus on some of the key structural features that confer on rhodopsin the unique ability to function as a finely tuned dim-light photoreceptor.

The X-ray structure was determined from diffraction-quality crystals of detergent solubilized bovine ROS rhodopsin (Okada et al., 2000; Palczewski et al., 2001). A current refined (2.8 Å) model of rhodopsin (Teller et al., 2001) includes greater than 95% of the amino acid residues as well as post-translational modifications (Figure 4.4). One of the surprises from the crystal structure is a compact intradiscal arrangement, parts of which fold inwards to enclose the retinal moiety. This "lid" or "plug" over the chromophore binding pocket involves the second intradiscal loop and is comprised of an anti-parallel β-stranded motif. The disulfide forming Cys-187 residue of this loop anchors the latter to Cys-110, near the extracellular surface of TM III. The second part of this loop is proximal to the bound 11-*cis* retinal, whereas the first part of the loop lies on top of the second strand. As indicated in Section 4.3.3, the intradiscal domain contains two consensus *N*-glycosylation sites at Asn-2 and Asn-15. Neither of these sites appears to make structural contacts with the protein (Figure 4.4), although Asn-15 *N*-glycosylation has previously been implicated in rhodopsin function (Kaushal et al., 1994). The cytoplasmic domain, in contrast to intradiscal domain, is largely disordered. One exception is H8, which lies almost perpendicular to the carboxyl-terminal end of TM VII (Figure 4.4). The carboxyl-terminal cytoplasmic tail and portions of the third cytoplasmic loop are poorly resolved in the structure (Palczewski et al., 2000; Teller et al., 2001; Okada et al., 2002), while, quite interestingly, the palmitoyl chains attached to Cys-322 and Cys-323 display resolvable density (Figure 4.4).

The TM helices are the sites where a majority of the conserved amino acid residues are found in rhodopsin. For example, Asn-55 of TM I, Asn-78 and Asp-83 of TM II, and Asn-302, Pro-303, and Tyr-306 of TMVII (part of the NPXXY$(X)_{5,6}$F motif), are all

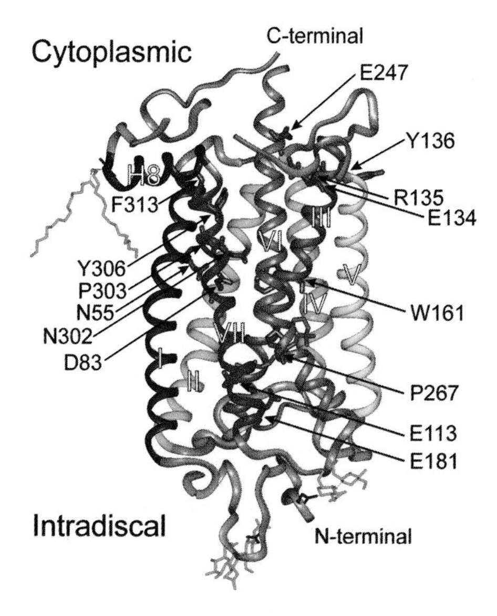

Figure 4.4 Three-dimensional structure of bovine rhodopsin. Ribbon drawing of the rhodopsin structure showing the seven TM helices and the cytoplasmic α-helix, H8, in various colors with the connecting segments in gray. The N-linked oligosaccharide chains attached to Asn-2 and Asn-15 in the N-terminal region and the thioester linked palmitoyl groups attached to Cys-322 and Cys-323 in the C-terminal region are shown in green. The 11-*cis* retinal chromophore is shown in blue. The side chains of key amino acid residues mentioned in the text that are important for maintaining the ground-state structure of rhodopsin or are involved in the mechanism of rhodopsin activation are shown in purple and indicated with arrows. The image was generated using InsightII and PDB entry 1HZX (A chain) and is based on work from Teller et al., 2001.

highly conserved amino acids across the GPCR family that appear to form contacts important for maintaining dark-state stability. While Asn-55 in TM I is hydrogen bonded to Asp-83 in TM II and Asn-78 in TM II is hydrogen bonded to Trp-161 in TM IV (Figure 4.4), conserved residues from the NPXXY(X)$_{5,6}$F motif form a network of interactions that stabilize the cytoplasmic domain (Palczewski et al., 2000; Teller et al., 2001; Fritze et al., 2003). Another important conserved element is the E(D)RY motif at the cytoplasmic end of TM III (Figure 4.4). Glu-134 and Arg-135 in this motif form ionic interactions with Glu-247 at the cytoplasmic end of TM VI and provide additional dark-state stability to rhodopsin (Fritze et al., 2003). It should be noted, however, that the electrostatic interaction between Glu-113 and Lys-296, residues which are largely conserved only among opsins, is considered to be the main factor in maintaining the dark-state conformation of rhodopsin (Rao et al., 1994).

4.5.2
Atomic Force Microscopy of Rhodopsin in the Disk Membrane

Oligomerization of GPCRs is now a widely accepted phenomenon. For most GPCRs, evidence for oligomerization is largely indirect and based on results from immunoprecipitation, chemical or disulfide crosslinking, size-exclusion chromatography, and fluorescence or bioluminescence resonance energy transfer studies (Terrillon and Bouvier, 2004). In the case of rhodopsin, however, direct evidence for oligomerization has recently been obtained using atomic force microscopy (AFM) (Fotiadis et al., 2003; Liang et al., 2003). The paracrystalline arrangement of rhodopsin dimers in the murine DM provides compelling evidence for how rhodopsin may be self-organizing, and provide a platform for signal propagation and/or quenching (Filipek et al., 2004). A current model for the rhodopsin dimer, termed the "IV–V model", where the dimeric interface is formed between helices IV and V (Fotiadis et al., 2003, 2004), proposes intermolecular contacts (hydrogen bonds) formed by Asn-199 and Ser-202 of both monomers. Asn-199 also extends the hydrogen bond network to Glu-196, and the adjacent Glu-197 residue points its carbonyl oxygen atom toward Asn-199, thereby contributing to the association of the two rhodopsin monomers. Despite the impressive nature of the AFM and modeling studies, it should be noted that these findings are at variance with the results of earlier biophysical measurements (Cone, 1972; Poo and Cone, 1974), which suggested that rhodopsin rapidly diffuses as a monomeric unit in the DM.

4.6
Activation Mechanism of Rhodopsin

Light activation of rhodopsin appears to rely on the disruption of dark-state contacts and replacing them with a new set of interactions. This process is directed towards creation of a highly specific and effective binding site for G_t. Since opsin exhibits negligible activity at neutral pH (Melia et al., 1997), the isomerized retinal chromophore is thought to play a major role in establishing these new interactions. In fact, studies with retinal analogs have shown that removal of the C19 methyl group from 11-*cis* retinal dramatically shifts the metarhodopsin I-metarhodopsin II equilibrium towards the former (Vogel et al., 2000; Meyer et al., 2000). An impairment of metarhodopsin II formation has also been observed when modifications are introduced in or around the β-ionone ring of retinal (Jager et al., 1994). Thus, it appears that both the C19 methyl group and β-ionone ring of retinal make important steric contributions to rhodopsin activation. The exact mechanism of rhodopsin activation is not fully understood but the general view is that there is an isomerization-induced separation of the cytoplasmic ends of TM III and TM VI relative to each other (Farrens et al., 1996; Hubbell et al., 2003). This is thought to be a consequence of charge separation and proton transfer from the retinylidene–Schiff base to the Glu-113 counterion, although a counterion switching mechanism at the metarhodopsin I stage involving Glu-181 has now been proposed (Yan et al., 2003). This latter mechanism, however, is at variance with recent NMR results indicating no evidence for any major structural reorganization around the chromophore binding site up to the metarhodopsin I stage (Spooner et al., 2003). Nonetheless, local conformational changes arising from retinal isomerization appear to be propagated to the cytoplasmic surface causing greater structural changes coupled to the reprotonation and rearrangement around the E(D)RY motif as well as coordinated changes in and around the NPXXY$(X)_{5,6}$F motif (Fritze et al., 2003). A strong distortion in TM VI imposed by Pro-267, one of the most conserved amino acid residues in the family of rhodopsin-like GPCRs (Sakmar, 2002), is also considered a key element of the activation mechanism.

4.7 Conclusions

Much has been revealed about the structure and function of mammalian rod rhodopsin over the past several years. As we begin the 21st century, it is likely that further molecular insights into this visual photosensory receptor will emerge. Having now established the basic structural features important for maintaining the dark-state of rhodopsin and having identified many of the key residues that contribute to function, future efforts will need to focus on placing these findings in the larger context of phototransduction. One immediate challenge, which should reveal the identity of specific steric interactions between the retinal chromophore and the protein, is the high-resolution structure determination of the rhodopsin photointermediates, and in particular, metarhodopsin II.

Acknowledgements

We thank Kris Palczewski for providing the template for Figure 4.2 and Eric DeJong for assistance in the preparation of Figure 4.4. This work was supported in part by U. S. Public Health Service grant EY13286 and a grant from the Karl Kirchgessner Foundation.

References

N. G. Abdulaev and K. D. Ridge, *Proc. Natl. Acad. Sci. USA.*, **1998**, 95, 12854–12859.

N. G. Abdulaev, *Trends Biochem. Sci.*, **2003**, 28, 399–402.

A. D. Albert and B. J. Litman, *Biochemistry*, **1978**, 17, 3893–3900.

C. Altenbach, K. Yang, D. L. Farrens, Z. T. Farahbakhsh, H. G. Khorana, and W. L. Hubbell, *Biochemistry*, **1996**, 35, 12470–12478.

V. Y. Arshavsky, T. D. Lamb TD, and E. N. Pugh, *Annu. Rev. Physiol.*, **2002**, 64, 153–187.

J. A. Ballesteros, L. Shi, and J. A. Javitch, *Mol. Pharmacol.* **2001**, 60, 1–19.

L. Bosch, E. Ramon, L. J. Del Valle, and P. Garriga, *J. Biol. Chem.*, **2003**, 278, 20203–20209.

M. E. Burns and D. A. Baylor, *Annu. Rev. Neurosci.*, **2001**, 24, 779–805.

T. M. Cabrera-Vera, J. Vanhauwe, T. O. Thomas, M. Medkova, A. Preininger, M. R. Mazzoni, and H. E. Hamm, *Endocr. Rev.*, **2003**, 24, 765–781.

R. A. Cone, *Nature New Biol.*, **1972**, 236, 39–43.

F. F. Davidson, P. C. Loewen, and H. G. Khorana, *Proc. Natl. Acad. Sci. USA.*, **1994**, 91, 4029–4033.

W. J. DeGrip, *Methods Enzymol.*, **1982**, 81, 256–265.

T. Doi, R. S. Molday, and H. G. Khorana, *Proc. Natl. Acad. Sci. USA.*, **1990**, 87, 4991–4995.

Z. T. Farahbakhsh, K. Hideg, and W. L. Hubbell, *Science*, **1993**, 262, 1416–1419.

D. L. Farrens, C. Altenbach, K. Yang, W. L. Hubbell, and H. G. Khorana, *Science*, **1996**, 274, 768–770.

S. Filipek, R. E. Stenkamp, D. C. Teller, and K. Palczewski, *Annu. Rev. Physiol.*, **2003**, 65, 851–879.

S. Filipek, K. A. Krzysko, D. Fotiadis, Y. Liang, D. A. Saperstein, A. Engel, and K. Palczewski, *Photochem. Photobiol. Sci.*, **2004**, 3, 628–638.

O. Fritze, S. Filipek, V. Kuksa, K. Palczewski, K. P. Hofmann, and O. P. Ernst, *Proc. Natl. Acad. Sci. USA.*, **2003**, 100, 2290–2295.

D. Fotiadis, Y. Liang, S. Filipek, D. A. Saperstein, A. Engel, and K. Palczewski, *Nature*, **2003**, 421, 127–128.

D. Fotiadis, Y. Liang, S. Filipek, D. A. Saperstein, A. Engel and K. Palczewski, *FEBS Lett.*, **2004**, 564, 281–288.

M. N. Fukuda, D. S. Papermaster, and P. A. Hargrave, *J. Biol. Chem.*, **1979**, 254, 8201–8217.

P. Garriga, X. Liu, and H. G. Khorana, *Proc. Natl. Acad. Sci. USA.*, **1996**, 93, 4560–4564.

P. A. Hargrave, J. H. McDowell, D. R. Curtis, J. K. Wang, E. Juszczak, S. L. Fong, J. K. Rao, and P. Argos, *Biophys. Struct. Mech.* **1983**, 9, 235–244.

P. A. Hargrave, J. H. McDowell, R. J. Feldmann, P. H. Atkinson, J. K. Rao, and P. Argos, *Vision Res.*, **1984**, 24, 1487–1499.

M. Heck, S. A. Schadel, D. Maretzki, F. J. Bartl, E. Ritter, K. Palczewski, and K. P. Hofmann, *J. Biol. Chem.*, **2003**, 278, 3162–3169.

W. L. Hubbell and B. K. Fung, *Soc. Gen. Physiol. Ser.*, **1979**, 33, 17–25.

W. L. Hubbell, C. Altenbach, C. M. Hubbell, and H. G. Khorana, *Adv. Protein Chem.*, **2003**, 63, 243–290.

S. J. Hug, J. W. Lewis, C. M. Einterz, T. E. Thorgeirsson, and D. S. Kliger, *Biochemistry*, **1990**, 29, 1475–1485.

J. B. Hurley, M. Spencer, and G. A. Niemi, *Vision Res.*, **1998**, 38, 1341–1352.

J. Hwa, J. Klein-Seetharaman, and H. G. Khorana, *Proc. Natl. Acad. Sci. USA.*, **2001**, 98, 4872–4876.

F. Jager, S. Jager, O. Krutle, N. Friedman, M. Sheves, K. P. Hofmann, and F. Siebert, *Biochemistry*, **1994**, 33, 7389–7397.

S. S. Karnik and H. G. Khorana, *J. Biol. Chem.*, **1990**, 265, 17520–17524.

S. S. Karnik, K. D. Ridge, S. Bhattacharya, and H. G. Khorana, *Proc. Natl. Acad. Sci. USA.*, **1993**, 90, 40–44.

S. Kaushal, K. D. Ridge, and H. G. Khorana, *Proc. Natl. Acad. Sci. USA.*, **1994**, 91, 4024–4028.

J. W. Lewis, F. J., van Kuijk, J. A. Carruthers, and D. S. Kliger, *Vision Res.*, **1997**, 37, 1–8.

J. Li, P. Edwards, M. Burghammer, C. Villa, and G. F. X. Schertler, *J. Mol. Biol.*, **2004**, 343, 1409–1438.

C. J. Liang, K. Yamashita, C. G. Muellenberg, H. Shichi, and A. Kobata, *J. Biol. Chem.*, **1979**, 254, 6414–6418.

Y. Liang, D. Fotiadis, S. Filipek, D. A. Saperstein, K. Palczewski, and A. Engel, *J. Biol. Chem.* **2003**, 278, 21655–21662.

B. J. Litman, *Methods Enzymol.*, **1982**, 81, 150–153.

J. Lugtenburg, R. A. Mathies, R. G. Griffin, and J. Herzfeld, *Trends Biochem. Sci.*, **1988**, 13, 388–393.

T. Maeda, Y. Imanishi, and K. Palczewski, *Prog. Retin. Eye Res.*, **2003**, 22, 417–434.

M. T. Mas, J. K. Wang, and P. A. Hargrave, *Biochemistry*, **1980**, 19, 684–691.

T. J. Melia, Jr., C. W. Cowan, J. K. Angleson, and T. G. Wensel, *Biophys. J.*, **1997**, 73, 3182–3191.

S. T. Menon, M. Han, and T. P. Sakmar, *Physiol Rev.*, **2001**, 81, 1659–1688.

C. K. Meyer, M. Bohme, A. Ockenfels, W. Gartner, K. P. Hofmann, and O. P. Ernst, *J. Biol. Chem.*, **2000**, 275, 19713–19718.

M. Michel-Villaz, H. R. Saibil, and M. Chabre, *Proc. Natl. Acad. Sci. USA.*, **1979**, 76, 4405–4408.

J. Nathans and D. S. Hogness, *Cell*, **1983**, 34, 807–814.

J. Nathans, *Biochemistry*, **1990**, 29, 9746–9752.

T. Okada, K. Takeda, and T. Kouyama, *Photochem. Photobiol.*, **1998**, 67, 495–499.

T. Okada, I. Le Trong, B. A. Fox, C. A. Behnke, R. E. Stenkamp, and K. Palczewski, *J. Struct. Biol.*, **2000**, 130, 73–80.

T. Okada, O. P. Ernst, K. Palczewski, and K. P. Hofmann, *Trends Biochem. Sci.*, **2001**, 26, 318–324.

T. Okada, Y. Fujiyoshi, M. Silow, J. Navarro, E. M. Landau, and Y. Shichida, *Proc. Natl. Acad. Sci. USA.*, **2002**, 99, 5982–5987.

D. D. Oprian, R. S. Molday, R. J. Kaufman, and H. G. Khorana, *Proc. Natl. Acad. Sci. USA.*, **1987** 84, 8874–8878.

Yu. A. Ovchinnikov, N. G. Abdulaev, M. Yu. Feigina, I. D. Artamonov, and A. S. Bogachuk, *Bioorg. Khim.* **1983**, 9, 1331–1340.

Yu. A. Ovchinnikov, N. G. Abdulaev, and A. S. Bogachuk, *FEBS Lett.*, **1988**, 230, 1–5.

K. Palczewski, T. Kumasaka, T. Hori, C. A. Behnke, H. Motoshima, B. A. Fox, I. Le Trong, D. C. Teller, T. Okada, R. E. Stenkamp, M. Yamamoto, and M. Miyano, *Science*, **2000**, 289, 739–745.

D. I. Papac, K. R. Thornburg, E. E. Bullesbach, R. K. Crouch, and D. R. Knapp, *J. Biol. Chem.*, **1992**, 267, 16889–16894.

M. Poo and R. A. Cone, *Nature*, **1974**, 247, 438–441.

R. Qanbar and M. Bouvier, *Pharmacol. Ther.*, **2003**, 97, 1–33.

E. Ramon, J. Marron, L. del Valle, L. Bosch, A. Andres, J. Manyosa, and P. Garriga, *Vision Res.*, **2003**, 43, 3055–3061.

V. R. Rao, G. B. Cohen, and D. D. Oprian, *Nature*, **1994**, 367, 639–642.

K. D. Ridge, Z. Lu, X. Liu, and H. G. Khorana, *Biochemistry*, **1995**, 34, 3261–3267.

K. D. Ridge, C. Zhang, and H. G. Khorana, *Biochemistry*, **1995**, 34, 8804–8811.

K. D. Ridge, T. Ngo, S. S. Lee, and N. G. Abdulaev, *J. Biol. Chem.*, **1999**, 274, 21437–21442.

K. D. Ridge, N. G. Abdulaev, M. Sousa, and K. Palczewski, *Trends Biochem. Sci.* **2003**, 28, 479–487.

R. P. Riek, I. Rigoutsos, J. Novotny, and R. M. Graham, *J. Mol. Biol.*, **2001**, 306, 349–362.

T. P. Sakmar, R. R. Franke, and H. G. Khorana, *Proc. Natl. Acad. Sci. USA.*, **1989**, 86, 8309–8313.

T. P. Sakmar, *Curr. Opin. Cell Biol.*, **2002**, 14, 189–195.

P. P. Schnetkamp and F. J. Daemen, *Methods Enzymol.*, **1982**, 81, 110–116.

Y. Shichida, Y. Matuoka, and T. Yoshizawa, *Photobiochem. Photobiophys.*, **1991** 7, 221–228.

R. W., Schoenlein, L. A. Peteanu, R. A. Mathies, and C. V. Shank, *Science*, **1991**, 254, 412–415.

L. Sipos and G. von Heijne, *Eur. J. Biochem.*, **1993**, 213, 1333–1340.

H. G. Smith, Jr., G. W. Stubbs, and B. J. Litman, *Exp. Eye Res.*, **1975**, 20, 211–217.

S. O. Smith, I. Palings, V. Copie, D. P. Raleigh, J. Courtin, J. A. Pardoen, J. Lugtenburg, R. A. Mathies, and R. G. Griffin, *Biochemistry*, **1987**, 24, 1606–1611.

P. J. R. Spooner, J. M. Sharples, M. A. Verhoeven, J. Lugtenburg, C. Glaubitz, and A. Watts, *Biochemistry*, **2002**, 41, 7549–7555.
P. J. R. Spooner, J. M. Sharples, S. C. Goodall, H. Seedorf, M. A. Verhoeven, J. Lugtenburg, P. H. Bovee-Geurts, W. J. DeGrip, and A. Watts, *Biochemistry*, **2003**, 42, 13371–13378.
I. Szundi, J. W. Lewis, and D. S. Kliger, *Biophys. J.*, **1997**, 73, 688–702.
D. C. Teller, T. Okada, C. A. Behnke, K. Palczewski, and R. E. Stenkamp, *Biochemistry*, **2001**, 40, 7761–7772.
S. Terrillon and M. Bouvier, *EMBO Rep.*, **2004**, 5, 30–34.
R. Vogel, G. B. Fan, M. Sheves, and F. Siebert, *Biochemistry*, **2000**, 39, 8895–8908.
G. Wald, *Science*, **1968**, 162, 230–239.
E. C. Yan, M. A. Kazmi, Z. Ganim, J. M. Hou, D. Pan, B. S. Chang, T. P. Sakmar, and R. A. Mathies, *Proc. Natl. Acad. Sci. USA.*, **2003**, 100, 9262–9267.
K. Yang, D. L. Farrens, C. Altenbach, Z. T. Farahbakhsh, W. L. Hubbell, and H. G. Khorana, *Biochemistry*, **1996**, 35, 14040–14046.
E. A. Zhukovsky and D. D. Oprian, *Science*, **1989**, 246, 928–930.

Since the preparation of this article, a new 2.2 Å crystal structure for bovine rhodopsin has been determined that resolves the complete polypeptide chain and provides further details about the configuration of the 11-*cis* retinal chromophore [Okada et al., *J. Mol. Biol.*, **2004**, 342, 571–583].

5
A Novel Light Sensing Pathway in the Eye: Conserved Features of Inner Retinal Photoreception in Rodents, Man and Teleost Fish

Mark W. Hankins and Russell G. Foster

Summary

Until recently the light sensing capabilities of the eye have been considered well understood. Rods and cones of the outer retina were thought to be the only photoreceptors of the eye, with the neurones of the inner retina providing the initial stage of visual processing. However, studies on mice lacking rod and cone photoreceptors have shown that these animals can adjust their circadian clocks, suppress pineal melatonin, modify locomotor activity, and regulate pupil size in response to environmental brightness. At least two of these responses, circadian regulation and pupillary constriction, are regulated by a previously uncharacterized opsin/vitamin A-based photopigment with a wavelength of maximum sensitivity λ_{max} near 480 nm (OP^{480}). The use of calcium imaging in wholemount retinal preparations has demonstrated the existence of a heterogeneous network of directly light-sensitive neurones within the ganglion cell layer of the retina. At least some of these light sensitive neurones express melanopsin. The ablation of this gene in mice with no rods or cones abolishes all known responses to light. Thus rods, cones and melanopsin neurones fully account for all photoreception within the eye, and melanopsin remains the primary candidate for OP^{480}. Studies in humans have identified an opsin/vitamin A based photopigment with a λ_{max}~480 nm, and hence a probable homolog of OP^{480}. This photopigment regulates both circadian responses to light and appears to act through a local retinal mechanism to drive diurnal changes in the primary visual pathway.

Whilst considerable attention has been paid to the inner retinal photoreceptors of mammals, little attention has been paid to the physiological function of these neurones in fish. Recent electrophysiological evidence suggests that one function of this inner retinal photoreceptor is to modulate the activity of retinal horizontal cells in response to environmental irradiance. The action spectrum for this depolarizing response fits a single opsin photopigment with a λ_{max} of 477 nm. Both VA-opsin and melanopsin appear to be expressed in these intrinsically photosensitive horizontal cells, but it remains unclear which of these opsins form the photopigment.

Handbook of Photosensory Receptors. Edited by W. R. Briggs, J. L. Spudich

ISBN 3-527-31019-3

Inner retinal photoreception in rodents, humans and fish is driven by opsin/vitamin A photopigments with markedly conserved spectral sensitivities (λ_{max} = 477–483 nm). This contrasts with the marked shifts in λ_{max} that have evolved in the accompanying cone sensitivities of these species. The reason for this remarkable level of conservation in inner retinal photosensitivity remains unclear.

5.1 Introduction

5.1.1 A Novel Photoreceptor within the Eye

The vertebrate eye has been the subject of serious study for some considerable time, with over 50 years of intracellular electophysiological analysis alone (Svaetichin, 1953). In broad terms the photosensory functions of the eye were thought to be well understood. Indeed, our understanding of the visual process is often heralded as one of the great success stories in neuroscience. The story was as follows: opsin/vitamin A-based photopigments located in the rods and cones of the outer retina transduce light, and the cells of the inner retina provide the initial stages of signal processing before signals travel down the optic nerve to specific sites in the brain for advanced visual processing. Whilst this view is certainly accurate, it ignores a quite separate light-detecting pathway within the eye that has only recently been identified within teleost fish and mammals (Soni et al., 1998; Freedman et al., 1999; Lucas et al., 1999; Berson et al., 2002; Jenkins et al., 2003; Sekaran et al., 2003). With hindsight it seems inconceivable that something as important as an unrecognised ocular photoreceptor could have been overlooked for so long. A partial explanation for this oversight is that these receptors do not mediate classical visual responses to light. Vision neuroscientists had no real reason to look for unidentified receptors as the responses they were monitoring could be more-or-less fully accounted for by the rods and cones. By contrast, the photoreceptor mechanisms mediating the effects of light on the biological clock were essentially unknown. The receptors were naturally assumed to be the rods and cones, but when this was assumption was tested empirically in the late 1980s and early 1990s the results were unexpected and ultimately led to the discovery of non-rod, non-cone ocular photoreceptors in the vertebrates. Although these receptors were initially discovered as a result of studies on the biological clock, it soon became clear that they do more than regulate the circadian system. Varied aspects of mammalian physiology, endocrinology and behavior respond to gross changes in environmental light. For example, pineal melatonin production (Lucas et al., 1999), pupil size (Lucas et al., 2001), adrenal cortisol secretion (Leproult et al., 2001), heart rate (Scheer et al., 1999) and even mood (Cajochen et al., 1999) are all affected by irradiance, and as discussed below, it now seems likely that novel receptors contribute to these and other light detection tasks.

5.1.2
Biological Clocks and their Regulation by Light

Almost all living creatures, including some prokaryotic forms of life (cyanobacteria), have been shown to possess an endogenous 24-hour biological clock. These clocks are generally called "circadian clocks" or "circadian pacemakers". The circadian system can be considered to fine-tune physiology and behavior to the varying demands of night and day. This advanced warning provided by the clock is necessary because different physiological or behavioral states cannot be switched instantly (Pittendrigh, 1993). For example, the transition from sleep to wake requires several hours in most mammals and involves the complete realignment of almost every biological system and process. If an organism waited to trigger its biology in response to a major change in environment, then valuable time would be wasted whilst physiology was realigned to generate an optimal performance. Peak performance is demanded by natural selection, where even modest advantages are translated into long-term evolutionary success. Thus almost every cellular, physiological or behavioral output is modulated to a greater or lesser extent by a chronobiological process (Rajaratnam and Arendt, 2001; Roenneberg and Merrow, 2002; 2003).

A clock is not a clock unless it can be set to local time. The ability to anticipate environmental events will only have survival value if biological time remains synchronized or entrained to the solar day, and the systematic daily change in the gross amount of light (irradiance) at dawn or dusk seems to provide the most robust, indicator or reference for the time of day. As a result, most organisms have evolved to use the twilight transition as the main *zeitgeber* (time giver) to adjust circadian time to local time – a process that has been termed "photoentrainment" (Roenneberg and Foster, 1997). Light detection for the image-forming visual system needs a measure of brightness in a particular region of space (radiance). By contrast, the circadian system needs a measure of the overall amount of light (irradiance) in the environment in order to make a reliable judgement about the phase of twilight. These marked differences in the nature of the sensory tasks associated with image-detection and irradiance detection seem to have given rise to very different sets of photoreceptor mechanism within the eye, the rods and cones of the outer retina and the irradiance detectors of the inner retina. In this review non-rod, non-cone ocular photoreception will be considered in the three groups of animals in which we have most information: Rodents (Section 5.2); Humans (Section 5.3) and Teleost fish (Section 5.4).

5.2
Non-rod, Non-cone Photoreception in Rodents

5.2.1
An Irradiance Detection Pathway in the Eye

It had been known since the 1980s that the circadian system of human and non-human mammals is relatively insensitive to light, needing both relatively bright and long duration exposure to light to bring about entrainment (Foster and Provencio, 1999). In the hamster, for example, animals can recognise simple images at light levels at least 200 times lower than the levels necessary to induce phase shifts in locomotor rhythms (Emerson, 1980). This relative insensitivity of the circadian system may be of considerable importance. It will effectively filter out light stimuli that do not provide reliable time of day information (Nelson and Takahashi, 1991). For example, the irradiance of starlight (around 1.5 n mol m^{-2} s^{-1}, 400–700 nm) and full moonlight (around 50 n mol m^{-2} s^{-1}, 400–700 nm) (Munz and McFarland, 1977), are both below the threshold for photoentrainment in the hamster (Nelson and Takahashi, 1991) and mouse (Foster et al., 1991). A reliable measure of environmental light, and hence time of day, will also need to compensate for local fluctuations in the light environment. This is a particular problem for mobile organisms which may experience marked changes in light exposure as a result of shading by plants or other structures (Lythgoe, 1979). A consistent feature of the vertebrate circadian system is that it is insensitive to light stimuli of a short duration, but can integrate light information over long periods of time. For example, the circadian system of the hamster is relatively insensitive to light stimuli shorter than 30 s, but can integrate/add photic stimuli over periods as long as 45 min. (Nelson and Takahashi, 1991). By contrast, integration times for image-forming visual responses are in the order of seconds. These features of the circadian system will again act to smooth out any local fluctuations in the light environment to provide a broad measure of environmental irradiance.

The circadian and visual systems also show a marked anatomical separation in their projections from the eye to the visual and circadian structures within the brain. The circadian pacemaker of mammals resides within the suprachiasmatic nuclei (SCN) (Ralph et al., 1990). The SCN receives its retinal projections from the retinohypothalamic tract (RHT) formed from a small number of morphologically distinct retinal ganglion cells (RGCs) (~1% of the total number of RGCs). The RGCs of the RHT tend to be distributed more-or-less evenly over the entire retina and send a topographically unmapped projection to the SCN. By contrast, the ganglion cells of the visual system send a highly mapped projection to the visual centres of the brain, such that a point on the retina maps precisely to a group of cells in the visual cortex (retinotopic mapping). The visual system is thus able to deduce both how much light and where it occurs in specific regions of space, whereas the SCN receives only information about the general brightness of environmental light. Thus the mammalian eye has parallel outputs, providing both radiance and irradiance information (Foster, 2002).

5.2.2 The Discovery of a Novel Ocular Photopigment in Mice (OP^{480})

Removal of the eyes in every mammal, including humans, abolishes photoentrainment (Nelson and Zucker, 1981; Foster, 1998). Because the rods and cones were the only known ocular photoreceptors, this led to the assumption that all light detection was mediated by these cells. However, studies on mice homozygous for the *retinal degeneration* gene (*rd/rd* mice), which lack functional rods and most of their cone photoreceptors, showed that although classical visual responses were absent in these animals, circadian responses to light appeared normal (Foster et al., 1991). There remained the possibility, however, that only a few degenerate "renegade" cone photoreceptors were required to sustain normal circadian responses to light (Foster et al., 1993a), and this led to the development of a mouse lacking all functional rod and cone photoreceptors. This was achieved by crossing rodless mice [*rd/rd* or *rdta* (McCall et al., 1996)] with transgenic mice (*cl*) lacking cones (Wang et al., 1992; Soucy et al., 1998). Circadian responses to light, and the capacity to regulate pineal melatonin by light, were preserved in these rodless/coneless (*rd/rd cl* and *rdta/cl*) animals, demonstrating that the eye must contain at least one additional class of photoreceptor (Freedman et al., 1999; Lucas et al., 1999). Furthermore, studies on *rd/rd cl* mice have also shown that non-rod, non-cone photoreceptors do more than regulate the circadian system. They also contribute to both pupil constriction (Lucas et al., 2001) and acute alterations in locomotor behavior (Mrosovsky et al., 2001), and may even be involved in a broad range of physiological and behavioral responses to light, including sleep (Gooley et al., 2003).

The *rd/rd cl* mouse has also provided a highly valuable model to characterize the novel ocular photopigment using action spectrum techniques. The first completed action spectrum was for pupil constriction and demonstrated the involvement of a novel opsin/vitamin A-based photopigment with a λ_{max} ~480 nm (OP^{480} – opsin pigment 480) (Lucas et al., 2001). The known photopigments of mice peak at ~360 nm (UV cone) (Jacobs et al., 1991), ~498 nm (rod) (Bridges, 1959), and ~508 nm (green cone) (Sun et al., 1997b), and do not show any significant fit to the pupil constriction action spectrum in *rd/rd cl* mice. Whether OP^{480} mediates all non-rod, non-cone ocular responses to light remains to be determined. This seems increasingly likely, however, as our recently completed action spectrum for phase-shifting circadian rhythms of locomotor behavior in *rd/rd cl* mice has identified an opsin/vitamin A-based photopigment with a λ_{max} at 481 nm (Hattar et al., 2003) (Figure 5.1a). In addition, these action spectra fail to provide any support for the involvement of a non-opsin photopigment such as the flavoprotein-based cryptochromes (Sancar, 2000; Van Gelder et al., 2003). For additional discussion see Lucas and Foster, (1999a; b).

Although rod and cone photoreceptors are not required for the regulation of the circadian system, this does not mean that these photoreceptors play no role. Indeed, the data emerging suggests that there is a complex interaction between rods, cones and novel photoreceptors in the regulation of circadian responses to light. For example, *rd/rd cl* mice fail to entrain to dim light/dark cycles with a normal phase, initiating their activity several hours before congenic wild-type controls (Foster et al., 2003a). In

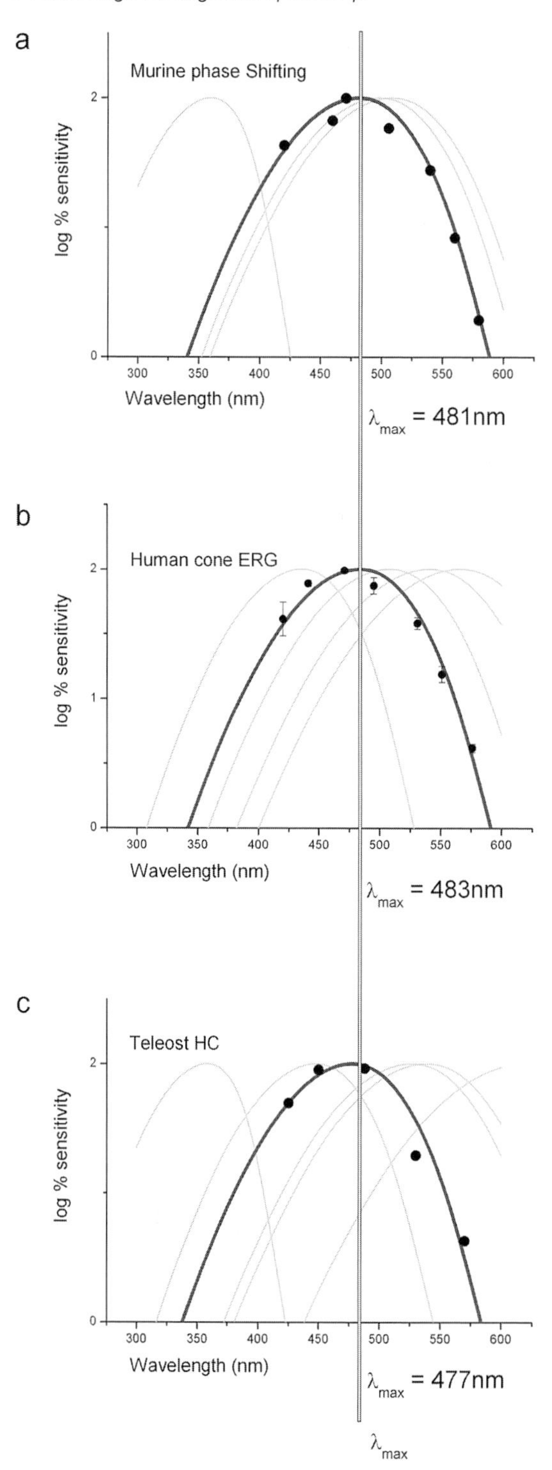

Figure 5.1 The action spectra for a range of irradiance detection tasks. (a) Action spectra for circadian phase shifting in the *rd/rd cl* mouse (Hattar et al., 2003). (b) Action spectra for the irradiance-dependent shift in the human ERG cone b-wave (Hankins and Lucas, 2002). (c) Action spectra derived for the novel depolarizing response of the teleost horizontal cell (*HC–RSD*) (Jenkins et al., 2003). In each case the derived action spectra (solid lines) are compared with the known rod and cone photopigments for each species (faint lines). The λ_{max} for each of the novel responses is shown. The data show a marked conservation in the spectral sensitivities for each of the irradiance detection systems. This contrasts with the marked shifts in λ_{max} that have evolved in the accompanying cone sensitivities of these species.

addition, action spectra for phase-shifting circadian rhythms in wild-type mice suggest the involvement of rods and/or cones (Thompson, Lucas, Hankins and Foster, unpublished). Perhaps the rods/cones provide information about acute transitions in the light environment, whilst the novel photoreceptors provide non-adapting irradiance information to the clock?

5.2.3
Melanopsin and Non-rod, Non-cone Photoreception

Action spectrum studies in *rd/rd cl* mice have defined a novel opsin/vitamin A-based photopigment with a λ_{max} close to 480 nm, and a number of candidate genes have been identified that might generate OP^{480}. By far the strongest candidate to emerge is melanopsin. The melanopsin gene family was first identified in *Xenopus* (Provencio et al., 1998b), and orthologs subsequently isolated from other vertebrate classes (e.g. teleosts (Bellingham et al., 2002) see 4.3 below) and several mammalian species including humans and mice (Provencio et al., 2000). The *Xenopus*, teleost and mammalian melanopsins have a relatively low identity to the vertebrate rod and cone opsins (~27%). They also show a different gene structure, and possess a tyrosine rather than a glutamate at the position of the putative counterion. In short they do not resemble the known photosensory opsins or even the inner retinal photosensory VA-opsins discussed below (Section 5.4.2) (Bellingham and Foster, 2002; Foster and Bellingham, 2002). Yet melanopsin has been shown to play a critically important role in non-rod, non-cone photoreception in mice.

In the murine retina, melanopsin is sparsely expressed within neurones of the ganglion cell layer and in a few cells in the amacrine cell layer (Provencio et al., 2000; Provencio et al., 2002). This distribution immediately suggested that melanopsin is expressed in those retinal cells that project to the suprachiasmatic nucleus (SCN) (Provencio et al., 1998a). This was confirmed by two independent studies. Using a combination of retrograde labeling and *in situ* hybridization, most of the retinal ganglion cells (RGCs) that project to the rat SCN where shown to express melanopsin (Gooley et al., 2001). The second study exploited the finding that pituitary adenylate cyclase-activating polypeptide (PACAP) is expressed in the retina exclusively within the RGCs of the retinohypothalamic tract (RHT). Melanopsin was then shown to be co-expressed with the PACAP-containing RGCs (Hannibal et al., 2002). Most recently melanopsin-projecting ganglion cells have been shown to project, in addition to the SCN, to other nuclei in the brain associated with processing irradiance information. These include the intergeniculate leaflet (IGL), olivary pretectal nuclei (OPN) and the ventrolateral preoptic nuclei (VLPO) (Hattar et al., 2002; Gooley et al., 2003; Hattar et al., 2003).

A series of publications has clearly demonstrated a functional link between melanopsin and non-rod, non-cone photoreception. As discussed below, two independent approaches were used to identify the non-rod, non-cone photoreceptor cell-type within the eye (Berson et al., 2002; Sekaran et al., 2003). In the rat and mouse these studies showed that at least some of the retinal ganglion cells that project to the SCN are directly light sensitive. Significantly, these light sensitive RGCs have been shown to express melanopsin. Additional approaches using *rd/rd cl* mice have strengthened the links between melanopsin and ganglion cell photosensitivity. Aged *rd/rd cl* mice show substantial transneuronal retinal degeneration. After ~18 months of age only the retinal ganglion cell layer and small patches of inner retina remain in these mice. Quantitative RT-PCR showed normal levels of melanopsin expression, and immunocytochemistry demonstrated both the presence and normal cellular appearance of melanopsin positive ganglion cells in these aged *rd/rd cl* mice. Despite the loss of the outer retina and most of the inner retina, circadian responses to light remain intact, even in *rd/rd cl* mice older than two years. These data provided the first positive correlation between the persistence of melanopsin-expressing ganglion cells and the maintenance of circadian responses to light (Semo et al., 2003b). In addition, light-induced expression of the immediate early gene *c-fos* was only observed in melanopsin-positive RGCs of *rd/rd cl* mice. Suggesting that Fos induction is linked to the intrinsic photosensitivity of these melanopsin-positive neurons (Semo et al., 2003a).

The final body of evidence showing that melanopsin plays a critical role in the transduction of light information of the directly photosensitive RGCs comes from melanopsin gene ablation studies. The first sets of experiments showed that melanopsin knock-out mice have attenuated phase shifting and pupil responses to light, and that the photosensitive RGCs fail to respond to light in these melanopsin knock-out animals (Panda et al., 2002; Ruby et al., 2002; Lucas et al., 2003). More recently, mice in which rods, cones and melanopsin have all been ablated fail to show any responses to light, arguing that these three classes of photoreceptor account for all light detection within the eye (Hattar et al., 2003). These data reinforce the argument that the cryptochromes are not involved in circadian photoreception. Indeed, Russell Van Gelder at Washington University School of Medicine, who has been the main proponent for a photosensory role for the cryptochromes stated recently: “It would thus appear that the primary photopigment in non-visual photoreception is melanopsin-dependent, but not cryptochrome-dependent.” (Panda et al., 2003).

Although highly suggestive, the melanopsin knock-out data do not confirm that melanopsin is the photopigment of the photosensitive RGCs. Gene ablations studies alone can only indicate that a gene is critical; biochemistry on the protein product is required to define its function. In an attempt to address the biochemical role of melanopsin, Phyllis Robinson's group at the University of Maryland have expressed melanopsin in COS cells and found that after reconstitution with 11-*cis*-retinal the pigment showed a maximal absorbance between 420–440 nm (Newman et al., 2003). This absorption maxima is in marked contrast to the action spectra for non-rod, non-cone photoreception in *rd/rd cl* mice which defined a photopigment close to 480 nm (OP^{480}). If ~430 nm is the true spectral absorbance maxima of melanopsin, then it is difficult to see how melanopsin could mediate circadian responses to light. A further complication is that melanopsin was originally thought to be expressed exclusively within a sub-set of RGCs of mammals. However, very recent studies have show that melanopsin mRNA and protein are expressed at high levels within the retinal pigment epithelium (RPE) (Peirson et al., 2004). This fact might complicate the interpretation of the melanopsin gene ablation studies. A final issue relating to the interpretation of the role of melanopsin is considered below (Section 5.4). Melanopsin appears to be co-expressed with known photosensory pigments in fish and amphibian photoreceptors.

5.2.4
A Functional Syncitium of Directly Light-sensitive Ganglion Cells

David Berson's laboratory, at Brown University, has recorded from retinal ganglion cells in the rat that were retrogradely labeled with fluorescent microspheres injected into the retino-recipient areas of the hypothalamus (Berson et al., 2002). These cells were demonstrated as directly light sensitive, as their light-evoked depolarizations persisted in the presence of a cocktail of drugs reported to block all retinal inter-cellular communication, and even continued when dissected from the surrounding retinal tissue. Using this approach Berson's group went on to generate an action spectrum for the light-evoked depolarization. The data suggested a best fit to an opsin/vitamin A-based photopigment with a λ_{max}~484 nm (Berson et al., 2002), and in this regard the results are strikingly similar to pupillary and circadian action spectra in mice (Lucas et al., 2001; Hattar et al., 2003) and the modulation of cone ERG responses in humans (Hankins and Lucas, 2002) (see Section 5.3.2 below and Figure 5.1b).

A parallel series of studies was undertaken to address which inner retinal neurones are directly photosensitive in the mouse. This approach used fluorescent imaging with FURA-2AM to study the role of Ca^{2+} in the generation of the light responses in the isolated retinae of *rd/rd cl* mice. This preparation has several distinct advantages: (a) intrinsic light responses can be detected in a physiological state without the need for potentially toxic pharmacological treatment or physical damage from mechanical isolation; (b) no assumptions are made concerning the retino-recipient targets of the intrinsically light sensitive cells; and (c) the approach can simultaneously image and screen several hundred neurones of the ganglion cell layer (GCL) permitting a real-time survey of the nature and extent of individual inner retinal photoreceptors. Using this approach a heterogeneous population of light-responsive neurones were identified in the *rd/rd cl* retina, constituting around 2.7% of the cells sampled (Figure 5.2). Significantly the light-induced Ca^{2+} fluxes could be blocked by application of 1 mM Cd^{2+} suggesting that the increases in FURA-2 fluorescence may reflect Ca^{2+} influx rather than intracellular Ca^{2+} mobilization (Sekaran et al., 2003).

Three discrete classes of light-induced Ca^{2+} change are apparent in these neurones: (a) *Sustained.* Cells continued to show increases in intracellular Ca^{2+} following termination of the light stimulus and generally recovered within 5 min; (b) *Transient.* The increase in intracellular Ca^{2+} quickly returned to baseline levels typically before the end of light stimulation; (c) *Repetitive.* Repeated firing without recovery. Continuous Ca^{2+} oscillations were observed which failed to show full recovery up to 20 min after cessation of the light stimulus (Figure 5.3). These different response sub-types may be associated with projections to different regions of the brain (e.g. SCN, IGL, VLPO and OPN). Alternatively, those brain regions involved in processing irradiance information could receive a more complex set of irradiance signals than previously as-

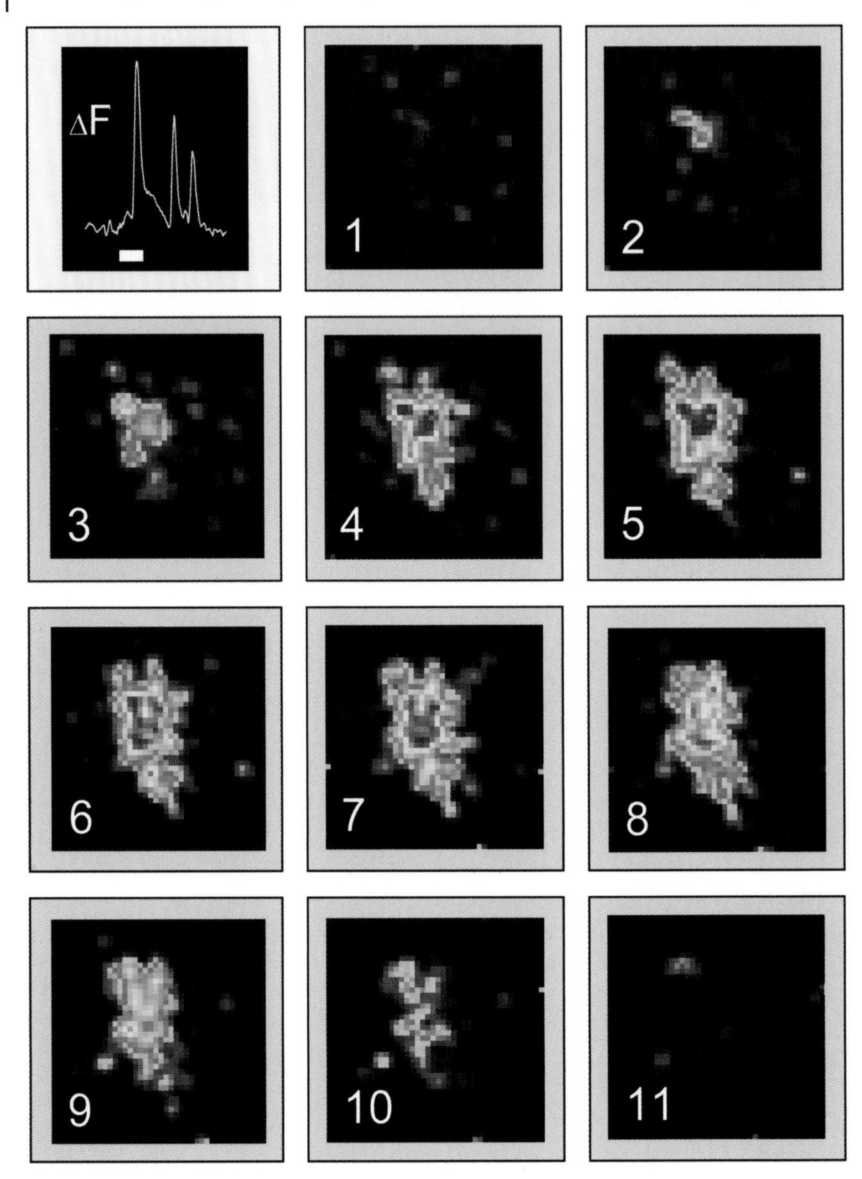

Figure 5.2 Calcium imaging of light-sensitive ganglion cells in the isolated *rd/rd cl* mouse retina. The retina was loaded with the Ca^{2+}-sensitive indicator FURA-2AM and stimulated with 470-nm light. In the first panel in the sequence the fluorescence intensity (ΔF) is plotted for this cell before, during, and after light stimulation (indicated by the bar). The plot shows successive transient waves in Ca^{2+} in response to the light stimulus. The peak changes in fluorescence were detected 44–54 s after lights on. During the recovery phase, secondary calcium oscillations were apparent. The sequential series of images (1–11) shows that illumination induced a significant increase in fluorescence of the Ca^{2+}-sensitive indicator. On average, light stimulation induced an increase in fluorescence in approximately 2.7% of the neurones in the ganglion cell layer.

sumed, receiving inputs from multiple populations of ganglion cells, either directly or through axon collaterals (Sekaran et al., 2003).

A particularly striking feature of the light responsive neurones is the synchronous nature of the individual light-induced Ca^{2+} wavelets in cells that are some 200–300 μm apart. This synchrony suggested that the light responsive neurones might be organized into a functional syncytium, perhaps linked by gap junction connections between directly photosensitive cells and neighbouring neurones in the GCL. The presence of the gap junction blocker carbenoxolone reduced the number of light responding cells by approximately 56%. This suggests that light responses from directly sensitive ganglion cells are processed and dispersed through additional non-photosensitive neurones via gap junction signalling. Thus intercellular gap junctional

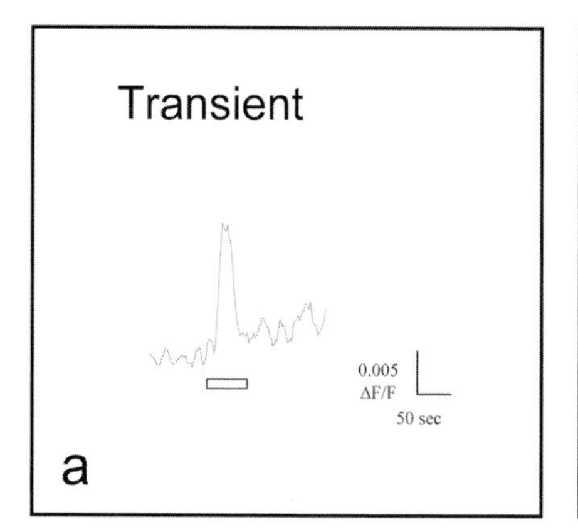

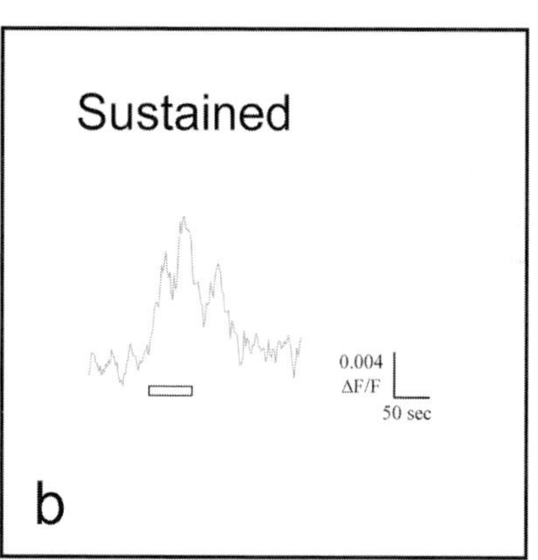

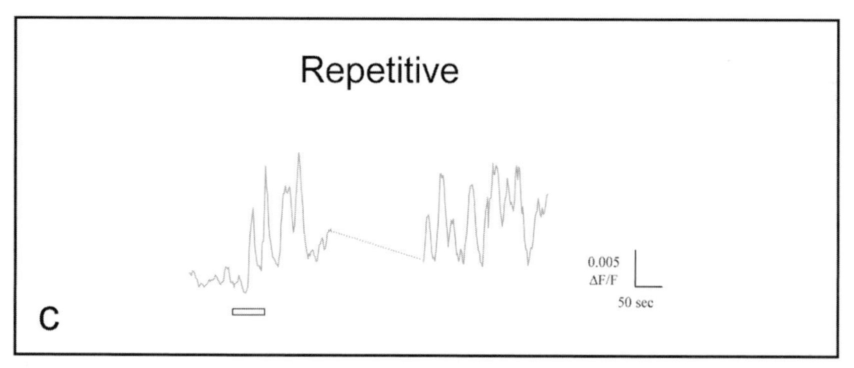

Figure 5.3 Multiple phenotypes of light-evoked Ca^{2+} responses. In response to 470-nm light stimulation (open bar; 6.7×10^{14}–1.3×10^{15} photons cm–2 s^{-1}) three distinct types of change in the normalized fluorescence intensity were observed: (a) transient response. The increase in fluorescence quickly recovered towards the baseline level, typically before the end of stimulation; (b) sustained response: the increase in fluorescence was maintained up to 1 min after termination of the light stimulus; (c) repetitive transients without full recovery: light stimulation induced successive transients that continued following termination of the stimulus. In the absence of any stimulation for a 10-min period (dashed lines) the cell continued to demonstrate repetitive Ca^{2+} transients.

communication may provide an important secondary population of light-driven inner-retinal neurones whose role remains to be established (Sekaran et al., 2003).

Combining the Ca^{2+} imaging with melanopsin immunocytochemistry confirmed that at least some of the directly light responsive cells express melanopsin (Sekaran et al., 2003). However, since we estimate that 2.7% of the neuronal population in the GCL respond to light, it is very unlikely that all of these neurones express melanopsin. This is because melanopsin is reported to be present in only ~1% of the ganglion cells (Hattar et al., 2002), and not 3.0%. It remains to be determined if the non-melanopsin light responding cells are directly photosensitive, or respond to light as a result of the gap-junctional connections with photosensitive RGCs.

The use of calcium imaging in wholemount *rd/rd cl* retinal preparations has demonstrated the existence of a heterogeneous network of light sensitive inner-retinal neurones that function in the absence of rod and cone inputs. Phototransduction in several invertebrate species is associated with an influx of Ca^{2+} through light-gated cation channels (Hardie and Minke, 1992; Hardie, 2001; Hardie and Raghu, 2001) and indeed, melanopsin does share some features in common with the invertebrate-like opsins (Provencio et al., 1998b; Provencio et al., 2000). Currently little is known about the signaling cascade linked to OP^{480}, but our results suggest that Ca^{2+} influx is a component part of the light-sensitive cells' physiological response to light. These data do not, of course, preclude the involvement of other ions.

5.3 Non-rod, Non-cone Photoreception in Humans

5.3.1 Introduction

The demonstration of non-rod, non-cone photoreception in rodents naturally suggested that such photoreceptors might also exist in humans. As in rodents, eye loss in humans blocks circadian responses to light (Czeisler et al., 1995; Lockley et al., 1995). Also like rodents, visual loss does not always result in circadian blindness. Some humans with retinal degenerative disease, and lacking conscious light perception, are still able to regulate their circadian physiology by light (Czeisler et al., 1995; Lockley et al., 1995). These data strongly suggest that, like rodents, humans possess non-rod, non-cone ocular photoreceptors. Before we proceed, some mention should be made of a report in *Science* which suggested that bright light applied to skin behind the knee can induce a shift in circadian rhythms of body temperature and melatonin (Campbell and Murphy, 1998). Several attempts have been made to support these findings e.g. (Lockley et al., 1998; Koorengevel et al., 2001; Wright and Czeisler, 2002). To date, however, all such attempts have failed, suggesting that some experimental artefact in the original study resulted in the apparent effects. The direct financial cost associated with attempts to duplicate these findings has been estimated as in excess of $5 million.

The variability in the pathology of retinal degeneration in humans, and the difficulty in assessing the extent of rod and cone photoreceptor loss in such patients, has not allowed access to the human equivalent of an *rd/rd cl* mouse. As a result, the unequivocal demonstration of non-rod, non-cone photoreceptions in our own species has been even more of a problem. However, the use of action spectrum techniques has provided clear support that such novel photoreceptors exist.

5.3.2 Novel Photoreceptors Regulate Melatonin

Two laboratories have independently determined an action spectrum for light-induced melatonin suppression in normally-sighted subjects (Brainard et al., 2001; Thapan et al., 2001). In both cases experiments were performed at night and determined the effect of long-term light exposure on plasma melatonin using a range of retinal irradiance/wavelength combinations. The resulting irradiance response curves were found to be univariant, and hence consistent with the involvement of a single photoreceptor driving the response. Both groups concluded that the data could be best described by a single novel opsin/vitamin A photopigment, quite distinct from the rods and cones, and maximally sensitive in the blue part of the spectrum, somewhere between 460–480 nm. Furthermore, a very recent study suggests that the same novel receptors not only suppresses melatonin, but also entrains the circadian driven rhythm of melatonin synthesis (Lockley et al., 2003). The similarity of these action spectra to mouse OP^{480}, suggests that humans possess an ortholog of the same novel opsin.

5.3.3 Novel Photoreceptors Regulate the Primary Visual Cone Pathway

Under natural ambient light cycles, the response of second-order retinal neurones to stimulation by cone photoreceptors is significantly slower at night than during the day (Hankins et al., 1998). This diurnal variation in the processing of cone signals by second order neurones is observed as a 20% increase in b-wave response (implicit) times in the middle of the night as compared to daytime values (Hankins et al., 1998; Hankins and Lucas, 2002). The function of this adaptation may be to overcome the temporal disparity in the response of rods and cones under mesopic light levels when both will be active. A primary drive for this change in cone response appears to be retinal light exposure. Thus, light delivered at night will shift the cone pathway towards the daytime state.

Significantly, long and relatively bright light exposure is required to drive human b-wave implicit time in cones from the night to the daytime state. This relative insensitivity to light is reminiscent of the light stimuli required for circadian entrainment, and led to speculation that the photoreceptors responsible for driving diurnal variations in cone b-wave implicit time might be the same as the circadian system. Detailed irradiance response functions for the reduction in b-wave implicit time were constructed by examining the effect of single 15-min monochromatic light exposure of varied irradiance and wavelengths. The light pulse was delivered at midnight, and its effects assessed by comparing ERG recordings made at 23:00 and 02:00. (Figure 5.4). Low irradiances of the experimental stimulus had no effect on b-wave implicit time. By contrast, at high irradiances it was possible to reduce b-wave implicit time by ~10 ms, driving it close to values observed in the middle of the day. The monochromatic irradiance response curves were then used to derive an action spectrum for the cone shifting ERG response. There is a strikingly poor fit between this action spectrum and the sensitivity curves for all four of the known human photoreceptors. When the action spectrum data are corrected for pre-retinal lens absorption (Stockman et al., 1999), and fitted to the standard absorbance template of an opsin/vitamin A photopigment the best fit (lowest sums of squares, $R^2 = 0.94$) is to a pigment template (Govardovskii et al., 2000) with a λ_{max} of 483 nm (Figure 5.1b). The closeness with which this template approximates the action spectrum for mouse OP^{480}, and melatonin suppression in humans, supports the hypothesis that humans possess an ortholog of mouse OP^{480} and that, like that in mice, there is a single novel pigment within the human eye regulating diverse responses to environmental irradiance.

The regulation of synaptic plasticity associated with the transition from scotopic to photopic vision is a critical component in the mixed rod/cone retina, and the demonstration in humans that this pathway is regulated by a novel opsin photopigment has important experimental and clinical implications. By what means the novel photoreceptors regulate this pathway remains a matter of speculation. One possibility is that

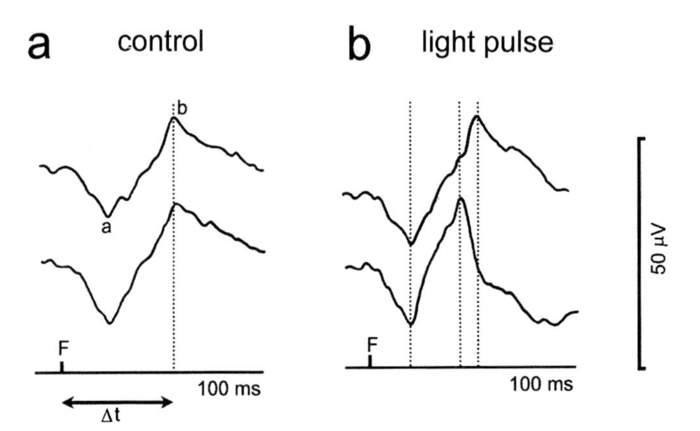

Figure 5.4 ERG waveforms recorded from an observer in the human cone ERG studies. The recordings represent 100-ms sweeps, with a flash delivered at the time marker "F". Components reflecting the hyperpolarization of photoreceptors (a-wave) and depolarization of second order neurones (b-wave) are denoted by "a" and "b" (a). The upper sweep was recorded at 21:00 GMT from an eye patched at dusk. The observer was then kept in darkness and the ERG recorded again at hourly intervals. The lower sweep was then taken at 03:00 GMT. There were no significant changes in either the amplitude or kinetics of the a-wave or b-wave over this period. (b) The upper sweep was recorded at 21:00 GMT from an eye patched at dusk. A 15-min light pulse (λ_{max} = 471 nm, 32 μW cm^{-2}) was then delivered at 23:00 and the ERG recorded again in the lower sweep at 03:00. The light pulse evoked a selective effect upon b-wave latency reducing it in this example by some 7 ms.

non-classical photoreceptors regulate retinal physiology by regulating melatonin and/or DA signals. Aspects of retinal plasticity associated with transitions between rod and cone vision have been associated with the local paracrine function of dopamine (Dowling, 1991; Witkovsky and Dearry, 1991) and melatonin (Dubocovich, 1989). In mammals, local retinal melatonin synthesis and release is high at night and suppressed by exposure to light or DA (Tosini and Menaker, 1996; 1998). By contrast, DA release is induced by light (Nir et al., 2000) and inhibited by melatonin (Dubocovich, 1989). Thus, retinal melatonin and DA concentrations may represent contra-regulatory signals of night and day respectively.

5.4 Non-rod, Non-cone Photoreception in Teleost Fish

5.4.1 Background

Thus far the discussion for the existence of a novel opsin-based photoreceptor within the inner retina has been limited to the mammals. However, parallel experiments on the retina of teleost fish provide independent support of the existence of a non-rod, non-cone ocular photoreceptor in the vertebrates. In contrast to the mammals, the non-mammalian vertebrates possess a diverse complement of photoreceptors, with at least five different types of photoreceptor organ developing from the forebrain (Shand and Foster, 1999). These photoreceptor organs are classified as: (1) eyes; (2) an intracranial pineal organ or pineal body (*epiphysis cerebri*) which contains photoreceptors in all non-mammalian vertebrates; (3) an intracranial parapineal organ, found in many teleost fish and lampreys; (4) an extracranial 'third eye', variously called a frontal organ (frogs) or parietal eye or parietal body (lizards); (5) deep brain photoreceptors, located in several sites in the brain and found in all non-mammalian vertebrates. Mammals are unique among vertebrates in that they appear to have lost extraretinal photoreceptors. Why this has occurred is not clear, but may be associated with the early evolutionary history of mammals and their passage through a "nocturnal bottleneck" (Foster and Menaker, 1993). The multiplicity of photoreceptor organs in the non-mammalian vertebrates often comes as a surprise to many, and is due undoubtedly to our perceptual bias. We show an extreme dependence on vision and lack extraocular photoreceptors. Extraretinal photoreceptors have generally been dismissed as either vestigial or redundant, akin to a "sensory appendix". Yet it is these non-eye photoreceptors that are critically important for circadian entrainment and other irradiance detection tasks in the non-mammals. For example, birds use photoreceptors located deep within the brain as their primary receptors for photoentrainment (Menaker and Underwood, 1976). The presence of such photoreceptors, located deep within the brain, seems at first absurd. Why have photoreceptors in regions where there is no light? Direct measurements, however, have shown that the skull and brain are remarkably permeable to light. Although photons are scattered and selectively filtered by neural tissue, large amounts of light penetrate deep into the brain (Foster and Follett, 1985). This light cannot not be used to generate an image of the world, but it can be used to deduce environmental irradiance, and hence time of day. The deep brain photoreceptors provided a striking example of the existence of specialized photoreceptors, dedicated to the task of irradiance detection, that do not resemble the rods and cones of the eye. The study of deep brain photoreceptors also led, accidentally, to the discovery of VA-opsin, and the first characterization of an inner retinal photopigment in the vertebrate eye.

5.4.2 Vertebrate Ancient (VA) Opsin and Inner Retinal Photoreception in Teleost Fish

The study of deep brain photoreceptors has a very long history, originating with the classical work of Carl Von Frisch in 1911 on the European minnow (Frisch, 1911). However, the nature of these photoreceptors has remained remarkable elusive (Foster and Soni, 1998). In an effort to characterize the extraocular photoreceptors of fish we developed degenerate PCR primers to amplify opsin cDNAs from the teleost CNS. To optimize our PCR procedures we first tested these degenerate primers using ocular cDNA derived from eye tissue of the Atlantic salmon (*Salmo salar*) salmon. An initial screen identified a cDNA whose conceptual translation shared only 37–42% identity with any of the known opsin families. This level of identity immediately isolated this opsin into a foundling, previously unrecognized, opsin family that we subsequently termed the Vertebrate Ancient (VA) opsins (Soni and Foster, 1997; Soni et al., 1998). The VA-opsins have now been described in many species of teleost fish, and critically, when expressed *in vitro* and reconstituted with 11-*cis*-retinal, VA-opsin from salmon and zebrafish forms a functional photopigment having a λ_{max} between 460–500 nm (Soni et al., 1998; Kojima et al., 2000). Perhaps the most remarkable finding about the VA-opsins are their sites of expression. They are expressed variously in a subset of neurones of the horizontal cell layer, amacrine cell layer, and ganglion cell layer of the retina (Figure 5.5). As a result, the VA-opsin family became the first characterized photopigment of the inner retina. Furthermore, VA-opsin was also found to be expressed in two extra retinal photoreceptor organs of fish, the pineal and the habenular region of the brain (Kojima et al., 2000; Philp et al., 2000). Both structures are either known to house, or implicated in housing, photoreceptors (Ekström et al., 1987; Foster et al., 1994; Ekström and Meissl, 1997; Yoshikawa and Oishi, 1998).

5.4.3 A Novel Light Response from VA-opsin- and Melanopsin-expressing Horizontal Cells

Whilst considerable attention has been paid to the role of novel opsins in the generation of direct light sensitivity in the RGCs of mammals, little attention has been paid to the physiological function of the opsins that are expressed within the teleost inner retina, including the horizontal cells (HCs) which act as local circuit neurones that have no direct projections outside the retina. In our recent studies we have used a combination of molecular biology, neuroanatomy and single cell electrophysiology to study these neurones in a teleost species (*Rutilus rutilus*) in which the physiology of

HCs has been well characterized. Our studies started with the isolation of roach VA-opsin and melanopsin.

Roach VA-opsin was isolated and shown to be expressed within a subset of neurones of the inner nuclear and ganglion cell layers (Figure 5.5b). The VA-opsin-positive neurones of the inner nuclear layer showed the characteristic morphology and location of horizontal cells. These horizontal cells were rare (<1.0% of all horizontal cells) and uniformly distributed across the retina. We also succeeded in isolating melanopsin from the roach. Melanopsin-positive cells had a strikingly similar distribution to VA-opsin, being expressed in a sub-set of neurones of the inner nuclear layer and ganglion cell layer of the retina (Figure 5.5a). The cytology, number, and location of VA-opsin- and melanopsin-expressing neurones in both the horizontal cells and the ganglion cell layer suggest these genes are co-expressed (Jenkins et al., 2003). This also seems to be the case in two other species: the zebrafish (Bellingham et al., 2002) and cod (Drivenes et al., 2003).

Intracellular recording, together with cytological examination and Lucifer Yellow injection of recorded neurones, identified a hitherto uncharacterized luminosity type of horizontal cell (Jenkins et al., 2003). These neurones showed a distinctive light response consisting of a secondary delayed depolarization following light-off (R*) (Figure 5.6). These units were termed HC–RSD (Horizontal Cell–Rod Secondary Depolarization) neurones. The novel response appears in a sub-population of cells that are rod-driven, and the novel component appears after the conventional hyperpolarizing S-potential response to light. HC–RSD cells represent <0.5% of the total HC population.

The conventional and novel components of the HC–RSD exhibited clear differences in their response to prolonged light (Figure 5.7a). While the amplitude of the conventional components saturated around 500–750 ms after stimulus onset, the novel component displayed a continued and near linear increase in amplitude, even when the stimulus duration exceeded 3–5 s. These experiments suggest that the novel component has a uniquely long integration time relative to those driven by rods and cones. Constant bright light was used to saturate the conventional rod input to the HC–RSD. Under these conditions the novel component was largely unaffected (Figure 5.7b), suggesting that the novel component can act independently of the conventional photoreceptor input to the cell (Jenkins et al., 2003).

5.4.4 Action Spectra for the HC–RSD Light Response Identify a Novel Photopigment

The action spectrum for the HC–RSD reveals that the novel depolarizing response fits a single opsin photopigment with a λ_{max} of 477 nm (Figure 5.1c). This λ_{max} lies well outside the range of the known photopigments for this species (UV-cone = 355 nm, SWS-cone = 447 nm, MWS-cone = 526 nm, LWS-cone = 619 nm and rod = 538 nm; Downing et al., 1986), none of which provides a significant fit to the action spectra. This novel action spectrum contrasts with the action spectra derived from other conventional components of HC light responses of these cells, all of which show the involvement of multiple photopigments (Jenkins et al., 2003).

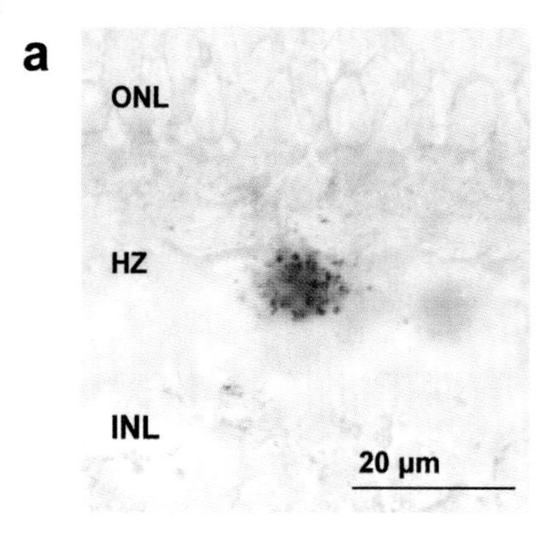

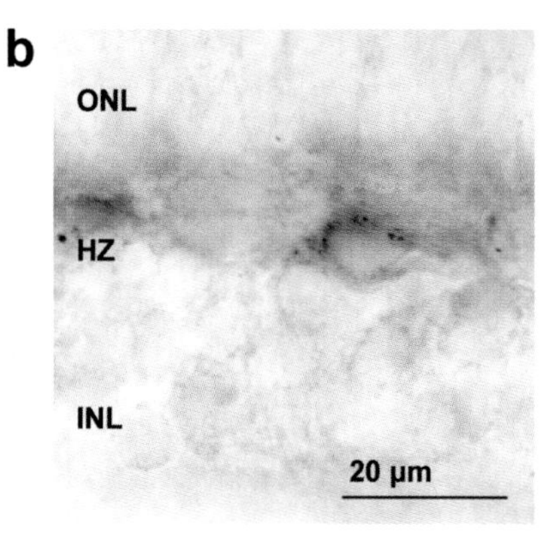

Figure 5.5 *In situ* localization of melanopsin and VA-opsin. Sections of roach retina were examined for VA-opsin and melanopsin expression using species specific probes. (a) Melanopsin expression in a neurone within the horizontal cell layer. In all cases, cells were located 1–2 cells below the outer plexiform layer (OPL). (b) An example of VA-opsin expression in the horizontal cell layer of the retina. As with melanopsin neurones, VA-opsin cells were always located 1–2 cells below the outer plexiform layer. The cytology, number, and location of VA-opsin and melanopsin cells in the horizontal cell layer may suggest that these genes are co-expressed. Abbreviations: outer nuclear layer (ONL), horizontal cell layer (HZ), inner nuclear layer (INL), outer plexiform layer (OPL).

On the basis of their cytology and distribution within the retina, the HC–RSD cells injected with Lucifer Yellow, VA-opsin, and melanopsin neurones all appear to belong to the same population of HCs. Whether the novel 477 nm photopigment is the prod-

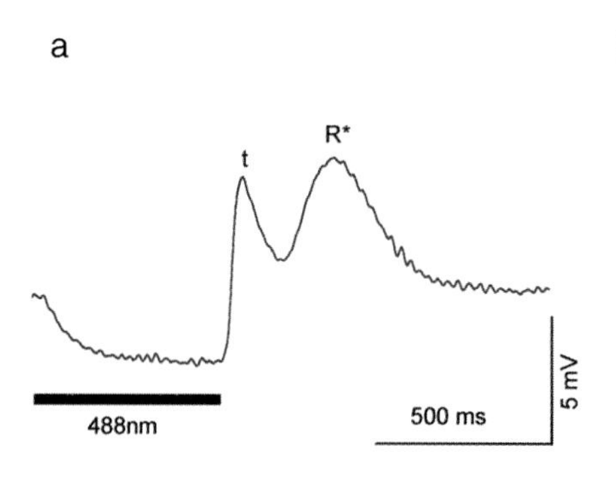

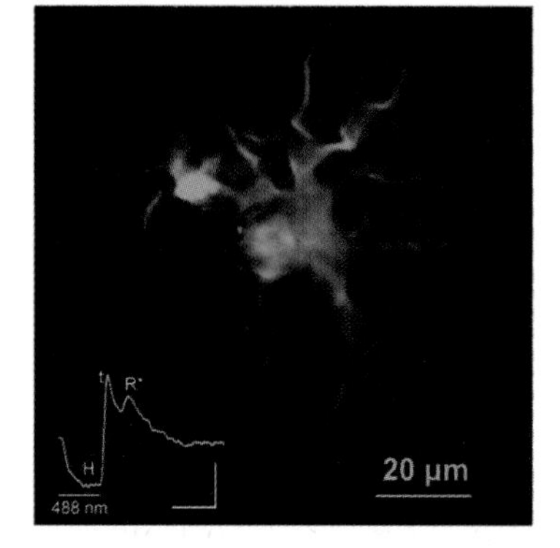

Figure 5.6 Anatomy and responses of HC–RSD cells. (a) Intracellular recording of the HC–RSD to a 488-nm light stimulus. This cell showed a marked novel component (sd*) that was iontophoretically marked with Lucifer Yellow, marked and photoconverted in panel b. (voltage and time scale bars as indicated). (b) Confocal image of a HC–RSD cell injected with Lucifer Yellow after electrophysiological recording. The light response of this cell is shown in the inset.

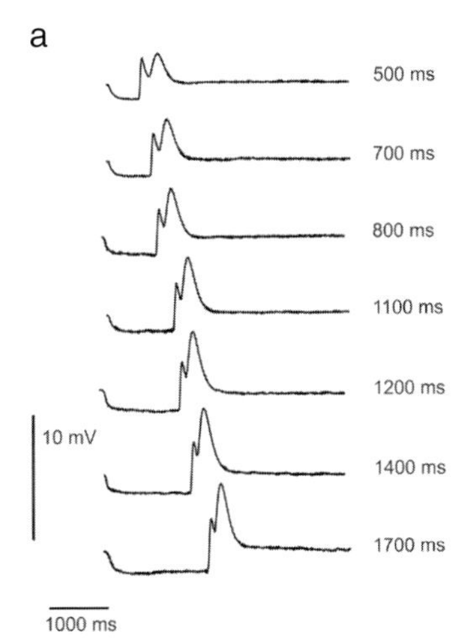

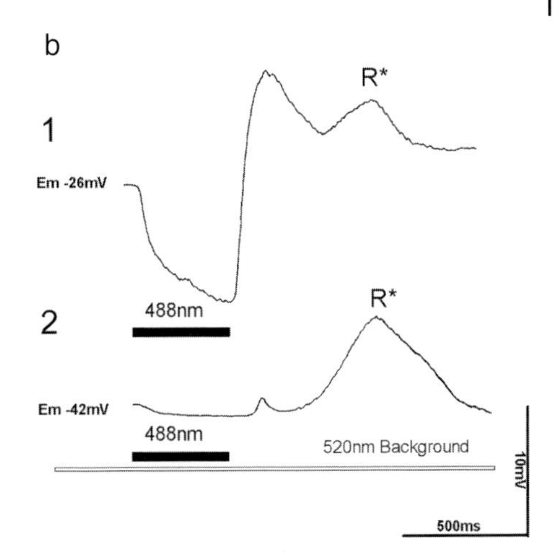

Figure 5.7 (a) Temporal summation of the novel response. Series of novel responses to increasing stimulus durations. The traces begin at the onset of light stimulation, with a small hyperpolarization sustained for the duration of light stimulation. At light off, there is an initial off response followed by a large novel off component that increases with the duration of light stimulation. (b) The effect of a rod-saturating background (520 nm) upon the light responses of an HC–RSD neurone. The control light response to 488 nm stimulation (8×10^{13} quanta cm–2 s^{-1}) is shown in the upper panel (1); note the clear conventional hyperpolarization and novel component (sd*). In the lower panel (2), the cell is hyperpolarized by a constant 520 nm background (5.3×10^{13} quanta cm–2 s^{-1}). Re-stimulation at 488 nm reveals that the conventional component is largely abolished, whilst the novel component is retained.

uct of roach VA-opsin or melanopsin remains uncertain. The spectral maxima of expressed VA-opsin from other species are between 460–500 nm, and so the novel 477 nm photopigment is within the projected range of the VA photopigments. There are no biochemical data that address the photopigment capacity of the teleost melanopsins. Indeed, it still remains uncertain whether the melanopsins in general form sensory photopigments, and there has been some speculation that they play a support function such as chromophore regeneration (Foster and Bellingham, 2002) (Section 5.5 below).

5.4.5
The Possible Function of HC–RSD Neurones

HCs are second order neurones that reside next to the OPL, and their traditional role is considered to be the processing of color and/or surround information. They have no clear direct pathway for signaling retinal light signals to the optic nerve. The extensive specific homologous coupling of given HC subtypes in the vertebrate retina has led to the concept that HCs act to signal broad retinal irradiance, essentially acting as a reference level with respect to bipolar cell signals. Given that cells within these syncytia express novel photopigments we have proposed that VA-opsin and/or melanopsin may play a role in setting these reference levels. Accordingly we suggest that the VA-opsin/melanopsin light responses we have described play a role in the intrinsic regulation of retinal physiology in relation to long-term changes in retinal illuminance.

5.5
Opsins can be Photosensors or Photoisomerases

The opsins probably perform several different tasks, but their known roles are as photosensors or photoisomerases (Foster and Bellingham, 2002). Photosensory opsins, such as the rod and cone opsins, and VA-opsin use light to activate a phototransduction cascade that ultimately results in a change in membrane potential of the photoreceptor cell. By contrast, photoisomerases are involved in photopigment regeneration. The best described photoisomerase is Retinal G-protein Coupled Receptor (RGR)-opsin. RGR is expressed in high concentrations in the retinal pigment epithelium (RPE), has the Lys296 retinal attachment site, a histidine at the Glu113 site, and acts by harvesting the energy of a photon to photoisomerase *all-trans*-retinal into the 11-*cis*-retinal isoform. 11-*cis* retinal is then transported to the outer segments of the rods and cones where it is associated with a rod or cone opsin to regenerate a photopigment. Consistent with RGRs non-photosensory role, RGR shares a relatively low level of amino acid identity (21–24%) and has a non-conserved genomic structure with the photopigment opsins (Hao and Fong, 1996; 1999). Another candidate photoisomerase, peropsin (Sun et al., 1997a), is likewise different from the photosensory opsins, sharing only ~27% amino acid identity with the photosensory opsins, a non-conserved genomic structure, and along with melanopsin, a tyrosine at the putative counterion position. This comparison of the opsins would argue that the functionally related photosensory opsins share a close phylogenetic relationship based upon both high levels of amino acid identity (~40%) and a largely conserved genomic structure (Bellingham and Foster, 2002).

RGR-opsin, Melanopsin and the photosensory opsins differ in their amino acid identitity, genomic structure, and in a number of critical residues. If melanopsins form photosensory opsins, then they represent a distinct line of photopigment evolution in the vertebrates. Alternatively melanopsin may function as a photoisomerase, acting to regenerate chromophore for an as yet unrecognised pigment. In this regard it is worth noting that there are recently discovered orphan opsins that have been isolated from mammals e.g. Opn5 (Tarttelin et al., 2003). It is additionally worth noting that VA-opsin and melanopsin appear to be co-expressed in the same photosensitive horizontal cells of the roach (see Figure 5.5 and Section 4.3). Furthermore, rod-opsin and melanopsin have been shown to be co-expressed in *Xenopus* melanophores (Miyashita et al., 2001). One interpretation of the latter finding is that rod-opsin mediates the aggregation of the melanosomes, and this would agree with the aggregation action spectrum which has a λ_{max} 500 nm, whilst melanopsin may be governing melanophore dispersion (Miyashita et al., 2001). Alternatively there may be a single rod-opsin photopigment in the melanophores, and melanopsin supports rod-opsin photopigment activity, perhaps acting as an RGR-like photoisomerase. Finally, of course, melanopsins may act as both photosensors and photoisomerases, and in this respect resemble the invertebrate photopigments, to which they share the highest level of amino acid identity (Provencio et al., 1998b; Bellingham and Foster, 2002).

5.6
Placing Candidate Genes and Photopigments into Context

Studies on teleost fish and mammals have demonstrated the existence of non-rod, non-cone ocular photoreceptors. In the case of the roach, electrophysiological evidence suggests that one function of this inner retinal photoreceptor is to modulate the activity of retinal horizontal cells, and this fits with the general role of horizonal cells in the teleost retina as regulators of retinal activity in response to environmental irradiance. In *rd/rd cl* mice and humans, non-rod, non-cone photoreceptors (which appear to be intrinsically light-sensitive RGCs) are used for a broad range of irradiance detection tasks. The candidate photopigments (melanopsin and VA opsin) mediating these responses to light have been variously linked to these photoreceptive neurones.

The relative ease with which genes can be isolated contrasts with the time it takes to determine their function. In the case of photopigment genes the assignment of function has been based traditionally on a number of criteria.

The primary criteria have been:

(1) The candidate protein should form a functional photopigment and alter its activity in response to photic rather than non-specific kinetic actions. For example, both infrared and ultraviolet energy can non-specifically activate a protein. In addition, it is often useful to show that an opsin can activate a transduction cascade to distinguish photosenory pigments from photoisomerases.

(2) The candidate photopigment should have an absorbance spectrum that matches the action spectrum of the response in question. As discussed above if mouse melanopsin forms a functional photopigment then its absorption spectrum should ideally match the action spectra for *rd/rd cl* responses to light.

(3) The candidate molecule should be expressed in areas/cells defined as photoreceptors using physiological assays. The expression of melanopsin within the intrinsically photosensitive RGCs, suggests that melanopsin is likely to play some role in the light-detecting capacity of these cells.

Secondary criteria for photopigment identification would include:

(4) Genetic ablation of the candidate molecule. Two broad results are possible in knock-out studies. If the candidate gene provides the only photosensory input then the response will be abolished. If, however, there are multiple photoreceptor inputs,

then gene ablation may result in an attenuated response or there may be no obvious phenotype. If attenuated, the action spectrum should be altered in a manner predicted by the absorbance spectrum of the photopigment. In the absence of the primary criteria (1)–(3) (above), gene ablation studies can only be used to correlate a gene with a light-dependent process, and will not distinguish between the loss of a photosensory pigment and/or loss of a critical element in the phototransduction process.

(5) Chromophore identification or depletion. For example, 11-*cis* retinaldehyde is only associated with opsin-based photopigments. 11-*cis* retinaldehyde can be readily identified using HPLC, and its identification (Foster et al., 1993b), apparent lack (Foster et al., 1989; Foster et al., 2003b) or depletion (Zatz, 1994) has been helpful in defining the nature of photoreceptive pathways. Some care should be exercised when using this approach however, as chromophore depletion is not the same as chromophore loss. Even visual responses may be only moderately affected after severe chromophore depletion (Zimmerman and Goldsmith, 1971). The identification of 11-cis retinal can be significant, but failure to identify 11-*cis*, like all negative results, may be misleading (Van Gelder et al., 2003).

(6) Homology to known photopigment molecules. Much of the discussion about the possible role of opsins is based on homology. While this can be informative, considerable care has to be exercised in the absence of criteria (1)–(3) as not all proteins classified as "opsins" perform the same task (Foster and Bellingham, 2002).

Melanopsin represents a promising candidate for the non-rod, non-cone photopigment in mammals, but we do not have definitive results showing whether melanopsin forms a functional photopigment, and should it do so, whether circadian action spectra will match its absorption spectra. As discussed above, preliminary data on expressed melanopsin suggest a pigment with a λ_{max} between 420 and 440 nm (Newman et al., 2003). This absorption maxima is 60 nm different from the action spectra for non-rod, non-cone photoreception in *rd/rd cl* mice, ~480 nm (OP^{480}). This difference is too great to be attributed to the variation normally seen between *in vitro* expressed pigments and their absorption spectra *in vivo*.

5.7
Conclusions

We have discussed three examples of novel irradiance photoresponses that appear to be driven through a group of non-rod non-cone photoreceptors in the inner retina. These are the novel horizontal cell light responses in the teleost retina, regulation of the human cone ERG, and the light-evoked activity of retinal neurones in the GCL of the mouse. These systems appear to involve long-term integration of the photoresponse, and provide irradiance signals both for the brain and/or for the local regulation of retinal physiology. The spectral sensitivity of these cells has been well defined through action spectroscopy. In Figure 5.1 a–c, we compare the action spectra for phase shifting in the *rd/rd cl* mouse with those obtained from the teleost horizontal cell studies and human ERG studies. What emerges is that all three systems are driven by opsin/vitamin A photopigments and that the spectral sensitivities of all these systems are markedly conserved (λ_{max} = 477–483 nm). This level of conservation tells us something fundamentally important about these novel light sensing pathways. It is unfortunate that we currently do not have the wit to appreciate what that might be!

References

Bellingham, J. and Foster, R.G. (2002) Opsins and mammalian photoentrainment. *Cell Tiss. Res.*, 309, 57–71.

Bellingham, J., Whitmore, D., Philp, A.R., Wells, D.J. and Foster, R.G. (2002) Zebrafish melanopsin: isolation, tissue localization and phylogenetic position. *Brain Res Mol Brain Res*, 107, 128–136.

Berson, D.M., Dunn, F.A. and Takao, M. (2002) Phototransduction by retinal ganglion cells that set the circadian clock. *Science*, 295, 1070–1073.

Brainard, G.C., Hanifin, J.P., Greeson, J.M., Byrne, B., Glickman, G., Gerner, E. and Rollag, M.D. (2001) Action spectrum for melatonin regulation in humans: evidence for a novel circadian photoreceptor. *J Neurosci*, 21, 6405–6412.

Bridges, C. (1959) The visual pigments of some common laboratory animals. *Nature*, 184, 727–728.

Cajochen, C., Khalsa, S.B., Wyatt, J.K., Czeisler, C.A. and Dijk, D.-J. (1999) EEG and ocular correlates of circadian melatonin phase and human performance decrements during sleep loss. *Am. J. Physiol.*, 277, R640–649.

Campbell, S.S. and Murphy, P.J. (1998) Extraocular circadian phototransduction in humans. *Science*, 279, 396–399.

Czeisler, C.A., Shanahan, T.L., Klerman, E.B., Martens, H., Brotman, D.J., Emens, J.S., Klein, T. and Rizzo III, J.F. (1995) Suppression of melatonin secretion in some blind patients by exposure to bright light. *N. Engl. J. Med.*, 332, 6–11.

Dowling, J.E. (1991) Retinal neuromodulation: the role of dopamine. *Vis Neurosci*, 7, 87–97.

Downing, J.E.G., Djamgoz, M.B.A. and Bowmaker, J.K. (1986) Photoreceptors of a cyprinid fish, the roach – morphological and spectral characteristics. *J Comp Physiol A*, 159, 859–868.

Drivenes, O., Soviknes, A.M., Ebbesson, L.O., Fjose, A., Seo, H.C. and Helvik, J.V. (2003) Isolation and characterization of two teleost melanopsin genes and their differential expression within the inner retina and brain. *J Comp Neurol*, 456, 84–93.

Dubocovich, M. (1989) Role of melatonin in the retina. *Progress in Retinal Research*, 8, 129–151.

Ekström, P., Foster, R.G., Korf, H.-W. and Schalken, J.J. (1987) Antibodies against retinal photoreceptor-specific proteins reveal axonal projections from the photosensory pineal organ of teleosts. *J Comp Neurol*, 265, 25–33.

Ekström, P. and Meissl, H. (1997) The pineal organ of teleost fishes. *Reviews in Fish Biology and Fisheries*, 7, 199–284.

Emerson, V.F. (1980) Grating acuity in the golden hamster: The effects of stimulus orientation and luminance. *Experimental Brain Research*, 38, 43–52.

Foster, R.G. and Follett, B.K. (1985) The involvement of a rhodopsin-like photopigment in the photoperiodic response of the Japanese quail. *J Comp Physiol A*, 157, 519–528.

Foster, R.G., Schalken, J.J., Timmers, A.M. and De Grip, W.J. (1989) A comparison of some photoreceptor characteristics in the pineal and retina: I. The Japanese quail (*Coturnix coturnix*). *J. Comp. Physiol. A*, 165, 553–563.

Foster, R.G., Provencio, I., Hudson, D., Fiske, S., DeGrip, W. and Menaker, M. (1991) Circadian photoreception in the retinally degenerate mouse (*rd/rd*). *J Comp Physiol [A]*, 169, 39–50.

Foster, R.G., Argamaso, S., Coleman, S., Colwell, C.S., Lederman, A. and Provencio, I. (1993a) Photoreceptors regulating circadian behavior: A mouse model. *J Biol Rhythms*, 8, Suppl. 17–23.

Foster, R.G., Garcia-Fernandez, J.M., Provencio, I. and DeGrip, W.J. (1993b) Opsin localization and chromophore retinoids identified within the basal brain of the lizard *Anolis carolinensis*. *J. Comp. Physiol. A*, 172, 33–45.

Foster, R.G. and Menaker, M. (1993) Circadian photoreception in mammals and other vertebrates. In Wetterberg, L. (ed.) *Light and biological rhythms in man.* Pergamon, pp. 73–91.

Foster, R.G., Grace, M.S., Provencio, I., Degrip, W.J. and Garcia-Fernandez, J.M. (1994) Identification of vertebrate deep brain photoreceptors. *Neurosci. Biobehav. Rev.*, 18, 541–546.

Foster, R.G. (1998) Shedding light on the biological clock. *Neuron*, 20, 829–832.

Foster, R.G. and Soni, B.G. (1998) Extraretinal photoreceptors and their regulation of temporal physiology. *Reviews of Reproduction*, 3, 145–150.

Foster, R.G. and Provencio, I. (1999) The regulation of vertebrate biological clocks by light. In Archer, S.N., Djamgoz, M.B.A., Loew, E.R., Partridge, J.C., Vallerga, S. (eds.) *Adaptive Mechanisms in the Ecology of Vision.* Kluwer Academic Publishers, Dordrecht, Netherlands, pp. 223–243.

Foster, R.G. (2002) Keeping an eye on the time. *Invest. Ophth. Vis. Sci.*, 43, 1286–1298.

Foster, R.G. and Bellingham, J. (2002) Opsins and Melanopsins. *Current Biology*, 12, 543–544.

Foster, R.G., Hankins, M., Lucas, R.J., Jenkins, A., Munoz, M., Thompson, S., Appleford, J.M. and Bellingham, J. (2003a) Non-rod, non-cone photoreception in rodents and teleost fish. *Novartis Found Symp*, 253, 3–23; discussion 23–30, 52–25, 102–109.

Foster, R.G., Provencio, I., Bovee-Geurts, P.H. and DeGrip, W.J. (2003b) The photoreceptive capacity of the developing pineal gland and eye of the golden hamster (Mesocricetus auratus). *J Neuroendocrinol*, 15, 355–363.

Freedman, M.S., Lucas, R.J., Soni, B., von Schantz, M., Munoz, M., David-Gray, Z.K. and Foster, R.G. (1999) Regulation of mammalian circadian behavior by non-rod, non-cone, ocular photoreceptors. *Science*, 284, 502–504.

Frisch, K.v. (1911) Beitrage zur Physiologie der Pigmentzellen in der Fischhaut. *Pfluegers Archiv fuer die Gesamte Physiologie des Menschen und der Tiere*, 138, 319–387.

Gooley, J.J., Lu, J., Chou, T.C., Scammell, T.E. and Saper, C.B. (2001) Melanopsin in cells of origin of the retinohypothalamic tract. *Nat. Neurosci.*, 12, 1165.

Gooley, J.J., Lu, J., Fischer, D. and Saper, C.B. (2003) A broad role for melanopsin in nonvisual photoreception. *J Neurosci*, 23, 7093–7106.

Govardovskii, V.I., Fyhrquist, N., Reuter, T., Kuzmin, D.G. and Donner, K. (2000) In search of the visual pigment template. *Vis Neurosci*, 17, 509–528.

Hankins, M.W., Jones, R.J. and Ruddock, K.H. (1998) Diurnal variation in the b-wave implicit time of the human electroretinogram. *Vis Neurosci*, 15, 55–67.

Hankins, M.W. and Lucas, R.J. (2002) The primary visual pathway in humans is regulated according to long-term light exposure through the action of a non-classical photopigment. *Current Biology*, 12, 191–198.

Hannibal, J., Hindersson, P., Knudsen, S.M., Georg, B. and Fahrenkrug, J. (2002) The photopigment melanopsin is exclusively present in pituitary adenylate cyclase-activating polypeptide-containing retinal ganglion cells of the retinohypothalamic tract. *J Neuroscience*, 22, RC191: 191–197.

Hao, W. and Fong, H.K. (1996) Blue and ultraviolet light-absorbing opsin from the retinal pigment epithelium. *Biochemistry*, 35, 6251–6256.

Hao, W. and Fong, H.K. (1999) The endogenous chromophore of retinal G protein-coupled receptor opsin from the pigment epithelium. *Journal of Biological Chemistry*, 274, 6085–6090.

Hardie, R.C. and Minke, B. (1992) The trp gene is essential for a light activated Ca^{2+} channel in *Drosophila* photoreceptors. *Neuron*, 8, 643–651.

Hardie, R.C. (2001) Phototransduction in *Drosophila melanogaster. J. Exp. Biol.*, 204, 3403–3409.

Hardie, R.C. and Raghu, P. (2001) Visual transduction in Drosophila. *Nature*, 413, 186–193.

Hattar, S., Liao, H.W., Takao, M., Berson, D.M. and Yau, K.W. (2002) Melanopsin-containing retinal ganglion cells: architecture, projections, and intrinsic photosensitivity. *Science*, 295, 1065–1070.

Hattar, S., Lucas, R.J., Mrosovsky, N., Thompson, S., Douglas, R.H., Hankins, M.W., Lem, J., Biel, M., Hofmann, F., Foster, R.G. and Yau, K.W. (2003) Melanopsin and rod-cone photoreceptive systems account for all major accessory visual functions in mice. *Nature*, 424, 75–81.

Jacobs, G.H., Neitz, J. and Deegan, J.F. (1991) Retinal receptors in rodents maximally sensitive to ultraviolet light. *Nature*, 353, 655–656.

Jenkins, A., Munoz, M., Tarttelin, E.E., Bellingham, J., Foster, R.G. and Hankins, M.W. (2003) VA opsin, melanopsin, and an inherent light response within retinal interneurons. *Curr Biol*, 13, 1269–1278.

Kojima, D., Mano, H. and Fukada, Y. (2000) Vertebrate ancient-long opsin: a green-sensitive photoreceptive molecule present in zebrafish deep brain and retinal horizontal cells. *J Neurosci*, 20, 2845–2851.

Koorengevel, K.M., Gordijn, M.C., Beersma, D.G., Meesters, Y., den Boer, J.A., van den Hoofdakker, R.H. and Daan, S. (2001) Extraocular light therapy in winter depression: a double-blind placebo-controlled study. *Biol. Psychiatry*, 50, 691–698.

Leproult, R., Colecchia, E.F., L'Hermite-Baleriaux, M. and Van Cauter, E. (2001) Transition from dim to bright light in the morning induces an immediate elevation of cortisol levels. *J. Clin. Endocrinol. Metab.*, 86, 151–157.

Lockley, S.W., Skene, D.J., Tabandeh, H., Bird, A.C., Defrance, R. and Arendt, J. (1995) Assesment of 6-sulphatoxymelatonin, sleep and activity rhythms in visually impared subjects. *Biological Rhythms Research*, 26, 413–.

Lockley, S.W., Skene, D.J., Thapan, K., English, J., Ribeiro, D., Haimov, I., Hampton, S., Middleton, B., von Schantz, M. and Arendt, J. (1998) Extraocular light exposure does not suppress plasma melatonin in humans. *J. Clin. Endocrinol. Metab.*, 83, 3369–3372.

Lockley, S.W., Brainard, G.C. and Czeisler, C.A. (2003) High sensitivity of the human circadian melatonin rhythm to resetting by short wavelength light. *J Clin Endocrinol Metab.* 88: 4502–4505.

Lucas, R.J. and Foster, R.G. (1999a) Circadian Rhythms: Something to *cry* about? *Curr. Biol.*, 9, 214–217.

Lucas, R.J. and Foster, R.G. (1999b) Mammalian photoentrainment: A role for cryptochrome? *J. Biol. Rhythms*, 14, 4–9.

Lucas, R.J., Freedman, M.S., Munoz, M., Garcia-Fernandez, J.M. and Foster, R.G. (1999) Regulation of the mammalian pineal by non-rod, non-cone, ocular photoreceptors. *Science*, 284, 505–507.

Lucas, R.J., Douglas, R.H. and Foster, R.G. (2001) Characterization of an ocular photopigment capable of driving pupillary constriction in mice. *Nat Neurosci*, 4, 621–626.

Lucas, R.J., Hattar, S., Takao, M., Berson, D.M., Foster, R.G. and Yau, K.W. (2003) Diminished pupillary light reflex at high irradiances in melanopsin-knockout mice. *Science*, 299, 245–247.

Lythgoe, J.N. (1979) *The Ecology of Vision.* Clarendon Press, Oxford.

McCall, M.A., Gregg, R.G., Merriman, K., Goto, N.S., Peachey, N.S. and Stanford, L.R. (1996) Morphological and physiological consequences of the selective elimination of rod photoreceptors in transgenic mice. *Exp. Eye Res.*, 63, 35–50.

Menaker, M. and Underwood, H. (1976) Extraretinal photoreception in birds. *Photochemistry and Photobiology*, 23, 299–306.

Miyashita, Y., Moriya, T., Yamada, K., Kubota, T., Shirakawa, S., Fujii, N. and Asami, K. (2001) The photoreceptor molecules in *Xenopus* tadpole tail fin, in which melanophores exist. *Zoological Science*, 18, 671–674.

Mrosovsky, N., Lucas, R.J. and Foster, R.G. (2001) Persistence of masking responses to light in mice lacking rods and cones. *J. Biol. Rhythms*, 16, 585–587.

Munz, F.W. and McFarland, W.N. (1977) Evolutionary adaptations of fishes to the photic environment. In Crescitelli, F. (ed.) *Handbook of Sensory Physiology*. Springer-Verlag, Berlin, pp. 193–274.

Nelson, D. and Takahashi, J. (1991) Sensitivity and integration in a visual pathway for circadian entrainment in the hamster (*Mesocricetus auratus*). *J Physiol*, 439, 115–145.

Nelson, R.J. and Zucker, I. (1981) Absence of extraocular photoreception in diurnal and nocturnal rodents exposed to direct sunlight. *Comparative Biochemistry and Physiology*, 69A, 145–148.

Newman, L.A., Walker, M.T., Brown, R.L., Cronin, T.W. and Robinson, P.R. (2003) Melanopsin forms a functional short-wavelength photopigment. *Biochemistry*, 42, 12734–12738.

Nir, I., Haque, R. and Iuvone, P.M. (2000) Diurnal metabolism of dopamine in the mouse retina. *Brain Res*, 870, 118–125.

Panda, S., Sato, T.K., Castrucci, A.M., Rollag, M.D., DeGrip, W.J., Hogenesch, J.B., Provencio, I. and Kay, S.A. (2002) Melanopsin (Opn4) requirement for normal light-induced circadian phase shifting. *Science*, 298, 2213–2216.

Panda, S., Provencio, I., Tu, D.C., Pires, S.S., Rollag, M.D., Castrucci, A.M., Pletcher, M.T., Sato, T.K., Wiltshire, T., Andahazy, M., Kay, S.A., Van Gelder, R.N. and Hogenesch, J.B. (2003) Melanopsin Is Required for Non-Image-Forming Photic Responses in Blind Mice. *Science*.

Peirson, S., Bovee-Geurts, P.H., Lupi, D., Jeffery, G., DeGrip, W.J. and Foster, R.G. (2004) Expression of the candidate circadian photopigment melanopsin (Opn4) in the mouse retinal pigment epithelium. *Mol. Brain Res.*, (in press).

Philp, A.R., Garcia-Fernandez, J.-M., Soni, B.G., Lucas, R.J., Bellingham, J. and Foster, R.G. (2000) Vertebrate Ancient (VA) opsin and extraretinal photoreception in the atlantic salmon (*Salmo salar*). *J Exp Biol*, 203, 1925–1936.

Pittendrigh, C.S. (1993) Temporal Organization: Reflections of a Darwinian Clock-Watcher. *Annu.Rev.Physiol.*, 55, 17–54.

Provencio, I., Cooper, H.M. and Foster, R.G. (1998a) Retinal projections in mice with inherited retinal degeneration: implications for circadian photoentrainment. *J. Comp. Neurol.*, 395, 417–439.

Provencio, I., Jiang, G., DeGrip, W.J., Hayes, W.P. and Rollag, M.D. (1998b) Melanopsin: An opsin in melanophores, brain and eye. *Proc Natl Acad Sci USA*, 95, 340–345.

Provencio, I., Rodriguez, I.R., Jiang, G., Hayes, W.P., Moreira, E.F. and Rollag, M.D. (2000) A novel human opsin in the inner retina. *J. Neurosci.*, 20, 600–605.

Provencio, I., Rollag, M.D. and Castrucci, A.M. (2002) Photoreceptive net in the mammalian retina. *Nature*, 415, 493.

Rajaratnam, S.M. and Arendt, J. (2001) Health in a 24-h society. *Lancet*, 358, 999–1005.

Ralph, M.R., Foster, R.G., Davis, F.C. and Menaker, M. (1990) Transplanted suprachiasmatic nucleus determines circadian period. *Science*, 247, 975–978.

Roenneberg, T. and Foster, R.G. (1997) Twilight times: light and the circadian system. *Photochem Photobiol*, 66, 549–561.

Roenneberg, T. and Merrow, M. (2002) Life before the clock: modeling circadian evolution. *J Biol Rhythms*, 17, 495–505.

Roenneberg, T. and Merrow, M. (2003) The network of time: understanding the molecular circadian system. *Curr Biol*, 13, R198–207.

Ruby, N.F., Brennan, T.J., Xie, X., Cao, V., Franken, P., Heller, H.C. and O'Hara, B.F. (2002) Role of melanopsin in circadian responses to light. *Science*, 298, 2211–2213.

Sancar, A. (2000) Cryptochrome: The second photoactive pigment in the eye and its role in circadian photoreception. *Annu. Rev. Biochem.*, 69, 31–67.

Scheer, F.A., van Doornen, L.J. and Buijs, R.M. (1999) Light and diurnal cycle affect human heart rate: possible role for the circadian pacemaker. *J. Biol. Rhythms*, 14, 202–212.

Sekaran, S., Foster, R.G., Lucas, R.J. and Hankins, M.W. (2003) Calcium imaging reveals a network of intrinsically light-sensitive inner-retinal neurons. *Curr Biol*, 13, 1290–1298.

Semo, M., Lupi, D., Peirson, S.N., Butler, J.N. and Foster, R.G. (2003a) Light-induced c-fos in melanopsin retinal ganglion cells of young and aged rodless/coneless (rd/rd cl) mice. *Eur J Neurosci*, 18, 3007–3017.

Semo, M., Peirson, S., Lupi, D., Lucas, R.J., Jeffery, G. and Foster, R.G. (2003b) Melanopsin retinal ganglion cells and the maintenance of circadian and pupillary responses to light in aged rodless/coneless (rd/rd cl) mice. *Eur J Neurosci*, 17, 1793–1801.

Shand, J. and Foster, R.G. (1999) The extraretinal photoreceptors of non-mammalian vertebrates. In Archer, S.N., Djamgoz, M.B.A., Loew, E.R., Partridge, J.C., Vallerga, S. (eds.) *Adaptive Mechanisms in the Ecology of Vision*. Kluwer Academic Publishers, Dordrecht, Netherlands, pp. 197–222.

Soni, B.G. and Foster, R.G. (1997) A novel and ancient vertebrate opsin. *FEBS Let.*, 406, 279–283.

Soni, B.G., Philp, A.R., Knox, B.E. and Foster, R.G. (1998) Novel retinal photoreceptors. *Nature*, 394, 27–28.

Soucy, E., Wang, Y., Nirenberg, S., Nathans, J. and Meister, M. (1998) A novel signaling pathway from rod photoreceptors to ganglion cells in mammalian retina. *Neuron*, 21, 481–493.

Stockman, A., Sharpe, L.T. and Fach, C.C. (1999) The spectral sensitivity of the human short-wavelength cones. *Vision Research*, 39, 2901–2927.

Sun, H., Gilbert, D.J., Copeland, N.G., Jenkins, N.A. and Nathans, J. (1997a) Peropsin, a novel visual pigment-like protein located in the apical microvilli of the retinal pigment epithelium. *Proc. Natl. Acad. Sci. (USA)*, 94, 9893–9898.

Sun, H., Macke, J.P. and Nathans, J. (1997b) Mechanisms of spectral tuning in the mouse green cone pigment. *Proc Natal Acad Sci USA*, 94, 8860–8865.

Svaetichin, G. (1953) The cone action potential. *Acta Physiol Scand*, 29(s106), 565–600.

Tarttelin, E.E., Bellingham, J., Hankins, M.W., Foster, R.G. and Lucas, R.J. (2003) Neuropsin (Opn5): a novel opsin identified in mammalian neural tissue. *FEBS Lett*, 554, 410–416.

Thapan, K., Arendt, J. and Skene, D.J. (2001) An action spectrum for melatonin suppression: evidence for a novel non-rod, non-cone photoreceptor system in humans. *J Physiol*, 535, 261–267.

Tosini, G. and Menaker, M. (1996) Circadian rhythms in cultured mammalian retina. *Science*, 272, 419–421.

Tosini, G. and Menaker, M. (1998) The clock in the mouse retina: melatonin synthesis and photoreceptor degeneration. *Brain Res*, 789, 221–228.

Van Gelder, R.N., Wee, R., Lee, J.A. and Tu, D.C. (2003) Reduced pupillary light responses in mice lacking cryptochromes. *Science*, 299, 222.

Wang, Y., Macke, J., Merbsl, S., Zack, D., Klaunberg, B., Bennett, J., Gearhart, J. and Nathans, J. (1992) A locus control region adjacent to the human red and green visual pigment genes. *Neuron*, 9, 429–440.

Witkovsky, P. and Dearry, A. (1991) Functional roles of dopamine in the vertebrate retina. *Progress in Retinal Research*, 11, 247–292.

Wright, K.P., Jr. and Czeisler, C.A. (2002) Absence of circadian phase resetting in response to bright light behind the knees. *Science*, 297, 571.

Yoshikawa, T. and Oishi, T. (1998) Extraretinal photoreception and circadian systems in nonmammalian vertebrates. *Comp Biochem Physiol*, 119B, 65–72.

Zatz, M. (1994) Photoendocrine transduction in cultured chick pineal cells: IV What do vitamin A depletion and retinaldehyde addition do to the effects of light on the melatonin rhythm? *J. Neurochem.*, 62, 2001–2011.

Zimmerman, W.F. and Goldsmith, T.H. (1971) Photosensitivity of the circadian rhythm and of visual receptors in carotenoid-depleted *Drosophila*. *Science*, 171, 1167–1169.

6
The Phytochromes

Shih-Long Tu and J. Clark Lagarias

6.1
Introduction

6.1.1
Photomorphogenesis and Phytochromes

Oxygenic photosynthetic organisms have evolved sophisticated mechanisms to adapt to their environment. Dependent upon light as an energy source, these organisms must also cope with too much light, which is especially challenging for highly pigmented species living in an aerobic environment. For this reason, plants have evolved light-receptor systems to recognize and respond to light quality, fluence rate, direction and duration from their environment. Among the many physiological processes under light control are seed germination, seedling growth, synthesis of the photosynthetic apparatus, the timing of flowering, neighbor detection, and senescence. Such light-regulated growth and developmental responses are collectively known as photomorphogenesis (Kendrick and Kronenberg 1994; Whitelam et al. 1998; Neff et al. 2000; Casal et al. 2003).

Phytochrome was the first of the photomorphogenetic photoreceptor families to be identified nearly 50 years ago (Butler et al. 1959). In the last decade, research on the physiological, biochemical and functional properties of phytochromes has grown exponentially (Sage 1992; Furuya 1993; Quail 1997; Fankhauser 2001; Nagy and Schäfer 2002). More recent advances in this field can be attributed to the impact of genomics, studies that have revealed phytochrome-related genes in representative organisms from all forms of life on earth except Archaea (Vierstra and Davis 2000; Montgomery and Lagarias 2002). The extended phytochrome family can be categorized into three subfamilies: plant phytochromes (Phy), cyanobacterial phytochrome 1 (Cph1) and cyanobacterial phytochrome 2 (Cph2) families (Figure 6.1). This review focuses on phytochromes that are found in oxygenic photosynthetic organisms. The Cph1-related bacteriophytochrome (BphP) family of biliverdin-binding proteins that are found in non-oxygenic photosynthetic bacteria, non-photosynthetic eubacteria and fungi are reviewed by Vierstra (Chapter 8).

Handbook of Photosensory Receptors. Edited by W. R. Briggs, J. L. Spudich

ISBN 3-527-31019-3

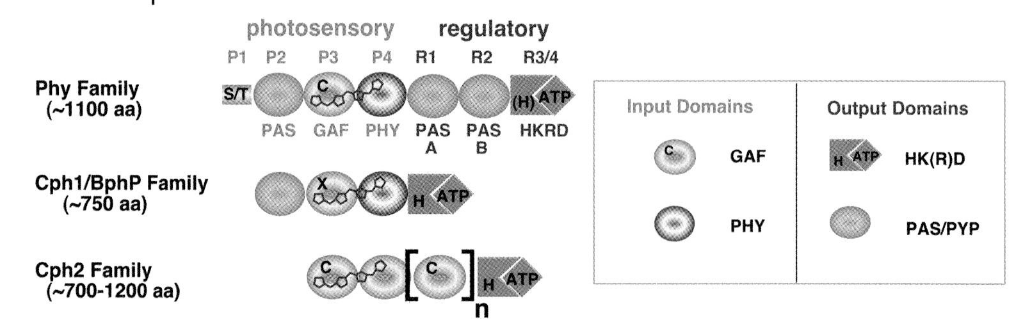

Figure 6.1 The phytochrome family. Orthologous photosensory input (GAF, PHY and PAS) and regulatory output (PAS and HKRD) subdomains are depicted in representative members of the phytochrome (Phy), cyanobacterial phytochrome 1/bacteriophytochrome (Cph1/BphP) and cyanobacterial phytochrome 2 (Cph2) families ($n = 0$–2). *The linear tetrapyrrole (phytobilin) chromophore is shown associated with the P3 GAF and P4 PHY domains. Adapted from Montgomery and Lagarias (2002).*

6.1.2
The Central Dogma of Phytochrome Action

Synthesized in the red light-absorbing Pr form, all phytochromes are regulated by red light (R) absorption which initiates the photochemical interconversion to the far-red (FR) light-absorbing Pfr form (Figure 6.2 A). FR promotes the reverse conversion of Pfr to Pr – a process which typically abolishes the R-dependent activation of the photoreceptor. This R/FR photoreversibility is conferred by a linear tetrapyrrole (phytobilin) prosthetic group that is covalently attached to the phytochrome apoprotein. Supporting evidence for the central dogma of phytochrome action, that is that Pfr is the active form, has been accumulating for years. Much of this evidence reflects the strong correlation between the amount of Pfr produced by a given fluence of light and the magnitude of the biological response. While the central dogma appears to hold true for R-FR-reversible low-fluence responses (LFR) and for R-dependent very low fluence responses (VLFR) (Shinomura et al. 1996), FR high irradiance responses (HIR) do not conform to this simple view of phytochrome action (Furuya and Schäfer 1996; Casal et al. 1998; Shinomura et al. 2000). Such data indicate that Pr, Pfr, photocycled-Pr, as well as intermediates produced during Pfr to Pr photoconversion, may all function to transduce the light signal for different phytochromes (Figure 6.2 B). In addition to photochemical interconversion processes, non-photochemical Pfr-to-Pr dark reversion plays an important role to attenuate the signal output by altering the lifetime of the Pfr form (Figure 6.2 B). For a given quality and fluence rate of light in the environment, signal output from phytochrome therefore depends on holophytochrome synthesis, the two photochemical interconversion processes, dark reversion as well as protein turnover. It is on these topics that we will mainly focus our discussion.

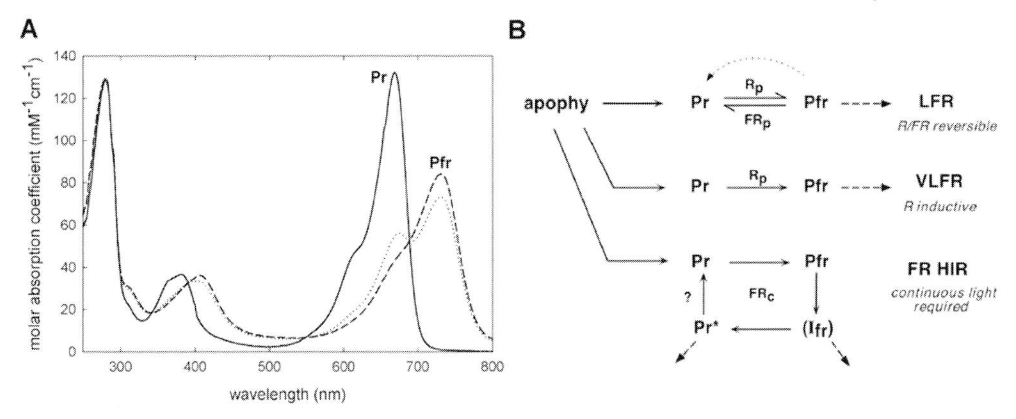

Figure 6.2 Spectral properties and modes of action of plant phytochromes. Panel A shows absorption spectra of purified oat phytochrome after saturating red (dotted line) and far-red (solid line) irradiation representing a mixture of 87% Pfr/13% Pr and 100% Pr, respectively. The deduced specturm of Pfr (dashed line) was obtained as described (Kelly and Lagarias 1985). Panel B depicts the three modes of phytochrome action in flowering plants that have been described by action spectroscopy.

6.2 Molecular Properties of Eukaryotic and Prokaryotic Phytochromes

6.2.1 Molecular Properties of Plant Phytochromes

In flowering plants, phytochromes are encoded by a small nuclear gene family reflecting repeated gene duplication of a eukaryote phytochrome progenitor during its evolution (Mathews et al. 1995). In the model plant *Arabidopsis thaliana*, the phytochrome family consists of 5 genes, denoted phyA–phyE (Clack et al. 1994), while monocot species (eg. rice or maize) appear to possess only representatives of the phyA–C families (Mathews and Sharrock 1997). The number of phytochrome species is not known for more primitive plants, that is gymnosperms, mosses, ferns, liverworts and algae. However the overall structure of the phytochrome photoreceptor has been preserved in all extant eukaryotic photosynthetic organisms (Figure 6.1). Although atypical phytochromes have been found in mosses and ferns, these organisms also possess conventional phytochromes (Thümmler et al. 1992; Nozue et al. 1998). For this reason, we will limit our discussion to those eukaryotic phytochromes which conform to the consensus structure. Our present understanding of phytochrome structure is mainly based on biochemical analysis of phyAs, but for those studied, other phytochromes appear to possess similar properties.

Eukaryotic phytochromes are soluble homodimeric proteins consisting of two ~120 kDa subunits. Small-angle X-ray scattering analyses indicate an overall protein fold with dimensions similar to mammalian immunoglobin Gs (Nakasako et al.

1990). Each subunit is composed of two domains that are separated by a protease-sensitive hinge region: the 60–70 kDa N-terminal 'photosensory' and 55 kDa C-terminal 'regulatory' domains. The photosensory domain functions to sense the light input and is composed of four sub-domains, that is the serine-rich N-terminal P1 domain, a PAS-related P2 domain, the bilin-binding P3 GAF domain, and the P4 PHY domain (Montgomery and Lagarias 2002). PAS and GAF domains have been shown to possess structurally-related protein folds (Ho et al. 2000), with PAS domains often comprising the binding site for small planar aromatic ligand molecules (Aravind and Ponting 1997; Ponting and Aravind 1997; Taylor and Zhulin 1999). PHY domains, evolutionarily related to PAS and GAF domains, may also adopt a similar three dimensional fold (Montgomery and Lagarias 2002). The phytobilin prosthetic group of plant phytochromes is covalently bound to a conserved cysteine residue found in the P3 GAF domain via a thioether linkage (Lagarias and Rapoport 1980). The presence of this conserved cysteine residue is one of the key structural features that distinguishes the phytochrome photoreceptors from members of the phytochrome-related BphP family.

Biochemical analyses have shown that the photosensory domain of plant phytochromes adopts a compact globular protein fold with the phytobilin chromophore mostly buried within the protein matrix (Yamamoto 1990). By contrast, the C-terminal regulatory domain of plant phytochromes is elongated and more sensitive to proteolytic degradation. Comprised of four sub-domains, including two PAS domains and two domains related to the transmitter 'output' domains found on prokaryotic two-component sensor proteins (Schneider-Poetsch et al. 1991), the regulatory domain of plant phytochrome contains the site(s) of homodimerization (Jones and Edgerton 1994). All four subdomains have been implicated in the high-affinity subunit–subunit interaction suggesting that no single subdomain of the C-terminus is necessary or sufficient (Jones et al. 1985; Vierstra 1993). By analogy to two-component sensor proteins, it is reasonable that the R4 H-ATPase-related domain on one subunit interacts with the R3 histidine phosphotransferase (HPT)-related domain on the other subunit (Stock et al. 2000). The potential for heterotypic interactions between R1 and R2 PAS domains on different subunits are also reasonable but has not been directly assessed to date.

The critical role of PAS domains to plant phytochrome function is underscored by the many loss-of-function alleles that map to these regions (Quail et al. 1995). The hypothesis that the PAS repeats play a central role in phytochrome signaling is also supported by localization of light-dependent conformational changes to this region (Lagarias 1985) and to the identification of nuclear targeting sequences within this domain (Chen et al. 2003; Matsushita et al. 2003). As depicted in Figure 6.3, light-induced conformational changes have been shown to occur in both photosensory and regulatory domains of phytochromes with the consensus view being that the P1 domain becomes less exposed while P4, R1, and R2 domains become more exposed upon Pr to Pfr phototransformation (Lagarias 1985). The surprising observation that the C-terminal domain is dispensable for phytochrome function, if it is replaced with a nuclear-targeted protein capable of homodimerization, indicates that R1–R4 subdomains regulate the subcellular localization, dimerization and overall topology of the

N-terminal domain (Matsushita et al. 2003). Together with structure-function studies using chimeric phytochromes (Wagner et al. 1996), these results indicate that phytochrome signaling in the nucleus is mediated by protein-protein interactions with its N-terminal photosensory domain.

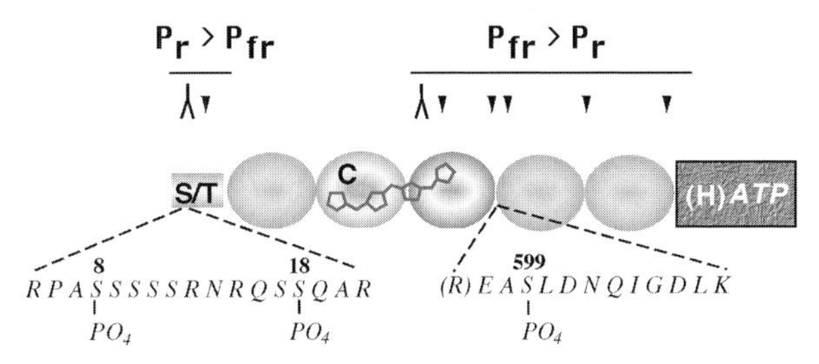

Figure 6.3 Light induced conformational changes of plant phytochromes. The light-dependent accessibility of various regions of plant phytochrome to modification/interaction by proteases (inverted arrows), protein kinases, and monoclonal antibodies (filled arrowheads) is depicted on the oat phyA subunit. Adapted from McMichael and Lagarias (1990).

Aside from homodimerization, the presence of transmitter kinase-related R3 and R4 'output' domains on all eukaryotic phytochromes implicates their role in ATP-dependent phosphotransferase activity. Phytochrome R4 domains possess all of the consensus residues found in H-ATPase domains of the transmitter kinase superfamily which is consistent with the observation that plant phytochromes bind ATP (Wong and Lagarias 1989). Plant phytochromes are also autophosphorylating, ATP-dependent serine-threonine protein kinases (Wong et al. 1989; Yeh and Lagarias 1998). These results, along with the lack of a conserved autophosphorylating histidine residue in the R3 HPT-related domain, indicate that plant phytochromes are atypical transmitter kinases that have gained a new function during evolution (Cashmore 1998). Resolution of an apparent enzymatic function for the transmitter-related domains of plant phytochromes with their dispensability remains an important, albeit unanswered question. The bonafide catalytic activity of transmitter domains of cyanobacterial phytochromes argue for an ATP-dependent enzymatic role for this region of plant phytochromes in light signaling (Yeh et al. 1997).

6.2.2 Molecular Properties of Cyanobacterial Phytochromes

The Cph1 family encompasses those prokaryotic phytochromes with the greatest similarity to plant phytochromes (Figure 6.1). All members of this family, which include the bacteriophytochromes, possess three of the four photosensory domains (P2-P4) found on plant phytochromes as well as the two conserved subdomains found in transmitter modules of two-component histidine kinases (Parkinson and Kofoid 1992). The two regulatory R1 and R2 PAS domains, as well as the serine-threonine rich P1 domain, are missing from all members of the Cph1 family. The first representative of this family to be identified was Cph1, a soluble 85 kDa protein from the cyanobacterium *Synechocystis* sp. PCC 6803 (Hughes et al. 1997; Yeh et al. 1997). Cph1-related genes are present in most cyanobacterial genomes examined; those with conserved cysteine residues in their P3 GAF domains are predicted to encode phytobilin-binding, R/FR photoreversible chromoproteins (Herdman et al. 2000). Evidence that the latter is true has been reported for *Synechocystis* Cph1 (Lamparter et al. 1997; Park et al. 2000b) and for CphA from *Calothrix* sp. PCC7601 (Jorissen et al. 2002). Many cyanobacterial species have two representatives of this family in their genome, the other being a bacteriophytochrome (Herdman et al. 2000). The bacteriophytochromes are discussed in more detail by Vierstra (Chapter 8).

Owing to the ability to express and reconstitute large quantities of recombinant Cph1s in bacteria, the biochemical properties of these proteins have been extensively investigated (Lamparter et al. 1997; Park et al. 2000b; Lamparter et al. 2001; Jorissen et al. 2002). Like plant phytochromes, Cph1s bind phytobilins via the conserved cysteine residue in the P3 GAF domain to generate R-FR photoreversible chromoproteins (Hubschmann et al. 2001a; Lamparter et al. 2001; Jorissen et al. 2002). Cph1 and CphA are also protein kinases, whose histidine autophosphorylation and aspartate phosphotransferase activities are both light dependent (Yeh et al. 1997; Hubschmann et al. 2001b). By contrast with plant phytochromes, both activities are Pr-dependent; thus light absorption leads to inhibition rather than activation of Cph1's catalytic function. These studies have shown that Cph1's enzymatic activities require both HPT and H-ATPase domains which function as phosphohistidine donor/acceptor and ATP-binding sites, respectively (Yeh et al. 1997).

The overall take-home lessons from biochemical studies on the Cph1 familiy of phytochromes are that 1) Cph1's P2-P3-P4 photosensory domain structure and phytobilin-binding environments are very similar to those of eukaryotic phytochromes, 2) Cph1's C-terminal HPT and H-ATPase domains are required for homodimerization, and 3) monomeric and dimeric forms of Cph1 are in dynamic equilibrium – a process that is regulated by both phytobilin binding and light. With regard to the third conclusion, size-exclusion chromatographic analysis has established that phytobilin binding to the Cph1 apoprotein promotes subunit–subunit association (Park et al. 2000b; Lamparter et al. 2001). Taken together, these results support the working hypothesis that phytobilin binding to apoCph1 activates the kinase activity by promoting homodimerization and trans-autophosphorylation. While additional experiments are needed to elucidate the structural basis of light inactivation of Cph1's kinase activities, this could be accomplished by increasing the subunit-subunit dissociation constant and/or by decreasing the affinity for ATP. These questions remain important topics for future investigation.

The second family of prokaryotic phytochromes, that is the Cph2s, also have been found exclusively in cyanobacteria (Montgomery and Lagarias 2002). With the exception of the founding member, that is Cph2 from *Synechocystis* sp. PCC6803, all members of the Cph2 family are composed of two to four GAF domains and terminate in a transmitter module containing both HPT and H-ATPase subdomains. The N-terminal GAF domains of Cph2s are most similar to the P3 GAF domains of plant and Cph1 phytochromes in which the conserved cysteine residue is located. Moreover, the GAF domain immediately adjacent to this conserved GAF domain possesses strong sequence similarity with P4 GAFs of both other phytochrome families (Montgomery and Lagarias 2002). To test the functional significance of this similarity, recombinant Cph2 was shown to encode a phytobilin-binding 1276 amino acid apoprotein that exhibited a characteristic R/FR photoreversible-phytochrome spectral signature (Park et al. 2000a). The phytobilin binding activity and spectroscopic properties of the full length protein were retained within the two N-terminal GAF domains of Cph2 (Wu and Lagarias 2000). These results provided compelling support for the hypothesis that the P3 GAF and P4 PHY subdomains of phytochromes delimit the region of phytochromes that is in direct contact with the phytobilin prosthetic group. Through expression of the N-terminal GAF domain of Cph2, it was also shown that the P3 GAF subdomain is sufficient for the phytobilin lyase activity (Wu and Lagarias 2000). Based on these results and the preserved GAF-PHY molecular architecture for members of all three phytochrome families, we hypothesize that the initial molecular changes that accompany light activation of the phytobilin prosthetic group will be shared by all three classes of phytochromes.

The widespread distribution of the Cph1 and Cph2 families of cyanobacterial phytochromes, together with their light-regulated enzymatic activities, suggests that these proteins perform important photosensory roles in these organisms. Interposon mutagenesis of Cph1 and Cph2 genes has so far failed to convincingly identify photoregulatory functions for their protein products although they appear to modulate phototaxis (Wilde et al. 2002). Owing to the presence of many phytochrome-related genes in cyanobacterial genomes, the potential for redundant function may account for the lack of a clear phenotype of knock-out mutants. Some of the potential functions of these proteins include both photosensory roles (i.e. phototaxis, light-intensity sensing, chromatic adaptation) as well as non-photosensory roles (i.e. regulators of tetrapyrrole and/or iron metabolism or oxygen sensors). There is clearly no dearth of interesting questions remaining to be answered on these two families of prokaryotic phytochromes.

6.3 Photochemical and Nonphotochemical Conversions of Phytochrome

6.3.1 The Phytochrome Chromophore

Based upon spectral measurements of phytochrome in 1959, Butler et al. proposed that the phytochrome chromophore was a linear tetrapyrrole similar to those found in phycobiliproteins – the major components of the phycobilisome light-harvesting antennae in cyanobacteria and red algae (Butler et al. 1959). Siegelman and colleagues later reported that the major pigment released from oat phytochrome by refluxing methanol was a linear tetrapyrrole (Siegelman et al. 1966). An oxidative degradation approach was also used to show that this compound was a 2,3-dihydrobiliverdin 'phytobilin' pigment similar to those derived from phycobiliproteins (Rüdiger and Correll 1969; Klein et al. 1977; Rüdiger et al. 1980). The structure of this phytobilin was subsequently confirmed by total chemical synthesis to be 2R, 3E-phytochromobilin (PΦB) (Weller and Gossauer 1980). Together with peptide-mapping studies (Fry and Mumford 1971), these data indicated that the phytobilin chromophore of oat phytochrome was covalently bound to the apoprotein via a thioether linkage – a hypothesis that was later confirmed by ^{1}H NMR spectroscopy (Lagarias and Rapoport 1980). The structure and linkage of the Pr chromophore of oat phytochrome based on these studies, and its phytobilin precursor, are shown in Figure 6.4.

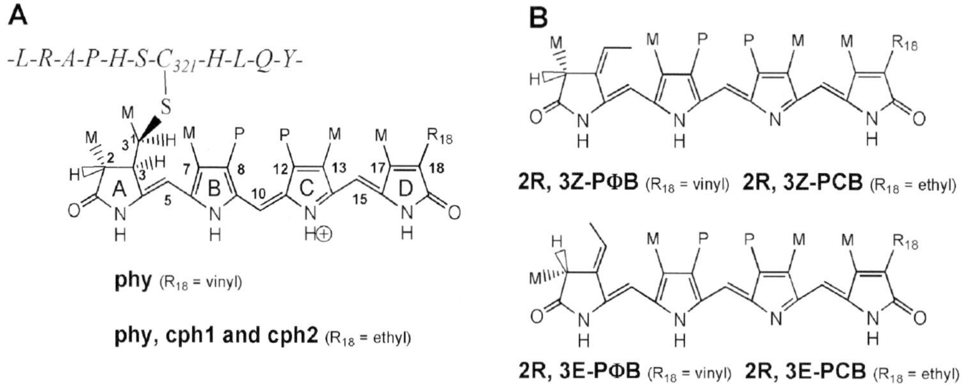

Figure 6.4 The chromophore and phytobilin precursors of plant, green algal and cyanobacterial phytochromes. Panel A illustrates the structure and linkage of the chromophore of oat phytochrome A as deduced by ^{1}H NMR spectroscopy (Lagarias and Rapoport 1980). The PCB-derived chromophores of green algal and cyanobacterial phytochromes are assumed to have identical linkages and stereochemistry by analogy. Panel B shows the phytobilin precursors of plant, green algal, and cyanobacterial phytochromes.

The chromophores of no other phytochrome have received direct chemical scrutiny to date. In this regard, it has been generally assumed that the structure and linkages of all flowering-plant phytochrome chromophores are the same. The situation is less clear for phytochromes from more primitive plant species, except for the evidence that the phytochrome from the green alga *Mesotaenium caldariorum* and Cph1 from the cyanobacterium *Synechocystis* both possess chromophores derived from the more reduced phytobilin prosthetic group, phycocyanobilin (PCB) (Wu et al. 1997; Hubschmann et al. 2001a).

Linear tetrapyrroles are flexible molecules, and their spectroscopic properties are strongly influenced by their conformation, protonation state, and chemical environment (Falk 1989). For this reason, the chemical structure of the phytochrome chromophore *in situ* is still the subject of debate. It is clear however that the phytobilin prosthetic group is protonated and adopts an extended configuration *in situ* – conclusions that are based on vibrational spectroscopic studies and the large ratio of phytochrome's red to near UV absorption maxima (Braslavsky 2003). This contrasts with the small ratio observed for bilins in free solution which assume more cyclic, porphyrin-like conformations. Based on vibrational spectroscopic analysis and semi-empirical vibrational energy calculations, the 4Z-*anti*, 10E-*anti*, 15Z-*syn* configuration was proposed for the Pr chromophore (Andel et al. 1996). This contrasts with the proposed 4Z-*anti*, 10Z-*syn*, 15Z-*anti* configuration, also found in chromophores of the phycobiliproteins, that was based on similar arguments (Kneip et al. 1999). This issue will likely remain controversial until a phytochrome crystal structure is determined. Regarding the Pfr chromophore structure, ^{1}H NMR analysis of chromopeptides derived from the Pfr form of oat phytochrome (Rüdiger et al. 1983), combined with vibrational spectroscopic studies, strongly implicate the Z to E isomerization of the C15 double bond upon Pr to Pfr photoconversion (Fodor et al. 1990; Hildebrandt et al. 1992; Mizutani et al. 1994; Andel et al. 1996; Kneip et al. 1999; Andel et al. 2000). This hypothesis is further supported by the intense fluorescence of covalent adducts between apophytochromes and phycoerythrobilin (PEB), a PΦB analog that lacks this double bond (Murphy and Lagarias 1997). Taken together, these results indicate that the phytobilin chromophore of phytochrome is tightly tethered to the apoprotein in a manner enabling Z to E isomerization of the C15 double bond to be the only efficient mode of radiationless de-excitation.

6.3.2
Phytochrome Photointerconversions

The photochemical interconversion processes of phytochromes have been extensively studied for many years (Butler 1972; Kendrick and Spruit 1977; Rüdiger 1992; Braslavsky et al. 1997; Braslavsky 2003). Since Pr and Pfr have overlapping absorption spectra in most regions of the light spectrum, Pr-to-Pfr and Pfr-to-Pr photoconversion processes lead to the formation of a photoequilibium consisting of a mixture of Pr and Pfr forms under saturating illumination. The light fluence needed to produce this photoequilibrium depends on the intensity and wavelength of light used as well as the relative quantum yields for the Pr-to-Pfr and Pfr-to-Pr phototransformation processes (Butler 1972). Photoequilibrium is most rapidly achieved at the absorption maxima of Pr and Pfr – red light (R) producing a mixture of roughly 87% Pfr and 13% Pr for oat and rye phyA (Kelly and Lagarias 1985; Lagarias et al. 1987). Owing to the lack of Pr absorption in the far-red (FR) region of the light spectrum, FR irradiation can convert >99% of phytochrome to the Pr form (Mancinelli 1986). This information has been used to calculate the amount of Pfr produced by a given fluence of light providing the basis of support for the central dogma(s) of phytochrome action described previously.

From time-resolved absorption and low-temperature trapping spectroscopic techniques, a number of intermediates accompaning the Pr-to-Pfr and Pfr-to-Pr photointerconversion processes have been identified. Two nomenclatures have been used to identify these intermediates – one based on the rhodopsin system and the other based upon the wavelength maxima of the intermediates. The former nomenclature will be used herein and in Figure 6.5. While the following discussion focuses on plant phytochromes (and mostly phyAs), comparative studies on Cph1 suggest that the initial photochemical interconversions are qualitatively similar (Remberg et al. 1997; Sineshchekov et al. 1998; Foerstendorf et al. 2000; Sineshchekov et al. 2002).

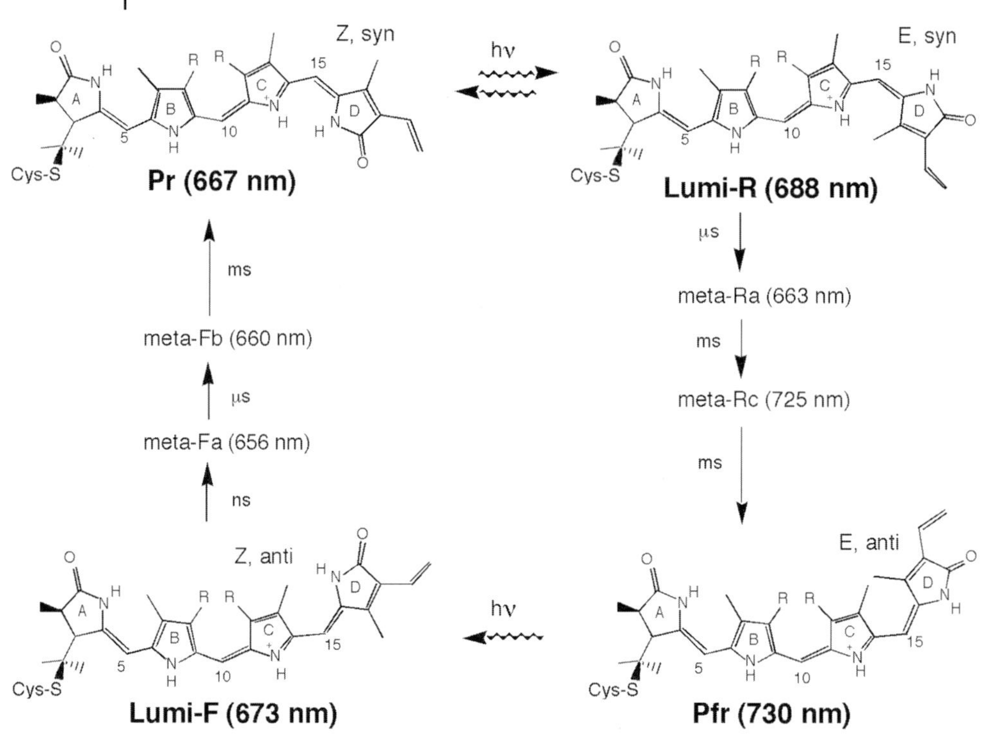

Figure 6.5 Phytochrome photointerconversion processes. The Pr-to-Pfr photoconversion involves the light-dependent 15Z-to-15E isomerization, followed by a light-independent rotation about the C14–C15 bond and changes in chromophore-protein interactions. The Pfr-to-Pr photoconversion is envisaged to involve a concerted in-plane photoisomerization of the C15 double bond, followed by a series of light-independent chromophore-protein relaxation steps. The lifetime and absorption maxima of the various spectroscopically detectible intermediates are shown. Adapted from Andel *et al,* (2000).

Photoexcitation of Pr yields Lumi-R, the first 'stable' photoproduct that absorbs near 700 nm. This conversion occurs very rapidly; the half-life of the Pr excited state(s) lying between 25 and 50 ps (Holzwarth et al. 1984; Andel et al. 1997; Bischoff et al. 2001). Based on resonance Raman analyses, it is generally accepted that Z, *syn* to E, *syn* isomerization of the C15 double bond occurs during this initial photoconversion (Mizutani et al. 1994; Andel et al. 1996). This mechanistic interpretation is consistent with 1) low-temperature absorption kinetic measurements (Eilfeld et al. 1986), 2) the intense fluorescence of PEB-apophytochrome adducts which lack the C15 double bond (Murphy and Lagarias 1997), and 3) the small deuterium isotope effect on Pr fluorescence yield – results that all but rule out intramolecular proton transfer as the primary mechanism of Pr excited-state decay (Brock et al. 1987).

Lumi-R decay occurs in the microsecond timescale to meta-Ra, followed by its conversion into other kinetically distinguishable intermediates (eg. meta-Rc) within the microsecond to millisecond range, ultimately yielding Pfr. The temperature- and solvent-dependence of these non-photochemical interconversions implicate significant changes in phytobilin-apoprotein interactions. At least one of these processes involves the rotation about the C14–C15 single bond to yield the final C15 E,anti conformation of the Pfr chromophore (Figure 6.5). The chemical structures of these intermediates is subject of debate, with some investigators favoring a deprotonation-reprotonation mechanism while others assert that the chromophore remains protonated throughout (Foerstendorf et al. 2001; Braslavsky 2003). The hypothesis that the phytobilin chromophore is moderately planar in both Pr and lumi-R but becomes more distorted from planarity upon conversion to Pfr is presently the most widely accepted interpretation of the available spectroscopic data (Braslavsky 2003). The overall Pr-to-Pfr photochemical quantum yield (Φr) has been determined to be in the range of 0.15–0.17 for plant phytochromes (Lagarias et al. 1987). This quantum yield is in agreement with that determined by femtosecond spectroscopy (Andel et al. 1997) and by optoacoustic spectroscopy (Gensch et al., 1996).

The reverse Pfr-to-Pr photointerconversion is less well characterized for technical reasons but clearly this also proceeds through multiple intermediates (Chen et al. 1996). It is notable that Pfr fluorescence and Pfr-to-Pr photoconversion quantum yields are both lower than those of Pr and Pr-to-Pfr phototransformations, respectively. In this regard, the fluorescence quantum yield for Pfr is vanishingly small, that is less than 10^{-6} at room temperature (Wendler et al. 1984), and Φfr at 0.06–0.08 is less than half of that of Φr (Lagarias et al. 1987). Together with the results of resonance Raman analyses and theoretical arguments (Andel et al. 2000), these observations have been interpreted to support a mechanism for the Pfr-to-Lumi-F photochemical interconversion that involves a concerted configurational and conformational isomerization about the C15 double bond (Figure 6.5). The spectroscopic evidence for strong nonbonded interactions between the C- and D-ring methyl groups in the Pfr chromophore indicate that the excited state of Pfr and its photoproduct are strongly coupled, which may be a major structural feature which favors the concerted C15 E,anti to C15 Z,syn isomerization mechanism (Andel et al. 2000). The small value for Φfr suggests that one or more of the Pfr-to-Pr intermediates can thermally dark-revert back to Pfr-like meta-R intermediates produced in the forward reaction. It also is conceivable that a second reversible photochemical reaction (i.e. proton transfer, amino acid adduct formation) had occurred in parallel with the photoisomerization mechanism to quench the Pfr excited state. Other than their absorption-spectroscopic signatures and thermal stabilities, little is known about the chemical structures of these intermediates and much still remains unknown regarding the reverse photointerconversion process.

6.3.3
Dark Reversion

From the first spectroscopic measurement of phytochromes, the process of non-photochemical Pfr-to-Pr dark reversion became evident (Franklin 1972). Dark reversion of phytochrome has not only been measured *in vitro,* but has been observed *in vivo,* which lead to the hypothesis that this process is of physiological significance (Butler et al. 1963). The bulk of these studies were performed using dark-grown plant seedlings to reduce chlorophyll absorption which strongly overlaps with phytochrome and because these seedlings accumulate high levels of phyA, considerably improving the sensitivity of phytochrome photoassays. Spectrophotometric surveys of etiolated plant tissues revealed considerable variation in the amount of dark reversion for different plant species. While monocot plants showed considerably less dark reversion compared with dicots, the distinct morphology and cell types of these two classes of plants made it difficult to conclude whether this difference reflected the intrinsic properties or cellular environments of phytochromes from the two classes of plants (Franklin 1972).

The development of methods to isolate and purify phytochromes have facilitated analysis of the molecular basis of dark reversion – a process that has been shown to be modulated by temperature, pH, various denaturants, proteases and reductants (Franklin 1972). As improvements have been made in the method of phytochrome purification, the discrepancy between *in vivo* and *in vitro* measurements of phytochrome dark reversion has been resolved. Consistent with *in vivo* measurements, full length monocot phyAs display very little dark reversion *in vitro,* while proteolytic removal of 50–100 amino acid at phyA's N-terminus (i.e. the P1 domain) significantly promotes dark reversion (Vierstra and Quail 1982). The stabilizing influence of the P1 domain has been seen for most phytochromes examined – its removal invariably leading to a 10–15 nm blue shift of the Pfr absorption maximum (Vierstra and Quail 1986). Indeed, the lack of a P1 domain in Cph1 and Cph2 phytochromes may be responsible for the blue-shifted absorption spectra and enhanced dark reversion of these cyanobacterial photoreceptors compared with flowering plant phyAs (Lamparter et al. 1997; Yeh et al. 1997).

The low abundance of phyB-E in plants has made it difficult to measure their spectroscopic properties until recently. The development of methods to express and reconstitute recombinant holophytochromes has enabled investigation of the biochemical and spectroscopic properties of phytochromes whose genes have been cloned (see subsequent section). Studies on the Arabidopsis phytochrome family have shown that recombinant phyC and phyE exhibit significantly more dark reversion than recombinant phyA and phyB (Eichenberg et al. 2000). The importance of dark reversion to phytochrome function is further underscored by the observation that the loss-of-function phyB-101 mutant displays enhanced dark reversion (Elich and Chory 1997). Localization of the phyB-101 mutation to the R2 PAS repeat implicates intra- and inter-molecular interations within the phytochrome dimer to stabilize the 'active' Pfr form. The possibility that other loss-of-function alleles of phytochrome may represent enhanced dark reversion has not been carefully assessed in all instances. It is based on these and other biochemical observations that the regulatory interplay between the N- and C-terminal regions of phytochromes is the subject of active ongoing investigation (Song 1999). The fact that plant phytochromes are all homodimers also raises the possibility that dark reversion of a Pfr chromophore on one subunit may be affected by the chromophore state on the other subunit (Furuya and Schäfer 1996). While there is evidence to support this hypothesis, for example that Pfr-Pfr homodimers dark revert more slowly than Pfr-Pr heterodimers, this simple interpretation does not hold true for all phytochromes so far examined (Eichenberg et al. 2000). In summary, dark reversion plays an important role to regulate the phytochrome signal output – a role that future investigations will hopefully better resolve.

6.4
Phytochrome Biosynthesis and Turnover

Since light perception by all phytochromes depends upon the presence of the linear tetrapyrrole chromophore, the metabolic processes involved in the synthesis of its phytobilin prosthetic group and its assembly with apophytochrome are of fundamental importance to light-mediated plant growth and development. As phytobilin-binding proteins, phytochromes also influence tetrapyrrole metabolism in addition to their photoregulatory function. This section summarizes the current understanding of the pathways involved in the synthesis and assembly of phytobilins to apophytochromes.

6.4.1
Phytobilin Biosynthesis in Plants and Cyanobacteria

The finding that phytochromes possess linear tetrapyrrole prosthetic groups indicated that its synthesis would share common intermediates with heme and chlorophyll biosynthetic pathways. Early studies exploited the development of specific inhibitors of early steps in the tetrapyrrole pathway which could be overcome by feeding hypothetical intermediates after the blocked step. Oat seedlings treated with gabaculine or 5-aminohexynoic acid, potent inhibitors of glutamate-1-semialdehyde amino transferase (GSAT), were shown to induce both phytochrome and chlorophyll deficiencies (Gardner and Gorton 1985; Elich and Lagarias 1988). Supplementation of the medium with 5-aminolevulinic acid (ALA), protoporphyrin or biliverdin IXα (BV) restored phytochrome levels to those of un-inhibited plants, thereby establishing the intermediacy of these compounds in the pathway of phytochrome chromophore biosynthesis (Gardner and Gorton 1985; Jones et al. 1986; Konomi and Furuya 1986; Elich and Lagarias 1987; Elich et al. 1989). Based on these studies, it has been well established that the phytobilin biosynthesis shares intermediates with siroheme, protoheme, and chlorophyll biosynthetic pathways that are depicted in Figure 6.6.

All plant tetrapyrroles are derived from glutamate which is converted to ALA via three enzymes: glutamate tRNA synthetase, glutamate tRNA reductase (GluTR) and

GSAT (Beale 1999; Vavilin and Vermaas 2002). Found in plants, algae, most eubacteria, and many archaebacteria, the glutamate C5 pathway for ALA synthesis can be contrasted with the C3 Shemin pathway of alpha proteobacteria and non-photosynthetic eukaryotes that utilizes the enzyme ALA synthase (ALAS). In all chlorophyll-based photosynthetic organisms, eight molecules of ALA are metabolized in several steps to yield protoporphyrin IX (Figure 6.6). A key branch point in chlorophyll and heme biosynthesis, protoporphyrin IX is metalated with magnesium or iron by two distinct chelatase systems to produce Mg-protoporphyrin or iron-protoporphyrin (heme). The intermediacy of heme in phytochrome chromophore biosynthesis was implicated by the observation that Mg-protoporphyrin was unable to restore holophytochrome levels in gabaculine-treated plants. Unfortunately, exogenous heme was not incorporated into phytochrome under the conditions used for ALA, protoporphyrin and BV rescue (Terry et al. 1993b). Resolution of this impasse was made from two lines of investigation – development of an *in organello* assay system and molecular cloning of genes involved in phytochrome chromophore biosynthesis.

With the knowledge that all plant tetrapyrroles are derived from ALA made in plastids, an isolated plastid system was used to address the hypothesis that linear tetrapyrroles are synthesized in this compartment (Terry and Lagarias 1991). These studies took advantage of the ability of apophytochromes to assemble with phytobilin precursors to yield photochemically active holoproteins [see (Wahleithner et al. 1991) and later discussion]. By adding apophytochrome to the assay mixtures, these investigators were able to show that the entire pathway of phytochrome chromophore biosynthesis resides in plastids (Terry and Lagarias 1991) and later confirmed the intermediacy of ALA, heme, and biliverdin in its synthesis (Terry et al. 1993b). While the low abundance of the enzymes unique to the pathway of phytobilin biosynthesis has proven challenging to their purification and molecular cloning, the key breakthrough on this subject was made with the molecular cloning of HY1 and HY2 (Muramoto et al. 1999; Kohchi et al. 2001) – two loci known from previous studies to be involved in the synthesis of the phytochrome chromophore in Arabidopsis (Parks and Quail 1991). These studies have confirmed that the phytochrome chromophore is derived from heme via the intermediacy of BV and PΦB (see Figure 6.4 for structure). The properties of the plant-specific heme oxygenase and phytobilin synthases are described in the following section.

6.4.1.1 Ferredoxin-dependent Heme Oxygenases

Heme oxygenases (HO) from mammals were the first of this family of enzymes to be identified (Tenhunen et al. 1968). Microsomal enzymes, mammalian heme oxygenases catalyze the oxygen-dependent conversion of heme to BV, CO, and Fe^{2+} products (Ortiz de Montellano and Auclair 2003). The seven electrons required for this reaction are provided by microsomal NADPH-cytochrome P450 reductases (Tenhunen et al. 1969; Schacter et al. 1972). Although they catalyze the same reaction, plant HOs are soluble plastid-localized proteins that utilize soluble ferredoxin for reducing power (Muramoto et al. 2002). Despite their functional differences, the two families of HOs are structurally and evolutionarily related (Terry et al. 2002; Frankenberg and Lagarias 2003b). Ferredoxin-dependent HOs plays a pivotal role in the biosynthesis of

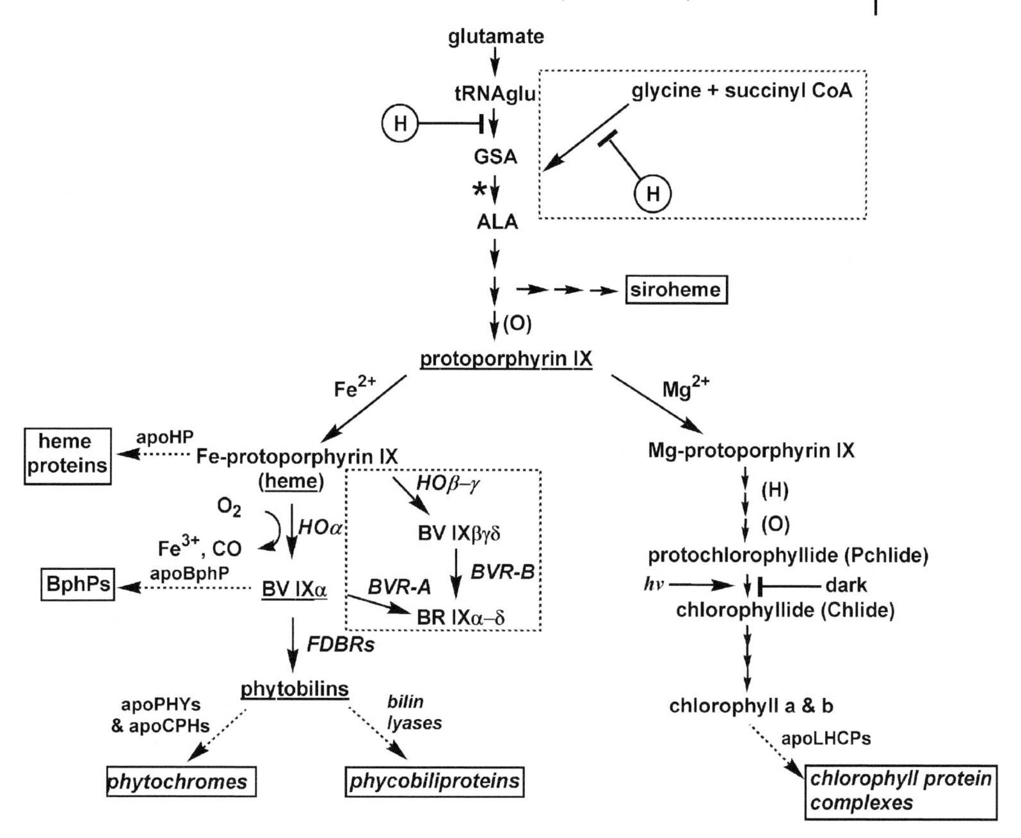

Figure 6.6 Phytobilin biosynthetic pathways in plants, cryptophytes, and cyanobacteria. Phytobilin biosynthesis shares common intermediates with siroheme, heme, and chlorophyll pathways up to the level of protoporphyrin IX. Iron insertion leads to heme, which is subsequently converted to BV by heme oxygenase and to phytobilins by a member of the ferredoxin-dependent bilin reductase (FDBR) family. Magnesium insertion leads to the chlorophylls. Heme, BV, phytobilins, and chlorophylls assemble with their respective apoproteins – processes that draw off these intermediates. Alternative pathways for ALA synthesis, heme, and BV metabolism found in animals and some bacteria are shown in the stipled boxes. Exogenous intermediates that were incorporated into the phytochrome chromophore are underlined. Metabolic steps that are feedback-inhibited by heme are shown with circled Hs. The symbols (H) and (O) indicate that oxidation and reduction steps occur. Abbreviations include: ALA, 5-aminolevulinic acid; apoBphP, apo-bacteriophytochromes; apoCph, apo-cyanobacterial phytochromes; apoHP, apo-hemoproteins; apoLHCP, apo-light harvesting chlorophyll binding proteins; apoPHYs, apo-phytochromes; FDBRs, ferredoxin-dependent bilin reductases; GSA, glutamate-1-semialdehyde; HO, heme oxygenase; tRNAglu, glutamate tRNA

phytobilins in all oxygenic photosynthetic organisms. In plants, BV is reduced by the enzyme phytochromobilin synthase to yield PΦB, the immediate precursor of the phytochrome chromophore of phytochromes that is critical for its photoactivity (Terry et al. 1993b). In cyanobacteria and algae, BV is converted into the phytobilin precursors of the chromophores of the light-harvesting phycobiliproteins for photosynthesis (Beale 1993). These include both PEB and PCB, the latter of which also serves as the immediate precursor of the cyanobacterial phytochromes, Cph1 and Cph2.

The first HO identified from a photosynthetic organisms was found in the red alga *Cyanidium caldarium* by Troxler and colleagues (Troxler et al. 1979). These researchers demonstrated that the phycobiliprotein chromophore precursor PCB was derived from heme in a manner similar to that in mammalian systems. More recently, Beale and Cornejo partially purified a *Cyanidium* HO and reported data establishing its ferredoxin dependence (Beale and Cornejo 1984; Cornejo and Beale 1988). Cyanobacterial HO activity was also described by these researchers (Cornejo and Beale 1997), who later succeeded to clone and functionally express a ferredoxin-dependent HO from the cyanobacterium *Synechocystis* sp. PCC 6803 (Cornejo et al. 1998). Other than *in organello* studies from a variety of plant species (Terry et al. 2002), there has been little published on the biochemistry of heme oxygenases isolated from plant sources. The cloning of HY1 has however lead to the identification of three other heme oxygenase-related genes in the Arabidopsis genome (Davis et al. 2001; Muramoto et al. 2002). From genetic and biochemical studies on recombinant proteins, it is clear that HY1 encodes a soluble, ferredoxin-dependent HO that is targeted to plastids (Muramoto et al. 2002). HO genes have been found in many other plant species, and like HY1, lesions in these genes also lead to phytochrome deficiencies (Terry 1997).

Aside from animals, plants, and cyanobacteria, novel HO genes have been identified in a wide variety of bacterial species (Wilks 2002). A soluble HO from the bacterial pathogen *Neisseria meningitides* encoded by the *hemO* locus was shown to convert heme to ferric-BV in the presence of ascorbate or NADPH-cytochrome P450 reductase (Zhu et al. 2000). Biochemical studies of a protein encoded by the HO-related *hmuO* locus from *Corynebacterium diphtheriae* suggests that it also encodes an HO enzyme similar to human HO (Schmitt 1997; Wilks and Schmitt 1998; Chu et al. 1999). Physiological studies on HmuO implicate a vital role in iron acquisition that is essential for survival and pathogenicity of this microorganism (Schmitt 1997). The HO-related PigA protein from the opportunistic human pathogen *Pseudomonas aeruginosa* was recently shown to convert heme to the IXβ/γ isomers of biliverdin (Ratliff et al. 2001). Perhaps the most interesting is the BphO family of HOs, whose genes are invariably found linked to BphPs in a wide variety of bacterial genomes (Bhoo et al. 2001). Although there is no direct biochemical evidence that shows that BphO is a heme oxygenase, the observation that *E. coli* cells overexpressing the *bphOP* operon exhibit green color strongly suggesting that BV is produced by this HO-related protein (Bhoo et al. 2001). These observations implicate a role for the bacteriophytochrome family in iron homeostasis (Montgomery and Lagarias 2002) – a hypothesis that will be addressed in the subsequent review by Vierstra (Chapter 8).

6.4.1.2 **Ferredoxin-dependent Phytobilin Synthases (Bilin Reductases)**

In oxygenic photosynthetic organisms exclusively, BV is converted to phytobilins by a family of ferredoxin-dependent bilin reductases (FDBRs) as illustrated in Figure 6.7. HY2, the founding member of this family, was shown to encode the enzyme PΦB synthase that mediates the conversion of BV to 3Z-PΦB (Kohchi et al. 2001). Unlike the NADPH-dependent biliverdin reductase (BVR) found in mammals (Maines and Trakshel 1993) and its cyanobacterial ortholog BvdR (Schluchter and Glazer 1997), biliverdin reduction by PΦB synthase appears to proceed via sequential one electron steps involving radical intermediates (Frankenberg and Lagarias 2003b). PΦB synthase isolated from etiolated oat seedlings as well as recombinant HY2 from Arabidopsis both catalyze the ferredoxin-dependent two electron reduction of BV to 3Z-PΦB (Kohchi et al. 2001; McDowell and Lagarias 2001). Formally a phytochromobilin:ferredoxin oxidoreductase (E.C. 1.3.7.4.), PΦB synthase possesses a submicromolar affinity for its BV substrate and also exhibits a relatively high turnover rate constant of >100 min^{-1} (McDowell and Lagarias 2001). A single-copy gene in Arabidopsis, mutations in HY2-related genes found in other plant species all lead to phytochrome deficiencies (Terry 1997). The paucity of HY2 cDNA clones in plant EST databases supports the observation that the HY2 protein is present in vanishingly low amounts in plant tissue (McDowell and Lagarias 2001). Together with its very high BV affinity, this observation may account for the high specific activity of PΦB synthase enabling low amounts of enzyme to produce sufficient PΦB precursor for holophytochrome synthesis.

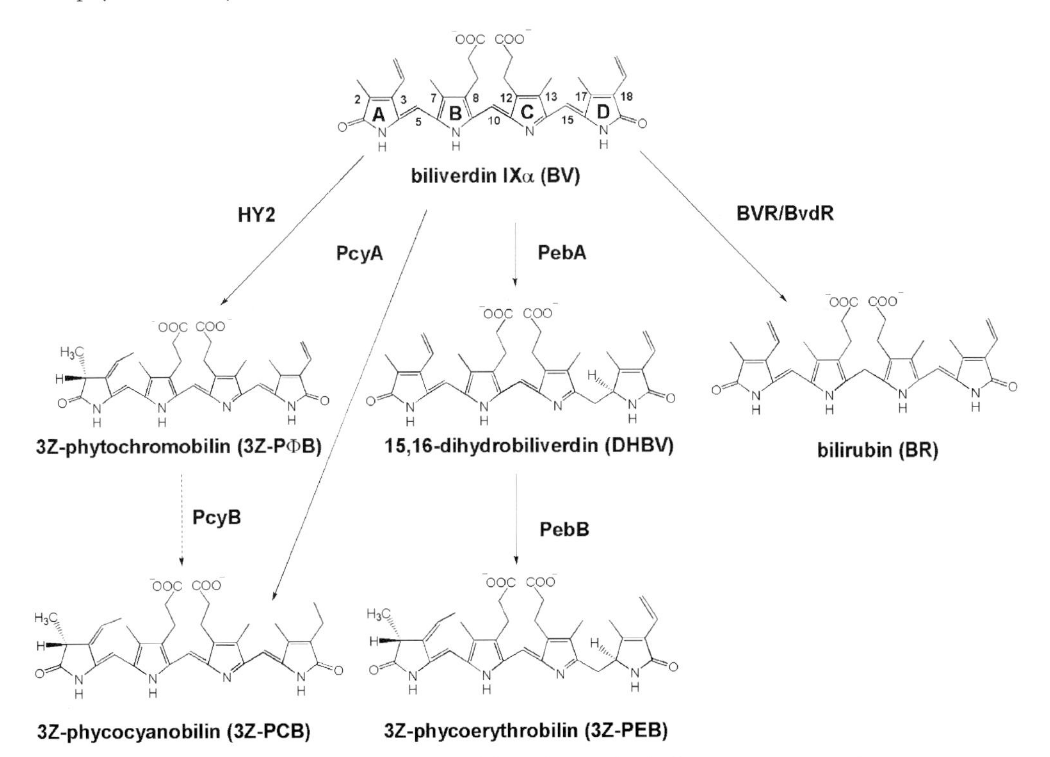

Figure 6.7 Bilin metabolism in plants, cyanobacteria and mammals. Phytobilins from plants, crytophytes and cyanobacteria are all derived from BV which is metabolized by various members of the ferredoxin-dependent bilin reductase family, that is HY2, PcyA, PcyB, PebA and PebB (Frankenberg et al. 2001). In mammals and cyanobacteria, BV is reduced to bilirubin (BR) by the NADPH-dependent enzymes BVR and BvdR, respectively.

The precursor of the chromophores of green algal and cyanobacterial phytochromes has been shown to be PCB – a phytobilin that is two electrons more reduced than PΦB (Wu et al. 1997; Hubschmann et al. 2001a). PCB is produced by the HY2-related enzyme PcyA, phycocyanobilin:ferredoxin oxidoreductase (E.C. 1.3.7.5.) (Frankenberg et al. 2001). Found in all phycobiliprotein-producing cyanobacteria as well as the oxygenic photobacterium *Prochlorococcus marinus*, *PcyA* genes encode the enzymes that catalyze the ferredoxin-dependent, four electron reduction of BV to a mixture of 3Z- and 3E-PCB (Frankenberg et al. 2001). Mechanistic studies on PcyA have established that the reduction of BV by PcyA proceeds via the two-electron reduced intermediate $18^1,18^2$-dihydrobiliverdin, rather than its isomer PΦB (Frankenberg and Lagarias 2003a). The distinct BV reduction regiospecificity of PcyA was proposed to ensure that cyanobacteria do not produce PΦB because its misincorporation into phycobiliproteins might alter their spectra and/or stability (Frankenberg and Lagarias 2003a). The observation that the green alga *Mesotaenium caldariorum* possess phytobilin synthase enzymes capable of the reductions of BV to PΦB and PΦB to PCB (Wu et al. 1997) indicates that an alterative enzyme that we have named PcyB can accomplish one or both of these conversions in this organism (Frankenberg and Lagarias 2003b). Although this enzyme has not yet been cloned, the recent demonstration that the PΦB chromophore of plant phytochromes can be substituted with PCB, and still yield a functional photoreceptor, raise many questions about chromophore selection during the evolution of higher plant phytochromes (Kami et al. 2004).

6.4.2 Apophytochrome Biosynthesis and Holophytochrome Assembly

Since phytochromes consist of two components, an apoprotein and a phytobilin pigment, phytochrome biosynthesis involves the convergence of two biosynthetic pathways which culminate in the assembly of functional holoprotein. Progress on the synthesis of the phytochrome apoprotein and its assembly with the phytobilin chromophore precursor is summarized below.

6.4.2.1 Apophytochrome Biosynthesis

Although they are encoded by a small gene family of 3–5 members (Clack et al. 1994), plant phytochromes can be classified into two groups based upon their light stability. Phytochromes encoded by the *phyA* gene family are responsible for the light-labile pool, while *phyB-E* genes encode the light-stable phytochromes (Furuya 1993). The pronounced light-lability of phyA proteins is due to two processes – light-dependent transcriptional repression of the *phyA* gene (Quail 1991) and light-dependent protein turnover (see Section 6.4.3). PhyA protein accumulates to very large levels in dark grown seedlings due to the elevated level of transcription and the pronounced stability of the phytochrome apoprotein. In this regard, the *phyA* promoter has been shown to be even stronger than the 35S promoter from the cauliflower mosaic virus when used to express heterologous proteins in dark-grown Arabidopsis plants (Somers and Quail 1995a). It is well established that the light-dependent repression of *phyA* gene expression is mediated by phyA itself (Quail 1991). The pattern of expression of the different phytochrome genes has been extensively studied in Arabidopsis (Somers and Quail 1995b; Goosey et al. 1997), as has the accumulation of the phytochrome apoproteins (Hirschfeld et al. 1998; Sharrock and Clack 2002). Taken together, these studies indicate that the five phytochrome apoproteins have an overlapping distribution with phyA and phyB being the dominant species of phytochrome that respectively accumulate in dark-adapted and light-adapted plant tissue. While similar results have been reported for phytochromes in other flowering plant species (Furuya 1993; Hauser et al. 1998), little is known about the regulation of apophytochrome expression in lower plants (Wada et al. 1997). In this regard, it has been reported that phytochrome accumulation in green algae and mosses is light-regulated, although these phenomena may reflect differential protein turnover rather than differential transcription (Morand et al. 1993; Esch and Lamparter 1998). Owing to their very low abundance, even less is known about the accumulation of the Cph1 and Cph2 apoproteins in cyanobacteria other than the observation that Cph1 mRNA accumulation is light-regulated (Garcia-Dominguez et al. 2000). The regulatory mechanism for this response is distinct from that of phyA in flowering plants however, and little is presently known how this affects apoCph1 accumulation in cyanobacteria.

6.4.2.2 **Holophytochrome Assembly**

Phytobilins are covalently bound to both phytochromes and phycobiliproteins via thioether linkages. Consequently, the process of holoprotein assembly has been the subject of extensive analysis (McDowell and Lagarias 2002). The assembly of plant phytochromes is autocatalytic and proceeds in two steps – noncovalent phytobilin binding to the apophytochrome followed by thioether linkage formation (Lagarias and Lagarias 1989; Li and Lagarias 1992; Li et al. 1995). Recent studies with the cyanobacterial phytochrome Cph1 have corroborated this autocatalytic mechanism (Borucki et al. 2003). The take-home lesson of these studies is that an A-ring ethylidene in the phytobilin precursor is required for covalent attachment – a substituent that enables Michael addition of the conserved cysteine thiol residue of apophytochromes to this electrophilic double bond. BVs and bilirubins, which lack the A-ring ethylidene moiety, do not form covalent linkages with apophytochrome (Li and Lagarias 1992). BVs can however non-covalently interact with apophytochrome and act as reversible competitive inhibitors for phytochrome assembly (Li et al. 1995). These results contrast with the observation that BV covalently binds to bacteriophytochromes (Bhoo et al. 2001; Lamparter et al. 2002; Lamparter et al. 2003; Lamparter et al. 2004). The autocatalytic mechanism of phytochrome assembly also contrasts with phycobiliprotein assembly which require lyase enzymes to mediate attachment of each phytobilin (Schluchter and Glazer 1999; Frankenberg and Lagarias 2003b).

In addition to the ethylidene moiety, both propionic acid moieties have been shown to be necessary for holophytochrome assembly (Bhoo et al. 1997). The requirement of a C10 methine bridge was based on the observation that rubin analogs of phytobilins fail to form covalent adducts with recombinant apophytochromes (Terry et al. 1993a). While most of the published reports on phytochrome assembly have utilized the 3E-phytobilin isomers, the observation that 3Z-phytobilin isomers are also capable of functional assembly with apophytochrome suggests that the 3Z to 3E isomerization is not critical for holophytochrome assembly (Terry et al. 1995). As was discussed in Section 2.1.2, 3Z-PΦB is the major product of PΦB synthase, hence its isomerization may not be necessary *in vivo*. Other studies have shown that bilins with modified substituents in the D-ring including PCB, PEB and isoPΦB can assemble with apophytochromes. However, the spectroscopic properties of the resulting phytobilin apophytochrome adducts are significantly altered (Elich et al. 1989; Li and Lagarias 1992; Lindner et al. 2000). PCB substitution has been shown to produce a functional, albeit blue-shifted plant phytochrome photoreceptor (Kami et al. 2004), while PEB adducts of apophytochromes, a.k.a. phytofluors, are intensely fluorescent (Murphy and Lagarias 1997). More recent studies using synthetic bilins indicate that, aside from the ethylidene moiety required for thioether linkage formation (Lagarias and Lagarias 1989; Li and Lagarias 1992; Li et al. 1995), modifications in both A- and D-rings can be tolerated (Hanzawa et al. 2001; Hanzawa et al. 2002). This work suggests that the phytobilin binding site on apophytochrome can accommodate structural variations, and that it should be possible to alter the spectral properties and photoregulatory function of phytochromes through engineering of new pathways of phytobilin biosynthesis in plants.

While the phytobilin specificity for holophytochrome assembly has been actively addressed, less is known regarding catalytic residues on apophytochrome that specify phytobilin attachment. The only absolute requirement for thioether linkage formation appears to be the conserved cysteine itself (i.e. Cys321 in the case of PHYA3) (Boylan and Quail 1989; Bhoo et al. 1997). Deletion analysis of recombinant phytochromes reveals that neither the N-terminal 68 amino acids nor the entire C-terminal regulatory domains are required for assembly (Deforce et al. 1991; Hill et al. 1994). Comparative domain analysis of phy, cph1 and cph2 families indicate that P1 and P2 domains are dispensable for phytobilin attachment, and in the same study, it was shown the P3 GAF domain delimits the phytobilin lyase domain of phytochromes (Wu and Lagarias 2000). A site-directed phytochrome mutant on the histidine residue adjacent to the conserved cysteine in the P3 GAF domain almost completely abolished phytobilin attachment, demonstrating a potential catalytic role for this residue (Bhoo et al. 1997; Remberg et al. 1999). Replacement of this histidine with asparagine did not prevent phytobilin attachment however, indicating that this residue cannot serve as a proton donor/acceptor for catalysis (Song et al. 1997). Mutagenesis of a conserved glutamate residue (E189) in the P3 GAF domain of Cph1 proved inhibitory to phytobilin attachment implicating its involvement in assembly – possibly as a proton donating residue to the phytobilin (Wu and Lagarias 2000). The detailed knowledge of the residues that participate in both noncovalent and covalent binding of phytobilins to apophytochromes is expected to provide insight into the molecular mechanism of holophytochrome assembly and its regulation, about which little is presently known.

6.4.3
Phytochrome Turnover

Aside from assembly, photoconversion and dark reversion, phytochrome localization and turnover are two important mechanisms that regulate phytochrome activity. The movement of phytochrome from the cytoplasm to the nucleus is controlled by the photoconversion between Pr and Pfr forms – a process that has been strongly linked to phytochrome signal transduction (see Nagy and Schäfer, Chapter 9). The following discussion focusses on the processes that regulate phytochrome stability.

Plant phytochromes have been classified into two groups, that is light-labile and light-stable. Accumulating to high levels in plants grown for prolonged periods in darkness, the light-labile phyA class of photoreceptors function to regulate gene expression and seed germination under very low light fluences and far-red light enriched canopy environments (Casal et al. 1997). Such conditions are encountered for seeds that germinate and develop underground. When dark-grown seedlings are exposed to light, phyA is rapidly degraded (Nagy and Schäfer 2002). The rate of phyA degradation increases 100-fold upon light exposure – from a half-life of over 100 h to less than 1 h after photoconversion to Pfr (Quail et al. 1973). Because of this phenomenon, light-stable phyB-Es predominate in light-grown plants (Sharrock and Clack 2002). Light-dependent phyA turnover is preceeded by the formation of sequestered areas of phytochromes or SAPs in the cytoplasm (Mackenzie et al. 1974; Speth et al. 1987). Since SAPs are only formed under R and are also detected in phyA preparations *in vitro*, the hypothesis that SAPs represent self-aggregated Pfr has been proposed (Hofmann et al. 1990; Hofmann et al. 1991). This proposal has received support from more recent studies showing that phyA must be translocated to the nucleus to initiate signal transduction (see Chapter 9 by Nagy and Schäfer). In this model, phyA that remains sequestered in the cytosol does not participate in signaling and is ultimately degraded.

In the late 1980s, the detection of ubiquitin-phyA conjugates (Ub-P) following light treatment provided evidence that phyA degradation utilizes the ubiquitin/26S proteasome pathway [for review see (Clough and Vierstra 1997)]. Ub-P conjugates appeared rapidly, that is within 5 min, following exposure to red light and disappeared with kinetics similar to those for the loss of Pfr; results that indicated that phyA is degraded in the Pfr form (Jabben et al. 1989; Cherry et al. 1991). This conclusion was more recently corroborated by the pronounced light-stability of phyA in dark-grown phytobilin-deficient plants where phyA accumulates as its apoprotein (Terry 1997). The role of phyA ubiquitination to both signal transduction and turnover has been the subject of great interest. Through deletion analysis of phyA and expression in transgenic plants, several regions important for Pfr degradation have been identified [see review of (Clough and Vierstra 1997)]. Key regions critical to the light-dependent turnover have been mapped to both N- and C-terminal domains of phytochrome. However, identification of key lysine residues that participate in the light-dependent ubiquitination of phyA remains an elusive challenge (Clough et al. 1999). The recent discovery of cytoplasm-to-nuclear translocation raises the question of in which plant cellular compartment phyA ubiquitination and turnover occurs. This issue will undoubtedly consume countless hours of research in the laboratory. The inability to identify a phyA-specific E3 ubiquitin ligase by both biochemical analysis and genetic screens suggests that a novel mechanism(s) for phyA turnover may be responsible.

6.5 Molecular Mechanism of Phytochrome Signaling: Future Perspective

In subsequent chapters of this book, biochemical, genetic, and cell biological approaches to identify the components of plant phytochrome signaling pathways and to unravel their role in the regulation of target genes will be presented by Quail (Chapter 7) and Nagy and Schäfer (Chapter 9). The emerging picture from these investigations is that phytochrome signaling requires both translocation to the nucleus and the maintenance of the Pfr form within the nucleus for a sufficient length of time to commit the light signal irreversibly. Based on the protein kinase activities of plant and cyanobacterial phytochromes, it is likely that protein phosphorylation plays a key role in the transmission of the light signal and/or in its regulation. The potential role of phytochromes to regulate tetrapyrrole metabolism is another emerging area of phytochrome signaling studies. Both topics are the subject of the following discussion.

6.5.1 Regulation of Protein–Protein Interactions by Phosphorylation

Previous studies have shown that the protein kinase activities of both Cph1 and eukaryotic phytochromes are bilin- and light-regulated (Yeh et al. 1997; Yeh and Lagarias 1998). Phytobilin binding to apoCph1 stimulates its histidine kinase activity, which is consistent with the observation that holoCph1 is mostly dimeric while apoCph1 is a monomer (Park et al. 2000b; Lamparter et al. 2001). Phytobilins are therefore apoCph1 ligands which regulate its trans(auto)phosphorylation – a mechanism that has been well described for other bacterial two-component histidine kinases (Stock et al. 2000). In contrast, the autophosphorylation activity of affinity-purified, recombinant oat phyA is inhibited by bilin attachment, indicating that the ser/thr kinase activity of a eukaryotic phytochrome is also bilin-regulated (Yeh and Lagarias 1998). As shown in Figure 6.8, the polarities of bilin- and light-regulation for plant and Cph1 phytochromes are opposite one another, with Pfr being more active than Pr for plant phytochromes and *vice versa* for Cph1. These results strongly implicate regulation of the monomer-dimer equilibrium by bilin and light for Cph1, while a different (allosteric) mechanism must be invoked for plant phytochromes which are are obligate dimers. The role of phytochrome phosphorylation activities in light signaling in plants is unknown. However, the observed bilin-stimulated and light-inhibited phosphotransferase activities of Cph1 implicate the regulation of the ATP/ADP-binding affinities, subunit-subunit dissociation constants, and/or ATP-to-histidine and phospho-histidine-to-aspartate equilibrium/rate constants. Little is presently known about the influence of bound ATP (and/or ADP) and of the histidine phosphorylation state on the thermodynamics and kinetics of the various protein–protein interactions. For higher plant phytochromes, it is reasonable that phytochrome-substrate interactions will be influenced by ATP/ADP binding as well as by the phosphorylation status of phytochrome itself and its substrate(s). We envisage that bilin, light, and ATP work together to regulate phytochrome–protein interactions which influence its subcellular localization, degradation, dark reversion, gene expression, and lead ultimately to changes in plant growth and development.

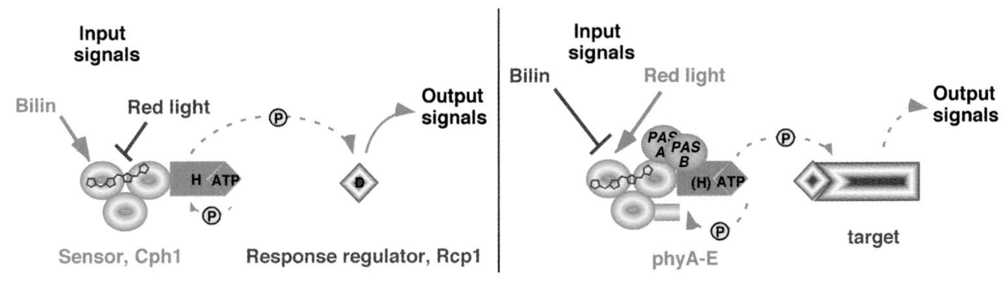

Figure 6.8 Models for prokaryotic (Cph1) and eukaryotic (phyA-E) phytochrome signal transmission. Bilin and light input signals are perceived by the photosensory domains leading to altered phosphotransfer activities of the regulatory domains. Biochemical studies have shown that Cph1 and eukaryotic phytochromes have opposite responses. The linear tetrapyrrole (bilin) chromophore is shown associated with the P3 and P4 domains as indicated in Figure 6.1 Potential downstream targets are indicated in red. The output signal(s) of the phytochrome signal transduction pathways are presently unknown.

6.5.2
Regulation of Tetrapyrrole Metabolism

The evidence that phytochromes are regulators of tetrapyrrole metabolism in plants has a rich history (Mohr and Oelze-Karow 1982). In addition to key enzymes of the chlorophyll pathway that are transcriptionally regulated by phytochrome, such as GluTR (McCormac et al. 2001) and protochlorophyllide oxidoreductase A (PORA) (Mösinger et al. 1985), phytochrome performs a metabolic regulatory role in the tetrapyrrole biosynthetic pathway that likely reflects a more ancestral role (Montgomery and Lagarias 2002). Phytochromes coordinate the expression of nuclear genes involved in the biogenesis of the photosynthetic apparatus, most notably the light harvesting chlorophyll a/b binding protein family, with the production of chlorophyll pigments. This is an extremely important function because misregulated chlorophyll synthesis can lead to production of toxic oxygen species via photosensitization (Cornah et al. 2003). Indeed, the combination of light, oxygen, and photosensitizing pigments is a deadly recipe that plants must avoid at all costs.

In flowering plants, chlorophyll synthesis does not occur in darkness since their protochlorophyllide oxidoreductases are all light-dependent (Willows 2003). This situation contrasts with cryptophytes, that is mosses, ferns and algae, which possess a light-independent protochlorophyllide oxidoreductase system enabling them to synthesize chlorophyll in darkness (Vavilin and Vermaas 2002; Fujita and Bauer 2003).

During prolonged dark periods, angiosperm seedlings therefore accumulate protochlorophyllide (Pchlide) which becomes bound to PORA. Upon light exposure, Pchlide bound to the ternary NADPH-PORA-Pchlide complex is photochemically converted to chlorophyllide (Chlide) which is subsequently metabolized to chlorophylls a and b (see Figure 6.6). The dark accumulation of Pchlide serves to prime chlorophyll biosynthesis and assembly of the photosynthetic apparatus when light becomes available; a process that is of adaptive significance to angiosperm seedlings developing underground (Casal et al. 1997). This program of etiolation can proceed as long as food reserves in the seed last or light becomes available. Highly expressed in dark-grown seedlings, the *PORA* gene is rapidly down-regulated upon light exposure; a response that has been shown to be mediated by phyA (Mösinger et al. 1985). The similar pattern of *PHYA* expression, that is elevated expression in darkness and rapid down regulation in light, suggests that the coordinate expression of PORA and PHYA contribute to this important adaptive process.

The metabolic consequence of elevated *PHYA* expression in dark-grown plants is elevated synthesis of Pchlide which in part is due to removal of heme-derived phytobilins. In this regard, it is well established that heme is a potent feedback inhibitor of ALA synthesis – and in plants, the major target for this inhibition is GluTR [see Figure 6.6; (Papenbrock and Grimm 2001; Cornah et al. 2003)]. For this reason, PORA expression must also be elevated to ensure that all of the Pchlide produced under these conditions is assembled into a PORA-protein complex. The consequence of insufficient PORA expression and the resulting accumulation of free Pchlide, seen in *porA* mutants and seedlings exposed to prolonged FR in which PORA gene expression is down-regulated via the action of phyA, is severe photooxidative damage upon transfer to light (Reinbothe et al. 1996). These results show that PORA and PHYA expression must be balanced during periods of prolonged etiolation, both to prime the system for rapid synthesis/assembly of the photosynthetic apparatus and to avoid photooxidative damage.

Beside contributing to the synthesis of light-modulated regulators of gene expression, the phytobilin-binding properties of apophytochromes, as well as the BV-binding properties of apobacteriophytochromes (see Vierstra, Chapter 8), represent one of many strategies found in nature to modulate heme levels and thereby to modulate tetrapyrrole synthesis. In animals and alpha protoeobacteria, ALA synthesis is also regulated by heme. However in these organisms, ALAS is the target of heme feedback inhibition. The regulation of heme is critical for all aerobic organisms since heme also has pro-oxidant properties. For this reason, nature has invented a wide variety of mechanisms to cope with heme accumulation aside from this feedback regulation. Heme oxygenases play a key role in this regulation. However the mammalian and plant enzymes are inefficient with poor catalytic turnover, unless their BV products are removed. To accomplish heme removal, plants and cyanobacteria utilize the FDBR family, apophytochromes and apophycobiliproteins, while mammals, fungi, pathogenic and nonpathogenic bacteria have evolved NADPH-dependent BV reductase (BVR-A), bacteriophytochromes, and unique heme oxygenase enzyme systems with distinct heme cleavage specificities [see Figure 6.6; (Frankenberg and Lagarias 2003b)]. In all of these organisms, removal of heme has the same effect to up-regulate new tetrapyrrole synthesis through metabolite-dependent de-repression of ALA synthesis. As for the fate of these newly synthesized tetrapyrroles, this varies from organism to organism – impinging on such processes as photosynthesis, nitrogen fixation, nitrate assimilation, oxygen-detoxification, synthesis/secretion of defense compounds, hormone synthesis/degradation, among many others. In addition to its well-publicized genetic regulatory function, phytochrome's role to regulate tetrapyrrole metabolism in plants and cyanobacteria is expected to receive increased experimental scrutiny in the years to come.

Acknowledgements

We gratefully acknowledge grant support from the National Institutes of Health (RO1 GM068552-01) and the United States Department of Agriculture (AMD-0103397) and to members of the Lagarias laboratory for critical reading of this manuscript.

References

Andel, F., K.C. Hasson, F. Gai, P.A. Anfinrud, and R.A. Mathies. 1997. *Biospectroscopy* 3, 421–433.

Andel, F., J.C. Lagarias, and R.A. Mathies. 1996. *Biochemistry* 35, 15997–16008.

Andel, F., J.T. Murphy, J.A. Haas, M.T. McDowell, I. van der Hoef, J. Lugtenburg, J.C. Lagarias, and R.A. Mathies. 2000. *Biochemistry* 39, 2667–2676.

Aravind, L. and C.P. Ponting. 1997. *TIBS* 22, 458–459.

Beale, S.I. 1993. *Chem. Rev.* 93, 785–802.

Beale, S.I. 1999. *Photosynth. Res.* 60, 43–73.

Beale, S.I. and J. Cornejo. 1984. *Arch. Biochem. Biophys.* 235, 371–384.

Bhoo, S.H., S.J. Davis, J. Walker, B. Karniol, and R.D. Vierstra. 2001. *Nature* 414, 776–779.

Bhoo, S.H., T. Hirano, H.Y. Jeong, J.G. Lee, M. Furuya, and P.S. Song. 1997. *J. Am. Chem. Soc.* 119, 11717–11718.

Bischoff, M., G. Hermann, S. Rentsch, and D. Strehlow. 2001. *Biochemistry* 40, 181–186.

Borucki, B., H. Otto, G. Rottwinkcl, J. Hughcs, M.P. Heyn, and T. Lamparter. 2003. *Biochemistry* 42, 13684–13697.

Boylan, M. and P. Quail. 1989. *Plant Cell* 1, 765–773.

Braslavsky, S.E. 2003. In *Photochromism, Molecules and Systems* (ed. D. H. and H. BouasLaurent), pp. 738–755. Elsevier Science BV, Amsterdam.

Braslavsky, S.E., W. Gärtner, and K. Schaffner. 1997. *Plant Cell Environ.* 20, 700–706.

Brock, H., B.P. Ruzicska, T. Arai, W. Schlamann, A.R. Holzwarth, S.E. Braslavsky, and K. Schaffner. 1987. *Biochemistry* 26, 1412–1417.

Butler, W.L. 1972. In *Phytochrome* (ed. K. Mitrakos and W. Shropshire), pp. 182–192. Academic Press, New York.

Butler, W.L., H.C. Lane, and H.W. Seigelman. 1963. *Plant Physiol.* 38, 514–519.

Butler, W.L., K.H. Norris, H.W. Siegelman, and S.B. Hendricks. 1959. *Proc. Natl. Acad. Sci. USA* 45, 1703–1708.

Casal, J.J., L.G. Luccioni, K.A. Oliverio, and H.E. Boccalandro. 2003. *Photochem. Photobiol. Sci.* 2, 625–636.

Casal, J.J., R.A. Sanchez, and J.F. Botto. 1998. *J. Exp. Bot.* 49, 127–138.

Casal, J.J., R.A. Sanchez, and M.J. Yanovsky. 1997. *Plant Cell Environ.* 20, 813–819.

Cashmorc, A.R. 1998. *Proc. Natl. Acad. Sci. USA* 95, 13358–13360.

Chen, E.F., V.N. Lapko, J.W. Lewis, P.S. Song, and D.S. Kliger. 1996. *Biochemistry* 35, 843–850.

Chen, M., R. Schwabb, and J. Chory. 2003. *Proc. Natl. Acad. Sci. USA* 100, 14493–14498.

Cherry, J.R., H.P. Hershey, and R.D. Vierstra. 1991. *Plant Physiol.* 96, 775–785.

Chu, G.C., K. Katakura, X. Zhang, T. Yoshida, and M. Ikeda-Saito. 1999. *J. Biol. Chem.* 274, 21319–21325.

Clack, T., S. Mathews, and R.A. Sharrock. 1994. *Plant Mol. Biol.* 25, 413–427.

Clough, R.C., E.T. Jordan-Beebe, K.N. Lohman, J.M. Marita, J.M. Walker, C. Gatz, and R.D. Vierstra. 1999. *Plant J.* 17, 155–167.

Clough, R.C. and R.D. Vierstra. 1997. *Plant Cell Environ.* 20, 713–721.

Cornah, J.E., M.J. Terry, and A.G. Smith. 2003. *Trends Plant Sci.* 8, 224–230.

Cornejo, J. and S.I. Beale. 1988. *J. Biol. Chem.* 263, 11915–11921.

Cornejo, J. and S.I. Beale. 1997. *Photosynth. Res.* 51, 223–230.

Cornejo, J., R.D. Willows, and S.I. Beale. 1998. *Plant J.* 15, 99–107.

Davis, S.J., S.H. Bhoo, A.M. Durski, J.M. Walker, and R.D. Vierstra. 2001. *Plant Physiol.* 126, 656–669.

Deforce, L., K.I. Tomizawa, N. Ito, D. Farrens, P.S. Song, and M. Furuya. 1991. *Proc. Natl. Acad. Sci. USA* 88, 10392–10396.

Eichenberg, K., I. Baurle, N. Paulo, R.A. Sharrock, W. Rüdiger, and E. Schäfer. 2000. *FEBS Lett.* 470, 107–112.

Eilfeld, P., P. Eilfeld, and W. Rüdiger. 1986. *Photochem. Photobiol.* 44, 761–769.

Elich, T.D. and J. Chory. 1997. *Plant Cell* 9, 2271–2280.

Elich, T.D. and J.C. Lagarias. 1987. *Plant Physiol.* 84, 304–310.

Elich, T.D. and J.C. Lagarias. 1988. *Plant Physiol.* 88, 747–751.

Elich, T.D., A.F. McDonagh, L.A. Palma, and J.C. Lagarias. 1989. *J. Biol. Chem.* 264, 183–189.

Esch, H. and T. Lamparter. 1998. *Photochem. Photobiol.* 67, 450–455.

Falk, H. 1989. *The Chemistry of Linear Oligopyrroles and Bile Pigments.* Springer-Verlag, Vienna.

Fankhauser, C. 2001. *J. Biol. Chem.* 276, 11453–11456.

Fodor, S.P., J.C. Lagarias, and R.A. Mathies. 1990. *Biochemistry* 29, 11141–11146.

Foerstendorf, H., C. Benda, W. Gärtner, M. Storf, H. Scheer, and F. Siebert. 2001. *Biochemistry* 40, 14952–9.

Foerstendorf, H., T. Lamparter, J. Hughes, W. Gärtner, and F. Siebert. 2000. *Photochem. Photobiol.* 71, 655–661.

Frankenberg, N. and J.C. Lagarias. 2003a. *J. Biol. Chem.* 278, 9219–9226.

Frankenberg, N., K. Mukougawa, T. Kohchi, and J.C. Lagarias. 2001. *Plant Cell* 13, 965–978.

Frankenberg, N.F. and J.C. Lagarias. 2003b. In *The Porphyrin Handbook. Chlorophylls and Bilins, Biosynthesis Structure and Degradation.* (ed. K.M. Kadish, K.M. Smith, and R. Guilard), pp. 211–235. Academic Press, New York.

Franklin, B. 1972. In *Phytochrome* (ed. K. Mitrakos and W. Shropshire), pp. 195–225. Academic Press, New York.

Fry, K.T. and F.E. Mumford. 1971. *BBRC* 45, 1466–1473.

Fujita, Y. and C.E. Bauer. 2003. In *The Porphyrin Handbook. Chlorophylls and Bilins, Biosynthesis Structure and Degradation.* (ed. K.M. Kadish, K.M. Smith, and R. Guilard), pp. 109–156. Academic Press, New York.

Furuya, M. 1993. *Ann. Rev. Plant Physiol. Plant Mol. Biol.* 44, 617–645.

Furuya, M. and E. Schäfer. 1996. *Trends Plant Sci.* 1, 301–307.

Garcia–Dominguez, M., M.I. Muro–Pastor, J.C. Reyes, and F.J. Florencio. 2000. *J. Bacteriol.* 182, 38–44.

Gardner, G. and H.L. Gorton. 1985. *Plant Physiol.* 77, 540–543.

Gensch, T., M.S. Churio, S.E. Braslavsky, and K. Schaffner. 1996. *Photochem. Photobiol.* 63: 719–725.

Goosey, L., L. Palecanda, and R.A. Sharrock. 1997. *Plant Physiol.* 115, 959–969.

Hanzawa, H., K. Inomata, H. Kinoshita, T. Kakiuchi, K.P. Jayasundera, D. Sawamoto, A. Ohta, K. Uchida, K. Wada, and M. Furuya. 2001. *Proc. Natl. Acad. Sci. USA* 98, 3612–3617.

Hanzawa, H., T. Shinomura, K. Inomata, T. Kakiuchi, H. Kinoshita, K. Wada, and M. Furuya. 2002. *Proc. Natl. Acad. Sci. USA* 99, 4725–4729.

Hauser, B.A., M.M. Cordonnier-Pratt, and L.H. Pratt. 1998. *Plant J.* 14, 431–439.

Herdman, M., T. Coursin, R. Rippka, J. Houmard, and N.T. de Marsac. 2000. *J. Mol. Evol.* 51, 205–213.

Hildebrandt, P., A. Hoffmann, P. Lindemann, G. Heibel, S.E. Braslavsky, K. Schaffner, and

B. Schrader. 1992. *Biochemistry* 31, 7957–7962.

Hill, C., W. Gärtner, P. Towner, S.E. Braslavsky, and K. Schaffner. 1994. *Eur. J. Biochem.* 223, 69–77.

Hirschfeld, M., J.M. Tepperman, T. Clack, P.H. Quail, and R.A. Sharrock. 1998. *Genetics* 149, 523–535.

Ho, Y.S.J., L.M. Burden, and J.H. Hurley. 2000. *EMBO J.* 19, 5288–5299.

Hofmann, E., R. Grimm, K. Harter, V. Speth, and E. Schäfer. 1991. *Planta* 183, 265–273.

Hofmann, E., V. Speth, and E. Schäfer. 1990. *Planta* 180, 372–377.

Holzwarth, A.R., J. Wendler, B.P. Ruzsicska, S.E. Braslavsky, and K. Schaffner. 1984. *Biochim. Biophys. Acta* 791, 265–273.

Hubschmann, T., T. Börner, E. Hartmann, and T. Lamparter. 2001a. *Eur. J. Biochem.* 268, 2055–2063.

Hubschmann, T., H.J. Jorissen, T. Börner, W. Gärtner, and N. Tandeau de Marsac. 2001b. *Eur. J. Biochem.* 268, 3383–9.

Hughes, J., T. Lamparter, F. Mittmann, E. Hartmann, W. Gärtner, A. Wilde, and T. Börner. 1997. *Nature* 386, 663.

Jabben, M., J. Shanklin, and R.D. Vierstra. 1989. *Plant Physiol.* 90, 380–384.

Jones, A.M., C.D. Allen, G. Gardner, and P.H. Quail. 1986. *Plant Physiol.* 81, 1014–1016.

Jones, A.M. and M.D. Edgerton. 1994. *Sem. Cell Biol.* 5, 295–302.

Jones, A.M., R.D. Vierstra, S.M. Daniels, and P.H. Quail. 1985. *Planta* 164, 505–506.

Jorissen, H., B. Quest, A. Remberg, T. Coursin, S.E. Braslavsky, K. Schaffner, N.T. de Marsac, and W. Gärtner. 2002. *Eur. J. Biochem.* 269, 2662–2671.

Kami, C., K. Mukougawa, T. Muramoto, A. Yokota, T. Shinomura, J.C. Lagarias, and T. Kohchi. 2004. *Proc Natl Acad Sci U S A* 101, 1099–104.

Kelly, J.M. and J.C. Lagarias. 1985. *Biochemistry* 24, 6003–6010.

Kendrick, R.E. and G.H.M. Kronenberg. 1994. *Photomorphogenesis in Plants,* pp. 828. Martinus Nijhoff Publishers, Dordrecht, The Netherlands.

Kendrick, R.E. and C.J.P. Spruit. 1977. *Photochem. Photobiol.* 26, 201–214.

Klein, G., S. Grombein, and W. Rüdiger. 1977. *Hoppe-Seyler's Z. Physiologie Chem.* 358, 1077–1079.

Kneip, C., P. Hildebrandt, W. Schlamann, S.E. Braslavsky, F. Mark, and K. Schaffner. 1999. *Biochemistry* 38, 15185–15192.

Kohchi, T., K. Mukougawa, N. Frankenberg, M. Masuda, A. Yokota, and J.C. Lagarias. 2001. *Plant Cell* 13, 425–436.

Konomi, K. and M. Furuya. 1986. *Plant Cell Physiol.* 27, 1507–1512.

Lagarias, J.C. 1985. *Photochem. Photobiol.* 42, 811–820.

Lagarias, J.C., J.M. Kelly, K.L. Cyr, and W.O. Smith, Jr. 1987. *Photochem. Photobiol.* 46, 5–13.

Lagarias, J.C. and D.M. Lagarias. 1989. *Proc. Natl. Acad. Sci. USA* 86, 5778–5780.

Lagarias, J.C. and H. Rapoport. 1980. *J. Am. Chem. Soc.* 102, 4821–4828.

Lamparter, T., M. Carrascal, N. Michael, E. Martinez, G. Rottwinkel, and J. Abian. 2004. *Biochemistry* 43, 3659–3669.

Lamparter, T., B. Esteban, and J. Hughes. 2001. *Eur. J. Biochem.* 268, 4720–4730.

Lamparter, T., N. Michael, O. Caspani, T. Miyata, K. Shirai, and K. Inomata. 2003. *J. Biol. Chem.* 278, 33786–33792.

Lamparter, T., N. Michael, F. Mittmann, and B. Esteban. 2002. *Proc. Natl. Acad. Sci. USA* 99, 11628–11633.

Lamparter, T., F. Mittmann, W. Gärtner, T. Börner, E. Hartmann, and J. Hughes. 1997. *Proc. Natl. Acad. Sci. USA* 94, 11792–11797.

Li, L. and J.C. Lagarias. 1992. *J. Biol. Chem.* 267, 19204–19210.

Li, L., J.T. Murphy, and J.C. Lagarias. 1995. *Biochemistry* 34, 7923–7930.

Lindner, I., S.E. Braslavsky, K. Schaffner, and W. Gärtner. 2000. *Angew. Chem. Intl. Ed.* 39, 3269–3271.

MacKenzie, J.M.J., Coleman, R.A., and Pratt, L.H. (1974). A specific reversible intracellular localization of phytochrome as Pfr. *Plant Physiol.* 53, Abstract No. 5.

Maines, M.D. and G.M. Trakshel. 1993. *Arch. Biochem. Biophys.* 300, 320–326.

Mancinelli, A.L. 1986. *Plant Physiol.* 82, 956–961.

Mathews, S., M. Lavin, and R.A. Sharrock. 1995. *Ann. Miss. Bot. Gard.* 82, 296–321.

Mathews, S. and R.A. Sharrock. 1997. *Plant Cell Environ.* 20, 666–671.

Matsushita, T., N. Mochizuki, and A. Nagatani. 2003. *Nature* 424, 571–574.

McCormac, A.C., A. Fischer, A.M. Kumar, D. Soll, and M.J. Terry. 2001. *Plant J.* 25, 549–561.

McDowell, M.T. and J.C. Lagarias. 2001. *Plant Physiol.* 126, 1546–1554.

McDowell, M.T. and J.C. Lagarias. 2002. In *Heme, Chlorophyll and Bilins, Methods and Protocols* (ed. A.G. Smith and M. Witty), pp. 293–309. Humana Press, Totowa, NJ.

McMichael, R.W. and J.C. Lagarias. 1990. In *Current Topics in Plant Biochemistry and Physiology* (ed. D.D.a.B. Randall, D.G.), pp. 259–270. University of Missouri–Columbia.

Mizutani, Y., S. Tokutomi, and T. Kitagawa. 1994. *Biochemistry* 33, 153–158.

Mohr, H. and H. Oelze-Karow. 1982. *Plant Physiol.* 70, 863–866.

Montgomery, B.L. and J.C. Lagarias. 2002. *Trends Plant Sci.* 7, 357–366.

Morand, L.Z., D.G. Kidd, and J.C. Lagarias. 1993. *Plant Physiol.* 101, 97–103.

Mösinger, E., A. Batschauer, E. Schäfer, and K. Apel. 1985. *Eur. J. Biochem.* 147, 137–142.

Muramoto, T., T. Kohchi, A. Yokota, I.H. Hwang, and H.M. Goodman. 1999. *Plant Cell* 11, 335–347.

Muramoto, T., N. Tsurui, M.J. Terry, A. Yokota, and T. Kohchi. 2002. *Plant Physiol.* 130, 1958–1966.

Murphy, J.T. and J.C. Lagarias. 1997. *Curr. Biol.* 7, 870–876.

Nagy, F. and E. Schäfer. 2002. *Annu. Rev. Plant Biol.* 53,329–55., 329–355.

Nakasako, M., M. Wada, S. Tokutomi, K.T. Yamamoto, J. Sakai, M. Kataoka, F. Tokunaga, and M. Furuya. 1990. *Photochem. Photobiol.* 52, 3–12.

Neff, M.M., C. Fankhauser, and J. Chory. 2000. *Gene Dev.* 14, 257–271.

Nozue, K., T. Kanegae, T. Imaizumi, S. Fukuda, H. Okamoto, K.C. Yeh, J.C. Lagarias, and M. Wada. 1998. *Proc. Natl. Acad. Sci. USA* 95, 15826–15830.

Ortiz de Montellano, P.R. and K. Auclair. 2003. In *The Porphyrin Handbook. The Iron and Cobalt Pigments, Biosynthesis, Structure and Degradation.* (ed. K.M. Kadish, K.M. Smith, and R. Guilard), pp. 183–210. Academic Press, New York City.

Papenbrock, J. and B. Grimm. 2001. *Planta* 213, 667–681.

Park, C.M., J.I. Kim, S.S. Yang, J.G. Kang, J.H. Kang, J.Y. Shim, Y.H. Chung, Y.M. Park, and P.S. Song. 2000a. *Biochemistry* 39, 10840–10847.

Park, C.M., J.Y. Shim, S.S. Yang, J.G. Kang, J.I. Kim, Z. Luka, and P.S. Song. 2000b. *Biochemistry* 39, 6349–6356.

Parkinson, J.S. and E.C. Kofoid. 1992. *Ann. Rev. Genet.* 26, 71–112.

Parks, B.M. and P.H. Quail. 1991. *Plant Cell* 3, 1177–1186.

Ponting, C.P. and L. Aravind. 1997. *Curr. Biol.* 7, R674–R677.

Quail, P.H. 1991. *Ann. Rev. Genet.* 25, 389–409.

Quail, P.H. 1997. *Plant Cell Environ.* 20, 657–665.

Quail, P.H., M.T. Boylan, B.M. Parks, T.W. Short, Y. Xu, and D. Wagner. 1995. *Science* 268, 675–680.

Quail, P.H., E. Schäfer, and D. Marme. 1973. *Plant Physiol.* 52, 128–131.

Ratliff, M., W.M. Zhu, R. Deshmukh, A. Wilks, and I. Stojiljkovic. 2001. *J. Bacteriol.* 183, 6394–6403.

Reinbothe, S., C. Reinbothe, K. Apel, and N. Lebedev. 1996. *Cell* 86, 703–705.

Remberg, A., I. Lindner, T. Lamparter, J. Hughes, C. Kneip, P. Hildebrandt, S.E. Braslavsky, W. Gärtner, and K. Schaffner. 1997. *Biochemistry* 36, 13389–13395.

Remberg, A., P. Schmidt, S.E. Braslavsky, W. Gärtner, and K. Schaffner. 1999. *Eur. J. Biochem.* 266, 201–208.

Rüdiger, W. 1992. *Photochem. Photobiol.* 56, 803–809.

Rüdiger, W., T. Brandlmeier, I. Blos, A. Gossauer, and J.–P. Weller. 1980. *Z.Naturforsch.* 35, 763–769.

Rüdiger, W. and D.L. Correll. 1969. *Liebigs.Ann.Chem.* 723, 208–212.

Rüdiger, W., F. Thümmler , E. Cmiel, and S. Schneider. 1983. *Proc. Natl. Acad. Sci. USA* 80, 6244–6248.

Sage, L.C. 1992. *Pigment of the Imagination, A History of Phytochrome Research.* Academic Press, Inc., San Diego.

Schacter, B.A., E.B. Nelson, H.S. Marver, and B.S.S. Masters. 1972. *J. Biol. Chem.* 247, 3601–3607.

Schluchter, W.M. and A.N. Glazer. 1997. *J. Biol. Chem.* 272, 13562–13569.

Schluchter, W.M. and A.N. Glazer. 1999. In *The Photosynthetic Prokaryotes.* (ed. G.A. Peschek, W. Loffelhardt, and G. Schmetterer), pp. 83–95. Kluwer Academic/Plenum Press, New York.

Schmitt, M.P. 1997. *J. Bacteriol.* 179, 838–845.
Schneider-Poetsch, H.A., B. Braun, S. Marx, and A. Schaumburg. 1991. *FEBS Lett.* 281, 245–249.
Sharrock, R.A. and T. Clack. 2002. *Plant Physiol.* 130, 442–456.
Shinomura, T., A. Nagatani, H. Hanzawa, M. Kubota, M. Watanabe, and M. Furuya. 1996. *Proc. Natl. Acad. Sci. USA* 93, 8129–8133.
Shinomura, T., K. Uchida, and M. Furuya. 2000. *Plant Physiol.* 122, 147–156.
Siegelman, H.W., B.C. Turner, and S.B. Hendricks. 1966. *Plant Physiol.* 41, 1289–1292.
Sineshchekov, V., J. Hughes, E. Hartmann, and T. Lamparter. 1998. *Photochem. Photobiol.* 67, 263–267.
Sineshchekov, V., L. Koppel, B. Esteban, J. Hughes, and T. Lamparter. 2002. *J. Photochem. Photobiol. B.* 67, 39–50.
Somers, D.E. and P.H. Quail. 1995a. *Plant Physiol.* 107, 523–534.
Somers, D.E. and P.H. Quail. 1995b. *Plant J.* 7, 413–427.
Song, P.S. 1999. *J. Biochem. Mol. Biol.* 32, 215–225.
Song, P.S., M.H. Park, and M. Furuya. 1997. *Plant Cell Environ.* 20, 707–712.
Speth, V., V. Otto, and E. Schäfer. 1987. *Planta* 171, 332–338.
Stock, A.M., V.L. Robinson, and P.N. Goudreau. 2000. *Ann. Rev. Biochem.* 69, 183–215.
Taylor, B.L. and I.B. Zhulin. 1999. *Microbiol. Mol. Biol. Rev.* 63, 479–506.
Tenhunen, R., H.S. Marver, and R. Schmid. 1968. *Proc. Natl. Acad. Sci. USA* 61, 748–755.
Tenhunen, R., H.S. Marver, and R. Schmid. 1969. *J. Biol. Chem.* 244, 6388–6393.
Terry, M.J. 1997. *Plant Cell Environ.* 20, 740–745.
Terry, M.J. and J.C. Lagarias. 1991. *J. Biol. Chem.* 266, 22215–22221.
Terry, M.J., P.J. Linley, and T. Kohchi. 2002. *Biochem. Soc. Trans.* 30, 604–609.
Terry, M.J., M.D. Maines, and J.C. Lagarias. 1993a. *J. Biol. Chem.* 268, 26099–26106.
Terry, M.J., M.T. McDowell, and J.C. Lagarias. 1995. *J. Biol. Chem.* 270, 11111–11118.
Terry, M.J., J.A. Wahleithner, and J.C. Lagarias. 1993b. *Arch. Biochem. Biophys.* 306, 1–15.
Thümmler, F., M. Dufner, P. Kreisl, and P. Dittrich. 1992. *Plant Mol. Biol.* 20, 1003–1017.
Troxler, R.F., A.S. Brown, and S.B. Brown. 1979. *J. Biol. Chem.* 254, 3411–3418.
Vavilin, D.V. and W.F. Vermaas. 2002. *Physiol. Plant.* 115, 9–24.
Vierstra, R.D. 1993. *Plant Physiol.* 103, 679–684.
Vierstra, R.D. and S.J. Davis. 2000. *Sem. Cell Dev. Biol.* 11, 511–521.
Vierstra, R.D. and P.H. Quail. 1982. *Planta* 156, 158–165.
Vierstra, R.D. and P.H. Quail. 1986. In *Photomorphogenesis in Plants* (ed. R.E. Kendrick and G.H.M. Kronenberg), pp. 35–60. Martinus Nijhoff, Dordrecht.
Wada, M., T. Kanegae, K. Nozue, and S. Fukuda. 1997. *Plant Cell Environ.* 20, 685–690.
Wagner, D., C.D. Fairchild, R.M. Kuhn, and P.H. Quail. 1996. *Proc. Natl. Acad. Sci. USA* 93, 4011–4015.
Wahleithner, J.A., L. Li, and J.C. Lagarias. 1991. *Proc. Natl. Acad. Sci. USA* 88, 10387–10391.
Weller, J.P. and A. Gossauer. 1980. *Chem. Ber.* 113, 1603–1611.
Wendler, J., A.R. Holzwarth, S.E. Braslavsky, and K. Schaffner. 1984. *Biochim. Biophys. Acta* 786, 213–221.
Whitelam, G.C., S. Patel, and P.F. Devlin. 1998. *Phil. Trans. Royal Soc. London Series B, Biological Sciences* 353, 1445–1453.
Wilde, A., B. Fiedler, and T. Börner. 2002. *Mol. Microbiol.* 44, 981–988.
Wilks, A. 2002. *Antioxid Redox Signal* 4, 603–14.
Wilks, A. and M.P. Schmitt. 1998. *J. Biol. Chem.* 273, 837–841.
Willows, R.D. 2003. *Nat Prod Rep* 20, 327–41.
Wong, Y.S. and J.C. Lagarias. 1989. *Proc. Natl. Acad. Sci. USA* 86, 3469–3473.
Wong, Y.S., R.W. McMichael, and J.C. Lagarias. 1989. *Plant Physiol.* 91, 709–718.
Wu, S.H. and J.C. Lagarias. 2000. *Biochemistry* 39, 13487–13495.
Wu, S.H., M.T. McDowell, and J.C. Lagarias. 1997. *J. Biol. Chem.* 272, 25700–25705.
Yamamoto, K.T. 1990. *Botan Mag* 103, 469–491.
Yeh, K.C. and J.C. Lagarias. 1998. *Proc. Natl. Acad. Sci. USA* 95, 13976–13981.
Yeh, K.C., S.H. Wu, J.T. Murphy, and J.C. Lagarias. 1997. *Science* 277, 1505–1508.
Zhu, W.M., A. Wilks, and I. Stojiljkovic. 2000. *J. Bacteriol.* 182, 6783–6790

7
Phytochrome Signaling

Enamul Huq and Peter H. Quail

7.1
Introduction

Perception, interpretation, and transduction of environmental light signals are critical for proper growth and development of higher plants throughout the life cycle. In the absence of light, young seedlings follow an etiolated growth program, termed skotomorphogenesis, characterized by long hypocotyls, small, appressed cotyledons, and unopened hooks. Light induces a dramatic switch to a de-etiolated growth program, termed photomorphogenesis, which is characterized by short hypocotyls, open hypocotyl hooks, and expanded, green cotyledons. Various facets of juvenile and adult plant development, including sculpturing of vegetational architecture and floral induction, are also regulated by light signals. To implement such modulation of growth and development in response to changes in ambient light conditions, plants deploy at least three sets of sensory photoreceptors: cryptochromes (crys) and phototropins to track the UV-A/blue region of the spectrum, an unidentified UV-B receptor, and phytochromes (phys) to track the red(R)/far-red(FR) region of the spectrum (Kendrick and Kronenberg, 1994; Smith, 2000; Quail, 2002a; Imaizumi et al., 2003). The diversity among these photoreceptors provides the capacity to track a number of parameters of the light environment, including the presence, absence, wavelength, intensity, direction and duration of the ambient light signals.

The phys are encoded by small multigene families in higher plants, consisting of five members, designated *PHYA* to *PHYE* in the dicot Arabidopsis, and three members, *PHYA*, *PHYB* and *PHYC* in the monocot rice (Mathews and Sharrock, 1997; Goff et al., 2002). This difference appears to represent a conserved divergence in the number of phys between monocot and dicot plants. The phy molecule is a soluble chromoprotein consisting of a linear tetrapyrrole chromophore covalently linked to a 125-kDa polypeptide. The molecule folds into two major domains: an amino-terminal domain containing the bilin chromophore responsible for sensing light signals, and the carboxy-terminal domain responsible for dimerization, and thought to function as a regulatory domain. phys exist as two reversibly-interconvertible, spectrally distinct forms: a red light absorbing Pr form and a far-red light absorbing Pfr form.

Handbook of Photosensory Receptors. Edited by W. R. Briggs, J. L. Spudich

ISBN 3-527-31019-3

The Pr form can be converted to the Pfr form by exposing it to red light, and subsequently, the Pfr form can be converted back to the Pr form by exposing it to far-red light. This inter-conversion is critical for biological function (Smith, 2000). The Pfr form is considered to be the biologically active form, as the newly synthesized Pr form is unable to induce photomorphogenesis in the absence of light activation. Light-induced Pfr formation triggers an intracellular signaling process that culminates in the altered gene expression which drives the overt growth and developmental changes observed in response to light (Figure 7.1 a). The phy system has been shown to control a variety of responses throughout a plant's life cycle, including seed germination, seedling de-etiolation, shoot and leaf development, shade avoidance, and flowering (Kendrick and Kronenberg, 1994). Intense research efforts have been devoted, in recent years, to defining the signaling and transcriptional networks that link the activated photoreceptor molecule to these biological responses.

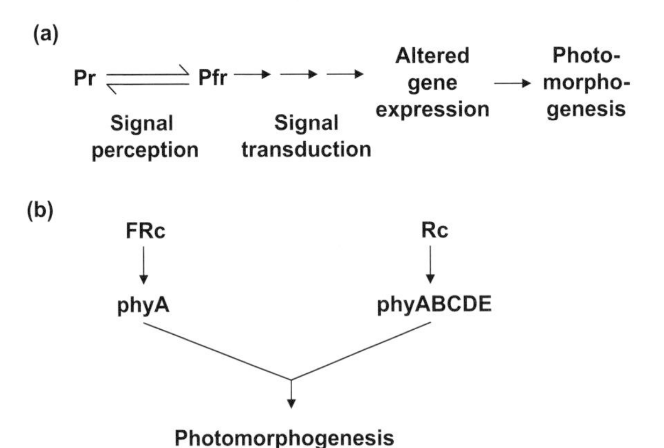

Figure 7.1 phy signaling pathway. (a) Simplified scheme for phy photoperception and signal transduction. Light-induced Pfr formation (signal perception) triggers an intracellular signal transduction process that culminates in altered expression of the genes that drive photomorphogenesis. (b) phy family members have differential photosensory specificity in response to continuous red (Rc) and continuous far-red (FRc) light.

7.2 Photosensory and Biological Functions of Individual Phytochromes

Since the discovery of five phys in the early 1990s in Arabidopsis, one of the challenges in the field was to define the photosensory specificity and biological functions of individual family members. With the advent of both genetic and reverse-genetic approaches, isolation and characterization of mutants for all five phys have recently been completed. Analysis of single and/or multiple mutants has provided important insight into their photosensory specificities and biological functions (Table 7.1). Although Table 7.1 shows that all five phys function in red light, individual phys have specific as well as overlapping photosensory and/or biological functions (Figure 7.1

b) (Quail et al., 1995, 2002a; Wang and Deng, 2002; Monte et al., 2003; Franklin et al., 2003). Specifically, phyA is the exclusive photoreceptor for continuous, monochromatic far-red light (FRc)-induced control of seedling de-etiolation, whereas phyB and phyC are involved in red light-induced seedling de-etiolation and control of flowering time. phyD and phyE contribute to these responses under red light in the absence of phyB, and phyE has been shown to control seed germination, internode elongation and flowering time (Devlin et al., 1998; Hennig et al., 1997; Aukerman et al., 1997).

Table 7.1 Photosensory specificity and biological functions of individual phy family members.

phy	Photosensory specificity	Biological functions	References
phyA	FR, R	seed germination hypocotyl elongation cotyledon expansion control of flowering time shade avoidance	Parks and Quail., 1993; Whitelam et al., 1993; Nagatani et al., 1993
phyB	R	seed germination hypocotyl elongation cotyledon expansion control of flowering time shade avoidance	Koorneef et al., 1980; Reed et al., 1993
phyC	R	hypocotyl elongation cotyledon expansion control of flowering time	Monte et al., 2003; Franklin et al., 2003
phyD	R	hypocotyl elongation cotyledon expansion	Aukerman et al., 1997
phyE	R	seed germination internode and petiole elongation	Devlin et al., 1998; Hennig et al., 2002

One of the intriguing findings from recent characterization of *phyC* mutants is that the phyC protein is unstable in the *phyB* mutant background (Monte et al., 2003; Hirshfield et al., 1999). Therefore, the phenotypes described for *phyB* monogenic mutants may in fact inevitably represent the phenotypes of the *phyB* single mutant with reduced levels of phyC. These results demonstrate that either phyB signaling or phyB protein is required for the stability of phyC, suggesting complex cross talk among the photoreceptors themselves.

7.3
phy Domains Involved in Signaling

Biochemical as well as spectroscopic data suggested that the phy polypeptide is folded into two major domains separated by a hinge region (Quail, 1997a). The N-terminal domain folds into a globular conformation cradling the bilin chromophore, and the C-terminal domain assumes a more extended conformation. Early domain-swap experiments using chimeric molecules between phyA and phyB in transgenic plants provided evidence that the N-terminal globular domain is responsible for the photosensory specificity, while the C-terminal domain, apart from being responsible for dimerization, was proposed to execute signal transduction through direct binding with phy interacting factors (Quail et al., 1995; Quail, 1997a). Moreover, intragenic revertants of phyA and phyB overexpression phenotypes in transgenic plants have been used to map the domains proposed to be responsible for signal transduction. Strikingly, the majority of the missense mutants identified in these studies mapped to a discrete "core" segment of the phy molecules (amino-acid positions 681–838) in the proximal region of the C-terminus. These missense mutants showed no perturbation of spectral integrity, indicating normal folding of the molecules. However, their regulatory activity was reduced *in vivo*. These data were interpreted to support the conclusion that this C-terminal region is responsible for binding to signaling partners. These data also led to considerable effort by a number of laboratories to identify phy interacting factors, mainly by Yeast-two-hybrid screening, using the C-terminal domain of the phy molecules (see below). Molecular genetic and biochemical approaches provided evidence that the some of the factors identified in this way appear to function in phy signaling, supporting the conclusion that the C-terminal domain of phys is involved in signal transduction (Quail, 2000).

However, more recent results have dramatically changed our view about the phy domains involved in signaling. Matsushita et al. (2003) showed that the N-terminal domain of phyB is sufficient to induce photomorphogenesis in response to red light when dimerized via a GUS-fusion at the C-terminus and targeted to the nucleus. Moreover, missense mutants in the abovementioned C-terminal "core region" (position 681–838) of phyA and phyB previously thought to be involved in signal transduction have been shown to be either deficient in light-dependent nuclear translocation ($phyB^{G767R}$) (Matsushita et al. 2003), or defective in nuclear speckle formation ($phyA^{A776V}$, $phyA^{E777K}$, $phyB^{G788E}$, $phyB^{E838K}$) (Kircher et al., 2002; Yanovsky et al., 2002). Since, nuclear translocation of phyB is necessary for its biological function (Huq et al., 2003), these data suggest that the major function of the C-terminal domain may be to provide dimerization and nuclear localization with a minor role in target binding. These data also provide evidence that the N-terminal domain is sufficient for both functionally important selection and interaction with signaling partners. Such interaction has been shown in the case of PIF3, one of the first phy-interacting factors identified. PIF3 has higher affinity for the N-terminus of phyB in a conformer-specific manner than for the C-terminal domain of phyB in a conformer-independent manner (Ni et al., 1999). Taken together, these data suggest that the N-terminal domain of the phys is responsible for target selection and conformer-specific interaction, and that the C-terminal domain is responsible for dimerization, nuclear localization and modulation of the interaction with signaling intermediates.

Apart from the photosensory and regulatory activities of the different phy domains, recent results have shown that certain regions of the molecule are involved in accumulation in nuclear bodies (often called speckles) in response to light (see Schäfer and Nagy, Chapter 9). However, the biological significance of speckle formation is still not clear. Certain nuclear speckles have been implicated in splicing, because they are enriched in splicing factors, transcription factors and RNA (Lamond and Spector, 2003). But the components of the phy speckles in the nucleus and their role in phy signaling have not yet been reported. Recently, Bauer et al. (2004) showed that PIF3 is transiently associated with phyB in so-called "early speckles" (formed within minutes), but not with "late speckles" (formed over hours). In addition, missense mutants in the core region of both phyA and phyB have been shown to be defective in speckle formation suggesting that nuclear speckles are important for phy signaling (see above) (Kircher et al., 2002; Yanovsky et al., 2002). Chen et al. (2003) have also identified intragenic determinants of phyB:GFP necessary for localization to nuclear speckles, and these missense mutations map to various regions of the phyB molecule suggesting that no discrete domain of the phy molecule is exclusively responsible for speckle formation. On the other hand, Matsushita et al. (2003) showed that the N-terminal domain of phyB, when dimerized and localized in the nucleus, was sufficient for signaling without nuclear speckle formation, suggesting that the nuclear speckles are not involved in phyB signaling. Clarification of the apparent discrepancies between these various studies must await further investigation. Chen et al. (2003) also reported isolation of extragenic *dsf* (defective speckle formation) mutants. Characterization of *dsf* mutants will provide more insight into the mechanism of nuclear speckle formation and their biological significance in phy signaling.

7.4 phy Signaling Components

7.4.1 Second Messenger Hypothesis

Prior to the mid-1990s, accumulated immunocytochemical and cell fractionation data had led to the widely accepted view that the phys were synthesized as Pr in the cytosol and remained there after light-induced Pfr formation. In the absence of any evidence of phy nuclear translocation, these earlier studies focused on identifying second messengers that might function in transducing phy signals from the cytosol to the nucleus for regulating gene expression. Pharmacological and microinjection approaches implicated G-protein, cGMP and Ca^{2+}/Calmodulin in phy signaling (Bowler et al., 1994; Neuhaus et al., 1997). Although a Ca^{2+}-binding protein, SUB1, has been shown to be involved in both phy and cry signaling pathways (Guo et al., 2001), no mutants in the putative cGMP or Ca^{2+}/Calmodulin pathway have been isolated as photomorphogenic mutants. Reverse genetic analyses of the wild-type and gain-of-function mutant versions of the alpha subunit of heterotrimeric G-protein were interpreted to show that phy signaling might involve heterotrimeric G-proteins (Okamota et al., 2001). However, recent studies using both loss-of-function and gain-of-function mutants showed a lack of involvement of G-protein in phy signaling (Jones et al., 2003). Moreover, using a glucocorticoid receptor-based fusion-protein system, Huq et al. (2003) showed that nuclear translocation and photoconversion combined are necessary and sufficient for the biological function of phyB, suggesting lack of a cytosolic pathway in phyB signal transduction. Although, cytosolic phy signaling has not been ruled out, there are currently no definitive data supporting this possibility. Consequently, a major focus at present is to understand how phys control gene expression in the nucleus.

7.4.2
Genetically Identified Signaling Components

7.4.2.1 Far-red Light Signaling Mutants

Since phyA and phyB were found to display differential photosensory specificity (Figure 7.1b) (Quail et al., 1995), efforts have been devoted to identifying signaling components specifically in the phyA (continuous far-red light) or phyB (continuous red light) signaling pathways. Such components might be expected to function early in the pathway. A number of putative signaling mutants have been isolated, primarily using forward genetic screening under continuous far-red light (Figure 7.2). Many of these have been shown to function as positively acting components (FAR1, HFR1, FHY1, FHY3, LAF1, LAF3, LAF6, PAT1, PAT3, *fin2*, *fin5*, FIN219 and AFR) (Soh et al., 1998; Quail, 2002a,b; Zeidler et al., 2001; Cho et al., 2003; Hare et al., 2003), whereas others perform as negatively acting components (SPA1, SPA3, SPA4, EID1, SUB1, PKS1 and PKS2) (Quail, 2002a,b; Lariguet et al., 2003; Laubinger and Hoecker, 2003). Of those that have been examined, the majority are wholly or partially nuclear localized (FAR1, HFR1, FHY3, LAF1, EID1, SPA1, FHY1, PAT3 and NDPK2), although others are localized in the cytosol (PAT1, LAF6, FIN219, SUB1 and PKS1). The nuclear localization of these factors along with light-dependent nuclear localization of phys suggests that early events in phy signaling occur in the nucleus (Quail, 2002 a,b).

While, the primary sequence of the latter two groups of these factors does not provide any indication about their biochemical mechanism of action, the nuclear-localized factors fall into two major classes of implied function: transcriptional regulation (FAR1, HFR1, FHY3 and LAF1) and protein degradation (SPA1-4 and EID1). Transcriptional activation activity of two of these factors has been shown either in yeast assays or in plants (FAR1, LAF1) (Ballesteros et al., 2001; Hudson et al., 2003). The biochemical role of EID1 is still under investigation. However, SPA1, SPA3 and SPA4 have been shown to interact with COP1 directly (Laubinger, and Hoecker, 2003; Hoecker and Quail, 2001). Moreover, SPA1 stimulates auto- and trans-ubiquitylation of COP1, facilitating the degradation of remaining LAF1 to downregulate photomorphogenesis in the dark (Seo et al., 2003) (see below). These results also suggest that an important regulatory event in phy signaling occurs in the nucleus.

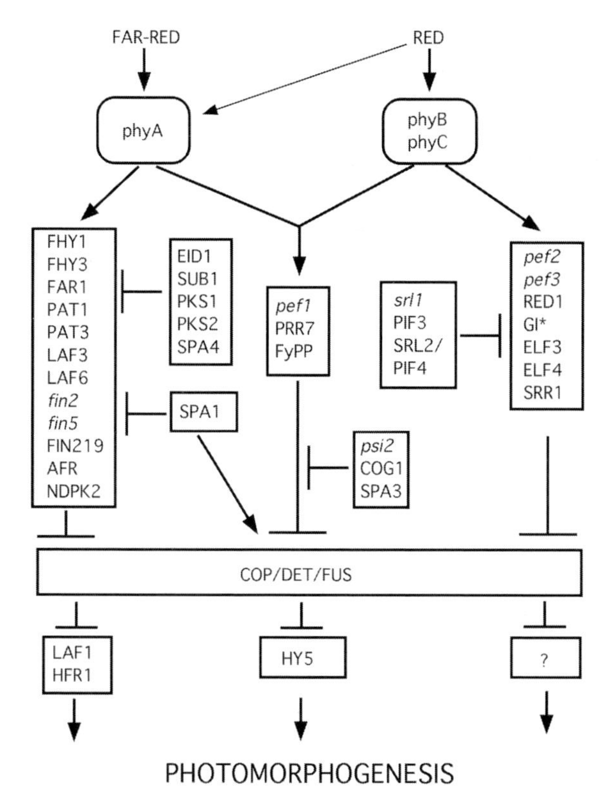

Figure 7.2 Simplified schematic overview of the phy signaling pathways based on genetically identified signaling components. Arrows indicate positive function/response, and the T-bars indicate negative function.

7.4.2.2 **Red Light Signaling Mutants**

Mutant loci displaying either long or short hypocotyls in de-etiolating seedlings compared to the wild-type have been identified by screening under continuous red light (Figure 7.2). Subsequent genetic and photobiological experiments were used to confirm their red-light specificity. This pathway, predominantly mediated by phyB and phyC, has been less investigated than the phyA pathway. However, as with the phyA pathway, both positively acting (*pef2*, *pef3*, RED1, GI, SRR1, ELF3 and ELF4), and negatively acting (*srl1*, SRL2/PIF4 and PIF3) components have also been identified using genetic and reverse genetic approaches (Quail, 2002a; Huq et al., 2000; Staiger et al., 2003; Kim et al., 2003; Hoecker et al., 2004; Khanna et al., 2004). All the molecularly characterized factors are localized in the nucleus (GI, SRR1, ELF3, ELF4, PIF3 and PIF4). The positively acting components (GI, SRR1, ELF3 and ELF4) are all novel proteins, and their mechanism of action is still unclear. However, the negatively acting components PIF3 and PIF4 are members of the well-characterized basic helix-loop-helix (bHLH) transcription-factor family, suggesting that they might function as transcriptional regulators. Moreover, PIF3 and PIF4 interact with each other, and therefore, potentially function as both homodimers and heterodimers (Toledo-Ortiz et al., 2003).

7.4.2.3 **Red/far-red Light Signaling Mutants**

Mutant loci defective in both red and far-red light signaling pathways have been identified using genetic approaches, supporting the hypothesis that common components might be shared by phyA and phyB signaling pathways (Figure 7.2). As found for the apparently separate red and far-red light pathways, both positively acting (*pef1* and PRR7) (Ahmad et al., 1996; Kaczarosky and Quail, 2003) and negatively acting (COG1, SPA3 and *psi2*) (Genoud et al., 1998; Park et al., 2003; Laubinger and Hoecker, 2003) components have been identified in this shared pathway. Molecular characterization of only one of these mutants, *prr7* has been reported. The data show that PRR7 is involved in both phyA and phyB signaling pathways as well as control of circadian rhythms. In addition, members of the PRR7 family have been shown to interact with bHLH factors involved in light signaling (Matsushika et al., 2002), suggesting a dual role of these proteins in both phy signaling and the circadian clock. Although, the genetic screening under red/far-red light conditions does not appear to be saturated, the existence of shared signaling intermediates among phys has also been supported by the identification of phy-interacting factors (PIF3, PKS1, NDPK2, FyPP) that interact with both phyA and phyB (see below).

7.4.2.4 **COP1/CSN**

Evidence for modulation of light signaling has been provided by isolation and characterization of a large group of pleiotropic mutants (*cop/det/fus*) that showed photomorphogenic phenotypes in the dark. These mutants show that the encoded proteins function as general repressors of photomorphogenesis as well as other pathways (Serino and Deng, 2003). One mutant among this group, *cop1*, encodes a ring finger protein with WD40 repeats, a structure similar to a certain class of E3 ubiquitin ligases (Deng et al., 1992; Serino and Deng, 2003; Saijo et al., 2003). The other mutants encode components of the COP9 signalosome complex (CSN), which is similar to the lid subunit of the 26S proteasome responsible for ubiquitin-mediated protein degradation. COP1 has been shown to have E3 ubiquitin-ligase activity, which can ubiquitylate itself as well as other interacting proteins, such as HY5, LAF1 and possibly others (Osterland et al., 2002; Seo et al., 2003; Saijo et al., 2003; Holm et al., 2002). This ubiquitylation provides a tag on the protein presumably enabling it to be recognized and degraded by the CSN proteasome complex. Strikingly, the coiled-coil domain of SPA1, a negative regulator of phyA signaling, has been shown to stimulate the E3 activity of COP1 in ubiquitylation of LAF1 (Seo et al., 2003). However, full-length SPA1 inhibited the E3 activity of COP1 to ubiquitylate HY5 (Saijo et al., 2003). These conflicting results might be from the use of a different SPA1 molecule in the two studies (full-length versus truncated protein) or due to differences in the conditions used in these assays. Although, these data need to be reconciled, they provide initial evidence for a mechanism by which COP1 and SPA1 function as repressors of phy signaling.

How are these activities controlled to induce photomorphogenesis? It turns out that light regulates subcellular distribution of the key regulatory protein, COP1. In darkness, COP1 is localized in the nucleus. COP1-mediated ubiquitylation of LAF1, HY5 and other COP1 interacting-proteins presumably provides a tag for these proteins to be recognized and degraded by the CSN complex-associated proteasome pathway in the dark. In prolonged light, COP1 is depleted from the nucleus using a nuclear exclusion mechanism that allows these target proteins to accumulate and induce photomorphogenesis (Subramanian et al., 2004). In addition to COP1's degradative role in darkness, it has been found to be required for accumulation of PIF3 in the dark (Bauer et al., 2003). Therefore, COP1 in conjunction with the CSN complex functions as a critical modulator that controls the level of these downstream transcription factors required to induce photomorphogenesis.

COP1 not only controls the transcription-factor levels in the dark, but also controls photoreceptor levels in the light. phyA, a light-labile phytochrome, has been shown to be degraded in light in part by ubiquitin mediated proteolysis (Vierstra , 1994; Clough et al., 1999). However, the factors responsible for degradation of phyA remained elusive until recently. Now Seo et al. (2004) have shown that COP1 is necessary for the degradation of phyA in continuous red light (Rc). However, COP10, a putative E2 enzyme, is not involved in the COP1-dependent phyA degradation. These data suggest that COP1 is also involved in the desensitization of incoming light signals by modulating the level of photoreceptors via the 26S proteasome-mediated degradation pathway.

7.4.3
phy-Interacting Factors

Identification of potential phy signaling partners has also been sought using either yeast two-hybrid screening or targeted yeast two-hybrid assays for interacting factors. These approaches have identified a group of unrelated factors that bind directly to either phyA or phyB or both, and the functional involvement of some of these factors in phy signaling has been shown by subsequent reverse genetic analyses (Quail, 2000; 2002ab). If taken at face value, the compiled results from these interaction studies would suggest that: i) there may be cross-talk between phy and cry signaling by direct interaction of phys with CRY1 and CRY2; ii) there may be cross-talk between phy and hormone signaling by direct interaction with IAA proteins; iii) phy signaling is modulated by phy interactors, i.e. PKS1, NDPK2, ARR4, FyPP, PIF1 and PIF4 ; iv) phy signaling involves interaction with circadian controlled factors, i.e. ELF3 and ZTL; and v) phy signaling involves direct interaction with a promoter element-bound transcription factor, PIF3 (Figure 7.3 a,c) (Kim et al., 2002; Quail, 2000; 2002ab; Huq and Quail, unpublished).

Of these various phy interactors, three (PIF1, PIF3, and PIF4) have been shown to have selective interaction with the Pfr form of phyA and/or phyB (Figure 7.3 a). These are all members of the well-characterized basic helix-loop-helix (bHLH) family of

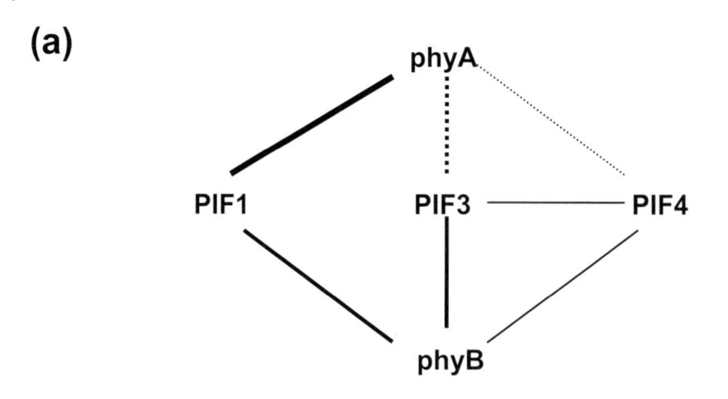
(a)
phyA
PIF1
PIF3
PIF4
phyB

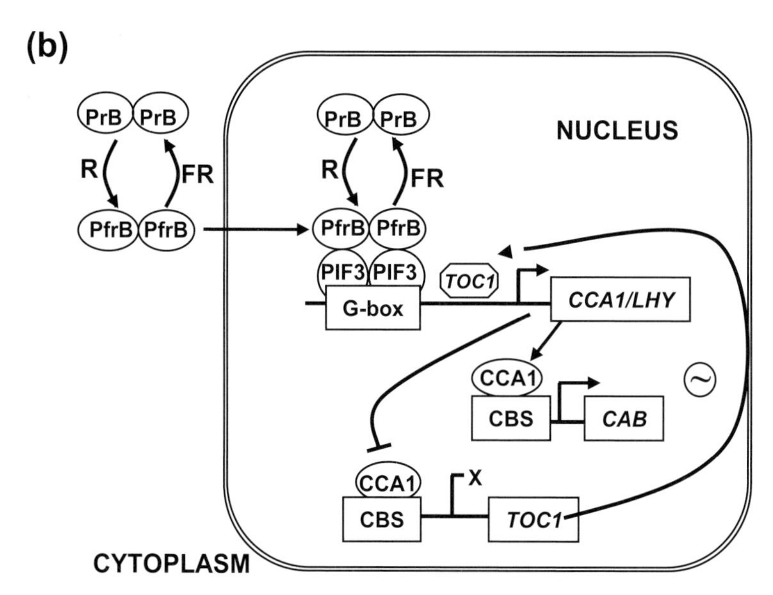
(b)
PrB PrB
R
FR
PfrB PfrB
NUCLEUS
PrB PrB
R
FR
PfrB PfrB
PIF3 PIF3
G-box
TOC1
CCA1/LHY
CCA1
CBS
CAB
CCA1
X
CBS
TOC1
CYTOPLASM

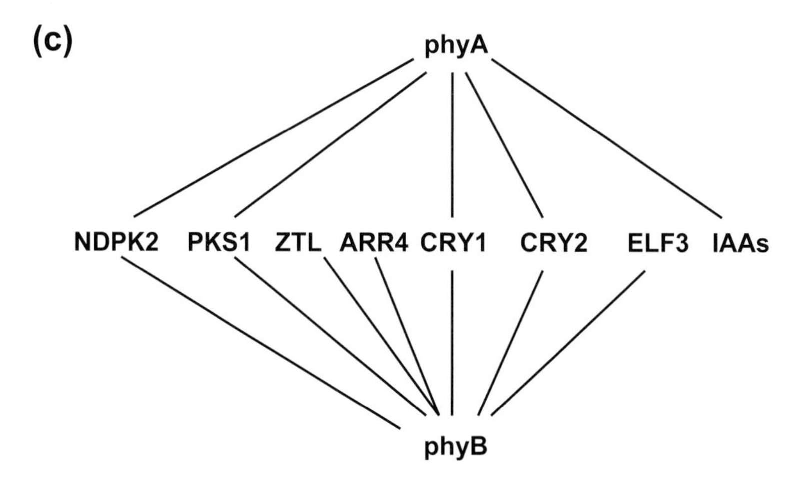
(c)
phyA
NDPK2
PKS1
ZTL
ARR4
CRY1
CRY2
ELF3
IAAs
phyB

transcription factors. Genetic and reverse-genetic analyses have established that they function in distinct pathways of phy signaling. PIF1 has by far the strongest interaction with phyA with more than 10-fold higher affinity for phyA than PIF3 and PIF4. PIF1 and PIF3 have similar affinity for phyB, while PIF4 has half the apparent affinity of PIF1 and PIF3 for phyB (Huq and Quail, 2002; Huq and Quail, unpublished). These results suggest that different bHLH factors have differential affinities for different phys, and may transduce light signals to different genes or the same genes with differential magnitude.

PIF1 has been shown to be a critical regulator of the chlorophyll biosynthetic pathway. In the absence of PIF1, young seedlings accumulate higher levels of the chlorophyll biosynthetic intermediate, protochlorophyllide, in the dark, and this accumulation is phototoxic to plants in light when present in the free form. This excess free protochlorophyllide causes photo-bleaching of dark-grown *pif1* seedlings upon exposure to light. Strikingly, the severity of the bleaching phenotype increases with increasing hypocotyl length of the seedlings before light exposure. *pif1* seedlings with hypocotyl lengths longer than 10 mm showed a strong bleaching phenotype, while wild-type seedlings with similar hypocotyl lengths develop normal green and expanded cotyledons. These results suggest that PIF1 may provide a selective advantage under ecological conditions where seed burial is crucial for seedling survival (Huq and Quail, unpublished).

PIF3, the first member of the bHLH family to be identified as a phy interactor, interacts with the biologically active Pfr form of both phyA and phyB (Ni et al., 1998, 1999; Zhu et al., 2000). It also binds to a G-box DNA sequence motif (CACGTG), a variant of the E-box motif (CANNTG) commonly recognized by bHLH factors. The G-box motif is found in many light regulated promoters (Quail, 2000; Hudson and Quail, 2003). Strikingly, G-box promoter-element bound PIF3 can interact with the Pfr form of phyB, which together with the well-established light-induced translocation of phys to the nucleus, has been interpreted as suggesting a direct pathway for controlling gene expression by phys in response to light (Figure 7.3b) (Quail, 2000; 2002ab; Martinez-Garcia et al., 2000). However, this behavior is in contrast to PIF1

◀ **Figure 7.3** phy-Interacting factors. (a) Molecular interactions between the bHLH factors (PIF1, PIF3 and PIF4) and phyA and phyB in a conformer-dependent manner. The strengths of the interaction are presented in a semi-quantitative way using different thickness connecting lines. Dashed lines indicate minor interactions. (b) Postulated direct targeting of light signals to primary light-response genes through promoter-bound basic helix-loop-helix transcription factor (PIF3) complexed with photoactivated phyB. The model proposes that phyB translocates into the nucleus in response to light as the Pfr form, and binds to the G-box-bound PIF3 to induce primary light-response genes such as *CCA1* and *LHY*. CCA1 and LHY bind to CCA1-binding sites (CBS) to down-regulate *TOC1* expression or activate downstream target genes such as *CAB*. TOC1 functions as a positively acting factor activating expression of *CCA1*, and thereby creating the negative feedback loop of the proposed circadian clock. CAB, chlorophyll a/b binding protein; PIF3, phytochrome-interacting factor 3; TOC1, timing of cab expression 1 protein. (c) Molecular interactions between phys and other putative signaling factors in a marginally conformer-dependent or conformer-independent manner.

and PIF4 which only bind to phyB when not complexed with DNA (Huq and Quail, 2002; Huq and Quail, unpublished).

Previous reverse-genetic analysis with antisense and overexpression lines of *PIF3* suggested that PIF3 is a positive regulator of phy signaling (Ni et al., 1998). This conclusion was supported by the characterization of another Rc-specific hypersensitive mutant, *poc1*, which has a T-DNA insertion in the promoter region of *PIF3* (Halliday et al., 1999). Recent results with loss-of-function alleles of *pif3* suggested that PIF3 functions as a negative regulator of phyB signaling under prolonged Rc irradiations (Kim et al., 2003). This phenotype is similar to another member of this family, PIF4, which has been shown to function as a negative regulator of phyB signaling (Huq and Quail, 2002). In addition, Bauer et al. (2004) showed that PIF3 protein is undetectable in the *poc1* mutant, arguing that *poc1* might in fact be a null allele of PIF3. Taken together, these data strongly suggest that PIF3 is a negative regulator of phyB signaling under prolonged Rc irradiations. By contrast, recent evidence suggests that PIF3 acts positively in early responses, such as greening, following initial exposure of seedlings to light (E. Monte and P.H. Quail, unpublished).

Bauer et al. (2004) also showed using fusion proteins that PIF3 colocalizes with phyA and phyB in response to light in vivo, consistent with previous in vitro physical interactions between these proteins. PIF3 is accumulated constitutively in the nucleus in the dark, but is then degraded rapidly in light with a short half-life of about 10 min. COP1 is required for the accumulation of PIF3 in the dark, which is in contrast to other transcription factors such as HY5 and LAF1 where COP1 is responsible for their degradation in the dark (Seo et al., 2003; Saijo et al., 2003). Light-dependent degradation of PIF3 requires phyA under continuous far-red light, and phyA, phyB, and phyD under continuous red light. Bauer et al. (2004) conclude that PIF3 functions transiently, during early stages of the dark-to-light developmental transition. These data are consistent with the observed role for PIF3 in the early greening process following initial light exposure (E. Monte and P.H. Quail, unpublished].

7.4.4
Early phy-Responsive Genes

Identification and characterization of early phy-responsive genes is expected to provide insight into the initial events in phy signal transduction. Various techniques have been successfully used to this goal. Using a fluorescence differential display technique, Kuno et al. (2000) identified a number of transcripts that are regulated by phys. One of these differentially regulated genes was recently identified and designated as encoding early phytochrome-responsive 1, EPR1, a single MYB-domain protein homologous to early phy-responsive components of the central oscillator genes, CCA1 and LHY. EPR1 has been shown to function as a slave oscillator in circadian regulation (Kuno et al., 2003).

Microarray-based expression profile analyses have also been used to identify genes that are regulated by phys or by phy signaling components (Tepperman et al., 2001; 2004; Ma et al., 2001; Wang et al., 2002a; Hudson et al., 2003). Using this approach, Tepperman et al. (2001) have shown that approximately 10% of the genes represented on an 8000-gene Affymetrix Chip are regulated by light in a phyA-dependent manner. Among the early-response genes, 44% represent members of already established or putative transcription factor genes. One of these early-repressed transcription factors encodes a member of the bHLH family, recently designated PIF3-like (PIL1), which is involved in rapid shade-avoidance responses related to the circadian clock in *Arabidopsis* (Salter et al., 2003; Yamashino et al., 2003). One of the early-induced genes, recently designated as encoding ELF4, a novel protein, was also found to have a role in phyB signaling during seedling de-etiolation (Doyle et al., 2002; Khanna et al., 2003). Interestingly, many of these transcription-factor genes have a G-box in their promoter sequences, suggesting that they might be regulated by phys through interaction with DNA-bound PIF3 or other bHLH family members (Tepperman et al., 2001; Quail, 2002a,b; Hudson and Quail, 2003).

In contrast to the striking exclusive role of phyA under continuous far-red light (FRc), phyB has been shown to play a more minor role in regulating gene expression under Rc conditions (Ma et al., 2001; Tepperman et al., 2004). These authors showed that the majority of Rc-regulated genes continue to be substantially induced/repressed in a *phyB* null mutant under Rc. Surprisingly, only 2% of the Rc-regulated genes are regulated in a robustly phyB-dependent manner, suggesting that the other photoreceptors, such as phyA, phyC, phyD, and/or phyE either separately or additively play a major role in regulating gene expression under Rc (Tepperman et al., 2004). Thus, the molecular phenotype assayed under these conditions is apparently not inconsistent with the morphological phenotype induced by the *phyB* mutant. Rc induces significant photomorphogenic development in the RNA-rich apical zone of these seedlings, visible as partial cotyledon expansion and greening, despite a complete lack of responsiveness in the hypocotyls. These data may reflect potential partial organ-specific function of different phy family members.

Microarray analysis of signaling mutants was also performed using mostly phyA-pathway signaling mutants (Wang et al., 2003; Hudson et al., 2003). However, the results showed that only a limited number of genes are differentially regulated in the mutants compared to wild type, and the differences are quantitatively relatively modest. Expression of a number of genes was also different in the mutant compared to wild-type in darkness. The limited effect on normal light-induced control of gene expression in the signaling mutants suggests that phyA might control multiple parallel signaling pathways to regulate seedling de-etiolation.

Microarray analysis of gene expression under circadian conditions also yielded phy signaling mutants. Harmer et al. (2000) identified over 400 circadian regulated genes using this approach. Reverse genetic analysis of a sub set of circadian regulated F-box containing genes showed that the F box protein AFR is involved in the phyA-mediated far-red light signaling pathway (Harmon and Kay, 2003). Because, phys are involved in the light-dependent entrainment of the circadian clock, it is anticipated that a subset of the cycling genes might also function as components of phy signaling (see below).

7.5
Biochemical Mechanism of Signal Transfer

One of the most challenging questions in the field is to define the biochemical mechanism of signal transfer from photoactivated phys to their immediate signaling partners. phys were postulated some years ago to function as histidine kinases due to their limited homology to a group of bacterial histidine kinases (Schnieder-Posch et al., 1991; Schnieder-Posch and Braun, 1991; Quail, 1997ab). This hypothesis drew more attention after the discovery of a cyanobacterial histidine kinase that showed strong homology to plant phys and was shown to function as a R/FR reversible photoreceptor (Quail, 1997b; Hughes et al., 1997). Although, cyanobacterial phy has indeed been shown to have light-regulated histidine kinase activity (Yeh et al., 1997), no such activity has been detected for plant phys (Boylan and Quail, 1996; Krall and Reed, 2000). Instead, oat phyA preparations have been shown to have Ser/Thr protein kinase activity (Lagarias et al., 1987; Wong et al., 1989; McMichael and Lagarias, 1990). These results were initially supported by limited sequence similarity to various eukaryotic Ser/Thr/Tyr kinases (Lagarias et al., 1987; Thummler et al., 1995). However, none of the canonical motifs conserved in the majority of eukaryotic Ser/Thr/Tyr kinases are found in the eukaryotic phys.

Although, some data were presented suggesting that the biochemical results might be an artifact due to a contaminating kinase in the phyA preparation from oat extracts (Grimm et al., 1989; Kim et al., 1989), recombinant oat phyA purified from two yeast strains exhibited similar activity. These data were interpreted as additional evidence that these phyA preparations showed autophosphorylation of the photoreceptor itself, rather than being due to a contaminant (Yeh and Lagarias, 1998). In addition, these yeast preparations displayed phosphorylation of a phy-interacting factor, PKS1, interpreted as transphosphorylation catalyzed by the photoreceptor (Fankhauser et al., 1999). Both Pr and Pfr forms showed apparent autophosphorylation and transphosphorylation activities, and were slightly light regulated arguing for the proposition that these activities are biologically significant. In addition, a naturally occurring *phyA* allele (*Lm-2* allele) carrying a Met548Thr mutation showed reduced apparent autophosphorylation activity, increased phyA protein stability and reduced responsiveness to far-red light (Maloof et al., 2001). However, the precise role of these apparent auto and transphosphorylation activities in phy signaling is still unknown because of the absence of evidence that a phy molecule deficient in kinase activity lacks functional activity *in vivo*. Moreover, a recent report that the C-terminal domain considered responsible for the kinase activity is apparently completely dispensable for phy activity in vivo, provides evidence against this activity being directly involved in phy signaling (Matsushita et al., 2003). On balance, these conflicting results do not support the conclusion that the Ser/Thr/Tyr kinase activity of phy preparations represents the primary mechanism of phy signal transfer.

7.6 phy Signaling and Circadian Rhythms

Circadian rhythms are daily fluctuations in metabolic, physiological and behavioral activities on a 24-hour basis driven by an endogenous biological clock, which is entrained by environmental cues, such as light and temperature. Plants use different photoreceptors to gather information about the ambient light environment to reset and regulate the circadian clock oscillations (Devlin et al., 2001; Schulz and Kay, 2003; Miller, 2004). Importantly, Somers et al. (1998) have shown that phyA functions under low-intensity red light to control circadian oscillations. phyB is the major photoreceptor for high-intensity red light mediated control, and phyA and cry1 function together to exert blue-light mediated control of the circadian oscillations. phyD and phyE have also been shown to control circadian oscillations under red light (Devlin and Kay, 2003). Therefore, it is not surprising that some of the phy signaling components also function as components of the circadian clock (Figure 7.3 b; Figure 7.4). Circadian clock involvement in light-induced de-etiolation has also been demonstrated (Dowson-Day and Miller, 1999). Since phy signaling is required to initiate circadian-clock oscillations in dark-grown seedlings, modulation of the de-etiolation process by these oscillations is not unexpected. In addition, circadian clock controlled rhythmic expression of members of the photoreceptor family has been shown at the level of mRNA, albeit without apparent oscillations in the protein (McClung, 2001; Miller, 2004).

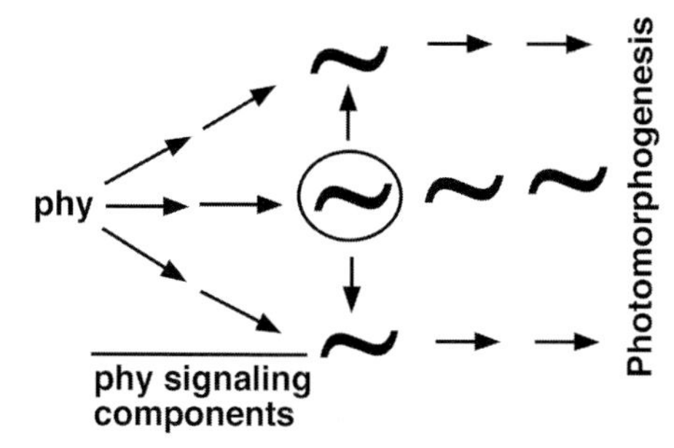

Figure 7.4 phys might control photomorphogensis in a circadian clock-independent and -dependent manner. Schematic overview of phy-mediated light-input pathway to the circadian clock, and phy signaling pathways. Some of the primary phy signaling components have been shown to be regulated by the circadian clock in a rhythmic manner as depicted in the scheme (above and below the clock symbol indicated by the circle in the middle).

Collectively, these results suggest that: i) phy signaling provides the information about the light environment to reset the circadian clock; ii) the clock components control their expression in a negative feedback manner to generate the circadian rhythm; and ii) the output genes in turn control the light input pathway providing rhythmic expression of the photoreceptor genes.

How do phys provide light information to initiate/entrain the circadian clock? The components with overlapping functions in both phy signaling and circadian clock oscillations include CCA1, LHY, TOC1, ELF3, ELF4, GI, SRR1, ZTL/ADO1, PRR7, PIL1, AFR and EPR1 (Quail, 2002 a,b; Doyle et al., 2002; Khanna et al., 2003; Mas et al., 2003; Kaczorowski and Quail, 2003; Somers et al., 2004; Kuno et al., 2003; Harmon and Kay, 2003; Salter et al., 2003; Staiger et al., 2003; Ahmad and Cashmore, 1996). Among these, three factors (CCA1, LHY and TOC1) have been proposed to function as components of the central oscillator (Alabadí et al., 2001). One model postulated direct targeting of light signals to regulate expression of these central oscillator components (*CCA1* and *LHY*) in young un-entrained seedlings possibly through DNA-bound PIF3 complexed with phyB (Martinez-Garcia et al., 2000). However, definitive evidence in favor of this hypothesis is still lacking. It remains to be tested whether *CCA1* and *LHY* expression is down regulated in *pif3* mutant as was the case with the *PIF3* antisense lines as previously reported (Martinez-Garcia et al., 2000).

Alternatively, GI, ELF3, and members of the PRR and ZTL families might function to mediate light signals to the central oscillator, as these components have been shown to function in both light signaling and the circadian clock (Mas et al., 2003; Kaczoroski and Quail, 2003; Somers et al., 2004; Miller, 2004). One member of the ZTL family, FKF1, has been proposed to function as a blue-light photoreceptor for controlling flowering time (Imaizumi et al., 2003). In addition, ZTL and ELF3 have been shown to interact with phyB in a conformer-independent manner, and TOC1 has been shown to interact with PIF3 (Jarrilo et al., 2001; Liu et al., 2001; Makino et al., 2002). However, the biochemical mechanisms by which these factors provide light signals to the circadian clock are still unknown.

7.7 Future Prospects

Although we have learned much in recent years, two of the most central questions in phy research remain unanswered. We still do not have a coherent picture of the molecular mechanism by which the phys induce changes in gene expression, and the biochemical mechanism of signal transfer from the activated photoreceptor to its primary signaling partner(s) remains undefined.

The earlier evidence that photoactivated phy molecules are translocated into the nucleus and are capable of interacting with at least a subset of specific transcription factors, strongly indicated that primary events in phy signaling are localized in the nucleus, and suggested the specific mechanism that phy molecules might act directly at the promoters of target genes to regulate transcription. However, although attention is still firmly focused on the nucleus as the locus of primary signaling events, robust supportive evidence for this specific mechanism of regulation of gene expression has not been forthcoming to date. On the contrary, phy-interacting bHLH factors other than PIF3, such as PIF1, PIF4, and PIF5, do not bind detectably to the phy molecule when DNA bound, despite a clear capacity to interact with the Pfr conformer of the photoreceptor when free in solution (i.e. not DNA bound). For these

PIFs, therefore, there is no evidence that they can recruit phy molecules to target genes. Moreover, the recent evidence that phy interaction with PIF3 in the nucleus of the living cell induces rapid degradation of the transcription factor raises the possibility of an alternative mechanism. These data suggest that the phy molecule may either directly or indirectly program the PIF3 molecule with which it interacts for degradation, potentially via the 26S proteasome system. In this case, the molecular mechanism of phy action would be to trigger the removal of transcriptional regulators from the cell, thereby abrogating their extant activity, and as such would represent indirect regulation of transcription of target genes. This possibility is readily testable.

The hypothesis that plant phys are unorthodox, light-regulated S/T protein kinases that transfer signaling information by transphosphorylation of primary signaling partner(s), has been driven for a number of years by the dual lines of evidence: i) that eukaryotic phy sequences are likely evolutionary descendents of the prokaryotic phys which exhibit biochemically verified, light-regulated histidine-kinase activity, and ii) the presence of S/T kinase activity in plant phy preparations. However, definitive evidence in support of this proposal has also been difficult to obtain. On the contrary, recent evidence that the kinase-like-sequence-containing C-terminal domain of phyB is completely dispensable for phy signaling in vivo argues against phy-catalyzed transphosphorylation as the mechanism of signal transfer. The evidence suggests instead that the C-terminal domain may act negatively on the signaling function which is localized in the N-terminal domain. This activity could involve autophosphorylation of the photoreceptor.

Collectively, current data suggest that a new direction in concepts of phy signaling is emerging. It appears possible that the light-activated phy molecule may act to facilitate tagging of signaling partners, such as PIF3, for degradation via the nuclear-localized ubiquitin-26S proteasome pathway. This action could be direct, through recruitment of components of this degradative machinery to such partners, perhaps manifested as nuclear "speckles", or indirect through facilitating covalent tagging of target proteins for subsequent recognition by the machinery. It may be anticipated that the combined power of functional genomics, in vivo imaging and biochemistry will continue to provide exciting new insights into this problem in the near future.

Acknowledgements

This work was supported by grants from NIH number GM47475, the Torrey Mesa Research Institute, San Diego, DOE Basic Energy Sciences number DE-FG03-87ER13742 and US Department of Agriculture Current Research Information Service number 5335-21000-010-00D to P.H.Q. and a set-up fund from The University of Texas at Austin to E.H.

References

Ahmad, M., A.R. Cashmore, *Plant J.* **1996**, *10*, 1103–1110.

Alabadí, D., T. Oyama, M.J. Yanovsky, F.G. Harmon, P. M s, S.A. Kay, *Science* **2001**, *293*, 880–883.

Aukerman, M.J., M. Hirschfeld, L. Wester, M. Weaver, T. Clack, R.M. Amasino, R. A. Sharrock, *Plant Cell* **1997**, *9*, 1317–1326.

Ballesteros, M., C. Bolle, L.M. Lois, J.M. Moore, J-P. Vielle-Calzada, U. Grossniklaus, N.-H. Chua, *Genes and Dev.* **2001**, *15*, 2613–2625.

Bauer, D., A. Viczian, S. Kircher, T. Nobis, R. Nitschke, T. Kunkel, K.C.S. Panigrahi, E. Adam, E. Fejes, E. Schafer, F. Nagy, *Plant Cell* **2004**, 16, 1433–1445.

Bowler, C., G. Neuhaus, H. Yamagata, N-H. Chua, *Cell* **1994**, *77*, 73–81.

Boylan, M.T., P.H. Quail, *Protoplasma* **1996**, *195*, 59–67.

Chen, M., R. Schwab, J. Chory, *Proc. Natl. Acad. Sci. USA* **2003**, *100*, 14493–14498.

Cho, D-S., S-H. Hong, H-G. Nam, M-S. Soh, *Plant Cell Physiol.* **2003**, *44*, 565–572.

Clough, R.C., E.T. Jordan-Beebe, K.N. Lohman, J.M. Marita, J.M. Walker, C. Gatz, R.D. Vierstra, *Plant J.* **1999**, *17*, 155–167.

Deng, X-W., M. Matsui, N. Wei, D. Wagner, A.M. Chu, K.A. Feldman, P.H. Quail, *Plant Cell* **1992**, 71, 791–801.

Devlin, P.F., S.A. Kay, *Annu. Rev. Physiol.* **2001**, *63*, 677–694.

Devlin, P.F., S.A. Kay, *Plant Cell* **2000**, *12*, 2499–2510.

Devlin, P.F., S.R. Patel, G.C. Whitelam, *Plant Cell* **1998**, *10*, 1479–1487.

Dowson-Day, M.J., A.J. Miller, *Plant J.* **1999**, *17*, 63–71.61.

Doyle, M.R., S.J. Davis, R.M. Bastow, H.G. McWatters, L. Kozma-Bognar, F. Nagy, A.J. Millar, R.M. Amasino, *Nature* **2002**, *419*, 74–77.

Fankhauser, C., K.-C. Yeh, J.C. Lagarias, H. Zhang, T.D. Elich, J. Chory, *Science* **1999**, *284*, 1539–1541.

Franklin, K.A., S.J. Davis, W.M. Stoddart, R.D. Vierstra, G.G. Whitelam, *Plant Cell* **2003**, *15*, 1981–1989.

Genoud, T., A.J. Millar, N. Nishizawa, S.A. Kay, E. Schäfer, A. Nagatani, N-H. Chua, *Plant Cell* **1998**, *10*, 889–904.

Goff *et al. Science* **2002**, *296*, 92–100.

Grim, R., D. Gast, W. Rudiger, *Planta* **1989**, *178*, 199–206.

Guo, H., T. Mockler, H. Duong, C. Lin, *Science* **2001**, *291*, 487–490.

Halliday, K.J., M. Hudson, M. Ni, M.-M. Qin, P.H. Quail, *Proc. Natl. Acad. Sci. USA* **1999**, *96*, 5832–5837.

Hare, P.D., S.G. Moller, L.F. Huang, N-H. Chua, *Plant Physiol.* **2003**, *133*, 1592–1604.

Harmer, S.L., J.B. Hogenesch, M. Straume, H-S. Chang, B. Han, T. Zhu, X. Wang, J.A. Kreps, S.A. Kay, *Science* **2001**, *290*, 2110–2113.

Harmon, F.G., S.A. Kay, *Curr. Biol.* **2003**, *13*, 2091–2096.

Hennig, L., W.M. Stoddart, M. Dieterle, G.C. Whitelam, E. Schafer, *Plant Physiol.* **2002**, *128*, 194–200.

Hirschfeld, M., J.M. Tepperman, T. Clack, P.H. Quail, R.A. Sharrock, *Genetics* **1998**, *149*, 523–535.

Hoecker, U. , P. H. Quail, *J. Biol. Chem.* **2001**, *276*, 38173–38178.

Hoecker, U., G. Toledo-Ortiz, J. Bender, P.H. Quail, *Planta* **2004**, *219*, 195–200

Holm, M., L.-G. Ma, L.-J. Qu, X-W. Deng, *Genes & Dev.* **2002**, *16*, 1247–1259.

Hudson, M.E., D.R. Lisch, P.H. Quail, *Plant J.* **2003**, *34*, 453–471.

Hudson, M.E., P.H. Quail, *Plant Physiol.* **2003**, *133*, 1605–1616.

Hughes, J., T. Lamparter, F. Mittman, E. Hartmann, W. Gartner, A. Wilde, T. Borner, *Nature* **1997**, *386*, 663.

Huq, E., B. Al-Sady, P.H. Quail, *Plant J.* **2003**, *35*, 660–664.

Huq, E., P.H. Quail, *EMBO J.* **2002**, *21*, 2441–2450.

Huq, E., Y. Kang, K.J. Halliday, M-M. Qin, P.H. Quail, *Plant J.* **2000**, *23*, 461–470.

Imaizumi, T., H.G. Tran, T.E. Swartz, W.R. Briggs, S.A. Kay, *Nature* **2003**, *426*, 302–306.

Jarillo, J.A., J. Capel, R-H. Tang, H-Q. Yang, J.M. Alonso, J.R. Ecker, A.R. Cashmore, *Nature* **2001**, *410*, 487–490.

Jones, A.M., J.R. Ecker, J-G. Chen, *Plant Physiol.* **2003**, *131*, 1623–1627.

Kaczorowski, K.A., P.H. Quail, *Plant Cell* **2003**, *15*, 2654–2665.

Kendrick, R.E., G.H.M. Kronenberg, *Photomorphogenesis in Plants*, 2nd ed.. Kluwer, Dordrecht, The Netherlands, **1994**.

Khanna, R., E.A. Kikis, P.H. Quail, *Plant Physiol.* **2003**, *133*, 1–9.

Kim, D-W., J-G. Kang, S-S. Yang, K-S. Chung, P.-S Song, C-M. Park, *Plant Cell* **2002**, *14*, 3043–3056.

Kim, I-S., U. Bai, P-S. Song, *Photochem. Photobiol.* **1989**, *49*, 319–323.

Kim, J., H. Yi, G. Choi, B. Shin, P-S. Song, G. Choi, *Plant Cell* **2003**, *15*, 2399–2407.

Kircher, S., P. Gil, L. Kozma-Bognar, E. Fejes, V. Speth, T. Husselstein-Muller, D. Bauer, E. Adam, E. Schafer, F. Nagy, *Plant Cell* **2002**, *14*, 1541–1555.

Koornneef, M., E. Rolff, C. Spruit, *Z. Pflanzenphysiol.* **1980**, *100*, 147–160.

Krall, L., J.W. Reed, *Proc. Natl. Acad. Sci. USA* **2000**, *97* 8169–8174.

Kuno, N., S.G. Moller, T. Shinomura, X. Xu, N-H. Chua, M. Furuya, *Plant Cell* **2003**, *15*, 2476–2488.

Kuno, N., T. Muramatsu, Hamazato, F. M. Furuya, *Plant Physiol.* **2000**, *122*, 15–24.

Lagarias, J.C., Y-S. Wong, T.R. Berkelman, D.G. Kidd, R.W. McMichael, Jr., Structure–function studies on Avena phytochrome. In *Phytochrome and Photoregulation in Plants* (ed. M. Furuya), pp. 51–62. Academic Press, Tokyo, **1987**.

Lamond, A.I., D.L. Spector, *Nature Rev.|Mol. Cell Biol.* **2003**, *4*, 605–611.

Lariguet, P., H.E. Boccalandro, J.M. Alonso, J.R. Ecker, J. Chory, J.J. Casal, C. Fankhauser, *Plant Cell* **2003**, *15*, 2966–2978.

Laubinger, S., U. Hoecker, *Plant J.* **2003**, *35*, 373–385.

Liu, X.L., M.F. Covington, C. Fankhauser, J. Chory, D.R. Wagner, *Plant Cell* **2001**, *13*, 1293–304.

Ma, L., J. Li, L. Qu, J. Hager, Z. Chen, H. Zhao, X-W. Deng, *Plant Cell* **2001**, *13*, 2589–2607.

Makino, S., A. Matsushika, A. Imamura, M. Kojima, T. Yamashino, T. Mizuno, *Plant Cell Physiol.* **2002**, *43*, 58–69.

Maloof, J.N., J.O. Borevitz, T. dabi, J. Lutes, R.B. Nehring, J.L. Redfern, G.T. Trainer, J.M. Wilson, T. Asami, C.C. Berry, D. Weigel, J. Chory, *Nat. Genet.* **2001**, *29*, 441–446.

Martinez-Garcia, J., E. Huq, P.H. Quail, *Science* **2000**, *288*, 859–863.

M s, P., D. Alabadí, M.J. Yanovsky, T. Oyama, S.A. Kay, *Plant Cell* **2003**, *15*, 223–236.

Mathews, S., R.A. Sharrock, *Plant Cell Environ.* **1997**, *20*, 666–671.

Matsushika, A., A. Imamura, T. Yamashino, T. Mizuno, *Plant Cell Physiol.* **2002**, *43*, 833–843.

Matsushita, T., N. Mochizuki, A. Nagatani, *Nature* **2003**, *424*, 571–574.118.

McClung, C.R. *Annu. Rev. Plant Physiol. Plant Mol Biol.* **2001**, 52, 139–162.119.

McMichael, Jr., R.W., J.C. Lagarias, *Biochem.* **1990**, *29*, 3872–3878.

Miller, A.J. *J. Exp Bot.* **2004**, *55*, 277–283.

Monte, E., J.M. Alonso, J.R. Ecker, Y. Zhang, X. Li, J. Young, S. Austin-Phillips, P.H. Quail, *Plant Cell* **2003**, *15*, 1962–1980.

Nagatani, A., J.W. Reed, J. Chory, J. *Plant Physiol.* **1993**, *102*, 269–277.

Neuhaus, G., C. Bowler, K. Hiratsuka, H. Yamagata, N-H. Chua, *EMBO J.* **1997**, *16*, 2554–64.

Ni, M., J.M. Tepperman, P.H. Quail, *Nature* **1999**, *400*, 781–784.

Ni, M., J.M. Tepperman, P.H. Quail, *Cell* **1998**, *95*, 657–667.

Okamota, H., M. Matsui, X-W. Deng, *Plant Cell* **2001**, *13*, 1639–1652.

Osterlund, M.T., C. Hardtke, N. Wei, X-W. Deng, *Nature* **2000**, *405*, 462–466.

Park, D.H., P.O. Lim, J.S. Kim, D.S. Cho, S.H. Hong, H.G. Nam, *Plant J.* **2003**, *34*, 161–71.

Parks, B.M., P.H. Quail, *Plant Cell* **1993**, *5*, 39–48.

Quail, P.H. *BioEssays* **1997b**, *19*, 571–579.

Quail, P.H. *Curr. Opin. Cell Biol.* **2002a**, *14*, 180–188.

Quail, P.H. *Nature Rev.(Mol. Cell Biol.* **2002b**, *3*, 85–93.

Quail, P.H. *Plant Cell Environ.* **1997a**, *20*, 657–665.

Quail, P.H. *Sem. Cell Dev. Biol.* **2000**,*11*, 457–466.

Quail, P.H., M.T. Boylan, B.M. Parks, T.W. Short, Y. Xu, D. Wagner, *Science* **1995**, *268*, 675–680.

Reed, J.W., P. Nagpal, D.S. Poole, M. Furuya, J. Chory, *Plant Cell* **1993**, *5*, 147–157.

Saijo, Y., J. A. Sullivan, H. Wang, J. Yang, Y. Shen, V. Rubio, L. Ma, U. Hoecker, X-W. Deng, *Genes and Dev.* **2003**, *17*, 2642–2647.

Salter, M.G., K.A. Franklin, G.C. Whitelam, *Nature* **2003**, *426*, 680–683.

Schneider-Poetsch, H.A.W., B. Braun, *J. Plant Physiol.* **1991**, *137*, 576–580.

Schneider-Poetsch, H.A.W., B. Braun, S. Marx, A. Schaumburg, *FEBS Lett.* **1991**, *281*, 245–249.

Schultz, T.F., S.A. Kay, *Science* **2003**, *301*, 326–328.

Schwechheimer, C., X-W. Deng, *Sem. Cell Dev. Biol.* **2000**, *11*, 495–503.

Seo, H.S., E. Watanable, S. Tokutomi, A. Nagatani, N.-H. Chua, *Genes and Dev.* **2004**, *18*, 617–622.

Seo, H.S., J-Y. Yang, M. Ishikawa, C. Bolle, M. Ballesteros, N.-H. Chua, *Nature* **2003**, *423*, 995–999.

Serino, G., X-W. Deng, *Annu. Rev. Plant Biol.* **2003**, *54*, 165–182.

Smith, H. *Nature* **2000**, *407*, 585–591.

Somers, D.E., P.F. Devlin, S.A. Kay, *Science* **1998**, *282*, 1488–1490.

Somers, D.E., W.Y. Kim, R. Geng, *Plant Cell* **2004**, *16*, 769–82.

Staiger, D., L. Allenbach, N. Salathia, V. Fiechter, S.J. Davis, A.J. Miller, J. Chory, C. Fankhauser, *Genes and Dev.* **2003**, *17*, 256–268.

Subramanian, C., B-H. Kim, N.N. Lyssenko, X. Xu, C.H. Johnson, A.G. von Arnim, *Proc. Natl. Acad. Sci. USA* **2004**, *101*, 6798–6802.

Tepperman, J.M., M.E. Hudson, R. Khanna, T. Zhu, S.H. Chang, X. Wang, P.H. Quail, *Plant J.* **2004**, *38*, 725–739.

Tepperman, J.M., T. Zhu, H-S. Chang, X. Wang, P.H. Quail, *Proc. Natl. Acad. Sci. USA* **2001**, *98*, 9437–9442.

Thummler, F., P. Algarra, G.M. Fobo, *FEBBS Lett.* **1995**, *357*, 149–155.

Toledo-Ortiz, G., E. Huq, P.H. Quail, *Plant Cell* **2003**, *15*, 1749–1770.

Vierstra, R.D. Phytochrome degradation: In *Photomorphogenesis in Plants*, 2nd edn. (eds R.E. Kendrick and G.M.H. Kronenberg), pp. 141–162. Kluwer, Dordrecht, The Netherlands. **1994**.

Wang, H., L.-G. Ma, J.-M. Li, H.-Y. Zhao, X.W. Deng, *Plant J.* **2003**, *32*, 723–733.

Wang, H., X-W. Deng, Phytochrome signaling mechanism in: *The Arabidopsis Book*. American Society of Plant Biologists (http://www.aspb.org/publications/arabidopsis/),**2002**.

Whitelam, G.C., E. Johnson, J. Peng, P. Carol, M.L. Anderson, J.S.,Cowl, N.P. Harberd,. *Plant Cell* **1993**, *5*, 757–68.

Wong, Y-S., R.W. McMichael, Jr., J.C. Lagarias, *Plant Physiol.* **1989**, *91*, 709–718.

Yamashino, T., A. Matsushika. T. Fujimori, S. Sato, T. Kato, S. Tabata, T. Mizuno, *Plant Cell Physiol.* **2003**, *44*, 619–629.

Yanovsky, J.M., P.J. Luppi, D. Kirchbauer, B.O. Ogorodnikova, A.V. Sineshchekov, E. Adam, J.R. Staneloni, E. Schaefer, F. Nagy, J.J. Casal, *Plant Cell* **2002**, *14*, 1591–1603.

Yeh, K.C., J.C. Lagarias, *Proc. Natl. Acad. Sci. USA* **1998**, *10*, 13976–13981.

Yeh, K-C., S-H. Wu, J.T. Murphy, J.C. Lagarias, *Science* **1997**, *277*, 1505–1508.

Zeidler, M., C. Bolle, N-H. Chua, *Plant Cell Physiol.* **2001**, *42*, 1193–1200.

Zhu, Y., J.M. Tepperman, C.D. Fairchild, P.H. Quail, *Proc. Natl. Acad. Sci. USA* **2000**, *97*, 13419–13424.

8
Phytochromes in Microorganisms

Richard D. Vierstra and Baruch Karniol

8.1
Introduction

Like plants, bacteria and fungi are profoundly influenced by their surrounding light environment (Hader, 1987; Armitage, 1997; Loros and Dunlap, 2001; Braatsch and Klug, 2004). For photoautotrophic species in which light supplies the energy that drives photosynthesis, various photosensory systems provide positional information to optimize light capture. These systems can control short-term responses that direct microorganism to move/grow toward more favorable light fluences. Sophisticated long-term adaptation responses are also evident that help these organisms adjust to the spectral quality of the light (sun versus shade) by modulating the complement of photosynthetic accessory pigments, or help entrain their growth and development to the diurnal and possibly seasonal cycles. For heterotrophic microorganisms, light is also monitored as an environmental cue. Although the roles of light are less obvious, it is likely that light directs preference/avoidance strategies similar to those used by photoautotrophic organisms, which in turn allow them to occupy more favorable ecological niches.

Whereas the microbial pigments involved in energy acquisition are known in exquisite detail, we have begun to appreciate only recently the repertoire of photoreceptors employed for sensory information. Much of this new understanding has emerged from the exponentially expanding wealth of genomic data that now provide a facile way to detect possible photoreceptors and associated sensory cascades by BLAST searches with signature motifs. Recent examples include the discovery of retinal-based bacteriorhodopsin, *p*-hydroxycinnamic acid-based xantopsin, and flavin-based LOV-type photoreceptors in a variety of prokaryotes [(van der Horst and Hellingwerf, 2004; Venter et al., 2004) and Chapters 13 and 15].

Phytochromes (Phys) are another example where genome analyses have greatly enhanced our understanding of light perception by lower organisms. This class of photoreceptors is defined by the use of a bilin (or linear tetrapyrrole) chromophore (Smith, 2000; Quail, 2002). Once bound to the apoprotein, the bilin enables detection of red (R) and far-red (FR) light by photointerconversion between two relatively stable conformations, a R-absorbing Pr form and a FR-absorbing Pfr form. Through

Handbook of Photosensory Receptors. Edited by W. R. Briggs, J. L. Spudich

ISBN 3-527-31019-3

their unique ability to photointerconvert between Pr and Pfr reversibly, Phys act as light-regulated switches by having one form behave as "active" and the other as "inactive". This photochromicity also can provide a crude form of color vision through measurement of the Pr/Pfr ratio generated by the R/FR ratio (Smith, 2000).

Members of Phy superfamily were first discovered over 50 years ago in higher plants based on the ability of R and FR to control many agriculturally important aspects of their life cycle (Smith, 2000; Quail, 2002). More recently, genetic analyses and BLAST searches have dramatically expanded their distribution to other kingdoms with the discovery of Phy-type photoreceptors in numerous cyanobacteria, eubacteria, actinobacteria, filamentous fungi, and possibly slime molds [(Wu and Lagarias, 2000; Vierstra, 2002) and B. Karniol and R.D. Vierstra, unpublished]. The purpose of this chapter is to review our current understanding of these Phys from lower organisms. As will be seen, they offer simple models to help unravel the biochemical and biophysical events that initiate signal transmission by these novel photochromic pigments. Microbial Phys also provide new clues concerning the evolution of what is now emerging as a superfamily of Phy-type pigments. Their widespread distribution implies that light has more important roles in microbial ecology than was previously appreciated, especially for heterotrophic species. Defining these roles is now an intriguing avenue of photobiological investigation.

8.2
Higher Plant phys

The best-characterized members of the Phy superfamily are those from higher plants (see Chapter 6). Each plant species contains a small collection of structurally similar but functionally distinct types. In *Arabidopsis thaliana* for example, five plant phy isoforms are present (PhyA-E) that have both overlapping and unique roles in light perception (Smith, 2000; Quail, 2002). The generic plant phy assumes a "Y-shaped" structure formed by the dimerization of two identical ~120-kDa polypeptides (Figure 8.1 A,B). The N-terminal half of each polypeptide functions as the sensory input module. It contains a bilin-binding pocket (BBP) that autocatalytically attaches a single 3(Z)-phytochromobilin (PΦB) chromophore (Figure 8.1 C). PΦB is synthesized by a three-step enzymatic cascade with the first committed step being the oxidative cleavage of heme by a heme oxygenase (HO) to form the linear bilin biliverdin IXa (BV) (Davis et al., 1999a; Muramoto et al., 1999). BV is then converted to PΦB by sequential reduction and isomerization reactions (Frankenberg et al., 2001). Free PΦB is linked to the apoprotein via a thioether bond to a positionally conserved cysteine in a signature c**G**MP phosphodiesterase/**a**denyl cyclase/**F**hlA (GAF) domain within the BBP (Wu and Lagarias, 2000). In addition to its role in sensory acquisition, it has been proposed that the BBP participates in sensory output (Matsushita et al., 2003). The C-terminal half of plant phys contains activit(ies) needed for both dimerization and sensory output. This half includes two **P**er/**A**rndt/**S**im (PAS) motifs that are essential for signal transmission and a more C-terminal region bearing dimerization contacts (Quail, 2002) (Figure 8.1 B).

A. Proposed Structure

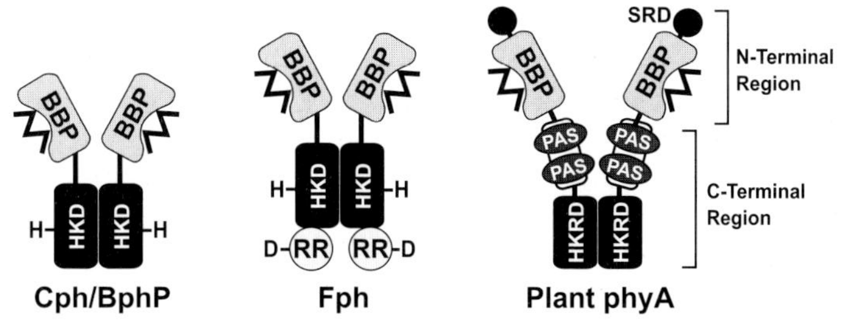

B. Linear Maps

C. Bilin Chromophores

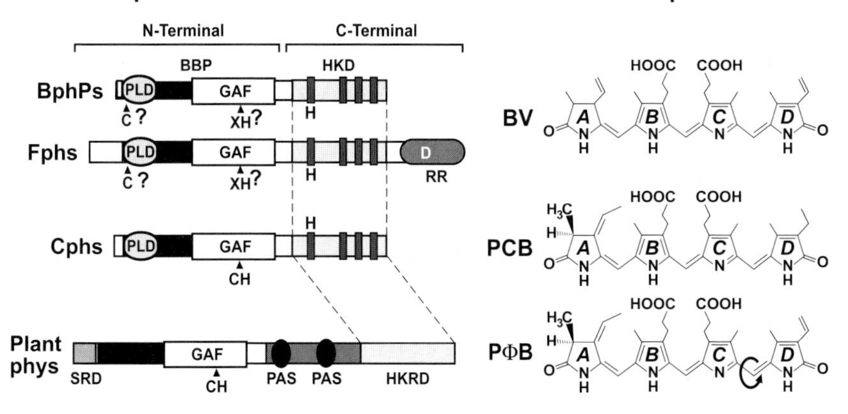

D. Absorbance Spectra

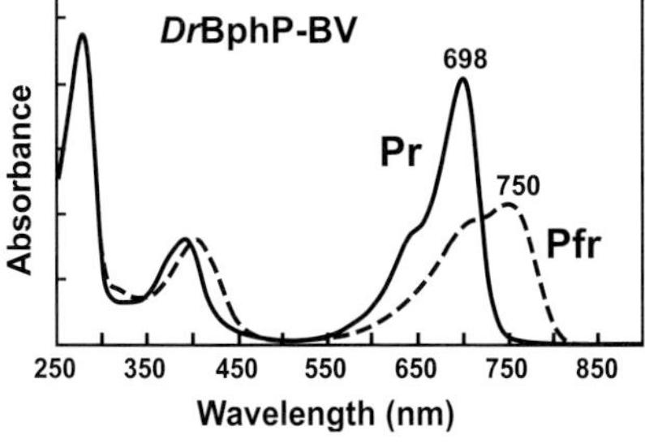

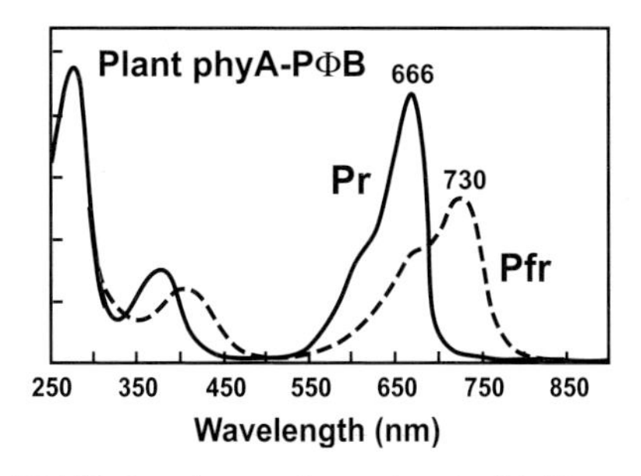

Figure 8.1 Organization, polypeptide structure, and spectral properties of Phy-type photoreceptors. (A) and (B): Predicted organization and polypeptide maps of the cyanobacterial Phy (Cph), bacterial Phy (BphP), and fungal Phy (Fph) families as compared with plant phys. BBP, bilin-binding pocket encompassing a PAS-like domain (PLD) and a c**G**MP phosphodiesterase/**a**denyl cyclase/**F**hlA (GAF) domain. HKD, histidine kinase domain. HKRD, histidine kinase-related domain. RR, response regulator. PAS, Per-Arndt-Sim domain. SRD, serine-rich domain. The histidine (H) and aspartic acid (D) residues that participate in TC-HK phosphotransfer as show in (A). The arrowheads in (B) identify the bilin attachment site with the key amino acids shown. The question marks identify the two possible sites in BphPs and Fphs. (C): Structure of the BV, PCB and PΦB chromophores. The curved arrow in PΦB identifies the C15 double bond that undergoes a *cis* to *trans* isomerization during Pr to Pfr photoconversion. (D): Absorption spectra of the BphP from *Deinococcus radiodurans* after saturating FR (Pr) and R (Pfr) irradiations as compared to plant phyA. The absorption maxima are indicated.

PΦB binding generates a Pr ground state. Once bound PΦB exhibits a dramatic red shift and an increase in absorption as compared to the free form, presumably caused by a network of chromophore/protein interactions between the BBP and the bilin (Wu and Lagarias, 2000). Upon excitation of Pr, the bound PΦB undergoes a *cis*-to-*trans* isomerization of the double bond between the C and D pyrrole rings (Figure 8.1 C) and a 31° reorientation of the bilin relative to the polypeptide, and the polypeptide undergoes multiple conformational changes (Quail, 2002). This photoconversion ultimately generates the relatively stable Pfr form, presumably with an altered biochemical output.

At present the nature of the output signal from Pfr is unclear, hindered, in part, by the lack of a robust activity and/or obvious sequence motif(s) that would conclusively define an enzymatic function. One attractive hypothesis since the 1980s is that plant phys act as light-regulated protein kinases. In particular, the realization that the C-terminus of plant phys is related to the histidine kinase domain (HKD) in two-component histidine kinases (TC-HKs) led to speculation that plant phys function in similar phosphorelays (Schneider-Poetsch, 1992). TC-HKs are a family of dimeric protein kinases commonly used by bacteria for environmental adaptation. They work by perceiving a signal through a sensor module, which then promotes an associated HKD to phosphorylate itself and then transfer the phosphate to a cognate response regulator (RR). The phosphorylated RR transmits the signal to appropriate effector pathways. Based on a similar architecture to TC-HKs, it appears plausible that plant phys works in an analogous fashion, using their BBP as a sensor module to activate the histidine kinase-related domain (HKRD) and thus begin a phosphorelay. Despite this homology, it remains unclear if plant phys are TC-HKs. Their HKRD is missing several important residues that typify a TC-HK, including the histidine residue that serves as the acceptor for the initial phosphorylation step. Furthermore, the assembled chromoprotein acts as a serine/threonine kinase (albeit weakly) and not as a HK, at least *in vitro* (Yeh and Lagarias, 1998). Following Pfr formation, important subsequent events include a redistribution of Pfr from the cytoplasm to the nucleus and a substantial alteration in gene expression, suggesting that plant phys work in close proximity to the nuclear transcriptional machinery (Smith, 2000; Quail, 2002).

8.3
The Discovery of Microbial Phys

Before 1996, the prevailing notion was that Phy-type pigments are present only in higher and lower plants and algae (Mathews and Sharrock, 1997). This perception changed radically with the identification by Kehoe and Grossman (Kehoe and Grossman, 1996) of *Response to Chromatic Adaptation E* (*RcaE*), a gene essential for complementary chromatic adaptation in the cyanobacterium *Fremyella diplosiphon*. Surprisingly, the encoded RcaE protein was discovered to contain a 150-amino acid N-terminal region with striking similarity to the signature BBP of plant phys followed by a prototypical HKD found in TC-HKs. Importantly, all the motifs and residues essential for phosphotransfer were evident in the HKD, including the H box that contains the histidine that becomes phosphorylated, and the N, F and G boxes that participate in ATP binding. Both its position within the complementary chromatic adaptation pathway and its similarity to plant phys suggested that RcaE behaves as a new bacterial Phy-type photoreceptor that initiates a TC-HK phosphorelay (Kehoe and Grossman, 1996, 1997).

Following the discovery of RcaE, five genes with varying degrees of relatedness to *Fremyella diplosiphon RcaE* and higher plant *PHYs* were detected in the completed genome of the cyanobacterium *Synechocystis* sp. PCC6803 (Kehoe and Grossman, 1996; Hughes et al., 1997; Wilde et al., 1997; Yeh et al., 1997; Park et al., 2000b). Like RcaE, three of the predicted polypeptides also have a C-terminal HKD, implicating them in TC-HK signaling. *Synechocystis* Cph1 in particular behaved like a true Phy. The recombinant apoprotein autocatalytically assembled with bilins like PΦB and 3Z-phycocyanobilin (PCB) to generate a dimeric R/FR photochromic chromoprotein *in vitro* (Hughes et al., 1997; Yeh et al., 1997). The resulting holoprotein displayed HK activity with the Pfr form being more active than the Pr form. These observations encouraged a number of investigators to perform similar searches with other bacterial genomes to define the limits of the prokaryotic Phy kingdom. Of the numerous cyanobacterial species surveyed, including *Calothrix* PCC7601, *Oscillatoria* PCC7821, and several *Anabaena, Pseudoanabaena,* and *Nostoc* species, all were found to contain one or more Phy-like protein sequences, indicating that these photoreceptors may be common to this phylogenetic group (Herdman et al., 2000; Wu and Lagarias, 2000). One Phy-like protein from *Synechococcus elongatus* PCC7942 (CikA) was identified genetically as required to reset the circadian clock, thus implicating this group in cyanobacterial photobiology (Schmitz et al., 2000). Multiple Phy sequences were also detected in the genomes of several purple bacteria, including *Rhodospirillum centenum, Bradyrhizobium* ORS278, *Rhodobacter sphaeroides,* and *Rhodopseudomonas palustris,* which further extended the distribution of prokaryotic Phys to photosynthetic eubacteria (Jiang et al., 1999; Wu and Lagarias, 2000; Bhoo et al., 2001; Giraud et al., 2002). As with RcaE and Cph1, these sequences often contained a canonical HKD appended to a BBP, strengthening the view that plant phys evolved from a prokaryotic progenitor TC-HK.

Parallel studies extended the range of Phys beyond photosynthetic organisms with the discovery of Phy-like proteins in numerous non-photosynthetic bacteria, including *Deinococcus radiodurans, Agrobacterium tumefaciens, Rhizobium leguminosarium,* and *Pseudomonas syringae* (Davis et al., 1999b; Bhoo et al., 2001). Like their photosynthetic bretheren, these eubacterial Phys contain the signature BBP, which in several cases was demonstrated to attach various bilins autocatalytically *in vitro*. The resulting chromoproteins display R/FR photochromic spectra typical of Phys (Davis et al., 1999b; Bhoo et al., 2001; Jorissen et al., 2002b; Lamparter et al., 2002; Karniol and Vierstra, 2003). Phy-like sequences were even evident in several filamentous fungi, including *Aspergillus nidulans* and *Neurospora crassa,* indicating for the first time the potential existence of Phy-type photoreceptors in the fungal kingdom (Bhoo et al., 2001; Catlett et al., 2003). Like cyanobacterial Phys, most of these heterotrophic bacterial and fungal sequences contain a HKD, and as a result also likely function in TC-HK cascades (Figure 8.1). However, one important distinction was immediately evident. The GAF domain in these Phys is missing the cysteine that was thought to be a prerequisite for bilin attachment (Bhoo et al., 2001). As a consequence these bacterial and fungal polypeptides must attach their bilin by a different mechanism if they are to act as photoreceptors (see below).

8.4 Phylogenetic Analysis of the Phy Superfamily

At present, more than 50 predicted proteins with the signature GAF domain of Phys can be identified in over 30 bacterial and fungal species. While defining their physiological functions awaits genetic analyses, it is clear from preliminary biochemical studies that these microbial polypeptides have extensively evolved, with many displaying physico-chemical properties different from their higher plant relatives (Figure 8.2). For example, chromophore-assembly studies with recombinant polypeptides indicated that some bind bilins other than PΦB and use different attachment sites, whereas others may not even be photoreceptors (e.g. Bhoo et al., 2001; Jorissen et al., 2002b; Mutsuda et al., 2003; Lamparter et al., 2004). A striking subset of Phys use the Pfr and not the Pr form as the ground state, indicating that they function backwards, requiring FR and not R to photoconvert the photoreceptor following assembly with the bilin (Giraud et al., 2002; Karniol and Vierstra, 2003). As a consequence, our long-held assumptions regarding characters that define a Phy may need to be relaxed to include these variants.

Unfortunately, the rapid and independent identification of these microbial polypeptides has led to a dizzying menagerie of nomenclatures in the literature. However by combining recent phylogenetic and biochemical characterizations, it appears that this superfamily of Phy polypeptides can be sorted into a few distinct clades. Using the GAF domain alone for sequence alignments, several families emerge that are clearly distinct from plant phys (Figure 8.2). Coupled with a grouping based on specific biochemical properties (e.g. identity of the bilin and its linkage site), we propose the formation of four major microbial Phy subdivisions that reflect their distribution within the bacterial and fungal kingdoms, their photobiological properties, and their possible modes of action.

These divisions encompass the **c**yanobacterial **Ph**ys (Cphs), the **b**acterio**ph**ytochrome **p**hotoreceptors (BphPs), the **f**ungal **Ph**ys (Fphs), and a collection of Phy-like sequences without apparent relationships. While we acknowledge that this preliminary nomenclature is still in flux given how little we know about many of these putative Phy sequences, these simple subdivisions do provide a starting point for discussing critical features of this photoreceptor superfamily. It should be emphasized that this classification does not imply common mechanisms of signal output, as microbial Phys with distinctly different C-terminal output modules are scattered among the four groups (Figure 8.3). Also evident from this classification is that some bacterial species contain members from the different divisions, indicating that a variety of light signaling systems can co-exist. For example, *Calorthrix* contains both a Cph (CphA) and a BphP type (CphB) (Jorissen et al., 2002b), while *Synechocystis* contains

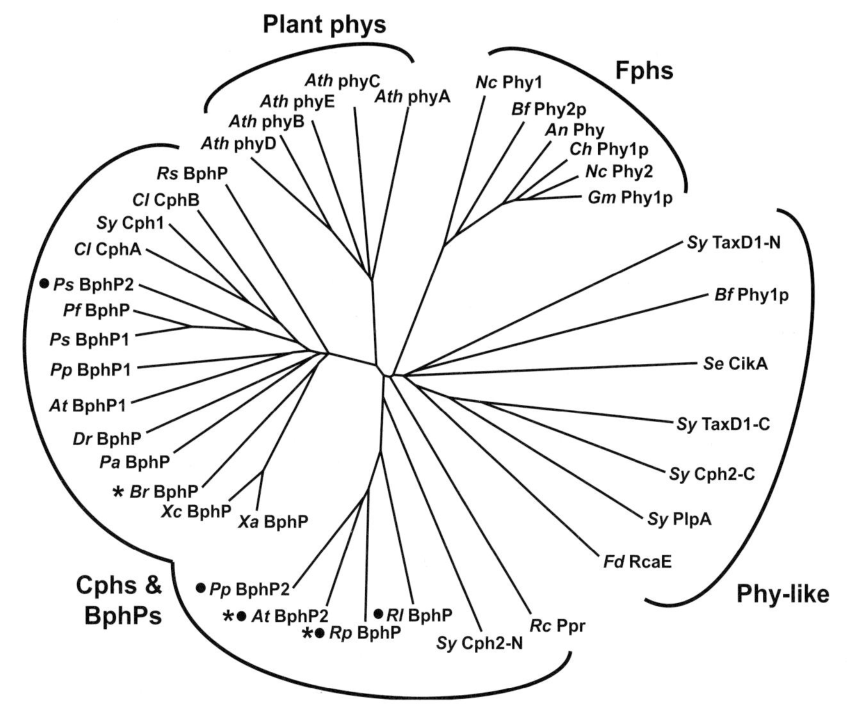

Figure 8.2 Phylogenetic organization of the Phy superfamily based on alignments of the GAF domain for representative members. The major plant phy, Cph, BphP, Fph and Phy-like families are indicated. Circles identify BphPs bearing the HWE-HK motif. Asterisks indicate known members of the bathyBphP subfamily. *Agrobacterium tumefaciens* (*At*), *Arabidopsis thaliana* (*Ath*), *Aspergillus nidulans* (*An*), *Botryotinia fuckeliana* (*Bf*), *Bradyrhizobium* ORS278 (*Br*), *Calorthrix* PCC7601 (*Cl*), *Cochliobolus heterostrophus* (*Ch*), *Deinococcus radiodurans* (*Dr*), *Fremyella diplosiphon* (*Fd*), *Gibberella moniliformis* (*Gm*), *Neurospora crassa* (*Nc*), *Pseudomonas aeroginosa* (*Pa*), *Pseudomonas putida* (*Pp*), *Pseudomonas syringae* (*Ps*), *Rhizobium leguminosarium* (*Rl*), *Rhodopseudomonas palustris* (*Rp*), *Rhodospirillum centenum* (*Rc*), *Synechococcus elongatus* PCC7942 (Se), *Synechocystis* PCC6803 (*Sy*), *Xanthomonas axonopodis* (*Xa*), and *Xanthomonas campestris* (*Xc*). Both the N-terminal (N) and C-terminal (C) GAF domains for *Synechocystis* TaxD1 and Cph2 were included in the analysis.

two Cphs (Cph1 and 2) and three Phy-like sequences (TaxD1, PlpA, and an RcaE-like) (Wilde et al., 1997; Yeh et al., 1997; Park et al., 2000b; Bhaya et al., 2001; Wilde et al., 2002; B. Karniol and R.D. Vierstra, unpublished). The Phy-like family encompasses an expanding collection of proteins that have been included in the superfamily based on the presence of a similar GAF domain. Coincidently, most are found in cyanobacteria. Several of these unorthodox proteins have been examined biochemically and found to be missing one or more attributes characteristic of *bona fide* Phys, suggesting that they function in a different manner. It is even possible that these Phy-like species are not R/FR photochromic bili-proteins but act as accessory components in

Phy-mediated transduction cascades [e.g. *Synechoccocus elongatus* CikA (Mutsuda et al., 2003)].

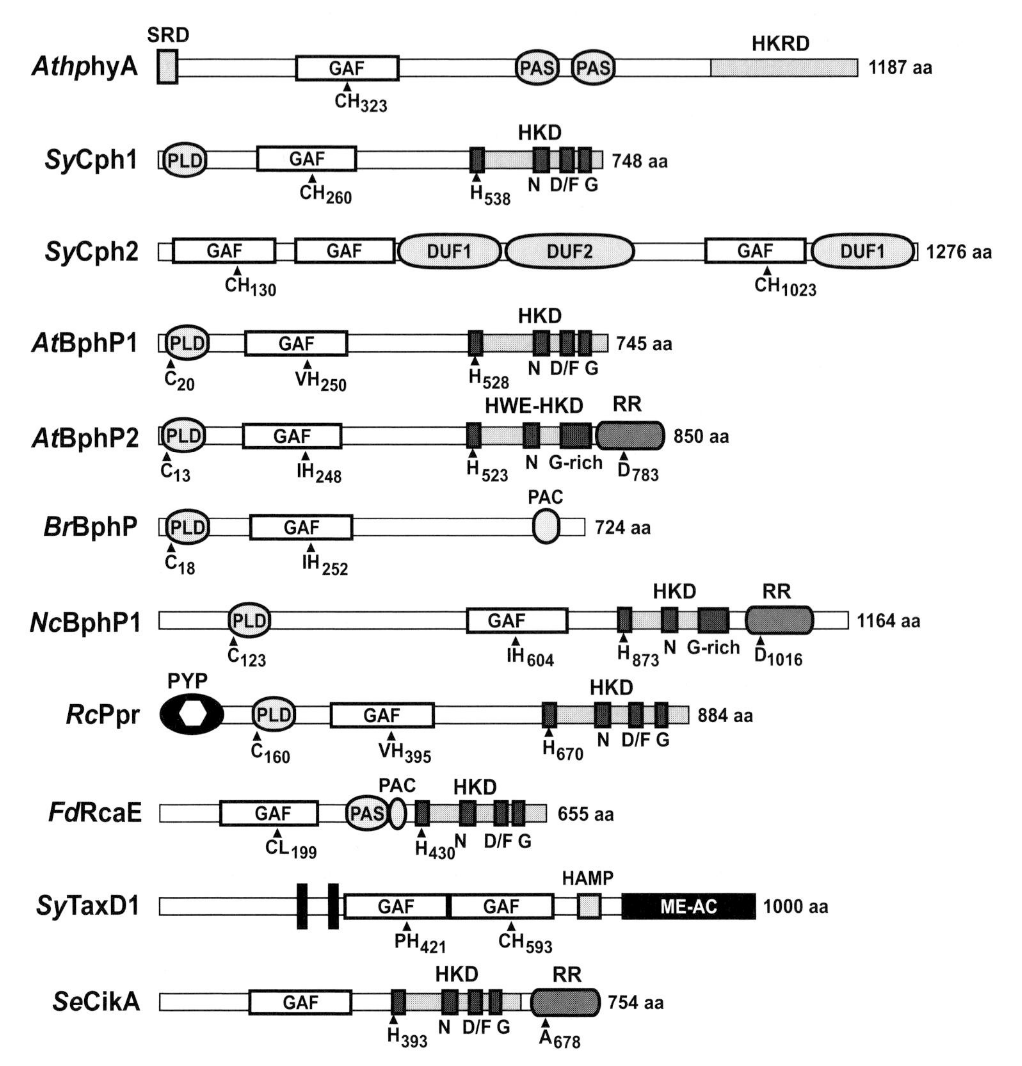

Figure 8.3 Structure of representative members of the Phy superfamily. Species designations can be found in the legend of Figure 8.2 DUF, domain of unknown function. GAF, cGMP phosphodiesterase/adenyl cyclase/FhlA. HKD, histidine kinase domain (the H, N, F and G boxes are indicated). HAMP, HKs/adenylcyclases/methyl-binding protein/phosphatases domain. HKRD, histidine kinase-related domain. Me-Ac, methyl-accepting chemotaxis protein domain. RR, response regulator. PAS, Per-Arndt-Sim domain. PAC, C-terminal PAS domain. PLD, PAS-like domain. PYP, photoactive yellow protein. SRD, serine-rich domain. The position of signature amino acids in the GAF, HKD and RR domains and the N-terminal cysteine that may bind bilins are indicated.

8.4.1
Cyanobacterial Phy (Cph) Family

The Cph family of Phys is spread among cyanobacterial species and appear most related to higher plant phys among the microbial forms. However, the cyanobacterial apoproteins likely use PCB as the chromophore instead of PΦB (Figure 8.1 C) (Hubschmann et al., 2001a; Jorissen et al., 2002a). PCB is synthesized in large quantities as a cyanobacterial photosynthetic accessory pigment and thus is in near unlimited supply for holopotein assembly. PCB is attached to the same positionally conserved cysteine within the GAF domain as employed by plant Phys to bind PΦB (Figure 8.4), and likely also uses its A ring ethylidene side chain to form a thioether linkage (Park et al., 2000b). The founding member of this group, *Synechocystis* Cph1 displays the characteristic R/FR photochromic spectra when assembled with PCB (Yeh et al., 1997; Park et al., 2000a). The slightly blue-shifted Pr and Pfr absorption maxima for holoprotein (654 and 706 nm) as compared to plant Phys (666 and 730 nm) are consistent with the loss of one double bond in the π-electron system of PCB versus PΦB. A similar bilin attachment site and R/FR spectra were confirmed for two other members of the Cph family, *Synechocystis* Cph2 (Park et al., 2000b; Wu and Lagarias, 2000) and *Calothrix* CphA (Jorissen et al., 2002b)), suggesting that the bilin-BBP interactions among this group have been conserved. *Synechocystis* Cph2 is unusual because it contains three predicted GAF domains (Figure 8.3). Binding studies with recombinant polypeptide showed that both the N-terminal and C-terminal GAF domains can bind bilins but only the N-terminal GAF domain generates a R/FR photochromic pigment (Park et al., 2000b; Wu and Lagarias, 2000).

Most Cphs have a PAS-like domain (PLD) within the N-terminal region of the BBP that is not evident in plant phys (e.g. *Synechocystis* Cph1 (Figure 8.3 and (Lamparter et al., 2004)). It can be distinguished from a similar PLD in BphPs and Fphs by the absence of a signature cysteine that may help the latter families bind bilins (see below). Most Cphs also contain a C-terminal HKD (e.g. *Synechocystis* Cph1 and *Calothrix* CphA) and thus likely function in TC-HK cascades (Yeh et al., 1997; Jorissen et al., 2002b). One exception is *Synechocystis* Cph2; instead of a HKD, Cph2 contains a pair of **d**omain of **u**nknown **f**unction (DUF)-1 motifs and one DUF2 motif in the C-terminal half (Figure 8.3). As implied from their names, the function(s) of these DUF sequences are unknown. Cph2 is also missing the N-terminal PLD but still retains normal R/FR photochromic absorption spectra, indicating that at least for Cph2, the PLD is not essential photochemically.

8.4.2
Bacteriophytochrome (BphP) Family

Like members of the Cph and plant phy families, BphPs typically can covalently bind various bilins and become R/FR photochromic (Bhoo et al., 2001; Lamparter et al., 2002; Karniol and Vierstra, 2003). Members of the BphP family are widely dispersed throughout the bacterial kingdom. While most known members are from non-photosynthetic eubacteria and photosynthetic purple bacteria (Davis et al., 1999b; Bhoo

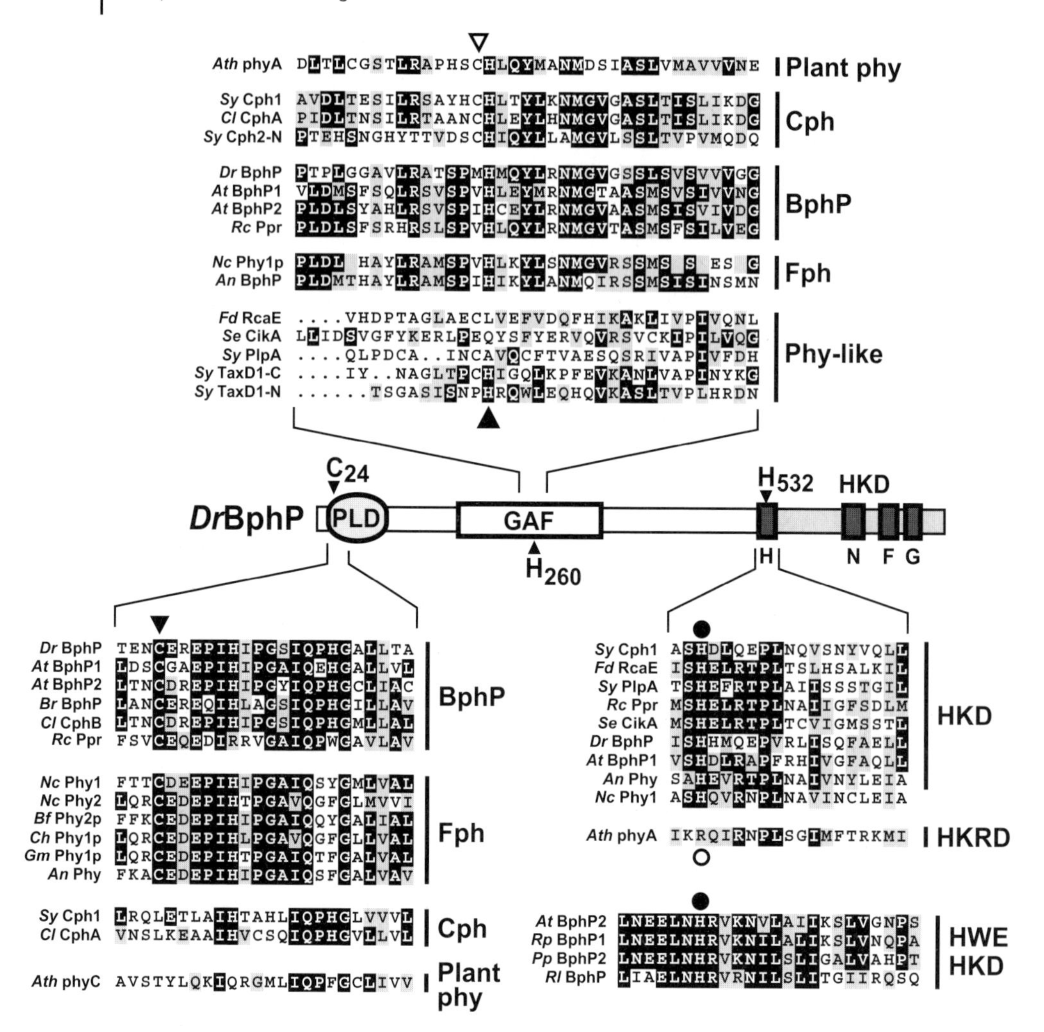

Figure 8.4 Amino acid sequence alignments of various domains within microbial Phy proteins. The position of each domain in Phys is shown using *Deinococcus radiodurans* BphP to help locate the region in the Phy polypeptide. Top alignment includes the region of the GAF domain that binds bilins via a cysteine thioether linkage in Cphs and plant phys or may bind bilins by a histidine Schiff base linkage in some BphPs. The cysteine and histidine residues are identified by the open and closed arrowheads, respectively. The N-terminal and C-terminal GAF domains are compared for *Synechocystis* TaxD1. Left bottom alignment compares the sequences surrounding the N-terminal PLD used by *Agrobacterium tumefaciens* BphP1 to covalently bind bilins by a cysteine thioether linkage. The cysteine is identified by the arrowhead. Right bottom alignment shows the H-box sequences from typical HKs and histidine/tryptophan/glutamic acid (HWE)-HKs. The histidine that serves as the phosphorylation site is identified by the closed circle. For plant phyA from the *Arabidopsis thaliana*, this residue is an arginine (open circle). Species designations can be found in the caption of Figure 8.2. Black and grey boxes denote identical and similar residues, respectively. Members of each of the Phy families are indicated by the brackets.

et al., 2001; Giraud et al., 2002), potential members are evident in some cyanobacteria (e.g. *Calothrix, Nostoc, Oscilatoria, and Geitlerinema* PCC9228 (Herdman et al., 2000; Bhoo et al., 2001)). These cyanobacterial species appear to contain members of both the Cph and BphP families (Herdman et al., 2000; Jorissen et al., 2002b) (Figure 8.2). Recently, we discovered a BphP sequence in the genome of *Kineococcus radiotolerans,* thus expanding the distribution of Phys to actinobacteria (B. Karniol and R.D. Vierstra, unpublished).

Important features that distinguish BphPs from Cphs include the type of bilin used and the nature of its attachment. Assembly studies indicate that BphPs use the PΦB/PCB precursor BV as the chromophore (Bhoo et al., 2001). Whereas plant phys and Cphs bind BV poorly or not at all, BphPs easily assemble with this bilin. The slightly red-shifted Pr and Pfr absorption maxima for BV-holoprotein (~698 and ~750 nm) as compared to plant phys are consistent with the addition of one double bond in the π-electron system of BV versus PΦB (Bhoo et al., 2001; Lamparter et al., 2002; Karniol and Vierstra, 2003). The use of BV is also supported by genomic analyses. Bacterial species that contain one or more BphPs invariably express a HO that converts heme to BV but do not appear to express the BV reductase activit(ies) needed to convert BV to PCB or PΦB (Bhoo et al., 2001). In several intriguing situations, the *HO* gene is physically linked to the *Bph* locus (Bhoo et al., 2001; Giraud et al., 2002). The most striking are the *Bph* operons of *Deinococcus. radiodurans, Pseudomonas aeruginosa, Pseudomonas syringae,* and *Rhizobium leguminosarium* that have a *HO* gene (*BphO*) included within the operon, and the *Bradyrhizobium* and *Rhodopseudomonas palustris* genomes where the *HO* gene (*HmuO*) is nearby (Figure 8.5). Presumably via such a genetic linkage, these bacteria can easily coordinate synthesis of the chromophore with the apoprotein.

Aligning the BBP sequence from members of the BphP family identified two other distinguishing features of this group. One feature is within the GAF domain. Instead of the conserved cysteine used by plant phys and Cphs to attach bilins, a small hydrophobic residue is often found in this position (Figure 8.4). Because of this substitution, BphPs must bind BV via a different mechanism and/or a different site than plant phys and Cphs (Davis et al., 1999b; Bhoo et al., 2001). The other feature is the presence of an N-terminal PLD containing an invariant cysteine (Figures 8.3 and 8.4), which appears to participate either directly or indirectly in chromophore binding (Lamparter et al., 2002; Lamparter et al., 2004). Although the situation is far from resolved, preliminary data indicate that members of the BphP family may bind bilins in several different ways. For the founding member of this group, BphP from *Deinococcus radiodurans,* mass spectroscopic analysis (MS) implicated a histidine within the GAF domain in bilin binding; it is immediately distil to the site where the cysteine is positioned in Cph and plant phys (Davis et al., 1999b). This histidine was previously shown to be important for the bilin lyase activity of plant phys (Bhoo et al., 1997) and has since been found to be conserved in all microbial Phys shown empirically to have bilin-binding activity (Figures 3 and 4). Its direct role in chromophore attachment would infer the use of a Schiff base-type linkage (Bhoo et al., 2001), which would be less stable than a thioether linkage involving the proximal cysteine. The only complication of these studies is that they were performed with PCB since the role

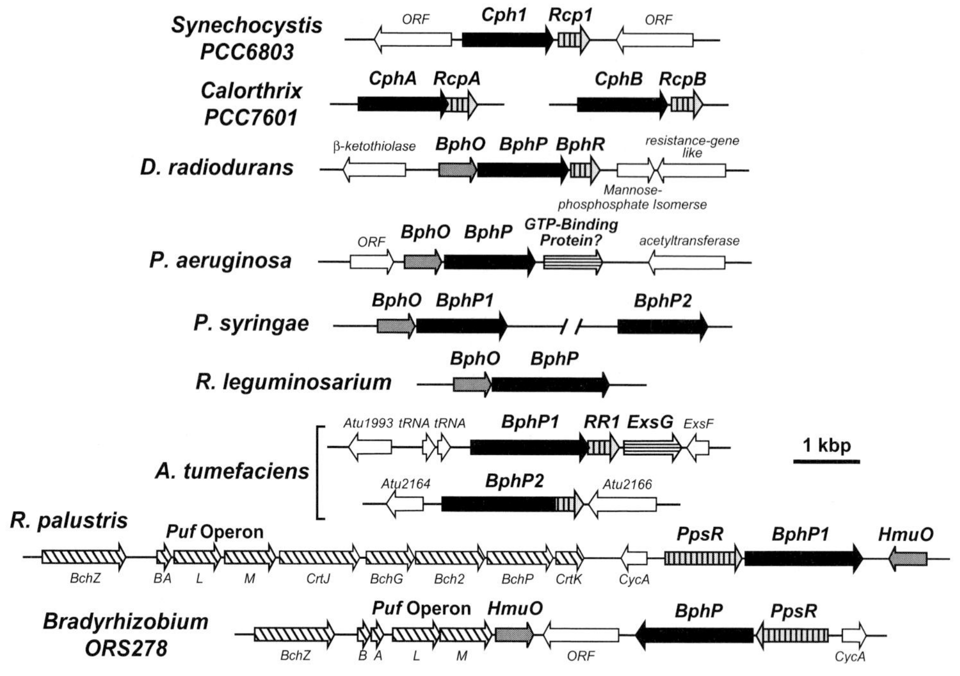

Figure 8.5 Organization of the genes surrounding the Phy apoprotein gene in various microbial species. Reading frames in black and grey encode the Phy apoprotein and a HO, respectively. Vertically hatched genes encode RRs genetically linked to the BphP. For *Agrobacterium tumefaciens* BphP1, the RR is appended to the Phy apoprotein to form a hybrid HK. Horizontal hatched genes indicate genes within the *BphP* operon. Diagonally hatched genes in the *Bradyrhizobium* and *Rhodopseudomonas palustris* chromosomes identify those linked genes predicted to be transcriptionally regulated by the Phy TC-HK cascade. ORF, open reading frame of unknown function. The direction of the arrow indicates the orientaion of the reading frame for each gene.

of BV as the natural chromophore was not yet known. Hence, it remains possible that BV binds via a different linkage/residue.

Studies with *Agrobacterium tumefaciens* BphP1 (or Agp1) identified a different site for BV attachment, at least for this photoreceptor. This alternative site was first suggested by the ability of cysteine modification reagents to block BV binding to the *At*BphP1 apoprotein (Lamparter et al., 2002). Assembly reactions with site-directed mutants and various BV derivatives implicated the conserved cysteine (Cys20) in the PLD and not a residue within the GAF domain (Lamparter et al., 2002; Lamparter et al., 2003). Despite this spatial change, BV appears to connect to the apoprotein via its A ring vinyl side chain, presumably through a thioether linkage similar to that used by plant phys and Cphs. Lamparter et al. (2004) have recently confirmed the involvement of the PLD cysteine by peptide mapping. Following digestion of the BV-*At*BphP holoprotein with trypsin, they identified by MS an N-terminal fragment containing the bilin bound to Cys20. Given the conservation of this cysteine in both BphPs and Fphs, it could be utilized by both of these families (Figure 8.4).

Recent analysis of *Calothrix* CphB indicates that a third mechanism is also possible. In this case, CphB appears to interact non-covalently with bilins even though both the PLD cysteine and the GAF-domain histidine are present (Jorissen et al., 2002b). The holoprotein retains R/FR photochromicity, suggesting that phototransformation between Pr and Pfr does not *a priori* require stable covalent attachment of the chromophore. Covalent binding of BV can be induced by introducing a cysteine into the GAF domain, indicating that BV fits into the CphB BBP in a similar arrangement as other BphPs. It has been proposed recently that CphB actually binds BV covalently via the histidine but that this Schiff-base linkage is unstable, thus allowing the chromophore to associate and dissociate readily (Quest and Gartner, 2004).

Spectroscopic studies indicate that most BphPs function as typical Phys, with the Pr form generated first following autocatalytic attachment of the bilin. Pfr is created only upon light absorption; Pfr can either photoconvert back to Pr by FR or revert nonphotochemically from Pfr back to Pr. For *Agrobacterium tumefaciens* BphP1, this dark reversion rate appears to be much faster than plant phys, suggesting that the Pfr conformation is less stable for some of these prokaryotic members (Lamparter et al., 2002; Karniol and Vierstra, 2003). Given the overlapping absorption spectra for Pr and Pfr, it is not possible to convert Pr to Pfr completely. Instead, a mixture is generated at photoequilibrium in R that still contains ~14% Pr. Surprisingly, analysis of several BphPs indicates that a novel subfamily of BphPs exists that works in reverse (Giraud et al., 2002; Karniol and Vierstra, 2003). These backwards (or bathy) BphPs rapidly assemble with BV to generate a Pr-like transient intermediate that quickly converts non-photochemically to a stable Pfr ground state. The absorption spectrum of the assembled bathyBphP at completion resembles the predicted spectrum for a Pfr without Pr contamination, strongly suggesting that all of the Phy pool becomes Pfr (Karniol and Vierstra, 2003). Pr is generated thereafter only by photoconversion with FR. This Pr is unstable and rapidly reverts to Pfr in the dark.

To date, members of the bathyBphP subfamily have been discovered in three species, *Agrobacterium tumefaciens*, *Bradyrhizobium*, and *Rhodopseudomonas palustris*, with more likely to be found (Giraud et al., 2002; Karniol and Vierstra, 2003). For example, sporulation of the slime mold *Physarum polycephalum* is controlled by a R/FR–absorbing photochromic pigment (Starostzik and Marwan, 1995). In contrast to typical Phy responses, this sporulaton is triggered by FR but not by darkness, R, or FR followed by R, indicating that the photoreceptor is initially synthesized as Pfr and requires FR for phototransformation to Pr. Phylogenetic clustering of the GAF domain does not place the three known bathyBphPs in a separate clade, suggesting that relatively subtle changes within the BBP account for their unusual spectral properties (Figure 8.2). The use of Pfr as the ground state appears to be physiologically relevant for *Bradyrhizobium* and *Rhodopseudomonas palustris* (Giraud et al., 2002). These species use their bathyBphPs to photoregulate expression of the photosynthetic apparatus by detecting FR, with R being without consequence. *Agrobacterium tumefaciens* is intriguing because it contains two BphPs; while *At*BphP1 behaves normally

(Pr as the ground state), the bathyBphP *At*BphP2 functions in reverse (Karniol and Vierstra, 2003). By simultaneously measuring R and FR, this pair may work either synergistically or antagonistically to detect fluctuating light quality.

Organization of the BphP proteins indicates that many function as TC-HKs. Often a canonical HKD is found C-terminal to the BBP. In several species, a RR domain is attached to the HKD (Figure 8.3). This organization creates a hybrid kinase in which a single polypeptide directs the phosphorylation of the HKD histidine as well as transfer of the phosphate to the asparate residue within the RR (West and Stock, 2001; Inouye and Dutta, 2003). Several BphPs that first appeared not to have a HKD were reexamined recently and found to have a new type of HKD, indicating that these BphPs also function in TC-HK relays. This new domain (designated HWE-HKD by the presence of conserved histidine, tryptophan and aspartic acid residues in the HKD) differs from the canonical HKD by substantial sequence alterations within the H, N and G boxes and the absence of an obvious F box [Figure 8.4 and (Karniol and Vierstra, 2003, 2004)]. In addition to BphPs from *Agrobacterium tumefaciens* (*At*BphP2), *Pseudomonas putida, Pseudomonas syringae, Rhodopseudomonas palustris* and *Rhizobium leguminosarium*, other potential sensor HKs contain this HWE-HKD, suggesting that it participates in a variety of signaling pathways besides light (Karniol and Vierstra, 2004). Not all members of the BphP family bear a recognizable HKD, implying that some may not act as protein kinases. The bathyBphP from *Bradyrhizobium* for example, contains a PAS-like domain called PAC that is often C-terminal to other PAS domains. This PAC likely functions in a light-regulated protein/protein interaction (Giraud et al., 2002).

Rhodospirillum centenum Ppr also appears to belong within the BphP family. It contains both a GAF domain (with the conserved histidine) and a recognizable cysteine-containing PLD upstream (Figures 8.3 and 8.4). However, it is distinguished from other BphPs by the presence of a **p**hotoactive **y**ellow **p**rotein (PYP) motif appended to the N-terminus of the BBP (Jiang et al., 1999). This 120-amino acid motif binds a single *p*-hydroxycinnamic acid chromophore to generate a photoreceptor capable of detecting blue light (B). Ppr has not yet been reported to bind bilins. As a consequence, Ppr may not be a true BphP but exploit the PYP motif to function as a B receptor. The fact that *ppr* mutants have altered expression of chalcone synthase gene in B supports this notion (Jiang et al., 1999).

8.4.3 Fungal Phy (Fph) Family

A surprising recent discovery was the identification of BBP-containing sequences in various filamentous fungi, suggesting for the first time that Phy-type pigments are present outside of the plant and bacterial kingdoms (Bhoo et al., 2001). The Fph family appears most related to BphPs. They contain a GAF domain with the conserved histidine preceded by a small hydrophobic residue and an N-terminal PLD with the signature cysteine similar to that in BphPs [Figures 8.3, 8.4 (Bhoo et al., 2001; Catlett et al., 2003)]. *Neurospora crassa* BphP2 can covalently assemble with bilins such as BV *in vitro* and become R/FR photochromic, suggesting that the Fphs are indeed Phy-type photoreceptors (B. Noh and R.D. Vierstra, unpublished). All the Fphs identified thus far contain a typical HKD followed by a RR, indicating that they function as hybrid kinases. The photobiological functions of Fphs remains to be determined.

8.4.4
Phy-like Sequences

Phylogenetic and biochemical studies have revealed a collection of Phy-like sequences that can be discriminated from the main Phy families by one or more criteria. As can be seen in Figure 8.2, most fall outside of the main Phy clades when the GAF domain alone is used for sequence comparisons. They are all devoid of an obvious PLD, which is one feature that may ultimately unify this collection (B. Karniol and R.D. Vierstra, unpublished). It is possible that these unorthodox polypeptides either do not bind bilins, bind them in different ways but are still photochromic, or bind bilins but are not photochromic. Many of these GAF domains contain a small deletion upstream of the putative chromophore-binding residue that may be photochemically significant (Wu and Lagarias, 2000). Included in the Phy-like group are *Fremyella diplosiphon* RcaE, *Synechocystis* TaxD1 and PlpA, *Synechococcus elongatus* CikA, and a sequence from the fungus *Botryotinia fuckeliana*. In several cases, biochemical studies have supported their delineation from the other main clades. *Fremyella* RcaE for example, has a recognizable GAF domain and can bind bilins both *in vivo* and *in vitro* but the resulting holoprotein is not photochromic (Terauchi et al., 2004). Although RcaE contains a cysteine near the expected site in the GAF domain, this residue is not essential for bilin attachment (at least *in vitro*) leaving the actual binding site unresolved. At present, it is unclear if RcaE is a photoreceptor despite genetic evidence connecting the corresponding locus to photoperception during complementary chromatic adaptation.

In a similar fashion to RcaE, *Synechococcus elongatus* CikA may not act as a typical Phy even though it appears to participate in photoperception by this cyanobacterium (Schmitz et al., 2000). CikA is missing the positionally conserved cysteine and histidine residues in the GAF domain and does not have a PLD (with its cysteine) (Figure 8.4). Despite these deficiencies, recombinant CikA can bind PCB and PΦB (but not BV) *in vitro*, but the resulting holoproteins are not R/FR photochromic (Mutsuda et al., 2003). The CikA protein extracted from *Synechococcus* cells is not a bili-protein, suggesting that the polypeptide does not assemble with these chromophores naturally. The GAF domain of CikA is followed by a typical HKD and a RR motif, implying that CikA functions in a TC-HK cascade (Schmitz et al., 2000). However, the appended RR motif is missing the aspartate residue necessary to receive the phosphate from the HKD. As a consequence, CikA cannot function as a *bona fide* hybrid HK. It has been proposed that CikA is a pseudo-RR that modulates a phosphorelay initiated by another photoreceptor (Mutsuda et al., 2003).

Synechocystis TaxD1 also clusters in the Phy-like group. It has two possible GAF domains with one containing the canonical cysteine-histidine sequence and the other a proline-histidine sequence at the putative bilin-binding site (Figure 8.4). Although its ability to bind bilins and become R/FR photochromic has not yet been reported, genetic analyses show that TaxD1 is essential for phototaxis toward low-fluence R and FR (Bhaya et al., 2001; Ng et al., 2003). TaxD1 does not contain a C-terminal HKD. Instead, a signaling domain similar to those found in methyl-accepting chemotaxis proteins is evident, suggesting that the phototactic response affected by TaxD1 is regulated by a methylation pathway similar to those utilized during halobacterial phototaxis and bacterial chemotaxis (Bhaya et al., 2001; Ng et al., 2003). *Synechocystis* PlpA and *Botryotinia fuckeliana* Phy1p also cluster phylogenetically in the Phy-like group but it remains to be determined if their photochemical properties are distinct from more typical Phys.

8.5
Downstream Signal Transduction Cascades

The structural organization of microbial Phys indicates that many, if not all, function either directly or indirectly in TC-HK phosphorelays. This connection is further supported by the presence of a RR domain, either translationally linked to the HKD thus creating a hybrid HK, or expressed as a separate polypeptide within the operon that encodes the Phy apoprotein (Figures 8.3 and 8.5). Even for *Bradyrhizobium* BphP, which does not contain a recognizable HKD, the RR PpsR is transcriptionally linked to the photoreceptor in an operon, suggesting that a TC-HK cascade is employed with PspR serving as the phosphoacceptor [Figure 8.5 and (Giraud et al., 2002)].

The HK activity of several Phys has been confirmed by *in vitro* phosphotransferase assays with recombinant proteins. Importantly, the HK activity (as measured by autophosphorylation of the HKD histidine) is regulated by the spectral form of the assembled holoprotein (Yeh et al., 1997; Bhoo et al., 2001; Karniol and Vierstra, 2003; Mutsuda et al., 2003). In some cases, the holoprotein is more active as Pr and, in other cases, the holoprotein is more active as Pfr. Whether this difference is an artifact of using recombinant proteins in purified systems or reflects intrinsic differences of the photoreceptors is not yet known. For the BphP pair from *Agrobacterium tumefaciens*, the typical Phy *At*BphP1 is a more active kinase as Pr, whereas the bathyBphP *At*BphP2 is a more active kinase as Pfr (Karniol and Vierstra, 2003). As a consequence, both photoreceptors would be maximally active in the dark and be simultanesously repressed by white light. Donation of the bound phosphate to the aspartate in the cognate RR has also been demonstrated for several Cphs and BphPs (Yeh et al., 1997; Bhoo et al., 2001; Hubschmann et al., 2001b; Karniol and Vierstra, 2003). This coupling appears to be specific as little phosphotransfer occurs between the HKD and an unrelated RR.

For most microbial Phys, the sequence of events that follows RR phosphorylation is not yet clear. Whereas many RRs have an appended output module (e.g. a DNA-binding or protein-interaction motif) that would be altered by the phosphorylation signal, these modules are often absent in the RRs associated with the microbial Phys (West and Stock, 2001; Inouye and Dutta, 2003). This absence suggests that microbial Phy RRs either interact directly with their targets (in a similar fashion to *Escherichia coli* CheY), or participate in a four-step His→Asp→His→Asp phosphorelay before signal output (e.g. *Escherchia coli* EnvZ).

A model based on the general rules of TC-HK cascades (West and Stock, 2001; Inouye and Dutta, 2003) is shown in Figure 8.6. Signaling begins with light-triggered conformational changes within the BBP altering the autophosphorylation activity of the HKD. This phosphorylation actually occurs in *trans*, using one HKD of the Phy dimer to direct phosphate addition to the HKD histidine of the second HKD. This cross phosphorylation is supported by the fact that all the microbial Phys tested thus far behave as homodimers (at least *in vitro*) (Yeh et al., 1997; Park et al., 2000b; Bhoo et al., 2001). The histidine phosphate is then transferred to the aspartate in a cognate RR. The phosphorylated RR could directly bind to an output apparatus (e.g. flagellar motor) to alter its function. Alternatively, the RR could donate the phosphate to the histidine of a histidine phosphotransferase (HPT). This HPT would continue the phosphorelay by transferring the phosphate to the aspartate in a second RR. The second RR could then interact with a separate output factor or contain an appended output domain to relay the signal further. While many HPTs and RRs with output do-

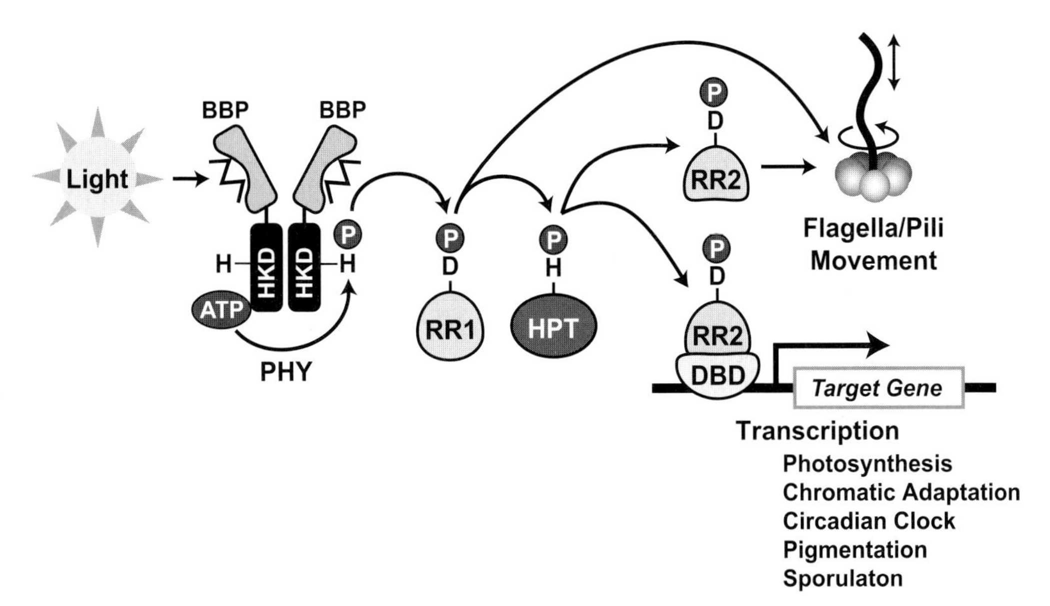

Figure 8.6 Proposed scheme for the function of BphPs in light perception. Absorption of light by the BphP homodimer triggers a conformational change in the sensor domain of the photoreceptor that activates (or inactivates) the histidine kinase activity in the HKD. Active BphP cross phosphorylates the conserved histidine (H) in the dimer and then transfers this phosphate to an aspartate residue (D) on an associated response regulator (RR1). RR1 can either directly affect an output (flagellar/pillus motor) or transfer the phosphate to a conserved histidine in a histidine phosphotransferase (HPT), which then donates the phosphate to a second RR (RR2). RR2 can interact with output factors or contains an appended output domain to initiate the response. In this example, RR2 is fused to a DNA-binding domain (DBD) and serves as a transcription factor that affects genes responsible for various photoresponses.

mains are evident in the various bacterial and fungal genomes [e.g, (Catlett et al., 2003)], those connected to Phy signaling are unknown in most cases. To date, the most complete pictures are from the signaling pathways involving *Fremyella diplosiphon* RcaE that directs complementary chromatic adaptation (Kehoe and Grossman, 1996, 1997), and *Bradyrhizobium* BphP that controls photosynthetic potential (Giraud et al., 2002). In each case, other members of the sensory cascade, including output modules and targets, can be inferred from genetic and genomic analyses (see below).

8.6 Physiological Roles of Microbial Phys

Given the varied structural, spectral, and biochemical properties of the microbial Phy families, we predict that they are involved in a diverse array of photosensory processes. Unfortunately, because most of these Phys systems have not yet been dissected at the physiological level, we currently have a rudimentary understanding of just a few. To date, one or more microbial Phys have been connected to the regulation of phototaxis, control of photosynthetic potential, entrainment of circadian rhythms, mitigation from high light fluences and damaging UV irradiation by producing protective pigments, and possibly sporulation. In most cases, the transcriptional control of gene expression is the ultimate output, thus implicating the associated TC-HK cascades in the regulation of DNA-binding proteins. Predictably, many of these same responses are also regulated by phys in higher plants (Smith, 2000; Quail, 2002). One potential complication to the study of Phy-mediated responses in microbes is that the responsible bili-proteins may have an unstable and thus transient Pfr state (Lamparter et al., 2002; Karniol and Vierstra, 2003). Consequently, it cannot be presumed that these responses will display the R/FR photoreversibility for induction that is often diagnostic for phy-mediated responses in higher plants (Smith, 2000; Quail, 2002).

8.6.1 Regulation of Phototaxis

As with the role of plant phys in directing growth and morphology, it is likely that microbial Phys also have important roles in regulating phototaxis and phototropism toward more favorable light environments. While few studies have directly implicated Phys in these movement/growth responses, action spectra showing a peak of activity centered around 700–730 nm suggest the involvement of Phy-type pigments. Examples in photosynthetic species include the phototactic behaviors of the cyanobacteria *Anabaena variabilis, Phormidium unicinatum, Synechococcus elongatus,* and *Synechocystis* (Hader, 1987; Ng et al., 2003). For non-photosynthetic organisms, similar roles are not yet clear. No studies have implicated a R-absorbing pigment in the taxis of non-photosynthetic prokaryotes and the phototropic response of many filamentous fungi is controlled by a B-absorbing pigment.

The best-understood phototactic response in cyanobacteria is that of *Synechocystis,* which displays a complex light response involving at least three photoreceptors that control positive phototaxis at low-fluence R and negative phototaxis at high-fluence B. The R phototaxis can be antagonized by simultaneous irradiation with FR (760 nm), supporting the participation of a Phy-type pigment. Genetic analyses have connected several Phys to both the R and B responses. The R response requires TaxD1, a member of the Phy-like clade (Bhaya et al., 2001; Ng et al., 2003). The organization of TaxD1 and its sensory cascade implies that R activates a phosphorelay involving a separate HK TaxAY1, with feedback methylation of TaxD1 possibly regulating the activity of the photoreceptor. Ultimately, the cascade directly impacts the motility motor of Type IV pili and its biogenesis. Although the B response likely requires another non-Phy receptor, the Phy Cph2 appears to modulate this response. Loss of Cph2 increases the sensitivity of *Synechocystis* to B, suggesting that the Cph2 protein functions to repress the B signaling system (Wilde et al., 2002). Loss of the Phy-like protein PlpA derepresses the autotrophic growth of *Synechocystis* in B, inferring a similar connection between a B-absorbing pigment and a Phy in this organism (Wilde et al., 1997).

Little is known about the phototactic response of non-photosynthetic eubacteria. *Agrobacterium tumefaciens* contains both flagella and Type IV pili and is motile but no photoresponses have been documented. Light does strongly promote the pathogenicity of this soil bacterium toward plant cells (Zambre et al., 2003) but the effective wavelengths have not been established. Since infection and subsequent transformation requires that the bacterium swims to the host, it is possible that a light-regulated motility response directed by *At*BphP1 and/or 2 is involved.

8.6.2 Enhancement of Photosynthetic Potential

As predicted, Phys from photoautotrophic microbes appear to have important roles in regulating the production, assembly, and modulation of the pigment/protein complexes needed for photosynthetic light harvesting. As in plants, the synthesis of the many components in the prokaryotic photosynthetic apparatus is likely regulated by Phys. In both the anoxygenic photosynthetic bacteria *Bradyrhizobium* and *Rhodopseudomonas palustris,* a bathyBphP is essential for this expression (Giraud et al., 2002). The corresponding apoprotein genes are close to the cluster of genes responsible for the synthesis of the light-harvesting complex (*Puf* operon), bacteriochlorophyll (*Bch*), and carotenoids (*Crt*) (Figure 8.5). In the same operon as *BphP* is *PpsR*; it encodes a RR with an appended DNA-binding motif. PspR represses the expression of photosynthetic genes in other bacteria and likely serves the same function in *Bradyrhizobium* and *Rhodopseudomonas palustris.* Disruption of *Bradyrhizobium BphP* blocks the light-induced activation of the adjacent photosynthetic gene cluster, whereas disruption of the *PpsR* gene constitutively induces their expression (Giraud et al., 2002). The simplest model is that both participate in a TC-HK cascade in which the BphP modulates a separate HK that represses transcriptional inhibition by PpsR, thereby stimulating the production of light-harvesting centers. The photoresponses of wild-type

Bradyrhizobium and *Rhodopseudomonas palustris* have a maximum at 750 nm, consistent with the Pr form of these bathyBphPs being the active form and the Pfr form being the inactive ground state (Giraud et al., 2002).

In numerous cyanobacteria, Phys are involved not only in producing the light-harvesting centers but also in altering their pigment composition in response to changing light quality. During complementary chromatic adaptation, reciprocal accumulation of the bili-proteins phycocyanin and phycoerythrin helps certain cyanobacteria optimize light capture in red- and green-rich light environments, respectively. This reversible process is mediated by differential expression of the apoproteins and associated bilins. In *Fremyella diplosiphon*, chromatic adaptation requires the Phy-like protein, RcaE (Kehoe and Grossman, 1996). Genetic dissection of the response has uncovered other components in the sensory cascade, including RcaC and RcaF (Kehoe and Grossman, 1996, 1997). RcaF is a RR without an obvious output motif. RcaC is a multi-domain protein containing an N-terminal RR followed by a potential DNA-binding motif, a HPT domain, and a second RR. The RRs and the HPT of RcaC contain the catalytic aspartic acid and histidine residues, respectively, indicating that they could be functional in a TC-HK cascade. Such an organization implies that complementary chromatic adaptation in *Fremyella diplosiphon* is driven by a four-step phosphorelay involving one or more Phys perceiving the light signal, which then initiate a phosphorelay from RcaF to RcaC. This relay culminates in the activation or repression of the DNA-binding motif in RcaC to effect transcription of appropriate phycobiliprotein and bilin biosynthetic genes. Even though RcaE participates genetically in the sensory cascade for complementary chromatic adaptation, it is not yet clear whether this Phy is the actual photoreceptor (Terauchi et al., 2004). RcaE can bind bilins but fails to demonstrate the characteristic R/FR photochromicity of typical Phys (see above).

Surprisingly, some unicellular microbes have circadian clocks whose entrainment by day length persists from one generation to another despite doubling times of less than one day. For several cyanobacterial species, this clock helps coordinate the synthesis of photosynthetic genes for maximum light use during the expected day. In the cyanobacterium *Synechococcus elongatus*, entrainment of the clock to the photoperiod requires input from the Phy-like protein, CikA (Schmitz et al., 2000). Both the period and amplitude of the clock as well as its phase entrainment are altered in *cikAΔ* mutants. How CikA resets the clock is not yet known. It binds bilins poorly and does not become R/FR photochromic (Mutsuda et al., 2003). While CikA does contain an appended RR, this motif is missing the phosphoacceptor aspartic acid, precluding its direct participation in a TC-HK. One possibility is that CikA functions in circadian entrainment but as an accessory factor that modulates the activity of the actual photoreceptor.

8.6.3
Photocontrol of Pigmentation

In several situations, Phys help regulate the synthesis of photoprotective pigments. In *Rhodospirillum centenum*, the BphP Ppr enhances the expression of chalcone synthase, the rate-limiting enzyme in the phenylpropanoid pathway used to synthesize flavonoid pigments (Jiang et al., 1999). In plants, flavonoids production is stimulated by various stresses including high fluence white light and UV light, implying that the flavoniods have a photoprotective role. The response in *Rhodospirillum centenum* is toward B and not R, indicating that the *p*-hydroxycinnamic acid chromophore bound to the PYP domain is responsible and not a bilin bound to its predicted BBP.

Deinococcus radiodurans BphP appears to have an important role in regulating carotenogenesis. Whereas dark grown cells accumulate carotenoids, the most abundant being a novel form called deinoxanthin, light stimulates this synthesis dramatically (Davis et al., 1999b). In fact, white light-grown *Deinococcus radiodurans* colonies are bright red. This light-induced increase is markedly attenuated in a *bphPΔ* mutant. The mutant is also slow growing under the high light conditions, suggesting a photoprotective role for the pigmentation. In higher plants, carotenogenesis is similarly regulated by phys via their ability to transcriptionally upregulate genes for the rate-limiting enzymes, phytoene synthase and phytoene desaturase. Notably, *Deinococcus radiodurans* counterparts to both genes are included within an operon, which would simplify their possible photoregulation by *Dr*BphP (Davis et al., 1999b).

8.7
Evolution of the Phy Superfamily

The recent discovery of the Phy-type pigments in cyanobacteria, eubacteria, and fungi now provides a potential route for the evolution of the Phy superfamily. Given their widespread presence in a number of non-photosynthetic and photosynthetic eubacteria and cyanobacteria, the BphP family likely represents the progenitor for all of the bilin-containing photochromic pigments. Their use of BV as the chromophore also represents the simplest way to obtain a linear bilin, requiring just one enzymatic step from the cyclic heme precursor (Bhoo et al., 2001). The incorporation of the HKD then provided a facile way to connect light to appropriate signaling networks. We presume that the bathyBphPs arose from the BphPs as a way to exploit FR as an environmental signal. However, it should be noted that BphPs are not universally present in eubacteria, being absent in many species (e.g. *Escherichia coli* and *Bacillus subtilis*), including those that are closely related to species that do contain a BphP (e.g. other *Pseudomonas* strains and *Thermus aquaticus*). Whether these species lost their BphPs over time or descended from lineages that arose before this photoreceptor is unknown.

Relatives of the BphPs then emerged during the evolution of cyanobacteria and fungi. For the Cph family, PCB was adopted as the chromophore, which was readily available given its synthesis as a photosynthetic accessory pigment. Such a change may have reflected the exploitation of Cphs to enhance light capture under competitive conditions. Whereas the absorption spectrum of the Pr form of BV-BphPs overlaps poorly with chlorophylls, the absorption spectrum of the Pr form of PCB-Cphs overlaps well. As a consequence, the Pr/Pfr ratio of PCB-Cphs at photoequilibrium is much more sensitive to shading by photosynthetic organisms than BV-containing forms. Such shading, which preferentially removes R versus FR, can then be monitored by a high Pr/Pfr ratio as opposed to a Pr/Pfr ratio close to one when R and FR are in equal proportions (e.g. full sunlight). Clearly the adoption of PCB as the chromophore would require a new method to link the bilin and a way to discriminate PCB from its precursor BV. Both problems may have been overcome by the use of the GAF cysteine to attach the chromophore covalently as opposed to the GAF histidine or the PLD cysteine. Various modifications of Cphs then created members of the Phy-like family. This process involved further changes in the BBP and replacement of the HKD with other signaling motifs.

Members of the Fph family most likely evolved from BphPs. They probably also use BV as the chromophore, consistent with the apparent inability of the fungi to synthesize other more complex bilins. At present, they have been found only in filamentous fungi with a preliminary action spectrum for light-induced sporulation, suggesting that one may be present in the slime mold *Physarum polycephalum* (Starostzik and Marwan, 1995). Given that Phy-type sequences are not evident in the genomes of single-cell fungi such as *Saccharomyces cerevisiae* and *Schizosaccharomyces pombe* or in another slime mold *Dictyostelium discoidium*, their distribution within the fungal kingdom may be limited. (Bhoo et al., 2001; Catlett et al., 2003). Likewise, BLAST searches of various *Archae* and animal genomes found no definitive evidence that Phys entered these kingdoms.

Plant phys then may have originated from a Cph precursor during the development of the chloroplast from a cyanobacterial endosymbiont. Plant phys became nuclear encoded and evolved further to help control the myriad of growth and developmental responses. This development also included the use of PΦB instead of PCB. However, the green alga *Mesotaenium caldorium* still employs PCB (Wu et al., 1997), suggesting that this substitution occurred later during the evolution of land plants. The major changes in plant phys relative to Cphs were the addition of the two internal PAS domains that help with signal output and modifications of the HKD. In typical TC-HKs, the HKD includes motifs not only for autophosphorylation and subsequent phosphotransfer but also for homodimerization. During the evolution of plant phys, it is possible that the HKRD has retained its dimerization contacts but either lost the HK activity or transformed the kinase domain to one with serine/threonine kinase activity.

8.8 Perspectives

The ongoing discovery of Phy-type pigments in microbes has impacted our understanding of Phys at many levels. In particular, their participation in TC-HK cascades has helped support the view that plant phys are light-regulated kinases. Whether plant phys use their serine/threonine kinase activity for signal transmission or for autoregulation is not yet clear (Quail, 2002). Whereas the physiological functions of plant phys are well documented, we know very little about the roles of their microbial counterparts. Preliminary data with photosynthetic bacteria suggest a common theme in optimizing light capture. This is accomplished either by directing movement toward more favorable light conditions or by regulating photosynthetic capacity with respect to light fluence, light quality, and/or photoperiod. For some microbes (e.g. *Synechocystis* and *Agrobacterium tumefaciens*), a complex interplay of signals emanating from multiple Phys is likely. In non-photosynthetic species, the roles of Phys are still unclear. Even though R is known to have important roles in conidiation, circadian rhythms, and sexual development in filamentous fungi, no aberrant phenotypes have been observed for *fphΔ* mutants in several species (Catlett et al., 2003 and A. Froehlich, J. Dunlap, and R.D. Vierstra, unpublished).

With respect to understanding how Phys function mechanistically as R/FR photochromic pigments, it is obvious that these microbial Phys now offer excellent opportunities to study this photoreceptor family. Genomic analyses indicate that several non-photosynthetic eubacteria encode only a single BphP and are void of other known photoreceptors, thus providing useful models to study Phy function in the absence of photosynthesis and other light sensing systems (Davis et al., 1999b; Bhoo et al., 2001; Lamparter et al., 2002; Karniol and Vierstra, 2003). They also offer the ability to produce an unlimited supply of homogeneous holoproteins for biochemical and biophysical studies. Several recombinant systems are now available that assemble Phy holoproteins by co-expressing the apoproteins with appropriate enzymes that synthesize the bilin chromophore (Bhoo et al., 2001; Gambetta and Lagarias, 2001). Furthermore, the availability of bathyBphPs that prefer Pfr as the ground state now affords the first opportunity to study Pfr without significant Pr contamination. Hopefully, biochemical, biophysical, and structural analyses of these pigments will reveal how Phys function as R/FR photochromic regulators of microbial and plant processes.

Acknowledgements

We thank various authors for providing information prior to publication and Allison Thompson and Brian Downes for helpful discussions. Our work was supported by grants from the U.S. Department of Energy and the National Science Foundation to RDV, and the Binational Research Development Fellowship to BK. This review is dedicated to my son Jeff.

References

Armitage, J.P. (1997) *Arch. Microbiol.* 168, 249–261.

Bhaya, D., Takahashi, A., and Grossman, A.R. (2001) *Proc. Natl. Acad. Sci. USA* 98, 7540–7545.

Bhoo, S.H., Davis, S.J., Walker, J., Karniol, B., and Vierstra, R.D. (2001) *Nature* 414, 776–779.

Bhoo, S.H., Hirano, T., Jeong, H.Y., Lee, J.G., Furuya, M., and Song, P.S. (1997) *J. Amer. Chem. Soc.* 119, 11717.

Braatsch, S., and Klug, G. (2004) *Photosyn. Res.* 79, 45–57.

Catlett, N.L., Yoder, O.C., and Turgeon, B.G. (2003) *Eukaryotic Cell* 2, 1151–1161.

Davis, S.J., Kurepa, J., and Vierstra, R.D. (1999a) *Proc. Natl. Acad. Sci. USA* 96, 6541–6546.

Davis, S.J., Vener, A.V., and Vierstra, R.D. (1999b) *Science* 286, 2517–2520.

Frankenberg, N., Mukougawa, K., Kohchi, T., and Lagarias, J.C. (2001) *Plant Cell* 13, 965–978.

Gambetta, G.A., and Lagarias, J.C. (2001) *Proc. Natl. Acad. Sci. USA* 98, 10566–10571.

Giraud, E., Fardoux, J., Fourrier, N., Hannibal, L., Genty, B., Bouyer, P., Dreyfus, B., and Vermeglio, A. (2002) *Nature* 417, 202–205.

Hader, D.P. (1987) In, *Cyanobacteria* (P. Fay and C. Van Baalen, eds) Elsevier, New York, NY, pp 325–345.

Herdman, M., Coursin, T., Rippka, R., Houmard, J., and Tandeau de Marsac, N. (2000) *J. Mol. Evol.* 51, 205–213.

Hubschmann, T., Borner, T., Hartmann, E., and Lamparter, T. (2001a) *Eur. J. Biochem.* 268, 2055–2063.

Hubschmann, T., Jorissen, H., Borner, T., Gartner, W., and de Marsac, N.T. (2001b) *Eur. J. Biochem.* 268, 3383–3389.

Hughes, J., Lamparter, T., Mittmann, F., Hartmann, E., Gartner, W., Wilde, A., and Borner, T. (1997) *Nature* 386, 663.

Inouye, M., and Dutta, R. (2003) *Histidine Kinases in Signal Transmission* Academic Press, NY.

Jiang, Z., Swem, L.R., Rushing, B.G., Devanathan, S., Tollin, G., and Bauer, C.E. (1999) *Science* 285, 406–409.

Jorissen, H., Quest, B., Lindner, I., de Marsac, N.T., and Gartner, W. (2002a) *Photochem. Photobiol.* 75, 554–559.

Jorissen, H., Quest, B., Remberg, A., Coursin, T., Braslavsky, S.E., Schaffner, K., de Marsac, N.T., and Gartner, W. (2002b) *Eur. J. Biochem.* 269, 2662–2671.

Karniol, B., and Vierstra, R.D. (2003) *Proc. Natl. Acad. Sci. USA* 100, 2807–2812.

Karniol, B., and Vierstra, R.D. (2004) *J. Bacteriol.* 186, 445–453.

Kehoe, D.M., and Grossman, A.R. (1996) *Science* 273, 1409–1412.

Kehoe, D.M., and Grossman, A.R. (1997) *J Bacteriol.* 179, 3914–3921.

Lamparter, T., Michael, N., Mittmann, F., and Esteban, B. (2002) *Proc. Natl. Acad. Sci. USA* 99, 11628–11633.

Lamparter, T., Michael, N., Caspani, O., Miyata, T., Shirai, K., and Inomata, K. (2003) *J. Biol. Chem.* 278, 33786–33792.

Lamparter, T., Carrascal, M., Michael, N., Martinez, E., Rottwinkel, G., and Abian, J. (2004) *Biochemisty* 43, 3659–3669.

Loros, J.J., and Dunlap, J.C. (2001) *Ann. Rev. Physiol.* 63, 757–794.

Mathews, S., and Sharrock, R.A. (1997) *Plant Cell Environ.* 20, 666–671.

Matsushita, T., Mochizuki, N., and Nagatani, A. (2003) *Nature* 424, 571–574.

Muramoto, T., Kohchi, T., Yokota, A., Hwang, I., and Goodman, H.M. (1999) *Plant Cell* 11, 335–348.

Mutsuda, M., Michel, K.P., Zhang, X.F., Montgomery, B.L., and Golden, S.S. (2003) *J. Biol. Chem.* 278, 19102–19110.

Ng, W.O., Grossman, A.R., and Bhaya, D. (2003) *J. Bacteriol.* 185, 1599–1607.

Park, C.M., Shim, J.Y., Yang, S.S., Kang, J.G., Kim, J.I., Luka, Z., and Song, P.S. (2000a) *Biochemistry* 39, 6349–6356.

Park, C.M., Kim, J.I., Yang, S.S., Kang, J.G., Kang, J.H., Shim, J.Y., Chung, Y.H., Park, Y.M., and Song, P.S. (2000b) *Biochemistry* 39, 10840–10847.

Quail, P.H. (2002) *Nat. Rev. Mol. Cell. Biol.* 3, 85–93.

Quest, B., and Gartner, W. (2004) *Eur. J. Biochem.* 271, 1117–1126.

Schmitz, O., Katayama, M., Williams, S.B., Kondo, T., and Golden, S.S. (2000) *Science* 289, 765–768.
Schneider-Poetsch, H.A. (1992) *Photochem. Photobiol.* 56, 839–846.
Smith, H. (2000) *Nature* 407, 585–591.
Starostzik, C., and Marwan, W. (1995) *FEBS Lett.* 370, 146–148.
Terauchi, K., Montgomery, B.L., Grossman, A.R., Lagarias, J.C., and Kehoe, D.M. (2004) *Molec. Microbiol.* 51, 567–577.
van der Horst, M.A., and Hellingwerf, K.J. (2004) *Acc. Chem. Res.* 37, 13–20.
Venter, J.C., Remington, K., Heidelberg, J.F., Halpern, A.L., Rusch, D., Eisen, J.A., Wu, D.Y., Paulsen, I., Nelson, K.E., Nelson, W., Fouts, D.E., Levy, S., Knap, A.H., Lomas, M.W., Nealson, K., White, O., Peterson, J., Hoffman, J., Parsons, R., Baden-Tillson, H., Pfannkoch, C., Rogers, Y.H., and Smith, H.O. (2004) *Science* 304, 66–74.
Vierstra, R.D. (2002) In *Histidine Kinases in Signal Transduction* (M. Inouya and R. Dutta eds.), pgs 273–295.
West, A.H., and Stock, A.M. (2001) *Trends Biochem. Sci.* 26, 369–376.
Wilde, A., Fiedler, B., and Borner, T. (2002) *Molec. Microbiol.* 44, 981–988.
Wilde, A., Churin, Y., Schubert, H., and Borner, T. (1997) *FEBS Lett.* 406, 89–92.
Wu, S.H., and Lagarias, J.C. (2000) *Biochemistry* 39, 13487–13495.
Wu, S.H., McDowell, M.T., and Lagarias, J.C. (1997) *J. Biol. Chem.* 272, 25700–25705.
Yeh, K.C., and Lagarias, J.C. (1998) *Proc. Natl. Acad. Sci. USA* 95, 13976–13981.
Yeh, K.C., Wu, S.H., Murphy, J.T., and Lagarias, J.C. (1997) *Science* 277, 1505–1508.
Zambre, M., Terryn, N., De Clercq, J., De Buck, S., Dillen, W., Van Montagu, M., Van Der Straeten, D., and Angenon, G. (2003) *Planta* 216, 580–586.

9
Light-activated Intracellular Movement of Phytochrome

Eberhard Schäfer and Ferenc Nagy

9.1
Introduction

Despite intensive efforts and significant progress, the primary processes of phytochrome-mediated responses are not yet fully understood. The three hypotheses most widely accepted describe the molecular functions of phytochrome as an enzyme, membrane effector, and transcription regulator, respectively. These molecular models and recent results clearly indicate that detailed knowledge about the intracellular localization of these photoreceptors is an essential pre-requisite for understanding early events in phytochrome-mediated signalling. Thus, in this chapter we describe data obtained about the distribution and localization of phytochrome by classical and contemporary methods. Beyond listing these observations, we also evaluate in detail some of the key findings and explain how they helped to develop novel molecular concepts for light-induced signalling.

9.2
The Classical Methods

The classical studies employed spectroscopic, immunocytochemical, and cell biological/ biochemical techniques to characterize the intracellular localization of phytochromes. For more detailed description of these studies and methods see Chapter 4 in *Plant Photomorphogenesis* (Kendrick and Kronenberg, 1994).

9.2.1
Spectroscopic Methods

Prior to the onset of the molecular era, micro-beam irradiation was a major tool for obtaining information about the intracellular localization of phytochromes. In their pioneering experiments Etzold (1965) and Haupt (1970) observed an action dichroism for photo- and polarotropism of the chloronemata of ferns and chloroplast ori-

Handbook of Photosensory Receptors. Edited by W. R. Briggs, J. L. Spudich

ISBN 3-527-31019-3

entation in the green alga *Mongeotia*. These findings suggested that the absorption dipole moment of Pr is parallel and that of Pfr perpendicular to the cell surface. The responses induced by a micro-beam pulse appeared to be local, since they could only be reversed by a subsequent far-red pulse given to the same spot. Thus it was concluded that the intracellular mobility of phytochrome in these cases is very limited.

The group led by M. Wada further refined these experiments and clearly demonstrated that the micro-beam must hit a region including the cell wall, the plasma membrane, and part of the cytosol to initiate the response. It was therefore concluded that phytochromes mediating these responses are not associated with plastids, mitochondria, or nuclei, but localized close to the plasma membrane. We note that although these experiments clearly indicated an ordered localization of phytochromes, the physical association of the photoreceptor molecules with the membrane could not be proved by this method.

Attempts to use similar techniques in higher plants failed primarily because light is scattered within the tissue and no strictly localized responses mediated by phytochromes are known in higher plants. In contrast, results obtained by Marmé and Schaefer, who used polarized light to induce photoconversion of phytochrome in vivo indicated partial action dichroism, i.e. an ordered localization of the photoreceptor (Marmé and Schaefer, 1972). Interpretation of these experimental data, however, was complicated since the role of differential light attenuation in causing the measured differences could not be rigorously excluded.

9.2.2
Cell Biological Methods

In addition to spectroscopic studies, cell fractionation was also considered as an efficient tool to determine whether phytochromes are associated with membranes (Kraml, 1994). In summary, these studies indicated that phytochrome could be associated with various organelles and also with the plasma membrane. The biological significance of these findings, however, has not yet been demonstrated and there is considerable doubt whether these observations indeed reflect localization of the phytochrome molecules in vivo. We note, however, that Quail et al. (1973), using the same method, reported red/ far-red reversible pelletability of phytochrome and this result was later confirmed using immunocytochemical methods when light-dependent formation of SAPs (sequestered areas of phytochrome) was reported (MacKenzie et al., 1975; Speth et al., 1986).

9.2.3
Immunocytochemical Methods

Because of the technical problems inherent in the cell fractionation method, the next approach, pioneered by the Pratt laboratory, was immunocytochemistry. McCurdy and Pratt (1986) showed that the immunodetectable phytochrome (phyA) in dark-grown oat coleoptiles is homogenously distributed throughout the cytoplasm. No association with organelles or membranes was observed and irradiation very rapidly – with a half-life of a few seconds – induced formation of SAPs. In darkness these SAPs disappeared with a half-life of about 30 minutes (Speth et al., 1986). Co-localization of SAPs and ubiquitin indicated that the SAPs might be the place of phyA degradation (Speth et al., 1987), a process mediated by the 26S proteasome. We should note, however, that this attractive hypothesis is still being debated and yet to be proved experimentally.

Ten years later Mösinger and Schäfer (1984) and Mösinger et al. (1985) demonstrated that transcription rates could be regulated by irradiating isolated nuclei. It is worth noting that these observations were ignored and forgotten for the following ten years, even though these data suggested that at least a fraction of phytochromes is localized in the nucleus during signal transduction.

9.3 Novel Methods

After the genes encoding phytochrome were cloned and protein sequences became available it was concluded that phytochromes are not integral membrane proteins and do not contain canonical nuclear localization signals (NLS). Thus, it became generally accepted that phytochromes are soluble cytosolic proteins, which probably associate with membranes only after binding to a molecule that itself is membrane-localized. Pioneering work performed by Sakamoto and Nagatani (1996) seriously challenged this view. These authors reported for the first time, the enrichment of phyB in nuclear extracts isolated from light-grown Arabidopsis seedlings. Moreover, the same authors demonstrated that a fusion protein consisting of the C-terminal part of *Arabidopsis thaliana* PHYB fused to the GUS reporter is constitutively localized into the nucleus in transgenic plants. These data obviously contradicted the membrane model but were overlooked for several years.

The situation, however, changed dramatically two years later, when Ni et al., (1998) reported interaction of phyA and phyB with PIF3 (phytochrome interacting factor 3), a transcription factor belonging to the family of bHLH (basic helix-loop-helix) proteins. This finding implied that phyA and phyB have to be localized in the nucleus in order to interact with this transcription factor at least temporarily, to mediate light-induced signal transduction. In 1999, independent of this hypothesis, Nagatani's group and we, ourselves, demonstrated beyond reasonable doubt, by analyzng the nucleo/cytoplasmic distribution of the PHYB:GFP (green fluorescent protein) fusion protein in transgenic plants, that light does indeed induce nuclear import of this photoreceptor in transgenic Arabidopsis plants (Yamaguchi et al., 1999; Kircher et al., 1999). The appearance of the characteristic PHYB overexpression phenotype of transgenic plants (Kircher et al., 1999) and complementation of an *Arabidopsis thaliana* (Yamaguchi et al., 1999) or a *Nicotiania plumbagenifolia* mutant lacking functional phyB (Gil et al., 2000) by the expression of the PHYB:GFP chimeric protein demonstrated that these fusion proteins represent photobiologically active photoreceptors.

Kircher et al. (1999) also studied nucleo/ cytoplasmic distribution of a chromophore-less mutant of phyB (the cysteine-encoding codon of the chromophore attachment site was mutated to code for an alanine) fused to GFP in transgenic plants. These authors found that this fusion protein is constitutively localized in the cytosol. Thus they concluded – based on the hypothesis that this mutant version has a conformation similar to the Pr form – that the Pr conformer of the photoreceptor is not compatible with nuclear import. The same holds for the N-terminal fragment fused to GFP (Matsushita et al., 2003), whereas the C-terminal half of PHYB must contain a functional NLS(s), since chimeric proteins containing the C-terminal part of PHYB fused to the GFP reporter showed constitutive nuclear localization (Yamaguchi et al., 1999; Nagy et al., 2000). With these various transgenic lines in hand that expressed an easily detectable, biologically functional PHYB:GFP photoreceptor, photobiological studies of the molecular mechanism regulating the intracellular localization of phyB and other phytochromes became feasible.

9.4 Intracellular Localization of PHYB in Dark and Light

In 6-day-old dark-grown seedlings the PHYB:GFP fusion protein is localized mainly in the cytosol. If the expression level of the transgene is high, occasionally a weak diffuse nuclear fluorescence is also observable, indicating nuclear localization of the fusion protein (Kircher et al., 1999; Yamaguchi et al., 1999; Kircher et al., 2002; Matsushita et al., 2003). Results obtained by Kircher et al. (2002) suggest that light treatment of imbibed seeds to promote homogenous germination can induce nuclear import of phyB. Thus it is conceivable that the weak nuclear staining detected in 6-day-old etiolated seedlings represents phyB molecules that were imported into the nucleus during this early phase of development. However, independently of the occasional diffuse staining, irradiation with either red or white light induces nuclear import of PHYB:GFP and subsequent accumulation of the photoreceptor in the nucleus. Nuclear-localized PhyB is not distributed homogenously in the nucleoplasm: it preferentially accumulates in characteristic structures termed speckles (Kircher et al., 1999, 2002).

Detailed studies showed that the nuclear import of PHYB:GFP as well as the formation of PHYB:GFP-containing speckles is a slow process, which saturates in about 4 h and shows a strong fluence-rate dependence (Gil et al., 2000). The wavelength dependence of these processes – tested under 6-h continuous irradiation – paralleled the described wavelength dependence of phyB-mediated seed germination (Shinomura et al., 1996). The almost complete lack of responsiveness to wavelengths longer than 695 nm establishing a Pfr/ Ptot ratio of ca. 40% was quite surprising. Tests with light pulses showed that a single light pulse was almost ineffective, but three consecutive 5-minute pulses given at hourly intervals induced import and formation of speckles containing the PHYB:GFP fusion protein. The inductive signal was reversible by a subsequent far-red light pulse, indicating that the nuclear import of phyB has the characteristics of a typical Low Fluence Response (LFR) (Kircher et al.,

1999; Nagy et al., 2000). Physiological experiments have shown that responsiveness to an inductive light pulse is often poor in etiolated seedlings, but could be strongly enhanced by pre-irradiation activating either phyB (red light), phyA (far-red light) or cry1, cry2 (blue light). Gil et al. (2000) reported that pre-irradiation with red and blue, but not with far-red light enhanced nuclear import and the formation of phyB-containing speckles. Moreover, the same authors also showed that the effectiveness of pre-irradiation with red light slowly decreased, thus the inductive effect of a 5-s red light treatment was completely lost after a 24-h dark period.

The PHYB:GFP fusion protein localized in the nucleus disappears slowly, with a half-life of about 6 h, in seedlings transferred back to darkness. First the speckles are dissolved. This stage is transient and the nuclei display diffuse staining. The next stage, i.e. the complete loss of nuclear staining takes about 10 h (Gil et al., 2000). Whether the slow disappearance of nuclear phyB is due to the slow export or slow turnover of the photoreceptor remains to be determined. We note, however, that the disappearance of nuclear staining could be accelerated by about 2 h by irradiating the seedlings with a FR pulse before the transfer to darkness, which is a typical end-of-day response.

Taken together, these data indicate that light-induced nuclear import of phyB exhibits the characteristics of a typical phyB-mediated physiological response. Namely, it displays low responsiveness to single pulses, red/ far-red reversibility of multiple pulses (LFR), sharp decline of responsiveness to wavelengths longer than 695 nm, fluence rate dependence, and responsiveness amplification. Moreover, it became evident that the light-induced import of phyB into the nuclei is followed by the rapid formation of large sub-nuclear complexes, termed speckles, which harbor the bulk of the photoreceptor detectable in the nuclei.

9.5 Intracellular Localization of PHYA in Dark and Light

The immunocytological experiments performed in the 1970s and 1980s characterized the localization of phyA primarily in monocotyledonous plants. They showed that light treatment results in a rapid rearrangement of cytosolic phyA and leads to the formation of phyA-containing cytosolic complexes (SAPs). With transgenic tobacco and Arabidopsis seedlings expressing the PHYA:GFP fusion protein on hand, the light-dependent intracellular localization of this photoreceptor was re-investigated (Kircher et al., 1999; Kim et al., 2000; Kircher et al., 2002). As in the case of PHYB:GFP, the functionality of the PHYA:GFP fusion protein was verified by successful complementation of a PHYA null mutant. These studies demonstrated that both, the rice PHYA:GFP and the Arabidopsis PHYA:GFP fusion proteins, were exclusively cytosolic in dark-grown transgenic tobacco and Arabidopsis seedlings, respectively. However, in contrast to PHYB:GFP, the intracellular distribution of the different PHYA:GFP fusion proteins showed a very rapid change after irradiation. Even a single far-red light pulse is sufficient in all these cases to induce a rapid formation of cytosolic spots, followed by translocation of the PHYA:GFP fusion protein to the nuclei. We note that the cytosolic PHYA:GFP spots are reminiscent of the SAPs previously described in monocotyledonous seedlings (MacKenzie et al., 1975; McCurdy and Pratt, 1986; Speth et al., 1986).

Nuclear import of PHYA:GFP was also followed by the formation of nuclear speckles, as was the case for PHYB:GFP. However, the PHYA:GFP-containing speckles appeared very rapidly and both, their size and number, were much reduced as compared to those of PHYB:GFP speckles (Kim et al., 2000; 2002). These data demonstrate that light-mediated nuclear import of phyA is a typical phyA-mediated Very Low Fluence Response (VLFR). PhyA can also mediate the far-red High Irradiance Response (HIR). In both, transgenic tobacco and Arabidopsis seedlings, continuous far-red light led to nuclear import of PHYA:GFP. The import process is fluence-rate and irradiance dependent (Kim et al., 2000). Thus it reflects a typical far-red HIR. A further characteristic of a far-red HIR is that it is diminished after a pre-treatment with red light (Beggs et al., 1980; Holmes and Schäfer, 1981). The nuclear import of PHYA:GFP could also be almost completely abolished by 24-h pre-treatment with red light (Kim et al., 2000). We note that similar results were obtained by Hisada et al. (2000), who analysed continuous far-red light- and light pulse-dependent intracellular localization of phyA in pea seedlings, using cytochemical methods.

In summary it can be concluded that i) PHYA:GFP is localized exclusively in the cytosol in dark grown seedlings, ii) irradiation initiates rapid formation of cytosolic SAPs and iii) nuclear import is followed by formation of nuclear speckles containing the PHYA:GFP fusion protein. These processes display complex dynamics and are mediated by VLFR and HIR.

9.6
Intracellular Localization of PHYC, PHYD and PHYE in Dark and Light

To complete the characterization of the nucleo/ cytoplasmic partitioning of all members of the phytochrome gene family, Kircher et al. (2002) produced transgenic Arabidopsis lines expressing PHYA–E:GFP fusion proteins under the control of the 35S cauliflower mosaic virus promoter. These authors found that in dark-grown seedlings the PHYD, PHYC and PHYE:GFP fusion proteins are primarily localized in the cytosol, as were PHYA:GFP and PHYB:GFP. Upon irradiation all phytochromes undergo nuclear import and speckle formation; however, both of these processes display phytochrome-specific kinetics and light-dependence. Nuclear transport of PHYD–E is red- and white-light inducible. Interestingly, although PHYB and PHYD are closely related genes, they showed the largest difference. PHYD:GFP displayed a very slow nuclear import and only one or two larger speckles per nucleus were detectable even after an 8-h irradiation by white light (Kircher et al., 2002). The speckle formation of all PHYs, except that of PHYD, showed a robust diurnal regulation under light/ dark cycles. The start of speckle formation even before the light-on signal indicates a circadian control. This phenomenon could be most clearly shown for PHYB:GFP (Gil et al., 2001; Kircher et al., 2002).

9.7
Intracellular Localization of Intragenic Mutant Phytochromes

Various approaches aimed at identifying signal transduction components for phytochrome-mediated responses resulted in the isolation of intragenic PHYA and PHYB mutants. These mutants can be classified as loss-of-function (hyposensitive, Yanovsky et al., 2002) and hypersensitive mutants (Kretsch et al., 2000; Casal et al., 2002). Dark- and light-dependent intracellular localization of some of these mutant photoreceptors has been examined by expressing them as PHYA:GFP and PHYB:GFP fusion proteins in transgenic Arabidopsis lines.

9.7.1
Hyposensitive, Loss-of-function Mutants

PhyA and phyB were both shown to interact with the transcription factor PIF3 in yeast (Ni et al., 1998). In vitro experiments demonstrated that the interaction of the photoreceptors with PIF3 is regulated by light, thus it is mediated by the biologically active conformer, namely the Pfr form of phyA and phyB (Ni et al., 1999). These authors also showed that the interaction of the transcription factor with photoreceptors encoded by a number of mutant alleles of phyA and phyB was significantly weakened. These mutants displayed hyposensitive phenotypes in vivo, thus the perturbed signalling was explained by the lack of interaction between the photoreceptor and PIF3. The majority of phyA and phyB mutants tested in these experiments carried missense point mutations in a specific region, termed the Quail box, of the photoreceptors. Kircher et al. (2002) investigated whether these point mutations affected just the interaction of the photoreceptors with PIF3 or also the nucleo/ cytoplasmic distribution of phyA and phyB. To this end they raised transgenic plants expressing the mutant PHYA and PHYB genes fused to GFP under the control of the 35S promoter and characterized the light-induced nuclear import of the fusion proteins in detail. The majority of the mutant photoreceptors were imported into the nuclei in a light-induced fashion, with no significant difference as compared to wild type phyA and phyB. The only exception observed so far was the G767R point mutant of PHYB. In this case the mutant protein accumulated to a significantly lower level in the nuclei of irradiated seedlings. We note that the insertion of an extra NLS into the G767R PHYB:GFP construct increased the accumulation of the fusion protein in the nuclei to levels similar to that of wild type PHYB:GFP and led to full complementation of a phyB deficient mutant (Matsushita et al., 2003). In contrast, to the seemingly normal nuclear import of the mutant photoreceptors, formation of speckles co-localizing with PHYA or PHY:GFP fusion proteins was almost completely absent or much reduced in all mutants, including the G767R point mutant. The Quail box of phyA and phyB had been shown in vitro to be essential for the interaction with PIF3. Thus the loss of nuclear speckles in the mutants was interpreted as an indication that these sub-nuclear complexes could be involved in mediating light-induced signalling and can be considered as molecular markers for physiologically active phyA and phyB (Kircher et al., 2002) (see Section 9.8 below for further discussion).

In an independent line of experiments Yanovsky et al. (2002) reported the isolation of a phyA mutant, which displayed complete loss of the HIR but retained the VLFR. This phyA-302 mutant was shown to contain the E777R point mutation in the PAS2 domain of the photoreceptor (see Tu and Lagarias, Chapter 6). The PHYA-302:GFP fusion protein was expressed in wild-type and phyA null-mutant (phyA-201) backgrounds and in both cases showed normal translocation from the cytosol to the nucleus under continuous far-red light, but failed to produce nuclear speckles. These data again indicated that these sub-nuclear complexes are required for light-induced signalling. Moreover, these results suggested that they are specifically involved in regulating HIR signalling and/ or degradation of phyA, but not VLFR signalling.

9.7.2
Hypersensitive Mutants

In a screen designed to isolate mutants exhibiting loss of reversibility, Kretsch et al. (2000) obtained a hypersensitive mutant (phyB-401), that was identified as a point mutant, carrying a G564E substitution in the conserved hinge region of phyB. Transgenic lines expressing the mutant PHYB:GFP fusion protein under the control of the 35S promoter, expressed in a phyB-minus background, showed extreme hypersensitivity both in wild-type and phyB null (phyB-9) backgrounds. The kinetics of light-induced nuclear translocation of the mutant PHYB:GFP differed from those of the wild-type PHYB:GFP. In contrast to wild-type PHYB:GFP, the nuclear import of the mutant PHYB:GFP was induced by a single light pulse. Moreover, the translocation of the mutant PHYB:GFP to the nuclei was followed by an immediate, rapid formation of speckles containing the fusion protein. These speckles were stable for more than 48 h in darkness and a far-red pulse could still induce the disappearance of the speckles even after incubating the seedlings for 24 h in dark. Thus it can be concluded that the phyB-401 mutant is stable in its Pfr form. We note that in a recent screen, designated specifically to isolate mutants displaying aberrant nuclear import and/ or formation of phyB-containing nuclear speckles, two independent, novel phyB alleles were identified showing the same phenotype (Bauer, Essing, Kircher, Schaefer and Nagy, unpublished). These data also suggest that nuclear speckles contain the phyB photoreceptor in its biologically active Pfr conformation.

9.8
Protein Composition of Nuclear Speckles Associated with phyB

Different laboratories demonstrated that the nuclear import of the PHYB:GFP fusion protein is always followed by the rapid formation of PHYB:GFP-containing speckles. It was found that the size and number of the speckles depend on the quality and quantity of the inductive light treatment (Kircher et al., 2002). Thus it was suggested that these large nuclear structures, associated with or containing the photoreceptor, might play a role in mediating phyB-dependent, light-induced signal transduction (Nagy and Schaefer, 2002).

More recently, in an attempt to screen for mutants impaired in intracellular localization of phyB, the patterns of nuclear speckle formation were more precisely categorized (Chen et al., 2003). These authors identified four types of speckles and showed that the number and size of the phyB speckles can be correlated with the ratio of Pfr and Pr conformers of the photoreceptor. The majority of isolated mutants displayed hyposensitivity to red light and aberrant formation of speckles co-localizing with phyB. A significant proportion of the mutants was identified as intragenic phyB mutations, whereas some were mapped to chromosome regions that do not contain known genes involved in the regulation of light-induced signalling. A similar approach in our laboratory yielded comparable results (Bauer, Essing, Nagy and Schaefer, unpublished). The identification of these mutant genes is expected to shed light on the organization and function of these sub-nuclear complexes in phyB mediated signalling.

In an independent line of experiments we could show that changes of intracellular distribution of phyB are extremely dynamic. Earlier studies indicated that the nuclear import of phyB is a slow process and GFP fluorescence in the nuclei could not be detected in a reliable fashion within less than about 45–60 min after the beginning of the irradiation (Gil et al., 2000). Optimization of microscopic techniques, however, made it possible to obtain a better resolution using GFP fluorescence. Thus we were able to show that within a few seconds after the beginning of irradiation many small nuclear PHYB:GFP speckles, not detected in earlier studies, are formed. These are extremely transient: they already disappear after 10–20 min and additional irradiation results in the formation of the more stable speckles reported in the earlier studies (Bauer et al., 2004). Analysis of the localization of the phytochrome-interacting factor PIF3 (Ni et al., 1998) showed that PIF3 is co-localized with the early phyB speckles but not with the late ones appearing after prolonged irradiation (Figure 9.1). Moreover, we showed that in a PIF3 null background, phyB forms only the late but not the early speckles on irradiation. Western-blot analysis and microscopic studies clearly demonstrated that light induces rapid degradation of PIF3 (the half life of the PIF3 protein is about 10 nm) and this process is controlled by the concerted action of phyA phyB and phyD (Bauer et al., 2004). These data suggest that early speckles co-localizing with PIF3 and phyB might be the site of PIF3 degradation and/ or represent early events of phyB-mediated signalling.

However, to elucidate the exact function of the various types of phyB-associated speckles in phyB signal transduction, it will be essential to obtain information about their molecular composition. To this end our laboratory undertook the following experimental approach: nuclei from 4-week-old transgenic Arabidopsis plants expressing the PHYB:GFP fusion protein were isolated, disrupted and the intact speckles, representing the stable ones described above, were further purified by differential gradient centrifugation. The purified speckles were then solubilized, their components were separated on SDS PAGE and analysed by MALDI-TOF. It was found that about 80% of the more than 25 proteins identified in phyB-containing speckles display significant homology to proteins shown to be present in the interchromatin granual clusters (ICGs) of animal cells. We note that although phyA, like phyB, forms nuclear speckles on photo transformation, all attempts to obtain information about

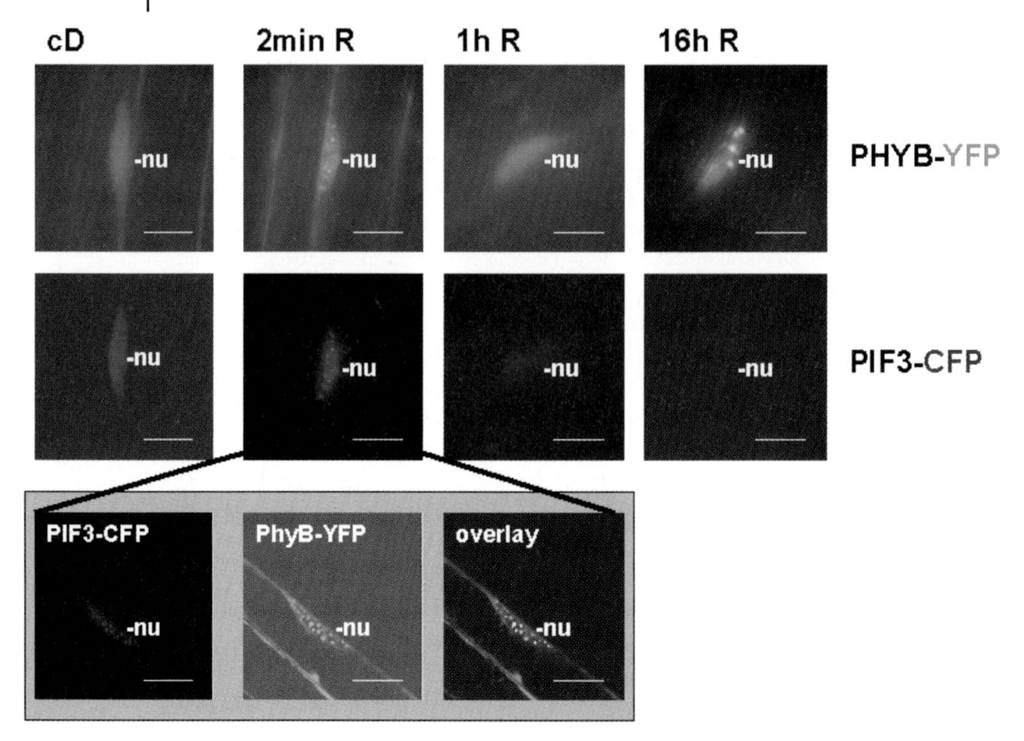

Figure 9.1 PHYB:YFP co-localizes transiently with PIF3:CFP in the nuclei of irradiated transgenic seedlings. Transgenic Arabidopsis lines expressing both PHYB:YFP and PIF3:CFP in *phyB⁻9* background were used to determine cellular distribution of these fusion proteins. Images taken during epifluorescence microscopic analysis of PHYB:YFP (upper row) and PIF3:CFP (lower row) are shown. 6-day-old etiolated seedlings (cD) were irradiated for 2 min, 1 h, and 16 h with red light (R). The insert shows confocal microscopic images of a cell expressing both fusion proteins after 2 nm R treatment. PIF3:CFP is displayed in red, PHYB:YFP is shown by green and overlay of the two signals is indicated by yellow. Scale bars = 10 μm. Positions of nuclei (nu) are indicated. Modified after Bauer et al. (2004).

the composition of this phyA-associated sub-nuclear protein complexes have so far proved unsuccessful.

Immunogold co-localization experiments clearly confirmed the co-localization of phyB with some of the proteins identified by MALDI-TOF. Taken together, these data indicate that the phyB-containing plant nuclear speckles are likely structural homologs of ICGs identified in animal cells. The exact biological role of ICGs is not yet fully understood even in animal cells. ICGs localize close to the actively transcribed regions and contain dozens of proteins involved in mRNA splicing. Thus they are considered to be involved in the storage, modification, and recruitment of factors necessary for transcription and splicing (Bubulya et al., 2002). In plant cells very little is known about these processes, but the association of the photoreceptor phyB with ICG-like complexes indicates that some of these molecular events could be regulated by light (Panihgrahi, Kunkel, Klement, Medzhradszky, Nagy, Schaefer, unpublished results). We note, however, that the N-terminal fragment of PHYB fused to GUS:NLS and the G767R point mutant PHYB fused to GFP:NLS did not show light-induced nuclear speckle formation, yet they successfully complemented the phenotype of phyB deficient mutants (for additional discussion see also Section 9.7.1 and 9.9). These data indicate that the formation of these nuclear complexes may not be essential for phyB-mediated signalling. Thus we conclude that, given their multiple forms and size, their transient and dynamically changing appearance, to understand the exact molecular function of the light-induced nuclear protein complexes, remains a challenging task.

9.9 The Function of Phytochromes Localized in Nuclei and Cytosol

Recent results provided compelling evidence that light quality- and quantity-dependent translocation of phytochromes into the nuclei represents a major regulatory step in light-induced signalling. This conclusion was further strengthened by the data recently reported by Huq et al. (2003). These authors expressed phyB fused to the glucocorticoid receptor (GR) in a phyB null background and investigated the cellular distribution of the fusion protein and the inhibition of hypocotyl elongation in the transgenic seedlings in red light in the presence or absence of the steroid hormone. They found that the fusion protein remained cytosolic in dark and in red light if no steroid hormone was added. Addition of the hormone allowed light-dependent transport of the fusion protein and complementation of the phyB null mutant phenotype as far as inhibition of hypocotyl elongation is concerned. (Huq et al., 2003).

In an independent line of experiments, Matsushita et al. (2003) inserted the SV40 NLS in the PHYB:GFP fusion protein and investigated the cellular distribution of PHYB:GFP:NLS in transgenic phyB null mutant Arabidopsis seedlings grown in light and dark. These authors reported that the fusion protein was constitutively localized in the nuclei, irrespective of the light conditions. Moreover, they found that transgenic plants did not exhibit an altered phenotype in the dark, but fully complemented the phyB null phenotype when grown in light. Using the same approach, our laboratory obtained somewhat different results. In our hands the expression of a similar fusion protein in Arabidopsis, although in a different genetic background, showed constitutive nuclear localization but resulted in pronounced hypersensitivity to red light (Kirchenbauer, Kircher, Nagy and Schäfer, unpublished). In contrast, the sub-cellular distribution of the same PHYB:GFP:NLS fusion protein in transgenic tobacco seedlings was not altered; the nuclear import of the chimeric photoreceptor remained light-inducible. Irrespective of the differences, these data clearly show that phytochrome localized in the nucleus is the functional phytochrome, light is still necessary for its activation and there is no obvious major contribution of cytosolic phyB to light-induced signalling underlying early steps of photomorphogenesis. In this context we point out that alteration of the nucleo/cytoplasmic distribution of PHYA:GFP by insertion of an additional NLS in the fusion protein has not yet been reported.

For phyB, the possibility of driving the photoreceptor constitutively into the nucleus even in darkness by fusing it with an extra NLS also allowed the functional analysis of truncated and mutated phyB molecules that otherwise would have been excluded from the nucleus. In a set of elegant experiments Matsushita et al. (2003) reported that the complete C-terminal domain of the phyB is dispensable for its function as a photoreceptor and it appears to be required to mediate the light-induced nuclear import of the full-length protein. It follows that the N-terminal domain fused to GUS (to facilitate dimerization) and NLS was constitutively imported into the nucleus. More importantly, transgenic seedlings expressing this fusion protein displayed hypersensitivity to red light, but showed no altered phenotype in dark, indicating that N-terminal domain alone can function as a biologically active photoreceptor. The main caveat of these exciting experimental results is that GUS was used as a dimerization domain although it is known to tetramerize. Thus GUS may provide an artificial platform to recruit signalling partners for the otherwise non-functional molecule. Therefore, additional experiments using other dimerization domains could be important to clarify this issue. However, if this is proven not to be the case, the results reported by Matsushita et al. (2003) indicate that phyB probably does not function as a kinase or the kinase function of the molecule is not essential for mediating light-induced signalling. This conclusion is based on the fact that the complete histidine kinase-like domain and also parts of the domain believed to be essential for the serine/threonine kinase function are absent from the truncated but biologically active fusion protein. The data described above convincingly prove that phyA and phyB localized in the nucleus are the functional form of these photoreceptors, but light is required to switch on signalling.

In contrast to the wealth of data obtained about the function of nuclear localization, our knowledge about the function of non-nuclear phytochromes in light-induced signalling is very limited. This is somewhat surprising, since a large fraction of phyA–E remains localized in the cytosol even in plants kept in constant light (Nagy and Schaefer, 2002). Even under saturating light conditions there is still phytochrome remaining in the cytosol. Its localization is preferentially in the periphery of the cytosol as detected by indirect immunocytochemical methods (Kunkel, Panigrahi and Schäfer, unpublished). To define the biological role of this pool of phytochromes, however, radically new methods and approaches are required since in the absence of reliable markers no specific screens aimed at the isolation of such novel mutants can be performed.

9.10 Concluding Remarks

Our view about the intracellular localization of phytochromes has changed remarkably during the last five years. It is obvious that there is a light-dependent nuclear transport of the photoreceptors and that the photoreceptors must be in the activated Pfr form to be functional. But there are still many, many questions awaiting answers. What retains phytochromes in the cytosol? Is there an NLS that is masked by folding in Pr and opened after photoconversion to Pfr? Do cytosolic phytochromes have a function not only in ferns, mosses, and some algae, but also in flowering plants? Is there a function of the phytochromes associated with the plasma membrane? What are the components of the different types of nuclear complexes and what is their function?

References

Bauer, D., Viczian, A., Kircher, S., Kunkel, T., Panigrahi, K., Adam, E., Fejes, E., Schäfer, E., Nagy, F. (2004). Spatial and temporal distribution of PIF3, a transcription factor required for light signaling, is controlled by the photoreceptors phytochromes. *Plant Cell* (in press).

Beggs, C. J., Holmes, M. G., Jabben, M., Schäfer, E. (1980). Action spectra for the inhibition of hypocotyl growth by continuous irradiation in light- and dark-grown *Sinapis alba* L. seedlings. *Plant Physiol.*, **66**, 615–618.

Bubulya, P.S. Spector, D. L. (2002). Dasassembly of interchromatin granule clusters alters the co-ordination of transcription and pre-mRNA splicing. *J. Cell Biol.*, **158**, 425–436.

Casal, J.J., Davis, S.J., Kirchenbauer, D., Viczian, A., Yanovsky,M.J., Clough· R.C., Kircher, S., Jordan-Beebe, E.T., Schäfer, E., Nagy, F., Vierstra, R.D. (2002). The serine-rich N-terminal Domain of oat phytochrome A helps regulate light responses and subnuclear localization of the photoreceptor. *Plant Physiol.*, **129**, 1127–1137.

Chen, M, Schwab, R., Chory, J. (2003), Characterization of the requirements for localization of phytochrome B to nuclear bodies. *Proc. Natl. Acad. Sci. USA.*, **25**, 100(24):14493–14498.

Etzold, H. (1965). Der Polarotropismus und Phototropismus der Chloronemen von *Dryopteris Filix-Mas* (L.) Schott. *Planta*, **64**, 254–280.

Gil, P. (2001). Analysis of the nucleo-cytoplasmic partitioning of phytochrome B and its differential regulation by light and the circadian clock. Fakultät für Biologie, Universität Freiburg.

Gil, P., Kircher, S., Adam, E., Bury, E., Kozma-Bognar, L, Schäfer, E., Nagy, F. (2000). Photocontrol of subcellular partitioning of phytochromeB:GFP fusion protein in tobacco seedlings. *Plant J.*, **22**, 135–145.

Haupt, W. (1970). Über den Dichroismus von Phytochrom$_{660}$ und Phytochrom$_{730}$ bei *Mougeotia Z. Pflanzenphysiol.*, **62**, 287–298.

Hisada, A., Hanzawa, H., Weller, J.L., Nagatani, A., Reid, J.B., Furuya, M. (2000). Light-induced nuclear translocation of endogenous pea phytochrome A visualized by immunocytochemical procedures. *Plant Cell*, **12**, 1063–1078.

Holmes, M. H., Schäfer, E. (1981). Action spectra for changes in the 'high irradiance reaction' in hypocotyls of *Sinapis alba* L. *Planta*, **153**, 267–272.

Huq, E., Al-Sady, B., Quail, P.H. (2003). Nuclear translocation of the photoreceptor phytochrome B is necessary for its biological function in seedling photomorphogenesis. *Plant J.*, **35**, 660–670.

Kendrick, R. E. and Kronenberg, G. H. M., eds. (1994). Photomorphogenesis in Plants. Dordrecht, The Netherlands: Kluwer Academic Publishers.

Kim, L. (2002). Analysen zur intrazellulären Lokalisation von Phytochrom A. Fakultät für Biologie, Universität Freiburg.

Kircher, S., Gil, P., Kozma-Bogn r, L., Fejes, E., Speth, V., Husselstein, T., Bury, E., dam, É., Schäfer, E., Nagy, F. (2002). Nucleo-cytoplasmic partitioning of the plant photoreceptors phytochrome A, B, C, D and E is differentially regulated by light and exhibits a diurnal rhythm. *Plant Cell*, **14**, 1541–1544.

Kircher, S., Kozma-Bognar, L., Kim, L., Adam, E., Harter, K., Schäfer, E., Nagy, F. (1999). Light quality-dependent nuclear import of the plant photoreceptors phytochrome A and B. *Plant Cell*, **11**, 1445–1456.

Kraml, M. (1994). Light direction and polarization. *Photomorphogenesis in Plants* (Kendrick R. E. and Kronenberg G.H.M eds), pp. 417–443.

Kretsch, T., Poppe, C., Schäfer, E. (2000). A new type of mutation in the plant photoreceptor phytochrome B causes loss of photoreversibility and an extremely enhanced light sensitivity. *Plant J.*, **22**, 177–186.

MacKenzie, J.M. Jr, Coleman, R.A., Briggs, W.R., Pratt, L.H. (1975). Reversible redistribution of phytochrome within the cell upon conversion to its physiologically active form. *Proc. Natl. Acad. Sci. USA*, **72**, 799–803.

Marmé, D., Schäfer, E. (1972). On the localization and orientation of phytochrome molecules in corn coleoptiles (*Zea mays* L.) *Z. Pflanzenphysiol.* **67**, 192–194.

Matsushita, T., Mochizuki, N. and Nagatani, A. (2003). Dimers of the N-terminal domain of phytochrome B are functional in the nucleus. *Nature*, **424**, 571–574.

McCurdy, D., Pratt, L.H. (1986). Immunogold electron microscopy of phytochrome in *Avena*: identification of intracellular sites responsible for phytochrome sequestering and enhanced pelletability. J. Cell Biol., **103**, 2541–2550.

Mösinger, E., Batschauer, A., Schäfer, E., Apel, K. (1985). Phytochrome control of in vitro transcirption of specific genes in isolated nuclei from barley (*Hordeum vulgare*). *Eur. J. Biochem.*, **147**, 137–142.

Mösinger, E., Schäfer, E. (1984). *In vivo* phytochrome control of *in vitro* transcription rates in isolated nuclei from oat seedlings. *Planta*, **161**, 444–450.

Nagy, F. and Schäfer, E. (2000). Control of nuclear import and phytochromes. *Curr. Op. Plant. Biol.*, **3**, 450–454.

Nagy, F., Schäfer, E. (2002) Phytochromes control photomorphogenesis by differentially regulated, interacting signalling pathways in higher plants. In: *Annu. Rev. Plant Biology*, **53**, 329–355 (Eds: Delmer, D., Bohnert, H.J., Merchant, S.).

Ni, M., Tepperman, J. M. and Quail, P. H. (1998). PIF3, a phytochrome interacting factor necessary for normal photoinduced signal transduction, is a novel basic helix-loop-helix protein. *Cell*, **95**, 657–667.

Ni, M., Tepperman, J.M., Quail, P. H. (1999). Binding of phytochrome B to its nuclear signalling partner PIF3 is reversibly induced by light. *Nature*, **400**, 784–784.

Quail, P.H., Marmé D., Schäfer, E. (1973). Particle-bound phytochrome from maize and pumpkin. *Nature*, **245**, 189–191.

Sakamoto, K. and Nagatani, A. (1996). Nuclear localization activity of phytochrome B. *Plant J.*, **10**, 859–868.

Shinomura T, Hanzawa H, Schäfer E, Furuya M. (1998). Mode of phytochrome B action in the photoregulation of seed germination in Arabidopsis thaliana. *Plant J.*, **13**, 583–590.

Speth, V., Otto, V., Schäfer, E. (1986). Intracellular localization of phytochrome in oat coleoptiles by electron microscopy. *Planta*, **168**, 299–304.

Speth, V., Otto, V., Schäfer, E. (1987). Intracellular localization of phytochrome and ubiquitin in red-light-irradiated oat coleoptiles by electron microscopy. *Planta*, **171**, 332–338.

Yamaguchi, R., Nakamura, M., Mochizuki, N., Kay, S.A. and Nagatani, A. (1999). Light-dependent translocation of a phytochrome B-GFP fusion protein to the nucleus in transgenic Arabidopsis. *J Cell Biol.*, **145**, 437–445.

Yanovsky, J.M., Luppi, P.J., Kirchbauer, D., Ogorodnikova, B.O., Sineshchekov, A.V., Adam, E., Staneloni, J.R., Schaefer, E., Nagy, F., Casal, J.J. (2002). Missense mutation in the PAS2 domain of phytochrome A impairs subnuclear localization and a subset of responses. *Plant Cell*, **14**, 1591–1603.

10
Plant Cryptochromes: Their Genes, Biochemistry, and Physiological Roles

Alfred Batschauer

Summary

Cryptochromes (cry) are sensory photoreceptors operating in the UV-A and blue-light region of the electromagnetic spectrum. They were first discovered in the plants *Arabidopsis thaliana* and *Sinapis alba* one decade ago, and afterwards identified not only in many other plant species but also in animals, humans, and bacteria. Therefore, cryptochromes are ubiquitous UV-A/blue light receptors of prokaryotes and eukaryotes. Cryptochromes are related in their sequence to DNA repair enzymes, the DNA photolyases, and share with them the same cofactor chromophores. In addition, cryptochromes seem to have conserved the ability of photolyases to bind DNA, although light absorption is used for different purposes, namely for catalysis (DNA repair) in the case of photolyase and for signaling in the case of cryptochromes. Since their discovery, a large quantity of data on the biological functions, the signaling mechanisms, and the biochemistry of cryptochromes has been accumulated and the reader is referred to several recent reviews on these topics (Ahmad, 1999; Cashmore et al., 1999; Briggs and Huala, 1999; Briggs and Olney, 2001; Lin, 2000a, 2000b, 2002; Lin and Shalitin, 2003; Sancar, 2003; Santiago-Ong and Lin, 2003; Thompson and Sancar 2002, Van Gelder, 2002). This chapter covers the biological roles and biochemistry of plant cryptochromes, whereas Chapter 11 (Cashmore) deals with the signaling of cryptochromes and Chapter 12 (Sancar and Van Gelder) with animal cryptochromes.

Handbook of Photosensory Receptors. Edited by W. R. Briggs, J. L. Spudich

ISBN 3-527-31019-3

10.1 Cryptochrome Genes and Evolution

10.1.1 The Discovery of Cryptochromes

Two different strategies led to the molecular cloning of cryptochrome genes in 1993. Margaret Ahmad and Anthony Cashmore (Ahmad and Cashmore, 1993) screened for T-DNA-tagged *Arabidopsis* mutants having the same phenotype as the *hy4* mutant isolated by Maarten Koornneef and coworkers in 1980 (Koornneef et al., 1980). The *hy4* mutant, in contrast to the wild type, has a long hypocotyl when the seedlings are grown under white or blue light whereas hypocotyl growth inhibition is normal under red and far-red light (Ahmad and Cashmore, 1993; Jackson and Jenkins, 1995). This result indicated that a gene encoding either a blue-light photoreceptor or a component in blue light signaling is affected in *hy4*. The insertion of a T-DNA facilitated the molecular cloning of the *HY4* gene (Ahmad and Cashmore, 1993). It turned out that HY4 has striking sequence similarity to class I CPD photolyases. These enzymes use the energy of photons in the UV-A/blue region of the spectrum to catalyze the repair of cyclobutane pyrimidine dimers (CPDs) which result from UV-B treatment of DNA (for review see Sancar, 2003). The fact that HY4 has homology with photolyase but lacks photolyase activity (Lin et al., 1995b; Malhotra et al., 1995) and many further findings described below led to the conclusion that *HY4* encodes a UV-A/blue light receptor and not a component in blue-light signaling. Therefore, HY4 was renamed cryptochrome 1 (Lin et al., 1995a), a term used formerly (Gressel, 1979; Senger, 1984) for unknown blue light receptors with action spectra that better resemble the absorption spectrum of phototropins (phot). Phototropins are another type of UV-A/blue-light receptors that regulate growth in response to the direction of light as well as movement of chloroplasts, stomatal opening, and other responses. Phototropins are reviewed in Chapters 13 (Christie and Briggs), 14 (Swartz and Bogomolni), and 15 (Crosson).

PCR amplification of DNA fragments from genomic DNA of white mustard (*Sinapis alba* L.) with degenerate oligonucleotides that resembled conserved regions in the few class I CPD photolyases known at that time, and further screening of genomic libraries with the PCR products as probes led to the isolation of another cryptochrome gene (Batschauer, 1993). Since no mutant of white mustard, similar to *hy4*, was available, it was not evident from the beginning that the isolated gene encodes a cryptochrome and not a DNA-photolyase. Later studies proved that this white mustard gene does not encode photolyase and thus most likely has a cryptochrome function (Malhotra et al., 1995). In retrospect it may not be too surprising that cryptochromes are related to DNA photolyase, and that photolyase itself as well as its chromophores were discussed as models for blue light receptors before the cryptochromes had been identified (Galland and Senger, 1988, 1991; Lipson and Horwitz, 1991).

10.1.2
Distribution of Cryptochromes and their Evolution

After cryptochromes were identified in *Arabidopsis* and *Sinapis*, they were found in many other plant species, animals, and bacteria either by using heterologous probes to screen for cryptochrome genes or by identifying such sequences in the growing databases of genome or EST projects. So far in the plant kingdom, cryptochromes have been described for the angiosperms *Arabidopsis thaliana* (Ahmad and Cashmore, 1993; Hoffman et al., 1996; Kleine et al., 2003; Lin et al., 1996b), *Sinapis alba* (Malhotra et al., 1995; Batschauer, 1993), tomato (Ninu et al., 1999; Perrotta et al., 2000; Perrotta et al., 2001; Weller et al., 2001), rice (Matsumoto et al., 2003), barley (Perrotta et al., 2001), for the fern *Adiantum capillus veneris* (Kanegae and Wada, 1998; Imaizumi et al, 2000), the moss *Physcomitrella patens* (Imaizumi et al., 1999; Imaizumi et al., 2002) and the green alga *Chlamydomonas rheinhardtii* (Small et al., 1995). In animals (see also Sancar and Van Gelder, Chapter 12) cryptochromes were found in mice (Kobayashi et al., 1998; Miyamoto and Sancar, 1998), rats (Eun et al., 2001), chicken (Yamamoto et al, 2001), Japanese quail (Fu et al., 2002), zebrafish (Kobayashi et al., 2000), *Xenopus* (Zhu and Green, 2001), bullfrog (Eun and Kang, 2003), the fly *Drosophila melanogaster* (Emery et al., 1998; Stanewsky et al., 1998), and in humans (Adams et al, 1995; Hsu et al., 1996; Todo et al., 1996; Van der Spek et al., 1996; Kobayashi et al., 1998). The only examples for prokaryotic cryptochromes are presently cry DASH (see also below) from the cyanobacterium *Synechocystis* PCC6803 (Brudler et al., 2003; Hitomi et al., 2000; Ng and Pakrasi, 2001), and VcCry1 and VcCry2 from *Vibrio cholerae* (Worthington el al., 2003). This list is not complete since cryptochrome sequences deposited in EST databases are not included.

Most plants seem to possess more than one cryptochrome. *Arabidopsis thaliana* contains two well-characterized cryptochromes (Ahmad and Cashmore, 1993; Hoffman et al., 1996; Lin et al., 1996b) and a third one (cry3 or cry DASH) for which the biological function is not yet defined (Kleine et al., 2003; Brudler et al., 2003). Tomato has three cryptochromes, CRY1a, CRY1b, CRY2 (Perrotta et al., 2000; Perrotta et al., 2001), and a putative cry DASH (G. Giuliano, personal communication), rice has two, CRY1a, CRY1b (Matsumto et al., 2003), *Adiantum* has at least five cryptochromes (Kanegae and Wada, 1998; Imaizumi et al., 2000) and the moss *Physcomitrella patens* at least two (Imaizumi et al., 1999; Imaizumi et al., 2002). Similarly, animals regularly contain more than one cryptochrome (see Sancar and Van Gelder, Chapter 12). The presence of more than one cryptochrome gene in most plants and animals is indicative of gene duplication events, whereas the presence of cryptochromes in prokaryotes indicates that eukaryotes may have received their cryptochromes by horizontal gene transfer from former endosymbionts that gave rise to mitochondria (α-proteobacteria) and chloroplasts (cyanobacteria).

The situation is probably even more complex since the plant cryptochromes do not group together with animal cryptochromes in phylogenetic trees (see Figure 10.1). Whereas plant cryptochromes are more closely related to the above-mentioned class I DNA-photolyases that repair cyclobutane pyrimidine dimers, and are mostly found in microbial organisms including the yeast *Saccharomyces cerevisiae* and other fungi, animal cryptochromes group with (6–4) photolyases. (6–4) photolyases repair another type of DNA photoproduct, the pyrimidine–pyrimidone (6–4) photoproduct, and are found exclusively in eukaryotes (for review see Sancar, 2003). One hypothesis for the evolution of the cryptochrome/photolyase family is that several gene duplication events gave rise to the present-day photolyases and cryptochromes (Kanai et al., 1997; Todo 1999). The most ancestral gene (possibly a CPD photolyase) duplicated to give rise to class I CPD photolyase and to class II CPD photolyase, the latter now present in metazoans including plants (Ahmad et al., 1997; Petersen et al., 1999; Taylor et al., 1996), marsupials (Kato et al., 1994; Yasui et al., 1994), fish (Yasuhira and Yasui, 1992; Yasui et al., 1994), and insects (Todo et al., 1994; Yasui et al., 1994), as well as in some bacteria (O'Connor et al., 1996; Yasui et al., 1994) and animal viruses (Afonso et al., 1999; Sancar, 2000; Srinivasin et al., 2001; Todo, 1999). The class I CPD photolyase duplicated again to give rise to the more recent class I CPD photolyases and the progenitor of cryptochromes and (6–4) photolyases. The latter duplicated again to give rise to the present-day cryptochromes in plants as well as the (6–4) photolyases and the cryptochromes in animals. More recent duplications led to an increase in the number of cryptochrome genes in plants and in animals such as *CRY1* and *CRY2* in *Arabidopsis*, or additional duplications of one of the cryptochrome members to give rise, for example, to *CRY1a* and *CRY1b* in tomato.

However, our phylogenetic analysis shown in Figure 10.1 allows for an alternative point of view concerning the evolution of plant cryptochromes after the discovery of a third cryptochrome (*CRY3* or *cryDASH*) in *Arabidopsis* (Kleine et al., 2003; Brudler et al., 2003). *CRY3* is closely related to *Synechocystis cryDASH* suggesting that plants may have received this cryptochrome from the cyanobacterial endosymbiont that gave rise to chloroplasts. The *CRY1/CRY2* members group closer to sequences of α-proteobacteria that are considered as the progenitors of mitochondria. Although the bootstrap value for grouping plant cryptochromes of the *CRY1/CRY2* family with α-proteobacteria sequences is only 63%, there is 0% probability for grouping *CRY3* with *CRY1* and *CRY2* (Kleine et al., 2003). Surely, further phylogenetic studies are required to gain deeper insight into the evolution of the photolyase/cryptochrome family.

10.2
Cryptochrome Domains, Cofactors and Similarities with Photolyase

As mentioned above, plant cryptochromes have significant sequence similarity with class I CPD photolyases. For example, a stretch of 500 amino acids within the N-terminal region of the cryptochromes CRY1 and CRY2 from *Arabidopsis* show about 30% sequence identity at the protein level with *E. coli* photolyase. Since photolyases are far better characterized at the molecular and even atomic level than cryptochromes, and since photolyase can be considered as a model molecule for cryptochrome, despite its difference in function, the current knowledge of photolyase structure and function will briefly be summarized.

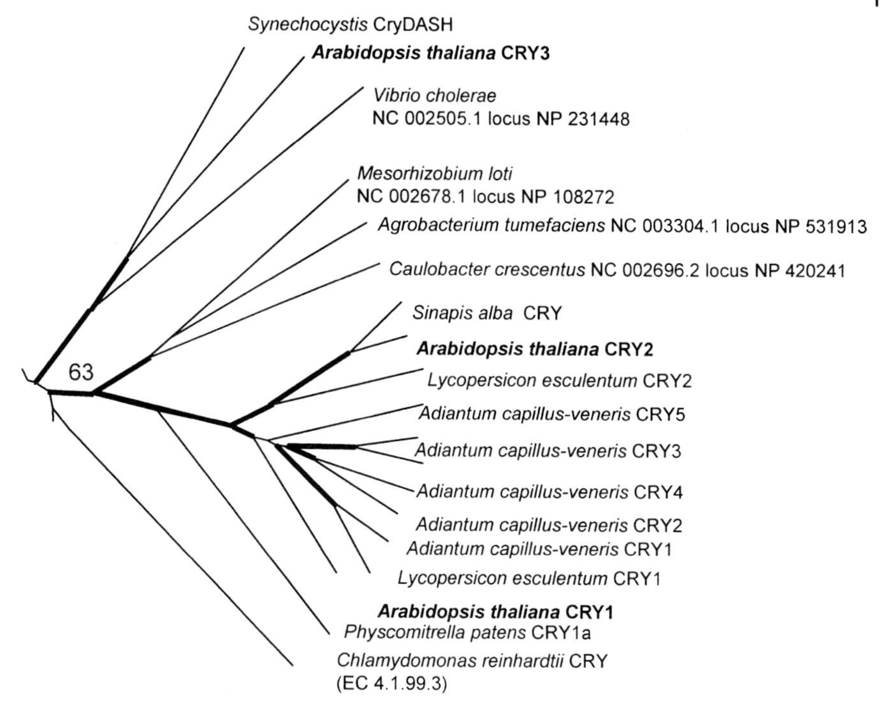

Figure 10.1 Phylogenic tree of the cryptochrome/photolyase family, showing the part of the tree that contains the plant cryptochromes. Our phylogenetic studies (Kleine et al., 2003) indicate that *Arabidopsis CRY3* groups with *cryDASH* from the cyanobacterium *Synechocystis*, whereas the family members represented by *CRY1* and *CRY2* group closer to genes from α-proteobacteria. As α-proteobacteria are considered to be the progenitors of mitochondria and cyanobacteria as the progenitors of chloroplasts one may speculate that two independent horizontal gene transfer events from the endosymbionts to the genome of the host gave rise to the present-day cryptochromes in plants. Although the bootstrap value for grouping sequences of the *CRY1/CRY2* family with α-proteobacteria sequences is only 63%, there is 0% support to group them with *CRY3*.

The protein structure of three class I CPD photolyases (*Escherichia coli, Anacystis nidulans, Thermus thermophilus*) have been solved (Komori et al., 2001; Park et al., 1995; Tamada et al., 1997). They, as well as all other photolyases, contain as a cofactor the flavin FAD in an U-shaped conformation, where the isoalloxaxine ring is in close proximity to the adenine ring (for review, see Sancar, 2003). FAD is essential for catalysis and active in its two-electron reduced deprotonated form ($FADH^-$). In addition, class I CPD photolyases contain a second cofactor that absorbs light and transfers the energy to $FADH^-$. Thus the second chromophore acts as an antenna. The second cofactor is not required for catalysis but it increases the rate of repair under limiting light conditions, because it has a higher extinction coefficient and absorbs at longer wavelengths than $FADH^-$. In most species the second chromophore is the pterin 5,10-methenyltetrahydropteroylpolyglutamate (methenyltetrahydrofolate, MTHF) having 3–6 glutamate residues. In other species such as *Anacystis nidulans* the second cofactor is the deazaflavin-type chromophore 8-hydroxy-7,8-didemethyl-5-deazariboflavin (8-HDF) also named F_o, that can contain γ-glutamates linked to the ribityl phosphate group of 8-HDF, and is then named F_{420}. Both, FAD and the second cofactor are present in stoichiometric amounts in photolyase and neither cofactor is covalently bound to the apoenzyme.

The reaction mechanism of CPD photolyase, as schematically shown in Figure 10.2, includes light-driven electron transfer from the reduced flavin ($FADH^-$) to the CPD creating an instable CPD radical anion and a neutral flavin radical ($FADH^\bullet$). The CPD radical causes a spontaneous cleavage of the carbon bonds within the cyclobutane ring and transfers the electron back to $FADH^\bullet$, thus completing the reaction cycle. The efficiency of energy transfer between the antenna and $FADH^-$ and the electron transfer between $FADH^-$ and CPD are high with values of 62% and 89%, respectively for *E. coli* photolyase and even higher for the deazaflavin-type *Anacystis nidulans* photolyase (for review, see Sancar, 2003).

Besides its high affinity for the substrate, photolyase also has low affinity for undamaged DNA. For example, the binding constant of *E. coli* photolyase to the thymine dimer in DNA is about 10^{-9} M, whereas the binding constant for undamaged DNA is about 10^{-4} M (Husain and Sancar, 1987; Sancar, 2003). In vitro binding and enzyme assays have shown that plant cryptochromes have neither detectable photolyase activity (Lin et al., 1995b; Malhotra et al., 1995; Hoffman et al., 1996) nor show significant binding to pyrimidine dimer-containing DNA (Malhotra et al., 1995). Recent data (see Section 10.5.3) have demonstrated, however, that cryptochromes, including those from plants, may well have DNA-binding activity.

Although no co-crystal structure is so far available of photolyase with its substrate, showing the contact sites of the enzyme with the pyrimidine dimer and the flanking bases, there is very good evidence that the hydrophobic amino acids along one side of the binding pocket as well as the polar amino acids lining at the opposite side are critical for substrate binding (Park et al., 1995; Tamada et al., 1997; Komori et al., 2001). Indeed, mutation of Trp277 in *E. coli* photolyase, which sits in the binding pocket, to a non-aromatic residue abolishes binding of the substrate (Li and Sancar, 1990). This Trp is not conserved however in *Arabidopsis* CRY1 and CRY2, where a Leu is found instead.

The highest conservation between plant cryptochromes and class I CPD photolyase is within the flavin-binding pocket. 9 of the 13 amino acids that make contact with FAD in *E. coli* photolyase are conserved in *Arabidopsis* CRY1 and CRY2 as well as in other plant cryptochromes. Since cryptochromes are not highly abundant proteins in plants and because of technical difficulties, they have not yet been purified from their endogenous source in amounts sufficient for the determination of their cofactors. However, heterologous expression of *Arabidopsis* CRY1 and CRY2 and of the white mustard cryptochrome, which has 89% amino acid sequence identity with *Arabidopsis* CRY2, in *E. coli* or insect cells has shown, that plant cryptochromes indeed bind FAD noncovalently and in stoichiometric amounts (Lin et al., 1995b; Malhotra et al., 1995). The *E. coli*-expressed *Arabidopsis* cry1 and *Sinapis* cry2 also contained non-covalently bound MTHF (Malhotra et al., 1995), although the amino acids that make contact in *E. coli* photolyase with the second cofactor are not as well conserved in

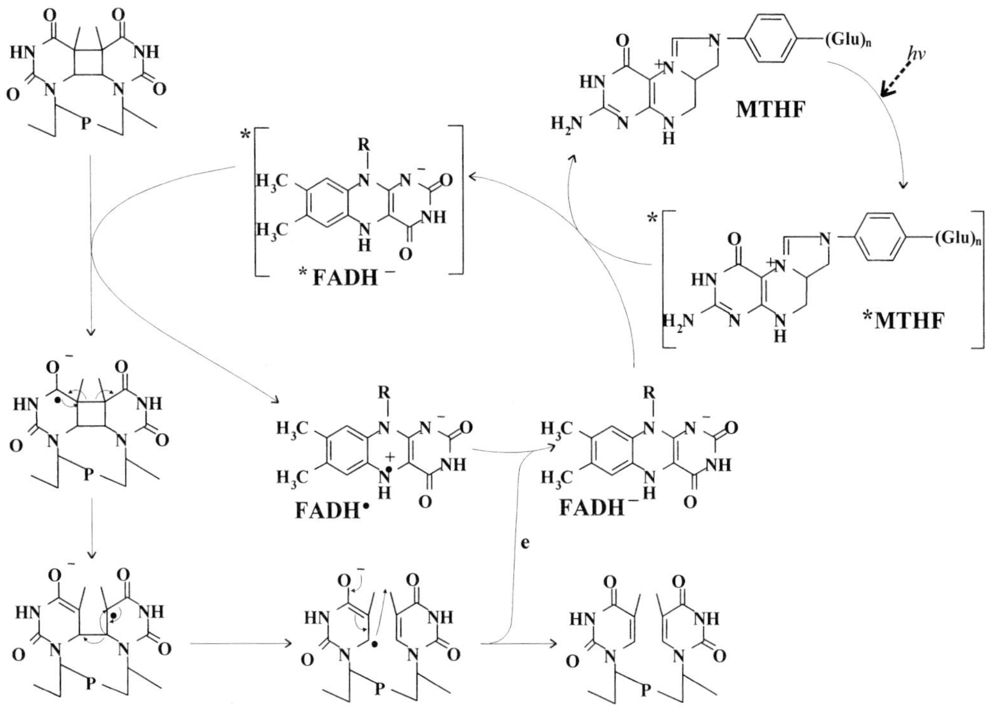

Figure 10.2 Reaction mechanism of CPD photolyase, showing the cofactors of photolyase (MTHF, FADH), the cyclobutane pyrimidine dimer (CPD) substrate, and the excitation and electron transfer events relevant for catalysis as established for the *E.coli* enzyme. For further details see Sancar, (2003). Light is absorbed by the second cofactor (MTHF) and excitation energy transferred to the catalytic cofactor ($FADH^-$). The excited $FADH^-$ transfers an electron to the CPD, which then undergoes cycloreversion. Electron back transfer from the pyrimidine radical anion to the flavin neutral radical FADH• restores the catalytic activity of flavin in its two-electron-reduced form ($FADH^-$). This scheme is also taken as model for cryptochrome signaling. Energy transfer from MTHF to FAD and electron transfer from FAD to an external electron acceptor has not been demonstrated for cryptochromes. However it was shown for recombinant *Arabidopsis* cry1 that electrons are transported from an artificial electron donor to fully oxidized FAD after its excitation. This electron transfer is mediated by aromatic amino acids of the cry1 protein (Giovani et al., 2003), a process also described for photolyase where it is used to gain the fully reduced flavin ($FADH^-$) that is required for catalytic activity (photoactivation). Figure adapted from Sancar, (2003).

these cryptochromes. Purification of the holoproteins from plant tissue and characterization of their cofactors is the only method in which the second cofactor, present in cryptochromes, can be conclusively identified.

In contrast to photolyase, most of the plant cryptochromes, as well as animal cryptochromes, carry extensions of varying length at their C terminus. This is schematically shown in Figure 10.3. The longest extension is 367 amino acids in case of *Chlamydomonas* cryptochrome (total length 867 amino acids) (Small et al., 1995), whereas AcCRY5 of *Adiantum capillus veneris* is lacking such an extension (Kanegae and Wada, 1998; Imaizumi et al., 2000). Although the sequences of plant cryptochromes are mostly conserved in the N-terminal region spanning about 500 amino acids (the so called photolyase or PHR domain) where the cofactors are bound, there is some conservation in the C-terminal extensions as well. The first motif described to be conserved between *Arabidopsis* CRY1 and CRY2 in this region is the so-called STAES (Ser-Thr-Ala-Glu-Ser) motif (Hoffman et al., 1996) followed by the sequence GGXVP (together called the S motif, X stands for a non-conserved amino acid). Further upstream are two other conserved regions in plant cryptochromes, one of which contains a varying number of acidic residues (Asp and Glu, A motif), and close to the start of the C-terminal extension the sequence DQXVP (D motif). Together these motifs are named the DAS domain (Lin, 2002). The conservation of the DAS domain in cryptochromes of moss, fern and seed plants indicates that it must already have been present in early land plants. Since mutations in *CRY1* leading to stop codons or amino acid changes directly before or within the D motif (*hy4-3* and *hy4-9* alleles) cause phenotypic changes (Ahmad and Cashmore, 1993, Ahmad et al., 1995), this region must be essential for the biological function of plant cryptochromes. The importance of the C-terminal extensions has also been shown by domain-switch experiments where the C-terminal extension of Arabidopsis CRY1 was fused to the PHR domain of CRY2 and *vice versa* (Ahmad et al., 1998a). Both combinations of fused photoreceptors were biologically active. The role of the C-terminal extension in cryptochrome signaling will be discussed by Cashmore (Chapter 11), whereas its role in

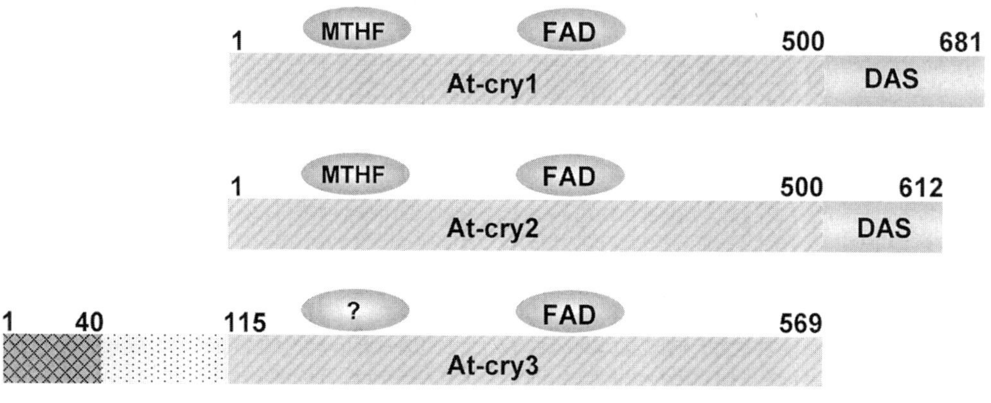

Figure 10.3 The three plant cryptochromes domains identified in *Arabidopsis* and their cofactors are shown schematically. The highest conservation among the protein sequences is found in the photolyase (PHR) related part (hatched part). This region is about 500 amino acids long and binds the FAD cofactor non-covalently. For cry1 and cry2 MTHF was also found to be bound to this region whereas no cofactor other than FAD was identified in cry3. Cry1 and cry2 carry an additional domain at the C-terminal end that varies in length and sequence. However, the C-terminal extension contains three motifs conserved in all land plant cryptochromes and named together the DAS domain. For cry2 it was demonstrated that the C-terminal domain is required for nuclear import and contains a bipartite nuclear localization signal (Kleiner et al., 1999; Guo et al., 1999). In contrast to cry1 and cry2, cry3 carries an extension at the N-terminal end, part of which (amino acids 1–40) is required for the import of cry3 into chloroplasts and mitochondria (Kleine et al., 2003).

the subcellular localization and biochemistry of cryptochromes is described in Sections 10.4 and 10.5.

Unfortunately, the production of full-length plant cryptochromes in large quantities was hampered for a long time by the fact that expression in *E. coli*, Pichia, or yeast cells resulted in preparations which contained mostly aggregated protein that could not be refolded (M. Müller, O. Kleiner, A. Batschauer, unpublished data). Consequently, efforts to crystallize plant cryptochromes were doomed to failure and even spectroscopic and biochemical studies were limited. However, more recent efforts to produce plant cry1 and cry2 in SF9 or SF21 insect cells were more effective (Lin et al., 1995b; Giovani et al., 2003; Bouly et al., 2003; Shalitin et al., 2003; R. Banerjee, A. Batschauer, unpublished data) and allowed studies on electron transfer within cry as well as some biochemical characterization that is described in Section 10.5.

In contrast to plant and animal cryptochromes, *Synechocystis* cry could be expressed in high quantities in *E. coli*, crystallized, and the structure solved at the atomic level (Brudler et al., 2003). This cryptochrome (cryDASH) contained FAD but no second cofactor, although the structure showed sufficient space for it. Its overall structure is very similar to the *E. coli* photolyase but has some differences that could explain its lack of photolyase activity. In particular, the pocket in photolyase that binds the pyrimidine dimer is wider and flatter in cryDASH because of the replacement of two amino acids and the rotation of a Tyr residue out of the pocket. One of the amino-acid changes affects the electronic structure of FAD and thus probably alters its ability to transfer an electron, whereas the second exchange in the pocket (Trp to Tyr) together with other exchanges on the surface of the protein that are also important for substrate binding, probably reduces the binding affinity for the photolyase substrate (Brudler et al., 2003).

10.3 Biological Function of Plant Cryptochromes

In plants nearly all growth and differentiation processes are affected by light, and many light responses are under control of different classes of photoreceptors and/or different types of photoreceptors from one class (Kendrick and Kronenberg, 1994; Gyula et al., 2003; Sullivan and Deng, 2003; Batschauer, 2003). For example, the light-regulated inhibition of stem growth in dicot plants at the seedling stage (hypocoytl growth inhibition) is under control of several phytochromes (see Quail, Chapter 7) that operate in the red and far-red region and are encoded by five genes in *Arabidopsis* (Sharrock and Quail, 1989), as well as under the control of cryptochromes and phototropins that operate in UV-A and blue light (Briggs and Christie, 2002; Briggs and Huala, 1999; Lin, 2002; Lin and Shalitin, 2003). It is not too surprising that plants, for which light is essential, have evolved mechanisms to sense their light environment precisely and to safeguard themselves against the loss of light sensing by possessing many photoreceptors. However, the redundancy of photoreceptors and their cross talk makes it more difficult to identify the specific function of a single photoreceptor in the presence of all the others. Despite this redundancy, the genes and the function of their encoded photoreceptors could be identified by isolating mutants that showed phenotypic changes either under white light or more specific light conditions such as red, far-red, or blue lights with different fluence rates, etc. The classical screen performed by Maarten Koornneef and coworkers (Koornneef et al., 1980) to identify *Arabidopsis* mutants that are altered in their light-responsiveness was done in white light by measuring the hypocotyl growth inhibition. Most of the isolated mutants (*hy* mutants) were shown later to be affected in the synthesis of either photoreceptor apoproteins such as phytochrome B (*hy3*) and cryptochrome 1 (*hy4*) or the phytochrome chromophore phytochromoblin (*hy1*, *hy2*).

10.3.1
Control of Growth

As already mentioned, *cry1* was identified in a screen for *Arabidopsis* mutants with reduced hypocotyl growth inhibition in white light. Because of the above-mentioned redundancy of photoreceptor action in white light, the effects of mutation of *CRY1* are less pronounced in white than in blue-light (Ahmad and Cashmore, 1993; Jackson and Jenkins, 1995). In contrast to blue light, the lack of cry1 and of cry2 seems to have no effects in darkness (Ahmad and Cashmore, 1993; Jackson and Jenkins, 1995; Lin et al., 1998) (see Figure 10.4 A). However, there are some reports describing red-light effects in the *cry1* mutant (see below). A red-light effect in the *cry1* mutant is surprising if one assumes that a flavin-type photoreceptor does not normally absorb light above 500 nm. However, the absorption properties of a flavin-type photoreceptor depend on the redox state of the flavin (and/or the second chromophore) which can be fully reduced, semi-reduced, or fully oxidized. The fully oxidized flavin does not absorb above 500 nm and has absorption maxima at around 470 nm and 390 nm. The fully reduced flavin absorbs strongly in the UV-A region but only weakly above 400 nm, whereas the flavosemiquinone radical form (semi-reduced) absorbs in the UV-A, blue and red region (Ehrenberg and Hemmerich, 1968).

There is some evidence from work on the fungus *Phycomyces*, based on action spectroscopy, that the chromophore of the photoreceptor, which mediates phototropism, can be in the semi-reduced form (Galland and Tölle, 2003). In addition, a significant amount of *Arabidopsis* cry1, when expressed in insect cells, was isolated in the semi-reduced form (Lin et al., 1995b), and there is clear evidence from *in vitro* studies on *Arabidopsis* cry1 that the flavin can be photoreduced (Giovani et al., 2003) (see also Section 10.5.4). However, further analysis is required to make definite conclusions about the redox state(s) of the chromophore(s) in cryptochromes *in planta* under different light conditions. The red-light effects seen in the *cry1* mutant cannot only be explained by the cryptochrome's role in red-light perception but also as an indirect effect where cryptochrome acts as a component in phytochrome signaling (Devlin and Kay, 2000) [see also Section 10.3.2 and Chapter 7 (Quail)].

Whereas the lack of cry1 has a strong effect on hypocotyl growth inhibition under high fluence rates of blue light (see Figure 10.4 A), the *cry2* mutant shows essentially no difference from wild type under these conditions (Figure 10.4 A). However, under lower fluence rates (1 μmol m^{-2} s^{-1} and less) the lack of cry2 becomes clearly vis-

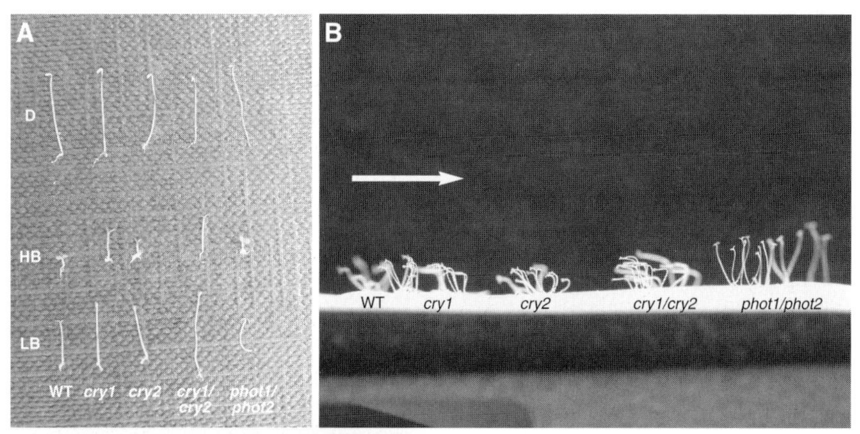

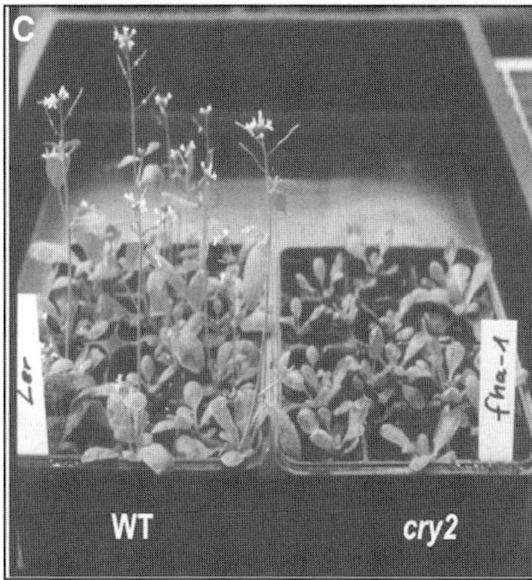

Figure 10.4 De-etiolation and flowering time are regulated by cryptochromes. In *Arabidopsis*, cry1 and cry2 regulate most of the blue-light-specific developmental programs. Shown are some phenotypic effects caused by mutations in *cry1* or *cry2*. A) Photomorphogenesis (deetiolation) of *Arabidopsis* wild type (WT) and photoreceptor mutants grown for 3 days in darkness (D) or continuous blue light of 30 μmol m^{-2} s^{-1} (HB) or 1 μmol m^{-2} s^{-1} (LB), respectively given from the top. In wild-type seedlings, stem (hypocotyl) growth is inhibited by blue light whereas opening of the hypocotyl hook and cotyledon opening and expansion is promoted by blue light. HB is more efficient than LB. The lack of cry1 (*cry1*) is most evident under HB conditions, whereas the lack of cry2 (*cry2*) becomes more evident under LB conditions, which is seen in particular for the double mutant (*cry1/cry2*). Defects in phototropin1 and phototropin2 (*phot1/phot2*) have no obvious effect on de-etiolation under these conditions. B) Mutations in *CRY1* and/or *CRY2* have no obvious effect on phototropism, the bending response in relation to the direction of the blue light source (indicated by the arrow), whereas the *phot1/phot2* double mutant does not respond to unilateral blue light. Wild-type and mutant seedlings were grown for three days in darkness before unilateral blue light (30 μmol m^{-2} s^{-1}) was given for 24 h. C) Flowering time is mainly regulated by cry2 in the facultative long-day (LD) plant *Arabidopsis thaliana*. Plants of the Landsberg erecta accession, either wild type (WT) or the *cry2* (*fha-1*) mutant were grown under LD (16 h light / 8 h darkness) photoperiods. Under these conditions, wild-type plants start to flower about 20 days after sowing whereas the *cry2* mutant plants require about 40 days to start flowering. Plants in this Figure are 30 days old.

ible and hypocotyls of the *cry2* mutant are longer that those of the wild type (Lin et al., 1998) (see Figure 10.4 A). Even more pronounced is this effect in the *cry1/cry2* double mutant, indicating that both cryptochromes act redundantly during the de-etiolation process where cry1 operates primarily under high light and cry2 under low light conditions. A similar situation was also observed for phytochromes (Quail, Chapter 7). The fact that cry2 does not operate under high fluence rates of blue light during seedling development was explained by the observation that under such conditions the cry2 protein is rapidly degraded (Lin et al., 1998). Cry2 degradation is further discussed in Section 10.5.1.

Although hypocotyl length is an easy measure to quantify the effects of photoreceptors on plant growth, the underlying molecular mechanisms that modulate hypocotyl growth are probably rather complex and involve many different mechanisms. Growth inhibition can be observed after only 30 seconds of blue light exposure. Within 30 minutes the growth rate decreases to almost zero and then reaches a rate for several days that is much lower than in dark-grown seedlings (Parks et al., 1998; Folta and Spalding, 2001; Parks et al., 2001). By analyzing *Arabidopsis* mutants deficient in *cry1, cry2* and *phot1,* it was shown that the early response (within 30 min) was similar to wild type in *cry* single and *cry1/cry2* double mutants but strongly reduced in the *phot1* mutant, demonstrating that phot1 but not the crys are responsible for the early growth inhibition. Although cry1 and cry2 are not required for the early response, they mediate a very fast membrane depolarization caused by the activation of anion channels which precedes the early inhibition response (Parks et al., 2001). Blocking the blue-light-regulated anion channels with 5-nitro-2-(3 phenylpropylamino)-benzoic acid has no effect on the phot1-regulated growth inhibition, but does affect the second phase (30–120 min of blue light) that is controlled by cry1 and cry2. Afterwards, growth inhibition in blue light seems to be controlled only by cry1, at least under high fluence rates where cry2 is degraded, and is independent of anion channels. The cry-regulated growth inhibition phase is however delayed in the *phot1* mutant, indicating that phot1 affects cry signaling (for review see Parks et al., 2001).

Besides hypocotyl growth inhibition, crys also regulate other processes during deetiolation, such as opening of the cotyledons (Lin et al., 1998), cotyledon expansion (Jackson and Jenkins, 1995; Weller et al., 2001), inhibition of petiole elongation (Jackson and Jenkins, 1995), anthocyanin formation (Ahmad et al., 1995; Jackson and Jenkins, 1995; Ninu et al., 1999; Weller et al., 2001), and gene expression, the latter probably involved in all these processes and discussed in more detail in Section 10.3.3.

As for hypocotyl growth inhibition, the loss of cry1 is most obvious for all the processes mentioned above under high fluence rates of blue light whereas the loss of cry2 becomes detectable under low fluence rates of blue light (Lin et al., 1998). The expression of either CRY1 or CRY2 under control of the constitutive and strong cauliflower mosaic virus 35S promoter in *Arabidopsis* or tobacco leads essentially to phenotypes opposite to loss of function mutations in *cry1* or *cry2,* resulting in exaggerated inhibition of hypocotyl and petiole growth, and enhanced cotyledon opening and anthocyanin production (Ahmad et al., 1998a; Lin et al., 1995a; Lin et al., 1996a; Lin et al., 1998). However, detailed inspection of expression profiles of blue-light regulated genes showed that overexpression of cry1 did not always have effects opposite to what was observed in the *cry1/cry2* double mutant (Ma et al., 2001) (see Section 10.3.3).

The cry1 overexpressor showed enhanced sensitivity for UV-A and blue-light, as expected, but also to green light (Lin et al., 1995a, 1995b; Lin et al., 1996a). The effect of green light could be due to the presence of a semireduced FAD chromophore in the cryptochrome, which has however not been demonstrated *in planta*. Surprisingly, there is no direct correlation between the amount of cry1 photoreceptor in *Arabidopsis* seedlings and their sensitivity to light. A comparison of the action spectra and threshold values for hypocotyl growth inhibition (end-point measurements) of wild type, *cry1* single and *cry1/cry2* double mutants and cry1 overexpressor showed that an increase in the amount of cry1 by a factor of ten leads to a shift in the threshold values of less than a factor of three. In addition, the shape of the action spectrum was altered by cry1 overexpression (Ahmad et al., 2002). Whereas limiting levels of signaling components or adverse effects causing 'light-stress' could explain the first observation, the effects on the shape of the action spectrum by cry1 overexpression are far from understood.

Besides *Arabidopsis*, the function of cryptochromes in plants was also studied in moss and fern plants (Suetsugu and Wada, Chapter 17) and to some extent in tomato. As mentioned above, tomato has three cryptochrome genes (CRY1a, CRY1b, CRY2) and a putative cryDASH. The function of cry1 was characterized by expression of antisense constructs (Ninu et al., 1999) and by mutant analysis (Weller et al., 2001). As with *Arabidopsis*, cry1 regulates the inhibition of hypocotyl growth and the induction of anthocyanin formation in tomato. However, phenotypic changes of the tomato *cry1* mutant and antisense plants were also found that have not been described for *Arabidopsis*, such as reduced chlorophyll content in seedlings and effects on stem elongation, apical dominance, and chlorophyll content in leaves and fruits of adult plants.

In addition to the growth and differentiation processes regulated by cryptochromes and described above, there are also some reports of cryptochromes having a role in phototropism (Ahmad et al., 1998b; Whippo and Hangarter, 2003).

10.3.2 Role of Cryptochromes in Circadian Clock Entrainment and Photoperiodism

10.3.2.1 Circadian Clock Entrainment

A detailed description of the components of the circadian clock in plants is beyond the scope of this chapter and the reader is referred to recent reviews on this topic (Devlin, 2002, 2003; Devlin and Kay, 2001; Fankhauser and Staiger, 2002; Hayama and Coupland, 2003; Millar, 2003; Somers, 1999) and to Schultz (Chapter 16). In brief, circadian clocks are self-sustained oscillators containing clock proteins that oscillate with a periodicity of about 24 h. The periodicity comes from rhythmic transcription of clock genes that encode clock proteins, which feed back to repress their own genes. In plants, there are at least three components in the oscillator, CIRCADIAN CLOCK ASSOCIATED 1 (CCA1), LATE ELONGATED HYPOCOTYL (LHY), both myb-domain DNA-binding proteins, and TIMING OF CAB1 (TOC1), a protein with a domain at the N terminus similar to receiver domains of two-component response regulators and a C-terminal domain present also in the CONSTANS protein family (see below). All three proteins cycle and the negative feedback loop is likely due to the repression of all three genes by the CCA1 and LHY proteins. When the levels of CCA1 and LHY proteins have decreased, the *TOC1* gene is released from the repression and the newly synthesized TOC1 protein promotes the transcription of *CCA1* and *LHY* starting another cycle (Alabadi et al., 2002; Mizoguchi et al., 2002). In addition to transcriptional control, there is regulation of clock gene transcripts and modification of clock proteins, which are also required to maintain the rhythm. Downstream of the circadian clock are so-called output pathways that are manifested by rhythmicity of many physiological processes including leaf movement, growth, and gene expression.

Circadian rhythms persist in the absence of external rhythms such as dark/light or temperature changes. However, under constant environmental conditions, the clock normally runs in a period that is not exactly 24 hours and therefore has to be entrained or reset by a *Zeitgeber* (time keeper). Daily dark/light changes are important *Zeitgeber* to set the phase of the circadian clock and to synchronize it with the environment. This light input to the clock is performed by sensory photoreceptors.

One of the important milestones in circadian clock research was the invention of reporter gene constructs that drive expression of the reporter under control of the regulatory region of a clock-controlled 'output' gene such as the promoter of the *CAB2* gene (Millar et al., 1992), encoding one of the chlorophyll a/b binding proteins of photosystem II. Use of the reporter luciferase (LUC) allowed high-through-put screens in *Arabidopsis* for the identification of mutant clock genes (Millar et al., 1995a; Somers et al., 1998a) as well as the study of the role different plant photoreceptors have in the entrainment of the circadian clock (Devlin and Kay, 2000; Millar et al., 1995b; Somers et al., 1998b).

As already illustrated for the de-etiolation process, light entrainment of the circadian clock is also regulated by several photoreceptors. Despite this redundancy in photoreceptor function it was possible to identify the role of single photoreceptors by using *Arabidopsis* mutants that are defective in one or several photoreceptors in combination with monochromatic light conditions. Light entrainment of the circadian clock can be analyzed under constant light conditions, since in diurnal species, as plants are, the period length becomes shortened when the fluence rates are increased. This phenomenon is known as Aschoff's rule (Aschoff, 1979). By analyzing light effects on the period length of *CAB2::LUC* expression in wild-type *Arabidopsis*, it was shown that both red- and blue-light entrain the circadian clock. This indicated that phytochromes and blue-light receptors are involved in clock resetting (Millar et al., 1998b). *Arabidopsis* mutants lacking single phytochromes showed that phyA is the major light input receptor under low fluence rate red light, whereas phyB operates under higher fluence rates of red light (Somers et al., 1998b). Consistent with the results obtained from the single mutants, the *phyA/phyB* double mutant showed a deficiency in the perception of red light at low and high fluence rates (Devlin and Kay, 2000). Mutations in other phytochromes such as *PHYD* and *PHYE* showed no effect on the period length of *CAB::LUC* expression under continuous red light. However, triple mutants with defects in *phyA/phyB* and in *phyD* or *phyE* showed stronger effects (longer periods) under high fluence rates of red light relative to the *phyA/phyB* double mutant (Devlin and Kay, 2000). This demonstrates functional redundancy of phyD and phyE with phyB in clock entrainment similar to their role in regulating the de-etiolation response.

The *phyA* mutant was not only affected under low fluence rate red light but also under low fluence rate blue light. The explanation could be either that phyA acts directly as a blue-light receptor, or that phyA acts as a signaling component of blue-light receptors under low light conditions (see below). The *phyB* mutant showed no defects in high or low fluence rates of blue light (Devlin and Kay, 2000).

In blue light, the *cry1* mutant was affected under low and high but not intermediate fluence rates, whereas the loss of cry2 had essentially no effect on clock entrainment under all tested fluence rates (Devlin and Kay, 2000; Somers et al, 1998b). However, the *cry1/cry2* double mutant showed stronger effects than the monogenic *cry* mutants (Devlin and Kay, 2000), demonstrating the redundancy of cry1 and cry2 at least under low and medium fluence rates of blue light. The lack of cry2 action in strong blue light can be explained by its degradation under such conditions. In contrast to double *cry1/cry2* knockout mice (Sancar, Chapter 12), the *cry1/cry2* double mutant of *Arabidopsis* still maintained rhythmicity demonstrating that in plants the cryptochromes are not integral components of the circadian clock.

Surprisingly, the *cry1* mutant also showed some deficiency in the entrainment of *CAB::LUC* rhythms under low fluence-rate red light (Devlin and Kay, 2000) similarly to the *phyA* and the *phyA/cry1* double mutant. Since cry1 is not considered to operate as a red-light photoreceptor it was concluded that cry1 might be acting as a component of phyA signaling (Devlin and Kay, 2000). Indeed, there is evidence for physical interaction of phyA with cry1 (Ahmad et al., 1998c) (see Section 10.5.2).

The phototropins do not seem to be involved in light input to the clock, since the *phot1* mutant showed the same period length as wild type under blue light (Devlin and Kay, 2001). Nevertheless, final proof for or against a role of phototropins as light input photoreceptors might require additional studies on mutants that are deficient in several photoreceptors.

The role of the ZEITLUPE/FKF/LKP family of putative photoreceptors in light entrainment is described by Schultz (Chapter 16).

Phytochromes and cryptochromes are not only part of the input pathway but also part of the output pathway of the circadian clock. This has been shown by measuring transcript levels and promoter activity of phytochrome and cryptochrome genes under constant light or dark conditions. The levels of *CRY1* and of *PHYB-E* transcripts show circadian fluctuations with peaks during the early hours of the subjective day, whereas the transcript levels of *CRY2* and *PHYA* fluctuate with peaks at the end of the subjective day (Bognár et al., 1999; Harmer et al., 2000; Toth et al., 2001). Very recent studies on the circadian regulation of *CRY2* transcript levels have shown, however, that *CRY2* peaks at the beginning of the light period (El-Assal et al., 2003). The cycling of the photoreceptors involved in the light input to the clock could control the effectiveness of clock resetting during subjective day and night cycles, a phenomenon called gating. However, the cycling of photoreceptor transcript levels is not in all cases reflected by a corresponding oscillation of photoreceptor protein levels. For example, the amount of phyB protein seems to be constant whereas its transcript level cycles (Bognár et al., 1999; Toth et al., 2001). In case of CRY2, there is oscillation of both transcript and protein levels, but under light/dark cycles the protein level of cry2 is primarily controlled by protein degradation in the light phase and depends on the photoperiod and not by changes in *de novo* synthesis (El-Assal et al., 2003; Mockler et al., 2003) (see Sections 10.3.2.2 and 10.5.1).

The nuclear transport of several plant photoreceptors is under light control thus providing another level of regulation of light input to the clock and for gating, as is discussed by Schäfer and Nagy (Chapter 9) for the phytochromes, and in Section 10.4 for the cryptochromes.

10.3.2.2 **Photoperiodism**

The ability of organisms to respond to changes in day length (or photoperiod) is of great importance because it allows their physiology and development to adjust to the annual seasons. Environmental factors such as photoperiod and temperature control the timing of flowering and seed-setting of plants. According to the inductive light conditions, plants are classified as short-day (SD) and long-day (LD) plants. Some species are day neutral and others depend on a defined order of short days followed by long days or *vice versa* (for reviews see Koornneef et al., 1998; Hayama and Coupland, 2003; Mouradov et al., 2002; Samach and Coupland, 2000; Samach and Gover, 2001; Thomas and Vince-Prue, 1997). *Arabidopsis* is a facultative LD plant and all ecotypes flower earlier under LD than under SD conditions. Analysis of mutants as well as identifying quantitative trait loci affecting flowering time led to the classification of flowering genes into at least three groups: i) the photoperiod response pathway (LD pathway); ii) the autonomous pathway; iii) the vernalization response pathway.

Mutants in the photoperiod pathway have delayed flowering under LD conditions but under SDs flowering time is normally not affected. Mutations that affect the autonomous pathway are delayed in flowering irrespective of the photoperiod. The vernalization response pathway is only found in such ecotypes of *Arabidopsis* (winter annual varieties) that germinate in summer, grow vegetatively during winter until spring and require a cold treatment during the winter season for flowering in spring.

It was already suggested by Bünning in 1936 (Bünning, 1936) that day-length measurement requires the circadian clock and two models were proposed to explain photoperiodism with clock function (Samach and Coupland, 2000; Samach and Gover, 2001; Thomas and Vince-Prue, 1997). The external coincidence model proposes that a endogenous rhythm has fluctuating sensitivity towards light at certain times of the day, promoting flowering when light is present during the sensitive phase in LD plants, and suppressing flowering in SD plants under the same conditions. The internal coincidence model suggests that under inductive conditions two endogenous rhythms would come into the same phase to promote flower induction whereas under non-inductive photoperiods these rhythms are out of phase.

One would assume that photoreceptors, if they are involved in flowering time regulation should operate in the LD pathway. Indeed, this is the case for phyA, cry1 and cry2 but not for phyB. That phyB does not operate specifically in day length perception was concluded from the observation that the *phyB* mutant flowers much earlier under LD (Goto et al., 1991) but not under SD conditions (Koornneef et al., 1995), suggesting an inhibitory role of phyB acting through the autonomous pathway. Such a role of phyB in the autonomous pathway is also supported by epistasis studies, which showed that FCA, a component of the autonomous pathway, is epistatic to phyB (Koornneef et al., 1995). However, the flowering-time pathways do not act independently from each other, and there is cross talk among them as well as through photoreceptors as outlined below.

The promoting role of cry1 in flowering-time regulation can only be detected in the absence of cry2 (Bagnall et al., 1996; Mockler et al., 1999) and therefore seems to be a minor contributor. Under certain light conditions such as far-red enriched white light, phyA is also found to be involved in day-length perception (Johnson et al., 1994; Más et al., 2000). At least in *Arabidopsis*, the most important photoreceptor for day length perception is cry2. The *cry2* mutant is severely delayed in flowering under LD but not in SD conditions (Guo et al., 1998) and it was found that cry2 is allelic to the photoperiod-insensitive flowering-time mutant *fha1* (Guo et al., 1998) that has been isolated by Koornneef and coworkers (Koornneef et al., 1991). As outlined above, photoperiodism relies on the circadian clock and the clock is entrained by photoreceptors. One could therefore assume that the role of cry2 in photoperiodism is caused by a direct effect on the circadian clock. This is however very unlikely because the *cry2* mutant shows normal clock function under a broad range of fluence rates and the role of cry2 in clock entrainment only becomes detectable when cry1 is also missing (Devlin and Kay, 2000; Somers et al., 1998b). It is therefore more likely that cry2 has a more direct effect on components of the photoperiod response pathway and this has very recently been demonstrated to indeed be the case (El-Assal et al., 2003; Suárez-López et al., 2001; Yanofsky and Kay, 2002).

The photoperiod response pathway contains at least four genes, *GIGANTEA* (*GI*), *CONSTANS* (*CO*), FLOWERING *LOCUS T* (*FT*) and *FWA* (for review see Hayama and Coupland, 2003; Koornneef et al., 1998; Mouradov et al., 2002; Samach and Coupland, 2000; Samach and Gover, 2001; Thomas and Vince-Prue, 1997). CO is a transcriptional activator of *FT* and shows circadian-controlled mRNA oscillations that are modified by the photoperiod with peak levels in darkness under SD and in light under LD (El-Assal et al., 2003; Suárez-López et al., 2001; Yanofsky and Kay, 2002). The peak expression of *CO* in light seems to be essential for the induction of FT. cry2 (and phyA) probably do not promote *CO* expression but modulate its ability to induce *FT* (El-Assal et al., 2003; Suárez-López et al., 2001; Yanofsky and Kay, 2002). The important role of cry2 in the LD pathway is not only evident from the phenotype of the mutant, that is late flowering under LD, but also from the identification of a quantitative trait locus of the Cape Verde Islands (Cvi) accession, which was shown to be a naturally occurring allele of *CRY2*. Plants with this *CRY2* allele are early flowering under LD and SD (El-Assal et al., 2001). The amino acid substitution important for Cvi-cry2 function was shown to be in position 367 where a Val present in other *CRY2* alleles is changed to methionine (El-Assal et al., 2001). This substitution causes a longer persistence of high cry2 protein levels in the light phase under SD conditions (El-Assal et al., 2001) (see also Section 10.5.1), and this increased protein stability of cry2 correlates with day-length insensitivity. By analyzing *Arabidopsis* mutants carrying either a loss-of-function *cry2* allele or the gain-of-function Cvi-*CRY2* allele in the background of other photoreceptor or flowering pathway mutants, it was shown that cry2 can act independently of cry1 and phyA (El-Assal et al., 2003). The same authors have also shown that the role of cry2 in promoting flowering does not exclusively depend on its inhibitory function on phyB as suggested earlier (Guo et al., 1998). Their conclusion was based on the observation that the flowering time phenotype of *cry2* is not detectable in continuous blue light, but in a mixture of blue with red light. As already mentioned above, *CRY2* transcription is under circadian control, and the protein level is strongly regulated by light and photoperiod. Interestingly, there is also control of *CRY2* transcription by FLC, a component of the autonomous flowering pathway (El-Assal et al., 2003), showing again the complexity in cross talk between the components of different flowering-promotion pathways.

10.3.3 Regulation of Gene Expression

As outlined above, cryptochromes regulate many physiological and developmental processes in plants and probably most of these processes involve differential gene expression at least in part. Before genome-wide expression profiling of gene expression was feasible, about one hundred plant genes had been identified by more classical methods such as Northern-blotting, primer extension and quantitative RT-PCR that show differential expression upon light-treatment. Among them a few are down-regulated by light such as *PHYA* (Bruce et al., 1989) and protochlorophyllide oxidoreductase (*PORA*) (Apel, 1981; Batschauer and Apel, 1984). However, most of these genes such as photosynthesis genes (*LHCPs*, *RubisCO*, etc.) and genes involved in the biosynthesis of flavonoids and anthocyanins (*CHS*, *CHI*, *DFR*), to mention only a few (for reviews see Fankhauser and Chory, 1997; Kuno and Furuya, 2000; Terzaghi and Cashmore, 1995), are upregulated. Besides nuclear genes, the expression of plastid-encoded genes is also under light-control. Only two of these genes for which cryptochromes play a major role in regulation will be considered here for further discussion, the nuclear encoded chalcone synthase (*CHS*) and the plastid encoded *psbD* gene encoding the D2 protein of photosystem II.

Chalcone synthase (CHS) is the first enzyme of the flavonoid and anthocyanin pathway. Flavonoids and anthocyanins serve many functions such as repellents of plant pathogens and herbivores, phytoalexins, mediators in plant-microbe interaction, and UV-protectants (Harbourne, 1993; Koes et al., 1994; Shirley, 1996). The red and purple anthocyanins may serve as sunscreens for etiolated seedlings to protect them against light damage as long as the photosynthetic apparatus is not completely assembled. According to their multiple functions, the synthesis of these compounds is strongly enhanced by environmental factors such as cold stress, pathogen attack, and light. Here, we will focus on the light-regulation of anthocyanin formation in *Arabidopsis* and *Sinapis* seedlings and how this regulation is mediated by differential expression of *CHS*.

Arabidopsis contains a single *CHS* gene (Feinbaum and Ausubel, 1988) that is strongly induced by UV-A, white, and blue light but only weakly by red light (Feinbaum et al., 1991). This result indicates that expression of this gene is primarily under control of a UV-A/blue-light receptor and that phytochrome(s) play only a minor role in up-regulating this gene. However, continuous far-red light acting on phyA also enhances *CHS* expression in etiolated *Arabidopsis* and *Sinapis* seedlings (Batschauer et al., 1991; Kaiser et al., 1995; Kunkel et al., 1996). The major role of cry1 in regulating *CHS* expression under blue-light was confirmed by analyzing the *cry1* (*hy4*) mutant. When *Arabidopsis* wild-type and *cry1* seedlings were grown for three days in red light and then transferred to blue light, only the wild-type seedlings responded with increased *CHS* mRNA levels (Ahmad et al., 1995), showing that cry1 is the major photoreceptor that regulates *CHS* gene expression in blue light. Similar results were obtained when plants were grown in dim white light for several weeks and then transferred to high-intensity blue light. In contrast to the *cry1* mutant, mRNA levels of *CHS* as well as of the subsequent enzymes in this pathway, chalcone isomerase (CHI) and dihydroflavonol reductase (DFR), were strongly increased in wild-type plants (Jackson and Jenkins, 1995; Kubasek et al., 1992). The reduced expression of these genes in the mutant corresponds quite well with lower anthocyanin formation (Ahmad et al., 1995; Jackson and Jenkins, 1995).

However, the regulation of anthocyanin formation in blue light is probably more complex since the *phyA* but not the *phyB* mutant has less pigment under these conditions (Neff and Chory, 1998). This difference indicates that phyA can either function as a blue light receptor at least under low fluence rates (Shinomura et al., 1996; Poppe et al., 1998) or that phyA modulates cry1 action. In contrast to anthocyanin formation, a lack of phyB affects the induction of *CHS* expression in blue light (Wade et al., 2001). Thus, there is probably some co-action of cry1 with phyA and phyB in blue light or some modulation of cry1 signaling by these phytochromes (discussed in Chapter 6). However, there is not a strict dependency on phyA and/or phyB for *CHS* regulation through cry1 because the expression of the β-glucuronidase (GUS) reporter under control of a white mustard *CHS*-promoter was similar to wild type in the *phyA*, *phyB* single and the *phyA*/*phyB* double mutants (Batschauer et al., 1996).

By using fusion gene constructs consisting of *CHS*-promoter fragments fused to the GUS reporter and analyzing their expression in stably transformed plants or transiently transformed protoplasts the *cis*-acting elements that mediate the light response could be identified for several *CHS* genes (Batschauer et al., 1996; Feinbaum et al., 1991; Hartmann et al., 1998; Kaiser et al., 1995; Rocholl et al., 1994). These light-responsive elements or units were also identified by *in vivo* footprinting and *in vitro* DNA-binding studies (Schulze-Lefert et al., 1989; Weisshaar et al., 1991), are well conserved among *CHS* genes of different species, and contain two binding sites for transcription factors, an ACGT element (G-box/E-box) and a myb-recognition element. At least for the ACGT element there is a clear idea of how the light signal is transduced to the gene. The bZIP factor HY5 binds to this element and enhances the transcription of the *CHS* gene (Ang et al., 1998). HY5 itself is regulated by differential stability in light and darkness. In darkness, HY5 binds to COP1 within the nucleus and is in this way guided to the proteasome, which degrades HY5. Degradation of HY5 is probably mediated by the COP9 signalosome (Hardtke et al., 2000; Osterlund et al., 2000). In light, COP1 is transported out of the nucleus (Osterlund and Deng, 1998; von Arnim and Deng, 1994), which rescues HY5 from degradation. Although physical interaction between cry1 and COP1 has been demonstrated (Wang et al., 2001; Yang et al., 2001) (see Cashmore, Chapter 11) it is not yet clear how light absorption by cry leads to the translocation of COP1.

Besides nuclear genes, chloroplast-encoded genes are also light-regulated. Whereas transcriptional control seems typically to dominate expression of nuclear genes, the situation is more complex for chloroplast-encoded genes where expression is regulated both by transcription and by posttranscriptional events, the latter often being more influential (Mayfield et al., 1995; Mullet, 1993; Sugita and Sugiura, 1996; Stern et al., 1997). One well-studied example of blue light-regulated chloroplast genes is *psbD*, encoding the reaction center core protein D2 of photosystem II. The expression of *psbD* is transcriptionally controlled by high-fluence blue light and the well-characterized light-responsive element of this gene is recognized by the plastid-encoded RNA polymerase (Gamble and Mullet, 1989; Kim et al., 1999; Thum et al., 2001), which is in general regulated by sigma-factors that mediate DNA-binding, and by other DNA-binding factors that regulate the activity of the polymerase (for review see Hess and Börner, 1999; Link, 1996; Maliga, 1998). Indeed it was found that a nuclear-encoded basic helix-loop-helix protein (PTF1) is transported into chloroplasts and binds to the light-responsive promoter of *psbD* (Baba et al., 2001).

An explanation for the blue-light regulation of *psbD* could come from the recent finding that the expression of one of the six sigma factors in *Arabidopsis* (SIG5), which are all encoded in the nucleus, is strongly induced by blue but not by red light (Tsunoyama et al., 2002). Together with results from mutant analysis, which showed that the blue light activation of *psbD* transcription depends on cry1 and cry2 (Thum et al., 2001), one could speculate that the expression of SIG5 is under cryptochrome control. This hypothesis has not yet been rigorously tested, however.

When the complete sequence of the *Arabidopsis* genome became available (Arabidopsis Genome Initiative, 2000), genome-wide expression profiling was feasible and was used to analyze light effects on gene expression in *Arabidopsis*. In one of these studies (Ma et al., 2001) long-term effects of light treatment on gene expression were analyzed by growing the seedlings for six days in continuous darkness or in white, blue, red or far-red light. In addition, seedlings grown for 4.5 days in darkness were treated with light for 36 h. Besides wild-type plants, mutants lacking phyA, phyB, or cry1/cry2 as well as overexpressors of phyA, phyB and cry1 were included in this study. It was shown that of the 9216 ESTs analyzed (representing about 6120 unique genes), 32% showed differential expression in white light of at least twofold. Under monochromatic light conditions, 73%, 57% and 40% of the genes expressed differentially in white light were affected by red, blue and far-red light, respectively. These numbers already show that the expression of most of these genes is affected by different wavelengths of light.

Although only a few genes had previously been identified as being down-regulated by light, the expression profiling showed that of the differentially expressed genes about 40% are repressed. This value is more or less the same in all light qualities. Although only 20% of the genes that are differentially expressed in white light could be functionally assigned based on experimental data or on their similarity with known genes from other organisms, at least 26 pathways seem to be coordinately up-regulated or down-regulated by light. The 11 pathways down-regulated by white light include those for the mobilization of stored lipids, enzymes which are probably no longer needed under these conditions; for ethylene and brassinosteroid synthesis, hormones known to be involved in repressing photomorphogenesis (see below); and for cell wall degradation and water transport across the plasma membrane and the tonoplast, that are probably involved in enhanced elongation growth of the hypocotyl in darkness.

In the context of this chapter it is of particular interest to discuss how many and which genes are affected by cryptochromes. This was addressed by Ma et al. (Ma et al., 2001) by studying the *cry1/cry2* double mutant and the cry1 over-expresser. Most of the genes that are up- or down-regulated in the wild type by blue light were not differentially expressed in the *cry1/cry2* double mutant under the same light conditions, demonstrating that cryptochromes are the major photoreceptors for regulation of gene expression in blue light and that other photoreceptors such as phototropins and phytochromes seem to play only a minor role under these conditions. However, over-expression of cry1 under control of the constitutive and strong 35S promoter of cauliflower mosaic virus resulted in reduced expression of 18% of the genes that are induced in wild type and 7% of the genes that were not up-regulated in wild type under blue light showed up regulation in the cry1 over-expresser. This result shows that the enhanced level of the cry1 photoreceptor does not lead only to quantitative but also to qualitative effects on gene expression and may explain, at least in part, why the increase in photoreceptor concentration does not result in a corresponding shift in the threshold response curve for hypocotyl growth inhibition (Ahmad et al., 2002) (see Section 10.3.1). However, driving the expression of the photoreceptor with a promoter, which causes ectopic expression, could also result in side effects resulting from the presence of the photoreceptor in cells where it is normally absent. Studies on the tissue- and cell-specific expression of CRY1 and CRY2 by using promoter-GUS fusions in transgenic *Arabidopsis* plants have however shown, that both genes seem to be expressed in all organs and tissues (Lin, 2002) (U. Grüne and A. Batschauer, unpublished data). Therefore, only the increased cry1 levels could cause the unexpected effects in transgenic CaMV35S::CRY1 plants.

In another DNA microarray study with Affimetrix gene chips, differences between wild type and the *cry1* mutant in gene expression under blue light were analyzed (Folta et al., 2003). Whereas the effects of blue-light treatment for at least 36 h were analyzed in the study of Ma et al. (Ma et al., 2001), Folta et al. (Folta et al., 2003) screened for differential effects 45 min after the onset of blue light, a time point when hypocotyl growth inhibition is already under cryptochrome control (Parks et al., 2001) (see Section 10.3.1). They found that 420 (5%) of the 8298 transcripts analyzed were differentially expressed in the *cry1* mutant, about half of them with higher and half of them with lower transcript levels compared to wild type. A possible explanation for the down-regulation of transcripts in blue light in the *cry1* mutant, which are up-regulated in wild type, is that their expression is regulated at the level of transcription and RNA stability, both of which are positively affected by cry1.

Among the pathways where gene expression is differentially affected soon after the onset of blue light are the cell cycle, auxin and gibberellin synthesis or signaling, and cell wall metabolism. All of the cell cycle genes are up-regulated in the *cry1* mutant, indicating that cry1 suppresses cell division at this developmental stage although hypocotyl growth inhibition is caused by reduced cell elongation. Most of the differentially expressed genes for auxin and gibberellin synthesis or signaling were up-regulated in the *cry1* mutant, indicating that blue light represses these pathways through cryptochromes very quickly resulting in reduced cell elongation. About half of the differentially expressed cell-wall genes are up-regulated and the other half down-regulated in the *cry1* mutant. Inspecting the known or putative functions of the encoded proteins, one can conclude that cry1 suppresses the expression of genes involved in cell wall loosening but enhances the expression of genes involved in cell wall strengthening (Ma et al., 2001; Folta et al., 2003).

Taken together the following can be concluded from the gene expression profiling studies:

1. Most of the blue light effects on gene expression are mediated by cryptochromes.
2. Cryptochromes have short- (minutes) and long-term (days) effects on gene expression.
3. Cryptochromes affect hormone biosynthesis and signaling by repressing auxin and gibberellin pathways at early stages and the brassinosteroid pathway in a later stage of development.
4. Genes involved in extension growth through cell-wall relaxation and increasing water transport through the plasma membrane and the tonoplast are suppressed by blue light via the cryptochromes.
5. Overexpression of cryptochrome 1 has not only quantitative but also qualitative effects on the gene expression pattern.
6. Although cryptochromes seem to affect transcription rates in most cases, there is also evidence for effects of cryptochromes on the stability of some transcripts.
7. Many of the genes that are regulated by cryptochromes are also controlled by phytochromes.

10.4
Localization of Cryptochromes

In the past few years significant progress has been made on elucidating the subcellular localization of plant photoreceptors. This was made possible in particular by using sensitive reporter proteins such as β-glucuronidase (GUS) (Jefferson, 1987), or the green fluorescent protein (GFP) (Tsien, 1998).

The localization of cry2 was studied in detail using GUS and GFP as reporters as well as immunological methods (Kleiner et al., 1999; Guo et al., 1999). All approaches showed consistently that cry2 is localized in the nucleus. In contrast to the phytochromes (see Chapter 6), there seems to be no light effect on nuclear targeting of cry2. The C-terminal extension of cry2 is required and sufficient for translocation into the nucleus, and this region contains a bipartite nuclear localization signal (NLS) (Kleiner et al., 1999; Guo et al., 1999). Interestingly, the localization studies of randomly fused *Arabidopsis* cDNAs with GFP led to the identification of a fusion protein that was associated with all chromosomes and contains the C terminus of cry2 (Cutler et al., 2000). It is not clear yet whether or not this association with chromosomes is of biological relevance but this topic will be discussed a bit further in Section 10.5.3.

Protoplasts transfected with cry2-GFP or cry2-RFP showed a homogenous signal within the nucleus when the cells were kept in darkness or red light. However, blue light treatment caused the rapid formation of so called nuclear speckles in tobacco (Más, et al., 2000) as well as in *Arabidopsis* and parsley protoplasts (M. Müller, A. Batschauer, unpublished data). These cry2 speckles co-localize with phyB (Más et al., 2000), further supporting the observation that cry2 and phyB interact (Más, et al., 2000) (see also Schäfer and Nagy, Chapter 9). The localization of *Arabidopsis* cry1 was studied in onion epidermal cells by bombarding them with cry1-GFP constructs (Cashmore et al., 1999) and in transgenic *Arabidopsis* plants as GUS fusions (Yang et al., 2000). The cry1-fusion proteins were also found in the nucleus. In contrast to cry2, there is a light effect described for the localization of cry1. The C-terminal extension fused to GUS was found to be enriched in the nucleus in dark-grown plants and to be cytosolic in light-treated plants (Yang et al., 2000). Since the C terminus of cryptochromes does not bind chromophores, the observed light effect on the localization of the fusion protein is not self-mediated. This does not however rule out that endogenous cryptochromes are involved in this process. Some of the cryptochromes of the fern *Adiantum capillus-veneris* are also transported to the nucleus (see Suetsugu and Wada, Chapter 17) as well as the cryptochromes in animals (Sancar, Chapter 12).

Very recently, a third cryptochrome (cry3, cryDASH) was identified in *Arabidopsis* (Brudler et al., 2003; Kleine et al., 2003) (see Section 10.1.2), which shows striking sequence similarity with the *Synechocystis* cryptochrome. Instead of having an extension at the C terminus, cry3 carries an extension at the N terminus that has significant similarities with targeting signals for import into chloroplasts and mitochondria. Using cry3-GFP fusion proteins and *in vitro* import studies it was shown, that cry3 is indeed transported into both organelles. Since the N terminus of cry3 is necessary and sufficient for the import into both organelles it must contain a dual targeting signal (Kleine et al., 2003). Which function cry3 fulfills in these organelles remains to be investigated.

10.5 Biochemical Properties of Cryptochromes

10.5.1 Protein Stability

The transcript levels of *Arabidopsis CRY1* and *CRY2* are not specifically regulated by light, but circadian control of both transcripts has been described (Bognár et al., 1999; El-Assal et al., 2003; Harmer et al., 2000; Toth et al., 2001). In other species, *CRY* transcript levels are affected by light. For example, induction of the *CRY2* gene by light has been reported for white mustard (Batschauer, 1993), and the mRNA levels of some fern *CRYs* are also regulated by light (see Chapter 17 of this volume). In contrast to the transcript level, the amount of the cry2 but not of the cry1 protein is strongly light-regulated in *Arabidopsis* seedlings (Ahmad et al., 1998a; Lin et al., 1998). Exposure of etiolated *Arabidopsis* seedlings to blue light leads to a rapid decrease in the amount of cry2. This effect is fluence-rate and wavelength dependent. After blue light treatment with 1 μmol m^{-2} s^{-1} for 6 h, essentially the same amount of cry2 protein as in dark samples can be detected and a significant fraction even after 24 h. However, with 20 times higher fluence rates of blue light only very small amounts are detectable after 1 h and nothing after 24 h (Ahmad et al., 1998a; Lin et al., 1998).

Whereas UV-A and green light also led to a decrease in the amount of cry2 protein, red light had no effect on the level of cry2 protein even when exposed for a long time and at high dose (Lin et al., 1998). The fact that the amount of cry2 protein is only affected by light with wavelengths below 500 nm and that this process is very similar for wild-type and *cry1* mutant plants (Ahmad et al., 1998a) suggests that cry2 could regulate its own down-regulation. However, the involvement of other blue light receptors in this process has not been tested rigorously so far. Interestingly, in etiolated tobacco and tomato seedlings cry1, but not cry2, is light labile (Ahmad et al., 1998a).

The rapid down-regulation of the cry2 protein in blue light together with the observation that its transcript level is not affected by light suggests that blue light either induces degradation of cry2 or blocks translation of its mRNA. To distinguish between these possibilities, dark-grown seedlings were incubated with the protein-synthesis inhibitor cycloheximide and then treated with blue light. Since no difference in the disappearance of cry2 was observed between inhibitor-treated and control plants it is very likely, that blue light induces the degradation of cry2 (Ahmad et al., 1998a).

In order to define the region of cry2 involved in its degradation, domain-switch experiments were performed in which different regions of *Arabidopsis* CRY1 and CRY2 were exchanged and the chimeric genes expressed under control of the CaMV 35S promoter in the *cry1* mutant of *Arabidopsis* (Ahmad et al., 1998a). The fusion proteins that contained either the C-terminal extension (amino acids 506–611) or the N-terminal region (amino acids 1–505) of cry2 were biologically active and showed significantly lower levels in blue than in red light (Ahmad et al., 1998a), indicating that both domains of cry2 can mediate degradation. Since chimeric proteins of the GFP or

GUS reporters and either the N-terminal or the C-terminal domain of cry2 are not reported to be light-labile, however (Kleiner et al., 1999; Guo et al., 1999), one may conclude that both cry2 domains can mediate degradation only when present in a functional cryptochrome photoreceptor. This result also supports the point of view that cry2 is likely the photoreceptor that induces its own degradation.

As outlined in Section 10.3.2.2, the naturally occurring *CRY2* allele in the Cape Verde Islands accession of *Arabidopsis* causes early-flowering and day-length insensitivity, and the relevant amino acid exchange is a substitution of valine for methionine at position 367 (El-Assal et al., 2001). When dark-grown seedlings were transferred to strong blue light (40 μmol m^{-2} s^{-1}) the depletion of Cvi-cry2 and cry2 from the accession Landsberg erecta was very similar. However, when plants were grown under SD conditions and the protein levels of Cvi-cry2 and Ler-cry2 compared after the lights were turned on, the Cvi-cry2 was degraded much more slowly than the Ler-cry2 and reaccumulated much faster in the following dark period. The same authors have also shown that the levels of Cvi-cry2 and Ler-cry2 in plants kept under LD conditions are very similar and do not oscillate significantly. Taken together, these data show that the level of cry2 protein is under photoperiodic control and that a single amino acid substitution within cry2 leads to some stabilization in light when plants are kept in SDs. This extended stability of Cvi-cry2 in SD is most likely the direct cause for the early-flowering phenotype of this accession under these conditions. The molecular mechanism of how the amino acid substitution at position 367 affects cry2 stability has not yet been investigated.

The photoperiodic effect on cry2 stability was also addressed by Chentao Lin and coworkers (Mockler et al., 2003). As Koornneef and coworkers found (El-Assal et al., 2003), the cry2 protein oscillates strongly under SD conditions with high levels at the end of the dark phase and low levels during the light phase. In LD conditions, the oscillation was very weak and the cry2 level was constitutively low, similar to the lowest level during the light phase of plants kept in SD. When white light was replaced by monochromatic light, the same oscillation in the cry2 level was observed under SD conditions with blue light but not in red light. Surprisingly, the cry2 level was very low in SD red light conditions. When plants were transferred to continuous light after the SD period, constitutively high levels were found in plants under continuous red light after being transferred from SD blue light conditions and constitutively low levels after transfer from SD red light conditions. After transfer from SD blue light to continuous blue the cry2 level remained low but showed some increase in the subjective night phase, probably caused by the circadian clock-controlled oscillation of *CRY2* transcript levels. Taking these results together the following can be concluded: 1. The cry2 protein level strongly oscillates under SD but not under LD conditions; 2. Oscillation of the cry2 protein level is mainly controlled by protein degradation and not by the circadian expression of the *CRY2* gene; 3. Blue light induces both oscillation and cry2 degradation.

The degradation of cry2 is reminiscent of the degradation of phyA, which is also rapidly broken down down upon light-treatment. In the case of phyA, ubiquitination has been shown to occur upon light treatment (Clough et al., 1999). Ubiquitination of proteins very often precedes their degradation by the proteasome (Vierstra, 1996). For cry2, ubiquitination has not been demonstrated yet but nevertheless there are some indications that cry2 could be degraded by the proteasome pathway. COP1 is a putative subunit of the E3 ubiquitin ligase complex, mediating the proteolytic degradation of the bZIP transcription factor HY5 in darkness. In light, HY5 is not degraded and activates the transcription of genes such as *CHS* (Ang et al., 1998; Hardtke et al., 2000; Osterlund et al., 2000), which were shown before to be active in light but not in the dark. In *Arabidopsis* seedlings carrying the weak *cop1–6* allele, the degradation of cry2 in blue light is impaired and the ratio between phosphorylated and unphosphorylated cry2 is increased (Shalitin et al., 2002) (see Section 10.5.2). This indicates that phosphorylated cry2 is the substrate for degradation, and requires functional COP1 for the process to be efficient.

In support of this conclusion are the results from yeast two-hybrid interaction studies which show that cry2 can physically interact with COP1 (Wang et al., 2001). Also cry1, which is not degraded in light interacts with COP1 (Wang et al., 2001; Yang et al., 2001) and from this association it was concluded that the interaction between COP1 and cry1 is involved in cry signaling (Yang et al., 2001) (see Cashmore, Chapter 11). Another problem in assuming COP1 involvement in cry2 degradation is the fact that COP1 is transported out of the nucleus in light (von Arnim and Deng, 1994; Osterlund and Deng, 1998), whereas cry2 seems to be located in the nucleus independent of the light conditions (see Section 10.4). However, since the degradation of cry2 seems to be much faster than the translocation of COP1, there could be enough COP1 present in the nucleus after dark-light transition to initiate cry2 degradation. Thus, further research is definitely needed to elucidate the molecular events in blue light-induced cry2 degradation.

10.5.2
Phosphorylation

Phosphorylation is a common principle for regulating protein activity or stability. It was found very recently that plant cryptochromes are phosphorylated and that the phosphorylation probably affects both their activity and stability. The first evidence that plant cryptochromes can be phosphorylated came from *in vivo* and *in vitro* studies, which showed that cry1 in dark-grown *Arabidopsis* seedlings incubated with ^{32}P-orthophosphate is not radioactively labeled. However, red light treatment leads to phosphorylation of cry1, an effect that could be reversed by a pulse of far-red light given after the red light pulse (Ahmad et al., 1998c). The red/far-red reversibility is indicative of the involvement of phytochrome in this process. Phytochrome A itself has Ser/Thr kinase activity (Yeh and Lagarias, 1998), and Ahmad et al. (Ahmad et al., 1998c) showed that cry1 is a substrate for the phyA kinase activity by incubating recombinant oat phyA with recombinant *Arabidopsis* cry1. With this assay, cry1 became phosphorylated only in the presence of phyA and this phosphorylation was light-dependent where red and blue light had very similar effects. In addition the authors demonstrated through yeast-two-hybrid interaction studies that the C-terminal regions of *Arabidopsis* phyA and cry1 physically interact. Based on these data a model was proposed in which the cryptochrome becomes fully active only when it is phosphorylated by phytochrome (Ahmad et al., 1998c).

The phosphorylation of *Arabidopsis* cry1 and cry2 was further investigated (Shalitin et al., 2003; Shalitin et al., 2002), and these studies showed that blue light treatment of dark-grown seedlings leads to a rapid phosphorylation of both photoreceptors, which can be detected by the incorporation of a ^{32}P-label within 5–10 min after the onset of light. This blue light effect on cry phosphorylation in vivo is most likely not mediated by phytochromes since it was not affected in the *phyA* and *phyB* single mutant, the *phyA/phyB* double mutant, the triple mutants *phyA/phyB/phyD* and *phyB/phyD/phyE* or in the *hy1* mutant, which is impaired in synthesizing the phytochrome chromophore phytochromobilin. The same authors could not repeat the above-mentioned red-light effect on cry1 phosphorylation *in planta* and concluded from all these results that the phosphorylation of cryptochromes is not mediated by phytochrome but instead by a blue light receptor, most likely cryptochrome itself. The involvement of a blue light receptor in cryptochrome phosphorylation is confirmed by an action spectrum for cry2 phosphorylation made with an *Arabidopsis* cell culture, which shows peaks in the blue and UV-A region and no effects at wavelengths above 500 nm (M. Müller and A. Batschauer, unpublished data), as well as by *in vitro* studies outlined below.

The phosphorylation of cryptochromes *in planta* can not only be detected by labeling with ^{32}P but also by a shift in mobility on SDS-PAGE (Shalitin et al., 2003; Shalitin et al., 2002). Upon blue light treatment several cryptochrome bands with lower mobility can be detected by Western-blotting with cry-specific antibodies. The shifted bands are phosphorylated and not otherwise modified forms of cryptochromes as shown by treating the native protein extracts with protein phosphatase, which causes a complete removal of the shifted bands. Using the shift in mobility on SDS-PAGE as an assay, the kinetics of cry phosphorylation and its fluence rate dependency was studied. The phosphorylation of cry1 and cry2 was fluence-rate dependent with strong differences in sensitivity and kinetics. Whereas for example 15 min of blue light treatment with a fluence rate of 5 μmol m^{-2} s^{-1} led to a significant phosphorylation of cry2, the same light treatment had almost no effect on cry1 phosphorylation. Cry1 required much higher fluence rates and exposure times to reach saturation of phosphorylation. Interestingly, the fraction of photoreceptor molecules that is phosphorylated under saturating light conditions seems to differ significantly between cry2 and cry1. About 75% of cry2 is phosphorylated after a 15 min blue light treatment with a fluence rate of 5 μmol m^{-2} s^{-1} (total fluence 4.5 mmol m^{-2}), whereas only 50% of the cry1 is phosphorylated after 30 min blue light treatment with a fluence rate of 60 μmol m^{-2} s^{-1} (total fluence 108 mmol m^{-2}). These data show that cry2 is significantly more sensitive towards phosphorylation than cry1.

As described in Section 10.3.1, cry2 operates under lower light intensities than cry1. One could therefore assume that the differences in sensitivity for phosphorylation are the reason why cry2 operates under low and cry1 under high intensity light. This would also imply that phosphorylation is required for the biological activity of cryptochromes, and there are good reasons to assume this requirement as described below. However, the situation may be more complex considering the differences in the stability of cry2 and cry1 under blue light. Cry2 activity is dependent on a balance between activation and degradation in an equilibrium that is fluence-rate dependent, whereas the activity of cry1 seems to depend only on blue light for activation.

Despite these differences, there are also similarities in the phosphorylation of cry1 and cry2. For example, under saturating light conditions only a fraction of photoreceptor molecules is phosphorylated, and transition to darkness causes dephosphorylation with little phosphorylated protein detectable after 15 min. The phosphorylation of cry1 and cry2 in blue light may indicate that phosphorylation is required for their biological activity. This hypothesis has been supported by the observation that all of the six analyzed *cry1* mutant alleles that are biologically inactive are not phosphorylated in blue light (Shalitin et al., 2003). Only one of these mutant alleles had an exchange of a residue (Ser66→Asn) that could be susceptible to phosphorylation. Whether this Ser is indeed phosphorylated upon blue light treatment has not been determined. How the other mutations in cry1 affect phosphorylation is not clear yet but the data show that there is a correlation between phosphorylation and the biological activity of cry1.

Because of the lack of appropriate mutant alleles of *CRY2*, another approach was used to test for the correlation between phosphorylation and biological activity (Shalitin et al., 2002). Expression of the C terminus of cry2 (or cry1), which can only be expressed in *Arabidopsis* to significant levels when fused to another protein such as GUS, causes a constitutive photomorphogenic phenotype (Yang et al., 2000), suggesting that the C terminus of plant cryptochromes mediates signaling (see Cashmore, Chapter 11). In contrast to full-length cry2, its C terminus is constitutively phosphorylated independent of the light conditions. This was taken as evidence that phosphorylation is required for the biological activity of cry2 (Shalitin et al., 2002). The data discussed above have shown that plant cryptochromes are rapidly phosphorylated in blue light, that phosphorylation correlates with biological activity, and that phytochromes are not required for this process. In addition it has been demonstrated that phosphorylation of cry2 occurs in the absence of cry1 and *vice versa* (Shalitin et al., 2003).

In vitro studies with recombinant *Arabidopsis* cry1 that has been expressed in insect cells and purified to apparent homogeneity have given insight into the molecular mechanism of cryptochrome phosphorylation (Bouly et al., 2003; Shalitin et al., 2003). These studies showed that cry1 autophosphorylates and that autophosphorylation is blue-light dependent. In the work by Bouly et al. (Bouly et al., 2003), autophosphorylation was analyzed in detail. It was shown that autophosphorylation depends not only on blue light but also on the presence of the FAD cofactor, and that flavin antagonists such as KI and oxidizing agents abolish the blue-light induced phosphorylation. Since cryptochromes do not have homology to known protein kinases, one may be concerned that the *in vitro* phosphorylation is caused by a copurified kinase and not by autophosphorylation. However, recombinant cry1 as well as cry1 purified from plant cells binds nearly quantitatively to ATP-agarose, and the binding affinity of cry1 for ATP (K_d = 20 μM) is in the same range as described for other ATP-binding proteins with high and specific affinity for ATP. The stoichiometry of ATP bound to recombinant cry1 was determined to be 0.4, indicating that cry1 contains one binding site for ATP. Several putative motifs for ATP and GTP binding (P-loop) are indeed present in cry1, one of which is well conserved between the cryptochromes of *Arabidopsis* and the fern *Adiantum* (Imaizumi et al., 2000). Consequently, it is very unlikely that the phosphorylation is caused by a copurified kinase.

The *in vitro* (Bouly et al., 2003) and *in vivo* (Shalitin et al., 2003) kinetic studies of cry1 phosphorylation give similar results with saturation being reached between 30 and 60 min after the onset of blue light. Interestingly, preillumination of cry1 with blue light, in the absence of the ATP substrate, followed by the addition of ATP in the absence of blue light nevertheless leads to the phosphorylation of cry1 (Bouly et al., 2003). This indicates that cry1 remains activated, at least for some time, after it is transferred to darkness. The identity of the amino acids phosphorylated in cry1 *in vitro* was determined and only serine was identified (Bouly et al., 2003). Since the autophosphorylated cry1 does not show the same shift in mobility on SDS-PAGE (Bouly et al., 2003) as the cry1 isolated from plant material (see above) one may assume, that *in planta* cry1 doesn't only autophosphorylate upon blue light treatment but is also phosphorylated by other kinases. The blue-light dependency of additional phosphorylation could be caused by a conformational change of cry after it has absorbed light, thus giving access to a kinase. Another explanation could be that blue light activates a kinase, which then phosphorylates the cryptochromes. In any case, the sites within cry1 and cry2 that are phosphorylated *in vivo* and *in vitro* have to be determined in order to address in greater detail the molecular mechanism of cry phosphorylation and the specific role of each phosphorylation site on the biological function of cryptochrome.

10.5.3
DNA Binding

As outlined in Section 10.2, photolyase has a high binding affinity for its substrate and a lower affinity for undamaged DNA, which is not sequence specific. Interestingly, it seems that the ability of photolyase to bind with low affinity to undamaged DNA is conserved in cryptochromes. Although not all known cryptochromes have been tested for DNA binding, there are at least several examples including animal and bacterial cryptochromes showing that cryptochromes bind to DNA. The only cryptochrome for which a crystal structure has been solved so far is cryDASH from the cyanobacterium *Synechocystis* (Brudler et al., 2003). This allowed comparison of its structure with already known structures from microbial (class I) CPD photolyases. Although no co-crystal structure of photolyase with substrate DNA is available, several amino acids are considered to be important for DNA recognition. Among them, there are five Arg residues (Arg226, Arg278, Arg342, Arg344, Arg397, numbering according to *E. coli* photolyase) on the surface of the protein and close to the substrate binding pocket that contribute to a positive electrostatic potential considered to be important for DNA binding (Park et al., 1995). Interestingly, all of these Arg residues are conserved in *Synechocystis* cryDASH for which binding to undamaged DNA has been demonstrated with an equilibrium dissociation constant of around 2 μM (Brudler et al., 2003), very similar to that described for *E. coli* photolyase. All of these Arg residues are also conserved in *Arabidopsis* cry3 (Kleine et al., 2003; Brudler et al., 2003) and it has been shown that cry3 also binds to DNA (Kleine et al., 2003).

For all the other plant cryptochromes, with the exception of cry2, DNA binding has not been analyzed or published. From random fusions of GFP with *Arabidopsis* cDNAs, a fusion protein was identified, which bound to chromatin. The fusion protein carried the C-terminal part of cry2 (Cutler et al., 2000). It is not clear from this study, however, whether the chromatin association was mediated by the interaction of the cry2 C-terminus with other proteins or by direct binding of cry2 to DNA. In vitro DNA-binding studies have indicated that the chromatin association of cry2 is at least in part caused by direct DNA binding (M. Müller and A. Batschauer, unpublished data). Arg278, 342, and 344 are fully conserved, and Arg226 is conserved in cry1 and substituted by Lys in cry2 which is, however, a conservative substitution. The Arg397 is just one position shifted in cry1 and cry2 in comparison to *E. coli* DNA photolyase. Since at least four of the five Arg residues are conserved in cry1 and cry2, and taking into consideration the strong argument mentioned above that the conserved Arg residues are responsible for DNA-binding activity, *Arabidopsis* cry1 and cry2 should indeed bind to DNA.

The question remains as to whether DNA binding by cryptochromes is regulated by light and what function this binding might have on the cellular response to light. In the case of *Synechocystis* it was concluded from a comparison of the gene-expression profiles from wild type and the *cry* mutant that cryDASH could act as a repressor of transcription (Brudler et al., 2003). Only a few genes were identified in this study that showed significantly higher expression in the mutant than in wild type and it is not clear yet how a protein with unspecific DNA binding can specifically regulate these genes. In addition, no comparison between light-grown and dark-grown (or dark adapted) cells was made and therefore does not allow conclusions to be drawn about a light-specific role of cryDASH in gene regulation. That plant cryptochromes have such a role has been demonstrated in detail (see Section 10.3.3). How they do this regulation at the molecular level, however, has still to be elucidated.

10.5.4 Electron Transfer

As described in Section 10.2, photolyases use light-driven electron transfer from the reduced flavin cofactor $FADH^-$ to the substrate for catalysis. In photolyase, only the fully reduced flavin is catalytically active, but not the semireduced or fully oxidized form (for review see Sancar, 2003). Photolyase containing semireduced or oxidized FAD can be transformed to the catalytically active form by photo-excitation of the FAD in the presence of reducing agents in the medium. This photoreduction involves conserved tryptophans and in some photolyases tyrosine residues in addition, which transfer electrons to the excited FAD. Owing to the similarities between photolyase and cryptochromes in amino acid sequence and cofactor composition it was speculated that cryptochromes might use light-driven electron transfer for signaling (Cashmore et al., 1999; Malhotra et al., 1995).

Indeed, it has recently been shown for *Arabidopsis* cry1 that electron transfer could be involved in cryptochrome signaling (Giovani et al., 2003). The photoreceptor used in this study was expressed and purified from baculovirus-transfected insect cells and contained fully oxidized FAD. After ns laser-flash excitation of the FAD, transient absorbance changes were monitored and the recovery kinetics indicated three components with half-lives of about 1 ms, 5 ms and >100 ms. From the kinetics and the observed spectral changes, it was concluded that upon excitation the semireduced radical FADH$^{\bullet}$ is formed concomitantly with a neutral tryptophan radical. There was further evidence for electron transfer from a tyrosine to the tryptophan radical (Trp$^{\bullet}$), as in *Anacystis nidulans* photolyase, that could compete with electron back-transfer from FADH$^{\bullet}$ to Trp$^{\bullet}$. Addition of β-mercaptoethanol as an external electron donor led to the reduction of the tyrosine radical and to accumulation of FADH$^{\bullet}$. Based on these studies one can conclude that the FAD cofactor can be photoreduced in cryptochromes, as in photolyase, involving Trp and Tyr radicals. In principle, all of these internal radicals as well as external electron donors that reduce the Tyr$^{\bullet}$ or electron acceptors, which can be reduced by FADH$^{\bullet}$ could mediate the signaling. However, physiological electron donors or acceptors of cryptochromes have not yet been identified. A final answer to the question whether electron transfer either inter- or intramolecularly is necessary for cryptochrome signaling therefore requires further studies.

10.6 Summary

Since cryptochromes were identified for the first time in plants in the year 1993, our knowledge about these photoreceptors has increased dramatically. Based primarily on genetic studies, the biological roles of cryptochromes have been identified and these studies show that cryptochromes affect many growth and differentiation processes, are important for photoperiodism, and regulate the expression of many genes. The discovery of cryptochromes in animals, humans, and bacteria broadened the field of cryptochrome research and furthermore allowed the identification of similarities and differences in the biological roles of cryptochromes from different kingdoms. Recent progress in understanding cryptochrome function at the molecular or even the atomic level was made possible by the production of these photoreceptors via heterologous expression. This progress has also laid the ground work for analyzing cryptochromes with various spectroscopic methods and one can thus expect tremendous progress in this research field within the next few years.

Acknowledgements

I wish to thank Erica Lyon (Max-Planck-Institute for Terrestrial Microbiology, Marburg, Germany) for critical reading of the manuscript, Birte Dohle for preparing the figures, and the Deutsche Forschungsgemeinschaft for the support of our research.

References

Afonso, C. L., Tulman, E. R., Lu, Z., Oma, E., Kutish, G. F., and Rock, D. L. (1999) *J. Virol. 73*, 533–552.

Adams, M. D., Kerlavage, A. R., Fleischmann, R. D., Fuldner, R. A., Bult, C. J. *et al.*, (1995) *Nature* 377, 3–174.

Ahmad, M. (1999) *Current Opinion Plant Biol. 2*, 230–235.

Ahmad, M., and Cashmore, A. R. (1993) *Nature 366*, 162–166.

Ahmad, M., Grancher, N., Heil, M., Black, R. C., Giovani, B., Galland, P., and Lardemer, D. (2002) *Plant Physiol. 129*, 774–785.

Ahmad, M., Jarillo, J., and. Cashmore, A. R (1998a) *Plant Cell 10*, 197–207.

Ahmad, M., Jarillo, J. A., Klimczak, L. J., Landry, L. G., Peng, T., Last, R. L., and Cashmore, A. R. (1997) *Plant Cell 9*, 199–207.

Ahmad, M., Jarillo, J. A., Smirnova, O., and Cashmore, A. R. (1998b) *Nature 392*, 720–723.

Ahmad, M., Jarillo, J. A., Smirnova, O., and Cashmore, A. R. (1998c) *Mol. Cell 1*, 939–948.

Ahmad, M., Lin, C., and Cashmore, A. R. (1995) *Plant J. 8*, 653–658.

Alabadi, D., Yanovsky, M. J., Mas, P., Harmer, S. L., and Kay, S. A. (2002) *Curr. Biol. 12*, 757–761.

Ang, L.-H., Chattopadhyay, S., Wei, N., Oyama, T., Okada, K., Batschauer, A., and Deng, X.-W. (1998) *Mol. Cell 1*, 213–222.

Apel, K. (1981) *Eur. J. Biochem. 120*, 89–93.

Aschoff, J. Z. (1979) *Tierpsychol. 49*, 225–249.

Baba, K,. Nakano, T., Yamagishi, K., and Yoshida, S. (2001) *Plant Physiol. 125*, 595–603.

Bagnall, D. J., King, R. W., and Hangarter, R. P. (1996) *Planta 200*, 278–280.

Batschauer, A. (1993) *Plant J. 4*, 705–709.

Batschauer, A. (2003) In *Photoreceptors and light signalling;* Batschauer A. (ed.), Comprehensive Series in Photochem. and Photobiol. Sciences, Royal Society of Chemistry, Cambridge UK.

Batschauer, A., and Apel, K. (1984) *Eur. J. Biochem. 143*, 593–597.

Batschauer, A., Ehmann, B., and Schäfer, E. (1991) *Plant Mol. Biol. 16*, 175–185.

Batschauer, A., Rocholl, M., Kaiser, T., Nagatani, A., Furuya, M., and Schäfer, E. (1996) *Plant J. 9*, 63–69.

Bognár, L. K., Hall, A., Adam, E., Thain, S. C., Nagy, F., and Millar, A. J. (1999) *Proc. Natl. Acad. Sci. USA* 96, 14652–14657.

Bouly, J.-P., Giovani, B., Djamei, A., Mueller, M., Zeugner, A., Dudkin, E. A., Batschauer, A., and Ahmad, M. (2003) *Eur. J. Biochem. 270*, 2921–2928.

Briggs, W. R., and Christie, J. M. (2002) *Trends Plant Sci. 7*, 204–210.

Briggs, W. R., and Huala, E., (1999) *Annu. Rev. Cell. Dev. Biol.* 15, 33–62.

Briggs, W. R., and Olney, M. A. (2001) *Plant Physiol. 125*, 85–88.

Bruce, W. B., Christensen, A. H., Klein, T., Fromm, M., and Quail, P. H. (1989) *Proc. Natl. Acad. Sci. USA 86*, 9692–9696.

Brudler, R., Hitomi, K., Daiyasu, H., Toh, H., Kucho, K., Ishiura, M., Kanehisa, M., Roberts, V. A., Todo, T., Trainer, A., and Getzoff, E. D. (2003) *Mol. Cell 11* 59–67.

Bünning, E. (1936) *Ber. Dtsch. Bot. Ges. 54*, 590–607.

Cashmore, A. R., Jarillo, J. A., Wu, Y. J., and Liu, D. (1999) *Science, 284*, 760–765.

Clough, R. C., Jordan-Beebe, E. T., Lohman, K. N., Marita, J. M., Walker, J. M., Gatz, C., and Vierstra, R. D. (1999) *Plant J. 17*, 155–167.

Cutler, S. R., Ehrhardt, D. W., Griffits, J. S., and Somerville, C. R. (2000) *Proc. Natl. Acad. Sci. USA, 97* 3718–3723.

Devlin, P. F. (2002) *J. Exp. Bot. 53*, 1535–1550.

Devlin, P. F. (2003) In *Photoreceptors and light signalling*, Batschauer A. (ed.), Comprehensive Series in Photochem. and Photobiol. Sciences, Royal Society of Chemistry, Cambridge UK.

Devlin, P. F., and Kay, S. A. (2000) *Plant Cell 12*, 2499–2509.

Devlin, P. F., and Kay, S. A. (2001) *Annu. Rev. Physiol. 63*, 677–694.

Ehrenberg, A., and Hemmerich, P. (1968) *Biological oxidations*, Singer T.P. (ed.) Interscience Publisher, New York, pp 239–262.

El-Assal, S. E.-D., Alonso-Blanco, C., Peeters, A. J. M., Raz, V., and Koornneef, M. (2001) *Nature Genetics 29*, 435–440.

El-Assal, S. E.-D., Alonso-Blanco, C., Peeters, A. J. M., Wagemaker, C., Weller, J. L., and Koornneef, M. (2003) *Plant Physiol. 133*, 1–13.

Emery, P. So, W. V. Kaneko, M., Hall, J. C., and Rosbash, M. (1998) *Cell 95*, 669–679.
Eun, B. K., and Kang, H. M. (2003) *Mol. Cell 16*, 239–244.
Eun, B. K., Lee, B. J., and Kang, H. M. (2001) *Mol. Cell 12*, 286–291.
Fankhauser, C., and Chory, J. (1997) *Annu. Rev. Cell Dev. Biol. 13*, 203–229.
Fankhauser, C., and Staiger, D. (2002) *Planta 216*, 1–16.
Feinbaum, R. L., and Ausubel, F. M. (1988) *Mol. Cell. Biol. 8*, 1985–1992.
Feinbaum, R. L., Storz, G., and Ausubel, F. M. (1991) *Mol. Gen. Genet. 226*, 449–456.
Folta, K. M., Pontin, M. A., Karlin-Neumann, G., Bottini, R., and Spalding, E. P. (2003) *Plant J.* 36, 203–214.
Folta, K. M., and Spalding, E. P. (2001) *Plant J. 28*, 333–340.
Fu, Z., Inaba, M., Noguchi, T., and Kato, H. (2002) *J. Biol. Rhythms 17*, 14–27.
Galland, P., and Senger, H. (1988) *Photochem. Photobiol. 48*, 811–820.
Galland, P., and Senger, H. (1991) In *Photoreceptor Evolution and Function;* Holmes M.G. (ed), Academic Press, London; pp. 65–124.
Galland, P., and Tölle, N. (2003) *Planta 217*, 971–982.
Gamble, P. E., and Mullet, J. E. (1989) *EMBO J. 8*, 2785–2794.
Giovani, B., Byrdin, M., Ahmad, M., and Brettel, K. (2003) *Nat. Struct. Biol. 10*, 489–490.
Goto, N., Kumagai, T., and Koornneef, M. (1991) *Physiol. Plant. 83*, 209–215.
Gressel, J. (1979) *Photochem. Photobiol. 30*, 749–754.
Guo, H., Duong, H., Ma, N., and Lin, C. (1999) *Plant J. 19*, 279–287.
Guo, H., Yang, H., Mockler, T. C., and Lin, C. (1998) *Science 279*, 1360–1363.
Gyula, P., Schäfer, E., and Nagy, F. (2003) *Curr. Opin. Plant Biol. 6*, 446–452.
Harbourne, J. B. (1993) *The flavonoids: Advances in research (1988–1991)* Chapman and Hall, London, UK.
Hardtke, C. S., Gohda, K., Osterlund, M. T., Oyama, T., Okada, K., and Deng, X.-W. (2000) *EMBO J. 19*, 4997–5006.
Harmer, S. L., Hogenesch, J. B., Straume, M., Chang, H. S., Han, B., Zhu, T., Wang, 10., Kreps, J. A., and Kay, S. A. (2000) *Science 290*, 2110–2113.
Hartmann, U., Valentine, W. J., Christie, J. M., Hays, J., Jenkins, G. I., and Weisshaar, B. (1998) *Plant Mol. Biol. 36*, 741–754.
Hayama, R., and Coupland, G. (2003) *Curr. Opinion Plant Biol. 6*, 13–19.
Hess, W. R., and Börner, T. (1999) *Int. Rev. Cytol. 190*, 1–59.
Hitomi, K., Okamoto, K., Daiyasu, H., Miyashita, H., Iwai, S., Toh, H., Ishiura, M., and Todo, T., (2000) *Nucleic Acids Res. 28*, 2353–2362.
Hoffman, P. D., Batschauer, A., and Hays, J. B. (1996) *Mol. Gen. Genet. 253*, 259–265.
Hsu, D. S., Zhao, 10. D., Zhao, S. Y., Kazatsev, A., Wang, R. P. Todo, T., Wei, Y. F., and Sancar, A. (1996) *Biochemistry 35*, 13871–13877.
Husain, I., and Sancar, A. (1987) *Nucleic Acids Res. 15*, 1109–1120.
Imaizumi, T., Kadota, A., Hasebe, M., and Wada, M. (2002) *Plant Cell 12*, 81–95.
Imaizumi, T., Kanegae, T., and Wada, M. (2000) *Plant Cell 12*, 81–96.
Imaizumi, T., Kiyosue, T., Kanegae, T., and Wada, M. (1999) *Plant Phys. 120*, 1205.
Jackson, J. A., and Jenkins, G. I. (1995) *Planta 197*, 233–239.
Jefferson, R. A. (1987) *Plant Mol. Biol. Rep. 5*, 387–405.
Johnson, E., Bradley, M., Harberd, N. P., and Whitelam, G. C. (1994) *Plant Physiol. 105*, 141–149.
Kaiser, T., Emmler, K., Kretsch, T., Weisshaar, B., Schäfer, E., and Batschauer, A. (1995) *Plant Mol. Biol. 28*, 219–229.
Kanai, S., Kikuno, R., Toh, H., Ryo, H., and Todo, T. (1997) *J. Mol. Evol. 45*, 535–548.
Kanegae, T., and Wada, M. (1998) *Mol. Gen. Genet. 259*, 345–353.
Kato, T., Jr., Todo, T., Ayaki, H., Ishizaki, K., Morita, T., Mitra, S., and Ikenaga, M. (1994) *Nucleic Acids Res. 22*, 4119–4124.
Kendrick, R. E., and Kronenberg, G. H. M. (1994) *Photomorphogenesis in Plants.* 2nd edn, Kendrick R.E., Kronenberg G.H.M. (eds.), Kluwer Academic Publishers, Dordrecht, The Netherlands.
Kim, M., Thum, K. E., Morishigi, D. T., and Mullet, J. E. (1999) *J. Biol. Chem. 274*, 4684–4692.
Kleine, T., Lockhart, P., and Batschauer, A. (2003) *Plant J. 35*, 93–103.
Kleiner, O., Kircher, S., Harter, K., and Batschauer, A. (1999) *Plant J. 19*, 289–296.

Kobayashi, Y., Ishikawa, T., Hirayama, J., Daiyasu, H., Kanai, S., Toh, H., Fukuda, I., Tsujimura, T., Terada, N. Kamei, Y., Yuba, S. Iwai, S., and Todo, T. (2000) *Genes Cells 5*, 725–38.

Kobayashi, K., Kanno, S., Smit, B., van der Horst, G. T. J., Takao, M., and Yasui, A. (1998) *Nucleic Acids Res. 26*, 5086–5092.

Koes, R. E., Quattrocchio, F., and Mol, J. N. M. (1994) *BioEssays 16*, 123–132.

Komori, H., Masui, R., Kuramitsu, S., Yokoyama, S., Shibata, T., Inoue, Y., and Miki, K. (2001) *Proc. Natl. Acad. Sci. USA 98*, 13560–13565.

Koornneef, M., Alonso-Blanco, C., Peeters, A. J. M., and Scoppe, W. (1998) *Ann. Rev. Plant Physiol. Plant Mol. Biol. 49*, 345–370.

Koornneef, M., Hanhart, C. J., and van der Veen, J. H. (1991) *Mol. Gen. Genet. 229*, 57–66.

Koornneef, M., Hanhart, C., van Loenen-Martinet, P., and Blankestijn-de Vries, H. (1995) *Physiol. Plant. 95*, 260–266.

Koornneef, M., Rolf, E., and Spruit, C. J. P. (1980) *Z. Pflanzenphysiol. 100*, 147–160.

Kubasek, W. L., Shirley, B. W., McKillop, A., Goodman, H. M., Briggs, W. R., and Ausubel, F. M. (1992) *Plant Cell* 4, 1229–1236.

Kunkel, T., Neuhaus, G., Batschauer, A., Chua, N.-H., and Schäfer, E. (1996) *Plant J. 10*, 625–636.

Kuno, N., and Furuya, M. (2000) *Cell & Developmental Biology 11*, 485–493.

Li, Y. F., and Sancar, A. (1990) *Biochemistry 29*, 5698–5706.

Lin, C. (2000a) *Plant Physiol. 123*, 39–50.

Lin, C. (2000b) *Trends Plant Sci. 5*, 337–342.

Lin, C. (2002) *Plant Cell 14 Suppl.*, S207–S225.

Lin, C., Ahmad, M., and Cashmore, A. R. (1996a) *Plant J. 10*, 893–902.

Lin, C., Ahmad, M., Chan J., and Cashmore, A. R. (1996b) *Plant Physiol. 110*, 1047.

Lin, C., Ahmad, M., Gordon, D., and Cashmore, A. R. (1995a) *Proc. Natl. Acad. Sci. USA 92*, 8423–8427.

Lin, C., Robertson, D. E., Ahmad, M., Raibekas, A. A., Schuman Jornes, M., Dutton, P. L., and Cashmore, A. R. (1995b) *Science 269*, 968–970.

Lin, C., and Shalitin, D. (2003) *Annu. Rev. Plant Biol. 54*, 469–496.

Lin, C., Yang, H., Guo, H., Mockler, T., Chen, J., and Cashmore, A. R. (1998) *Proc. Natl. Acad. Sci. USA 95*, 2686–2690.

Link, G. (1996) *BioEssays 18*, 465–471.

Lipson, E. D., and Horwitz, B. A. (1991) In *Sensory Receptors and Signal Transduction*, Spudich J.L., Satir B.H. (eds.). Wiley, New York, pp 64.

Ma, L., Li, J., Qu, L., Hager, J., Chen, Z., Zhao, H., and Deng, X.-W. (2001) *Plant Cell 13*, 2589–2607.

Malhotra, K., Kim, S.-T., Batschauer, A., Dawut, L., and Sancar, A. (1995) *Biochemistry 34*, 6892–6899.

Maliga, P. (1998) *Trends Plant Sci. 3*, 4–6.

Más, P., Devlin, P. F., Panda, S., and Kay, S. A. (2000) *Nature 408*, 207–211.

Matsumoto, N., Hirano, T., Iwasaki, T., and Yamamoto, N. (2003) *Plant Physiol. 133*, 1494–1503.

Mayfield, S. P., Yohn, C. B., Cohen, A., and Danon, A. (1995) *Annu. Rev. Plant Physiol. Plant Mol. Biol. 46*, 147–166.

Millar, A. J. (2003) *Biol. Rhythms 18*, 217–226.

Millar, A. J., Carre, I. A., Strayer, C. A., Chua, N.-H., and Kay, S. A. (1995a) *Science 267*, 1161–1163.

Millar, A. J., Short, R. S., Chua, N.-H., and Kay, S. A. (1992) *Plant Cell 4*, 1075–1087.

Millar, A. J., Straume, M., Chory, J., Chua, N.-H., and Kay, S. A. (1995b) *Science 267*, 1163–1166.

Miyamoto, Y., and Sancar, A. (1998) *Proc. Natl. Acad. Sci. USA 95*, 6097–6102.

Mizoguchi, T., Wheatley, K., Hanzawa, Y., Wright, L., Mizoguchi, M., Song, H.-R Carre, I. A., and Coupland, G. (2002) *Dev. Cell 2*, 629–641.

Mockler, T. C., Guo, H., Yang, H., Duong, H., and Lin, C. (1999) *Development 126*, 2073–2082.

Mockler, T. C., Yang, H., Yu, X., Parikh, D., Cheng, Y., Dolan, S., and Lin, C. (2003) *Proc. Natl. Acad. Sci. USA 100*, 2140–2145.

Mouradov, A., Cremer, F., and Coupland, G. (2002) *Plant Cell 14*: S111–S130.

Mullet, J. E. (1993) *Plant Physiol. 103*, 309–313.

Neff, M. M., and Chory, J. (1998) *Plant Physiol. 118*, 27–36.

Ng, W. O., and Pakrasi, H. B., (2001) *Mol. Gen. Genet. 264*, 924–930.

Ninu, L., Ahmad, M., Miarelli, C., Cashmore, A.R., and Giuliano, G. (1999) *Plant J. 18*, 551–556.

O'Connor, K. A., McBridge, M. J., West, M., Yu, H., Trinh, L., Yuan, K., Lee, T., and Zusman, D. R. (1996) *J. Biol. Chem. 271*, 6252–6359.

Osterlund, M. T., and Deng, X.-W. (1998) *Plant J. 16*, 201–208.

Osterlund, M. T., Hardtke, C. S., Wei, N., and Deng, X.-W. (2000) *Nature 405*, 462–466.

Park, H. W., Kim, S.-T., Sancar, A., and Deisenhofer, J. (1995) *Science 268*, 1866–1872.

Parks, B. M., Cho, M. H., and Spalding, E. P. (1998) *Plant Physiol. 118*, 609–615.

Parks, B. M., Folta, K. M., and Spalding, E. P. (2001) *Current Opinion in Plant Biology 4*, 436–440.

Perrotta, G., Ninu, L., Flamma, F., Weller, J. L., Kendrick, R. E., Nebuloso, E., and Giuliano, G. (2000) *Plant Mol. Biol. 42*, 765–773.

Perrotta, G., Yahoubyan, G., Nebuloso, E., Renzi, L., and Giuliano, G. (2001) *Plant Cell Environ. 24*, 991–997.

Petersen, J. L., Lang, D. W., and Small, G. D. (1999) *Plant Mol. Biol. 40*, 1063–1071.

Poppe, C., Sweere, U., Drumm-Herrel, H., and Schäfer, E. (1998) *Plant J. 16*, 465–471.

Rocholl, M., Talke-Messerer, C., Kaiser, T., and Batschauer, A. (1994) *Plant Science 97*, 189–198.

Samach, A., and Coupland, G. (2000) *BioEssays 22*, 38–47.

Samach, A., and Gover, A. (2001) *Curr. Biol. 11*, R651–R654.

Sancar, A. (2003) *Chem. Rev. 103*, 2203–2237.

Sancar, G. B. (2000) *Mutation Res. 451*, 25–37.

Santiago-Ong, M., and Lin., C. (2003) In *Photoreceptors and light signaling*; Batschauer A. (ed.), Comprehensive Series in Photochem. and Photobiol. Sciences, Royal Society of Chemistry, Cambridge, UK, pp 303–327.

Schulze-Lefert, P., Dangl, J. L., Becker-Andre, M., Hahlbrock, K., and Schulz, W. (1989) *EMBO J. 8*, 651–656.

Senger, H. (1984) *Blue Light Effects in Biological Systems;* Senger H. (ed), Springer Verlag, Berlin; pp. 72.

Shalitin, D., Yang, H., Mockler, T. C., Maymon, M., Guo, H., Whitelam, G. C., and Lin, C. (2002) *Nature 417*, 763–67.

Shalitin, D., Yu, X., Maymon M., Mockler, T., and Lin, C. (2003) *Plant Cell 15*, 2421–2429.

Sharrock, R. A., and Quail, P. H. (1989) *Genes Develop. 3*, 1745–1757.

Shinomura, T., Nagatani, A., Hanzawa, H., Kubota, M., Watanabe, M., and Furuya, M. (1996) *Proc. Natl. Acad. Sci. USA 93*, 8129–8133.

Shirley, B. W. (1996) *Trends Plant Sci. 1*, 377–382.

Small, G. D., Min, B. Y., and Lefebvre, P. A. (1995) *Plant Mol. Biol. 28*, 443–454.

Somers, D.E. (1999) *Plant Physiol. 121*, 9–19.

Somers, D. E., Devlin, P. F., and Kay, S. A. (1998b) *Science 282*, 1488–1490.

Somers, D. E., Webb, A .A. R., Pearson, M., and Kay, S. A. (1998a) *Development 125*, 485–494.

Srinivasan, V.,. Schnitzlein, W. M, and Tripathy, D. N. (2001) *J. Virol. 75*, 1681–1688.

Stanewsky, R., Kaneko, M., Emery, P., Beretta, B., Wagner-Smith, K., Kay, S. A., Rosbash, M., and Hall, J. C. (1998) *Cell 95*, 681–692.

Stern, D. B., Higgins, D. C., and Yang, J. (1997) *Trends Plant Sci. 2*, 308–314.

Suárez-López, P., Wheatley, K., Robson, F., Hitoshi, O., Valverde, F., and Coupland, G. (2001) *Nature 410*, 1116–1120.

Sugita, M., and Sugiura, M. (1996) *Plant Mol. Biol. 32*, 315–326.

Sullivan, J. A., and Deng, X.-W. (2003) *Dev. Biol. 260*, 289–297.

Tamada, T., Kitadokora, K., Higuchi, Y. Inaka, K., Yasui, A., de Ruiter, P. E., Eker, A. P. M., and Miki, K. (1997) *Nature Structural Biology 4*, 887–891.

Taylor, R., Tobin, A. K., and Bray, C. M. (1996) *Plant Physiol. 112*, 862.

Terzaghi, W. B., and Cashmore, A. R. (1995) *Annu. Rev. Plant Physiol. Plant Mol. Biol. 46*, 445–474.

The Arabidopsis Genome Initiative (2000) *Nature 408*, 796–815.

Thomas, B., and Vince-Prue, D. (1997) *Photoperiodism in plants*, 2nd edn; Academic Press, San Diego, CA, USA.

Thompson, C. L., and Sancar, A. (2002) *Oncogene 21*, 9043–9056.

Thum, K. E., Kim M., Christopher, D. A., and Mullet, J. E. (2001) *Plant Cell 12*, 2747–2760.

Todo, T. (1999) *Mutation Res. 434*, 89–97.

Todo, T., Ryo, H., Takemori, H., Toh, H., Nomura, T., and Kondo, S. (1994) *Mutation Res., DNA Repair 315*, 213–228.

Todo, T., Ryo, H., Yamamoto, K., Toh, H., Inui, T., Ayaki, H., Nomura, T., and Ikenaga, M. (1996) *Science 272*, 109–112.

Toth, R., Kevei, E., Hall, A., Millar, A. J., Nagy, F., and Kozma-Bognar, L. (2001) *Plant Physiol. 127*, 1607–1616.

Tsien, R. Y. (1998) *Annu. Rev. Biochem. 67,* 509–544.

Tsunoyama, Y., Morikawa, K., Shiina, T., and Toyoshima, Y. (2002) *FEBS Lett. 516,* 225–228.

Van der Spek, P. J., Kobayashi, K., Bootsma, D., Takao, M., Eker, A. P., and Yasui, A. (1996) *Genomics 37,* 177–182.

Van Gelder, R. N. (2002) *J. Biol. Rhythms 17,* 110–120.

Van Gelder, R. N. (2003) *Trends Neurosci. 26,* 458–461.

Vierstra, R. D. (1996) *Plant Mol. Biol. 32,* 275–302.

von Arnim, A. G., and Deng, X.-W. (1994) *Cell 79,* 1035–1045.

Wade, H. K., Bibikova, T. N., Valentine W. J., and Jenkins, G. I. (2001) *Plant J. 25,* 675–685.

Wang, H., Ma, L.-G., Li, J.-M., Zhao, H.-Y., and Deng, X.-W. (2001) *Science 294,* 154–158.

Weisshaar, B., Armstrong, G. A, Block, A., da Costa e Silva, O., and Hahlbrock, K. (1991) *EMBO J. 10,* 1777–1786.

Weller, J. L., Perrotta, G., Schreuder, M. E., van Tuinen, A., Koornneef, M., Giuliano, G., and Kendrick, R. E. (2001) *Plant J. 25,* 427–440.

Whippo, C. W., and Hangarter, R. P. (2003) *Plant Physiol. 132,* 1499–1507.

Worthington, E. N., Kavakli, I. H., Berrocal-Tito, G., Bondo, B. E., and Sancar, A. (2003) *Biol. Chem. 278,* 39143–39154.

Yamamoto, K., Okano, T., and Fukada, Y. (2001) *Neurosci. Lett. 313,* 13–16.

Yang, H. Q., Tang, R. H., and Cashmore, A. R. (2001) *Plant Cell 13,* 2573–2587.

Yang, H.-Q., Wu, Y.-J., Tang, R.-H., Liu, D., Liu, Y., and Cashmore, A. R. (2000) *Cell 103,* 815–827.

Yanovsky, M. J., and Kay, S. A. (2002) *Nature 419,* 308–312.

Yasuhira, S., and Yasui, A. (1992) *J. Biol. Chem. 267,* 25644–25647.

Yasui, A., Eker, A., Yasuhira, S., Yajima, H., Kobayashi, T., Takao, M., and Oikawa, A. (1994) *EMBO J. 13,* 6143–6151.

Yeh, K.-C., and Lagarias, J.C. (1998) *Proc. Natl. Acad. Sci. USA 95,* 13976–13981.

Zhu, H., and Green, C. B. (2001) *Mol. Vis. 7,* 210–215. Comprehensive Series in Photochem. and Photobiol. Sciences, Royal Society of Chemistry, Cambridge, UK, pp. 343–368.

11
Plant Cryptochromes and Signaling

Anthony R. Cashmore

11.1
Introduction

Cryptochromes were first characterized at the molecular level through the isolation of the *HY4* gene, now commonly referred to as *CRY1* (Ahmad and Cashmore, 1993). The *hy4* mutant was originally identified through the unusually long hypocotyl that it exhibits when grown under blue or UV-A light (Koornneef, et al., 1980). The mutant is indistinguishable from wild type when grown in darkness and it exhibits normal inhibition of hypocotyl cell growth when grown under either red or far-red light. In these respects, the *hy4/cry1* mutant is selectively attenuated in its response to blue light or UV-A light, and appears normal with respect to its response to red and far-red light.

11.2
Photolyases

The inability of the *cry1* mutant to respond normally to blue light could have resulted from a mutation in either a blue light photoreceptor or in a signaling component required for such a photoreceptor. It was the demonstration that the *CRY1* gene possessed strong sequence similarity to those of photolyases that provided the first indication that the encoded CRY1 protein was indeed a blue light receptor (Ahmad and Cashmore, 1993).

Photolyases are flavoproteins that mediate repair of pyrimidine dimers generated by exposure of DNA to UV-B light (Sancar, 2003). This DNA repair activity of photolyases requires irradiation with either UV-A or blue light – in this respect photolyases are photoreceptors. In addition to the primary catalytic chromophore, photolyases have a second light-harvesting chromophore; this is commonly a pterin, methenyltetrahydrofolate (MTHF), or in some cases a deazaflavin.

The catalytically active form of the flavin of photolyases is $FADH^-$, the anionic form of the fully reduced $FADH_2$. Excitation of $FADH^-$ occurs either by the direct ab-

Handbook of Photosensory Receptors. Edited by W. R. Briggs, J. L. Spudich

ISBN 3-527-31019-3

sorption of light or by excitation energy transfer from the light-harvesting chromophore. Following excitation, electron transfer occurs to bound pyrimidine dimers present in UV-B irradiated DNA, the latter then isomerizing to yield the monomeric repaired pyrimidines with the electron being returned to the flavin. Whereas no net electron transfer occurs, the reactions have the characteristics of redox reactions.

The flavin of the isolated *E. coli* photolyase exists in the redox state of the semiquinone. In order for this flavin to mediate pyrimidine dimer repair *in vitro* it must first be converted to the fully reduced form. This reduction is achieved through intramolecular electron transfer, the electron donor being a conserved Trp residue (Trp306). Photolyase containing a mutation of Trp306 exhibits no repair of pyrimidine dimers *in vitro* (Li, et al., 1991). However, *in vivo* the flavin exists in the catalytically active fully reduced form, and electron transfer from Trp306 is not required (Payne et al., 1987). In keeping with this conclusion, mutation of Trp306 has no noticeable effect on the ability of *E. coli* photolyase to mediate DNA repair *in vivo* (Li, et al., 1991). In view of these observations it is unlikely that the detailed findings that have been made concerning *in vitro* intramolecular electron transfer have any relevance to the functioning of photolyase *in vivo* (Sancar, 2003).

11.3
Cryptochrome Photochemistry

In contrast to isolated *E. coli* photolyase where the flavin is in the form of the semiquinone, *Arabidopsis* CRY1 is isolated with the flavin in the fully oxidized FAD form (Lin, et al., 1995). In anaerobic conditions, CRY1 undergoes photoreduction yielding first the semiquinone and then the fully reduced $FADH_2$. The semiquinone of *Arabidopsis* CRY1 is relatively stable, contrasting with *E. coli* photolyase, where photoreduction of the bound FAD yields $FADH_2$ without any detectable semiquinone intermediate. The biological significance of these differences is not known, although it has been speculated that the degree of sensitivity of *Arabidopsis* CRY1 to green light may reflect an *in vivo* role of the semiquinone that, in contrast to the other two redox forms of the flavin, absorbs in this region of the spectrum (Lin, et al., 1995).

As with *E. coli* photolyase, in vitro intraprotein electron transfer has been studied in *Arabidopsis* CRY1 (Giovani, et al., 2003). In this instance the reduction of FAD to the semiquinone was examined. This conversion was observed to involve electron transfer from a Trp (likely, Trp324) that in turn received an electron from one of four candidate Tyr residues. As these studies involved isolated cryptochrome it is difficult to know their relevance to the *in vivo* mode of action of cryptochrome. This conclusion follows from the observations reported above for *E. coli* photolyase where Trp306 was shown to form an essential role as the primary electron donor for the *in vitro* photoconversion of the semiquinone to the fully reduced flavin, yet mutation of this same Trp has no effect on the *in vivo* activity of the photolyase (Li, et al., 1991). Indeed, in the case of *E. coli* photolyase, there is no evidence that reduction of the semiquinone is required *in vivo*; the flavin apparently existing in the catalytically active fully reduced form. In the case of *Arabidopsis* CRY1, the catalytically active form of the flavin has not been determined and therefore the requirement or otherwise for the conversion of the fully oxidized FAD to the semiquinone is similarly unknown.

11.4 Cryptochrome Action Spectra

Action spectra can be useful in determining whether a photoreceptor is involved in mediating a particular photoresponse. However, such determinations are difficult at the best of times, and particularly so in the case of cryptochromes. The reasons for this are several: Firstly, as mentioned, in addition to the primary catalytic chromophore there is a second light-harvesting chromophore. The latter will not only contribute to the action spectra, indeed it is likely to dominate it; this follows from the fact that the extinction coefficient for pterins such as MTHF is substantially greater than that of flavins (Sancar, 2003). Thus any prediction of cryptochrome action spectra requires knowledge of the exact identity of the second chromophore. Whereas *Arabidopsis* CRY1 binds MTHF when expressed in *E. coli* (Malhotra, et al., 1995), the precise identity of the light harvesting chromophore in *Arabidopsis* is not known. An additional complication in predicting the identity of photoreceptors from action spectra involves the redox state of the chromophores, as the absorption properties, and therefore associated action spectra, will vary substantially according to the redox state. The biologically relevant *in vivo* redox states of either the flavin or the light harvesting chromophore are not known, for any cryptochrome. For all of these reasons it is not possible at the moment to make useful predictions concerning the action spectra for any process that might involve cryptochromes.

There is a further complication involving the interpretation of action spectra. In some instances a photoreceptor may play an essential role for a certain process; a temptation in such cases is to conclude that such a photoreceptor is the only photoreceptor mediating the process and therefore the corresponding action spectrum should mimic the absorption properties of the receptor. However, there may be multiple photoreceptors serving distinct and possibly essential roles. For instance, the *cry1 cry2* double mutant of *Arabidopsis* has been reported to lack any blue light induced shortening of the hypocotyl and in view of this finding the action spectrum of this response has been interpreted exclusively in terms of cryptochrome (Ahmad, et al., 2002). However, it is well established that PHYA plays a role in this blue light response (Neff and Chory, 1998, Poppe, et al., 1998) and therefore, even if the cryptochromes play an essential role, it does not follow that the corresponding action spectrum will exclusively reflect the cryptochrome photoreceptors.

This general topic of photoreceptors serving non-overlapping roles is of great importance in reference to arguments concerning what role if any cryptochromes play in the entrainment of mammalian circadian rhythms. Here, mice lacking both melanopsin and opsins are totally deficient in their capacity to undergo entrainment and for this reason it has generally been concluded that mice cryptochromes (CRY1 and CRY2) play no role in this process (Hattar, et al., 2003, Panda, et al., 2003). Whereas this conclusion may well be correct, the argument is not correct, for the reasons outlined above. It is conceivable that cryptochromes, serving as photoreceptors, do play a role in entrainment of the mammalian clock, even though either melanopsin or opsins are essential for this process. The quite unexpected role that cryptochromes play in the functioning of the central oscillator unfortunately precludes any easy determination of this question of whether or not these same cryptochromes function as photoreceptors for entrainment (van der Horst, et al., 1999; see Van Gelder and Sancar, Chapter 12). However, whereas there has been a willingness in the field to accept that CRY1 and CRY2 play an essential role in the central oscillator, distinct from the similarly essential role played by the PERIOD proteins, it is now the general consensus that cryptochromes play no role in entrainment (Hattar, et al., 2003, Panda, et al., 2003). As noted, whereas this conclusion may conceivably be correct, the logic of the argument is not correct. In reference to this debate it is of interest that the *Drosophila* CRY does function as a photoreceptor for entrainment (Helfrich-Forster, et al., 2001).

11.5
Cryptochromes and Blue Light-dependent Inhibition of Cell Expansion

Light-mediated inhibition of hypocotyl growth can be detected within 30 seconds of stem irradiation with a pulse of blue light (Parks, et al., 1998). With continuous irradiation, stem elongation continues at increasingly reduced rates for about 30 min after light exposure, whereupon growth continues at a markedly reduced rate for several days. Surprisingly, from mutant studies it appears that this early inhibition observed for the first 30 min of exposure to blue light is mediated not by cryptochrome but by PHOT1, the blue light photoreceptor responsible for phototropism (Folta and Spalding, 2001). However, after 30 min the response is largely mediated by CRY1 and CRY2 as, in contrast to the continued reduced growth rate of wild-type plants, normal growth is observed in both the *cry1* and *cry2* mutants after about 60 min of light exposure. Correlated with the rapid blue light-mediated inhibition of growth is an associated depolarization of the plasma membrane; anion channel blockers have a similar effect to that observed for the *cry1* and *cry2* mutants. From these studies it appears that blue light, acting through CRY1 and CRY2, activates an anion channel resulting in plasma membrane depolarization, which in turn inhibits cell expansion.

Related observations concerning the role of CRY1 in mediating blue light inhibition of cell expansion have come from studies with *Arabidopsis* protoplasts. Protoplasts isolated from hypocotyl tissue and kept under continuous red light undergo rapid and transient shrinkage over a period of five minutes subsequent to exposure to a pulse of blue light. This blue light induced protoplast shrinkage does not occur in protoplasts prepared from the *cry1* mutant, demonstrating a role for the CRY1 photoreceptor in this process (Wang and Iino, 1998). The observed responsivity to blue light requires previous exposure to red light and this response to red light is lost in protoplasts from the *phyA phyB* mutant.

11.6 Signaling Mutants

One way of identifying signaling partners for cryptochrome is to identify mutants that, like cryptochrome mutants, are deficient in blue light signaling. Several such mutants have been characterized, some of which are also altered with respect to phytochrome signaling.

One such mutant that is affected in both cryptochrome and phytochrome signaling is *sub1* (Guo, et al., 2001). SUB1 encodes a calcium binding protein and the *sub1* mutant shows a hypersensitive response to both blue and far-red light. Double mutant analysis shows that *sub1* is epistatic to both *cry1* and *cry2* at low intensity blue light, indicating under these conditions that the *sub1* phenotype is not dependent on either of these receptors. Conversely, at higher blue light intensities (10 μmol m^{-2} s^{-1}) *cry1* is epistatic to *sub1*, indicating that the *sub1* phenotype is dependent on CRY1 under these conditions. Similarly, under all fluence rates tested, the *sub1* phenotype is epistatic to *phyA*. The *sub1* mutant shows no phenotype in dark-grown seedlings. It has been proposed that SUB1 acts as a negative regulator downstream of CRY1 and CRY2 and that the *sub1* mutant acts constitutively in the absence of either photoreceptor. The absence of a dark phenotype for *sub1* is explained through the proposal that SUB1 signaling requires a downstream partner such as HY5, the activity of which is light-dependent. The situation with PHYA appears to be more straightforward with SUB1 apparently modulating PHYA signaling and the *sub1* mutation having no affect in the absence of PHYA. Similarly, under high blue light intensity, SUB1 activity shows a dependence on CRY1. Thus, SUB1 signaling is either CRY1 dependent or independent, depending on the light intensity.

Another mutant that affects both blue light and far-red light signaling is *hfr1*. This mutant (also known as *rsf1* and *rep1*) was identified through the long hypocotyl it possesses when grown under far-red light (Fairchild, et al., 2000, Fankhauser and Chory, 2000, Kim, et al., 2002). This hyposensitive response occurs under blue light as well and is also observed for cotyledon expansion and to a lesser extent, anthocyanin accumulation. Double-mutant analyses indicate that the activity of HFR1 involves both CRY1 as well as PHYA photoreceptors (Duek and Fankhauser, 2003). From analysis of the triple mutant *cop1 hy5 hfr1*, it was concluded that HFR1, like HY5, acts downstream of COP1 (Kim, et al., 2002).

11.7 Signaling by Cryptochrome CNT and CCT Domains

The simplest model of cryptochrome mode of action is based on the mechanism of action of photolyase. According to this model the C-terminal domain (CCT) of cryptochrome binds a signaling partner which is activated via a blue light-dependent redox reaction. In the simplest form of this model, CCT undergoes no change in response to light. In order to explore the activity of CCT isolated from the native CRY molecule, CCT was expressed as a GUS fusion protein in transgenic *Arabidopsis* plants (Yang, et al., 2000). The GUS:CCT fusion conferred a constitutive photomorphogenic (COP) phenotype on dark-grown *Arabidopsis* seedlings with the seedlings being distinguished by shortened hypocotyls, opened cotyledons, and enhanced anthocyanin production. Both CCT1 (the C-terminal domain of CRY1), as well as CCT2 (the C-terminal domain of CRY2), gave rise to this COP phenotype. Significantly, mutant CCT1 sequences corresponding to loss of function *cry1* mutants, did not give a COP phenotype.

In contrast to the model discussed above, these findings indicate that CCTs, on isolation from the native CRY molecules, contain sufficient information to mediate signaling. In the native CRY molecule CCT activity must be repressed by the action of the N-terminal domain (CNT), with this repression being alleviated through the action of light (Yang, et al., 2000).

In the initial studies of the activities of cryptochrome sub domains, no activity was observed for CNT. For these studies a fusion of CNT with GUS was employed. In subsequent studies the activity of CNT was re-examined employing GFP:CNT and cMyc:CNT fusion proteins. Expression of either of the respective N-terminal domains of CRY1 and CRY2 (CNT1 and CNT2) as GFP- or cMyc fusion proteins in transgenic light–grown *Arabidopsis* seedlings is now observed to result in enhanced suppression of hypocotyl growth. Furthermore, these CNT fusion proteins confer a COP phenotype, similar to that originally observed for the GUS:CCT fusions (Kang and Cashmore, unpublished). In a manner similar to that argued for CCT, these findings must mean that the potential signaling properties of CNT are repressed in the native CRY molecule, this repression being alleviated in response to light. That is, the repression of CNT must reflect the activity of CCT, in a manner analogous to the repression of CCT by CNT.

11.8
Arabidopsis Cryptochromes Exist as Dimers

The mutual repression of the potential signaling activity of the CNT and CCT domains of *Arabidopsis* cryptochromes raises the possibility that these two domains may physically interact with one another. Indeed, by yeast 2-hybrid studies such an interaction has been demonstrated (Kang and Cashmore, unpublished). This interaction could be either intramolecular or intermolecular, with the latter reflecting the existence of a dimer or some other oligomeric form of cryptochrome. By gel chromatographic studies it has been demonstrated that, at least *in vitro, Arabidopsis* CRY1 exists predominantly as a dimer (Wu and Cashmore, unpublished).

11.9
COP1, a Signaling Partner of *Arabidopsis* Cryptochromes

The phenotype of transgenic seedlings expressing GUS:CCT fusion proteins is similar to that of *Arabidopsis cop* mutants, in particular that of the *cop1* mutant. For this reason the possibility was entertained that COP1 may be a signaling partner of *Arabidopsis* cryptochromes. A physical interaction was demonstrated for both CRY1 and CRY2 with COP1 (Wang, et al., 2001, Yang, et al., 2001). These interactions were observed in the yeast 2-hybrid system, in studies *in vitro* and in *Arabidopsis*. Furthermore, from microarray hybridization studies it was observed that gene expression patterns for GUS:CCT1 dark-grown *Arabidopsis* seedlings were similar to those of the wild-type seedlings grown under blue light (Wang, et al., 2001).

The COP1 protein is a ring-finger zinc-binding protein containing WD-40 repeats and now known to function as a ubiquitin E3 ligase (Osterlund, et al., 2000, Seo, et al., 2003). COP1 protein facilitates COP9-dependent proteasome-mediated degradation of proteins such as the bZIP transcription factor HY5 that mediate the expression of genes that participate in photomorphogenesis. This proteasome-mediated degradation of HY5 is repressed in response to light and in blue light this activity requires cryptochrome. Dark-grown loss-of-function *cop1* mutant seedlings exhibit photomorphogenic properties including elevated levels of HY5 protein. Similarly, elevated levels of HY5 protein are observed in dark-grown seedlings expressing either GUS:CCT1 or GUS:CCT2 fusion proteins (Wang, et al., 2001), in keeping with the proposal that these *Arabidopsis* cryptochromes function through negatively impacting the interaction between COP1 and HY5 (Wang, et al., 2001, Yang, et al., 2001).

11.10
Cryptochrome and Phosphorylation

There are several reports concerning the phosphorylation of *Arabidopsis* cryptochromes. *Arabidopsis* CRY1 was reported to be a substrate for phosphorylation by the kinase activity associated with purified oat PHYA (Ahmad, et al., 1998). Conversely, CRY2 was shown to undergo blue light dependent phosphorylation *in vivo* (Shalitin, et al., 2002). This activity was not observed in red light or in far-red light, and phosphorylation occurred in several phytochrome-deficient mutants. Of interest, GUS:CCT2 was phosphorylated in a light-independent manner in transgenic seedlings. This correlation between constitutive phosphorylation and constitutive CCT signaling activity was interpreted as evidence in support of a model whereby phosphorylation is a necessary requirement for signaling. An alternative interpretation of the data would simply be that CCT exists in an active conformation, and that this active form of CCT is a substrate for phosphorylation. One possible role for phosphorylation is that it serves as a trigger for degradation, possibly via the activity of the proteasome. In keeping with this interpretation, the phosphorylated form of CRY2 is degraded more slowly in *cop1* mutants that in wild-type *Arabidopsis* (Shalitin, et al., 2002).

A similar series of observations was made for the phosphorylation of *Arabidopsis* CRY1 (Shalitin, et al., 2003). As in the case with CRY2, phosphorylation of CRY1 was blue light dependent and occurred in several phytochrome-deficient mutants. Several long hypocotyl *cry1* mutants were selected and all were found to be deficient in phosphorylation, consistent with a model whereby phosphorylation is a necessary prerequisite for cryptochrome activity.

An interesting finding is that *Arabidopsis* CRY1 protein isolated from insect cells is phosphorylated *in vitro* in a blue light-dependent manner (Bouly, et al., 2003, Shalitin, et al., 2003). Similar observations were made for human CRY1 and both the *Arabidopsis* and human cryptochromes were shown to bind to an ATP affinity column (Bouly, et al., 2003). These findings suggest that the observed phosphorylation of cryptochrome may reflect autophosphorylation. However, there is no obvious kinase domain within the *Arabidopsis* cryptochrome sequence and therefore it is difficult to eliminate the possibility that the observed phosphorylation simply reflects a contaminating kinase. This dilemma is remarkably similar to the related question concerning whether or not phytochrome is a kinase. Recombinant plant phytochrome purified from yeast cells possesses Ser/Thr protein kinase activity, similar in properties to the kinase activity demonstrated for phytochrome purified from plants (Yeh and Lagarias, 1998). From these observations it was concluded that plant phytochrome was a kinase, in spite of the fact that it lacked a traditional Ser/Thr protein kinase domain. Strongly supportive of the notion that cryptochrome is a kinase is a recent crystallographic study describing an ATP derivative bound to the CNT domain of Arabidopsis CRY1 (Brautigam, et al., 2004).

11.11 Cryptochrome and Gene Expression

There are many lines of evidence indicating that cryptochromes, activated by blue light, act in part through affecting gene expression. The induction of chalcone synthase gene expression by blue light in *Arabidopsis* is not observed in the *cry1* mutant (Ahmad, et al., 1995). This induction of gene expression occurs at the level of transcription and the required promoter elements have been identified (Batschauer, et al., 1996).

In microarray studies a large number of *Arabidopsis* genes have been shown to be regulated by either CRY1 or CRY2. It was shown that the majority of blue light regulated gene expression observed in *Arabidopsis* seedlings was mediated either by CRY1 and/or CRY2, and that this regulated expression was similar to that observed in dark-grown seedlings in the *cop1* mutant (Wang, et al., 2001). In an examination of seedlings expressing GUS:CCT1 or GUS:CCT2 it was observed that regulated gene expression for these two transgenic seedlings was very similar to one another and furthermore, the profile was similar, at least qualitatively, to that observed in the study of the *cry* mutants (Wang, et al., 2001). In a related study it was shown that many of the genes regulated by cryptochromes under blue light were similarly regulated by PHYA and PHYB during growth under far-red and red light respectively. The global importance of such regulation was exemplified by the estimate that approximately one-third of the *Arabidopsis* genome was similarly regulated by light (Ma, et al., 2001).

In a recent study, the requirement for CRY1 was determined for gene expression in *Arabidopsis* seedlings subsequent to irradiation for 45 min with blue light (Folta, et al., 2003). At this time point CRY1 is the primary photoreceptor mediating the light-induced inhibition of growth. Even after this relatively short exposure to light, more than 400 genes were differentially expressed in the *cry1* mutant compared with wild-type; these were approximately equally divided between those genes that were upregulated in the mutant and those that were down regulated. The types of CRY1-regulated genes included kinases, transcription factors, and genes involved in cell wall biosynthesis. Of interest, several genes involved in the response to the plant hormones auxin and the synthesis of gibberellic acid were identified. Inhibitor studies supported the notion that the effect of blue light acting through CRY1 is to repress gibberellic acid and auxin levels and/or sensitivity.

The observations that cryptochrome signaling includes changes in gene transcription raises a question concerning whether this activity involves association of cryptochrome with chromatin. Such an association could be either through direct interaction of cryptochrome with DNA or the interaction could be indirect through cryptochrome associating with DNA binding proteins. The fact that the activity of photolyases involves binding to DNA raises the possibility that cryptochromes may similarly function as DNA binding proteins. Indeed, there are reports that certain cryptochromes do bind to DNA although whether such an interaction is specific for a particular DNA sequence has not been determined. Mouse CRY1 and CRY2 negatively affect both their own transcription and that of the PER genes, this being achieved through an interaction with the positive transcriptional regulators CLOCK and BMAL1 (Kume, et al., 1999). In reference to plant cryptochromes, it is of interest that *Arabidopsis* CRY2 has been reported to be associated with chromatin (Cutler, et al., 2000).

11.12 Concluding Thoughts

Both the N-terminal (CNT) and the C-terminal (CCT) domains of *Arabidopsis* cryptochromes have the potential to activate signaling pathways individually in a light-independent fashion. As these constitutive signaling properties have been observed in multiple and distinct fusion proteins the observed activity must be a property of the respective CNT and CCT domains and not something conferred by the fusion partner. From this it follows that the signaling potential of CNT and CCT must be repressed in the native CRY molecule in the dark and this repression must be alleviated through the action of light.

The repression of CNT in the native CRY molecule must reflect the activity of CCT and similarly the repression of CCT must in turn reflect the activity of CNT. That is, these two domains of cryptochrome must negatively regulate one another in a light dependent fashion. This negative regulation could be direct or it could involve additional factors that are yet to be described. That the interaction is direct is suggested by yeast two hybrid studies where an interaction between CNT and CCT has been observed.

The signaling activity of CCT apparently results from a direct interaction of cryptochrome with COP1, this interaction being mediated by CCT (Figure 11.1). The partners of cryptochrome through which CNT signaling is mediated are yet to be determined. Other important questions that remain to be resolved include the photochemistry of the light activation of cryptochrome and how this effect of light leads to an activation of signaling by both CNT and CCT. A related question concerns the molecular details of how the action of CCT negatively impacts on COP1 activity. Yet to be

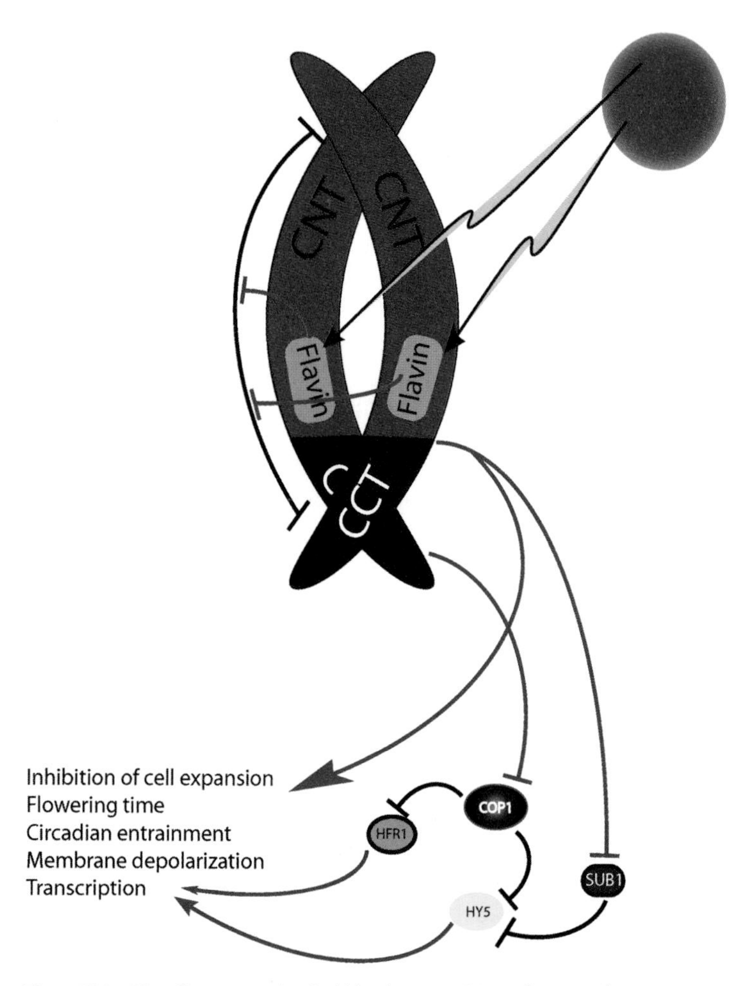

Figure 11.1 Signaling properties *Arabidopsis* CRY1. For sake of clarity only the catalytic chromophore of CRY1 is shown. The interaction of CRY1 with COP1 is mediated through CCT. Other signaling reactions may involve either CCT or CNT. Regulatory signaling interactions that are dependent on light are shown in blue.

identified are the cryptochrome signaling partners that interact with CNT. Another question of interest concerns the role of phosphorylation in cryptochrome signaling – is cryptochrome a kinase and is phosphorylation of cryptochrome required for signaling?

References

Ahmad, M and A. R. Cashmore (1993). *Nature* **366**, 162–6.

Ahmad, M., N. Grancher, M. Heil, R. C. Black, B. Giovani, P. Galland and D. Lardemer (2002). *Plant Physiol* **129**, 774–85.

Ahmad, M., J. A. Jarillo, O. Smirnova and A. R. Cashmore (1998). *Mol. Cell* **1**, 939–948.

Ahmad, M., C. Lin and A. R. Cashmore (1995). *Plant J* **8**, 653–8.

Batschauer, A., M. Rocholl, T. Kaiser, A. Nagatani, M. Furuya and E. Schafer (1996). *Plant J* **9**, 63–69.

Bouly, J. P., B. Giovani, A. Djamei, M. Mueller, A. Zeugner, E. A. Dudkin A. Batschauer and M. Ahmad (2003). *Eur J Biochem* **270**, 2921–8.

Brautigam, C. A., Smith, B. S., Ma, Z., Palnitkar, M., Tomchick, D. R., Machius, M., and Deisenhofer, J. (2004). *Proc Natl Acad Sci USA* 101, 12142–12147.

Cutler, S. R., D. W. Ehrhardt, J. S. Griffitts and C. R. Somerville (2000). *Proc Natl Acad Sci USA* **97**, 3718–23.

Duek, P. D. and C. Fankhauser (2003). *Plant J* **34**, 827–36.

Fairchild, C. D., M. A. Schumaker and P. H. Quail (2000). *Genes Dev* **14**, 2377–91.

Fankhauser, C. and J. Chory (2000). *Plant Physiol* **124**, 39–46.

Folta, K. M., M. A. Pontin, G. Karlin-Neumann, R. Bottini and E. P. Spalding (2003). *Plant J* **36**, 203–14.

Folta, K. M. and E. P. Spalding (2001). *Plant J* **26**, 471–8.

Giovani, B., M. Byrdin, M. Ahmad and K. Brettel (2003). *Nat Struct Biol* 10, 489–90.

Guo, H., T. Mockler, H. Duong and C. Lin (2001). *Science* **291**, 487–90.

Hattar, S., R. J. Lucas, N. Mrosovsky, S. Thompson, R. H. Douglas, M. W. Hankins, J. Lem, M. Biel, F. Hofmann, R. G. Foster and K. W. Yau (2003). *Nature* **424**, 75–81.

Helfrich-Forster, C., C. Winter, A. Hofbauer, J. C. Hall and R. Stanewsky (2001). *Neuron* **30**, 249–61.

Kim, Y. M., J. C. Woo, P. S. Song and M. S. Soh (2002). *Plant J* **30**, 711–9.

Koornneef, M., E. Rolff and C. J. P. Spruit (1980). *Z. Pflanzenphysiol. Bd.* **100**, 147–160.

Kume, K., M. J. Zylka, S. Sriram, L. P. Shearman, D. R. Weaver, X. Jin, E. S. Maywood, M. H. Hastings and S. M. Reppert (1999). *Cell* **98**, 193–205.

Li, Y. F., P. F. Heelis and A. Sancar (1991). *Biochem.* **30**, 6322–6329.

Lin, C., D. E. Robertson, M. Ahmad, A. A. Raibekas, M. Schuman Jorns, P. L. Dutton and A. R. Cashmore (1995). *Science* **269**, 968–970.

Ma, L., J. Li, L. Qu, J. Hager, Z. Chen, H. Zhao and X. W. Deng (2001). *Plant Cell* **13**, 2589–2607.

Malhotra, K., K. Sang-Tae, A. Batschauer, L. Dawut and A. Sancar (1995). *Biochemistry* **34**, 6892–6899.

Neff, M. M. and J. Chory (1998). *Plant Physiol.* **118**, 27–35.

Osterlund, M. T., C. S. Hardtke, N. Wei and X. W. Deng (2000). *Nature* **405**, 462–466.

Panda, S., I. Provencio, D. C. Tu, S. S. Pires, M. D. Rollag, A. M. Castrucci, M. T. Pletcher, T. K. Sato, T. Wiltshire, M. Andahazy, S. A. Kay, R. N. Van Gelder and J. B. Hogenesch (2003). *Science* **301**, 525–7.

Parks, B. M., M. H. Cho and E. P. Spalding (1998). *Plant Physiol.* **118**, 609–615.

Payne G., P. F. Heelis, B. R. Rohrs, and A. Sancar (1987). *Biochemistry* **26**, 7121–7127.

Poppe, C., U. Sweere, H. Drumm-Herrel and E. Schafer (1998). *Plant J* **16**, 465–471.

Sancar A. (2003). *Chem Rev* **103**, 2203–37.

Seo, H. S., J. Y. Yang, M. Ishikawa, C. Bolle, M. L. Ballesteros and N. H. Chua (2003). *Nature* **423**, 995–9.

Shalitin, D., H. Yang, T. C. Mockler, M. Maymon, H. Guo, G. C. Whitelam and C. Lin (2002). *Nature* **417**, 763–7.

Shalitin, D., X. Yu, M. Maymon, T. Mockler and C. Lin (2003). *Plant Cell* **15**, 2421–9.

van der Horst, G. T., M. Muijtjens, K. Kobayashi, R. Takano, S. Kanno, M. Takao, J. de Wit, A. Verkerk, A. P. Eker, D. van Leenen, R. Buijs, D. Bootsma, J. H. Hoeijmakers and A. Yasui (1999). *Nature* **398**, 627–630.

Wang, H., L. G. Ma, J. M. Li, H. Y. Zhao and X. W. Deng (2001). *Science* **294**, 154–8.

Wang, X. and M. Iino (1998). *Plant Physiol* **117**, 1265–79.

Yang, H. Q., R. H. Tang and A. R. Cashmore (2001). *Plant Cell* **13**, 2573–87.

Yang, H. Q., Y. J. Wu, R. H. Tang, D. Liu, Y. Liu and A. R. Cashmore (2000). *Cell* **103**, 815–27.

Yeh, K. C. and J. C. Lagarias (1998). *Proc Natl Acad Sci USA* **95**, 13976–81.

12
Animal Cryptochromes

Russell N. Van Gelder and Aziz Sancar

12.1
Introduction

Photolyase/cryptochrome blue-light photoreceptors are monomeric proteins of 50–70 kDa that contain two noncovalently bound chromophore/cofactors (Cashmore, 2003; Sancar, 2003). One of the cofactors is always flavin-adenine dinucleotide. The second chromophore is methenyltetrahydrofolate in most organisms and 8-hydroxy-5-deazariboflavin (8-HDF) in a few species that synthesize this cofactor. Although "cryptochrome" was originally used as a generic term for plant pigments of cryptic identity that mediated the physiological responses to blue light, now it is known that there are at least three flavoproteins (HY4, Nph1, FKF1) that mediate these responses (Briggs and Christie, 2002). Of these, the gene encoding apoprotein for HY4 in *Arabidopsis* was the first putative blue-light receptor gene identified (Ahmad and Cashmore, 1993) and it was named "*Cryptochrome*" (Lin et al., 1995). Now, the term cryptochrome has acquired a precise definition: a (putative) photoreceptor in plants, animals, and bacteria with a high degree of sequence and structural homology to DNA photolyase but with no repair function (Cashmore, 2003; Sancar, 2000). It is possible that photolyase and cryptochrome had a common progenitor that repaired the DNA damage caused by the high flux of UV present in the early atmosphere and that regulated the movement of an ancestral organism so as to optimize the beneficial effects of light (photosynthesis) while minimizing the harmful effects (DNA repair). Subsequently the pigments carrying out these two functions diverged to give rise to present day cryptochromes and photolyases. While it is interesting to speculate on whether the progenitor molecule was a photolyase or a cryptochrome it is quite conceivable that it was both and that it performed both functions poorly until further speciation and division in labor.

Handbook of Photosensory Receptors. Edited by W. R. Briggs, J. L. Spudich

ISBN 3-527-31019-3

12.2
Discovery of Animal Cryptochromes

Although photolyases are widespread in nature, they are not essential enzymes, and many species including such model organisms as *Bacillus subtilis, Scizosaccharomyces pombe*, and *Caenorhabditis elegans* do not have photolyase (see Sancar, 2000). Importantly, based on exhaustive biochemical data, it has been concluded that humans and other placental mammals do not have photolyase (Li et al., 1993). Therefore the report of a possible photolyase ortholog as an EST in the human genome database in 1995 (Adams et al., 1995) was unexpected and led to the re-evaluation of previous conclusions. Soon after this report of the first "human photolyase ortholog" Hsu et al., (1996) identified a second human potential photolyase ortholog and demonstrated that neither of these proteins had photolyase activity. They concluded that these proteins must perform other blue-light-dependent functions in human cells and named them human cryptochrome1 (hCRY1) and human cryptochrome2 (hCRY2), and suggested that these proteins may function as circadian photoreceptors. Shortly afterwards, cryptochrome homolog were found in many other animals including insects, fishes, amphibians, and birds. It must be noted, however, that these assignments are often based on sequence alignments and it is impossible to designate a member of the photolyase/cryptochrome family a photolyase or a cryptochrome unless assays specific for each group of the family have been performed.

12.3
Structure–Function Considerations

Currently, the photolyase/cryptochrome family encompasses 3 types of enzymes: cyclobutane pyrimidine dimer photolyase (or simply, photolyase); (6–4) photolyase; and cryptochrome. The first two repair the two major UV-induced DNA lesions, [the cyclobutane pyrimidine dimer and the pyrimidine-pyromidone (6–4) photoproduct], the third is either known or presumed to be a photosensory pigment (Cashmore, 2003; Sancar, 2003). It must be noted, however, that in all cases where detailed analysis is available both photolyases and cryptochromes perform light-independent functions as well (Sancar, 2000; 2003). Crystal structures of photolyases from *E. coli, A. nidulans*, and *T. thermophilus* (Sancar, 2003) and of cryptochromes from *Synecchococcus* (Brudler et al., 2003) and *A. thaliana* (Brautigam et al., 2004) have been solved. Remarkably, the structures of all these enzymes are very similar. However, most cryptochromes contain, in addition to the photolyase homology region, a unique carboxy terminal extension ranging from ~30 amino acids in Drosophila Cry to ~250 amino acids in *Arabidopsis thaliana* Cry1. The structure of this domain, which is thought to perform an effector function, is not known. Below, is a brief summary of the structure–function of *E. coli* photolyase, which is the best understood of this family of proteins, to set the stage for discussion of animal cryptochromes.

E. coli photolyase is a monomeric protein of 471 amino acids. The enzyme is composed of two domains (Figure 12.1): An N-terminal α/β domain (residues 1–113) and

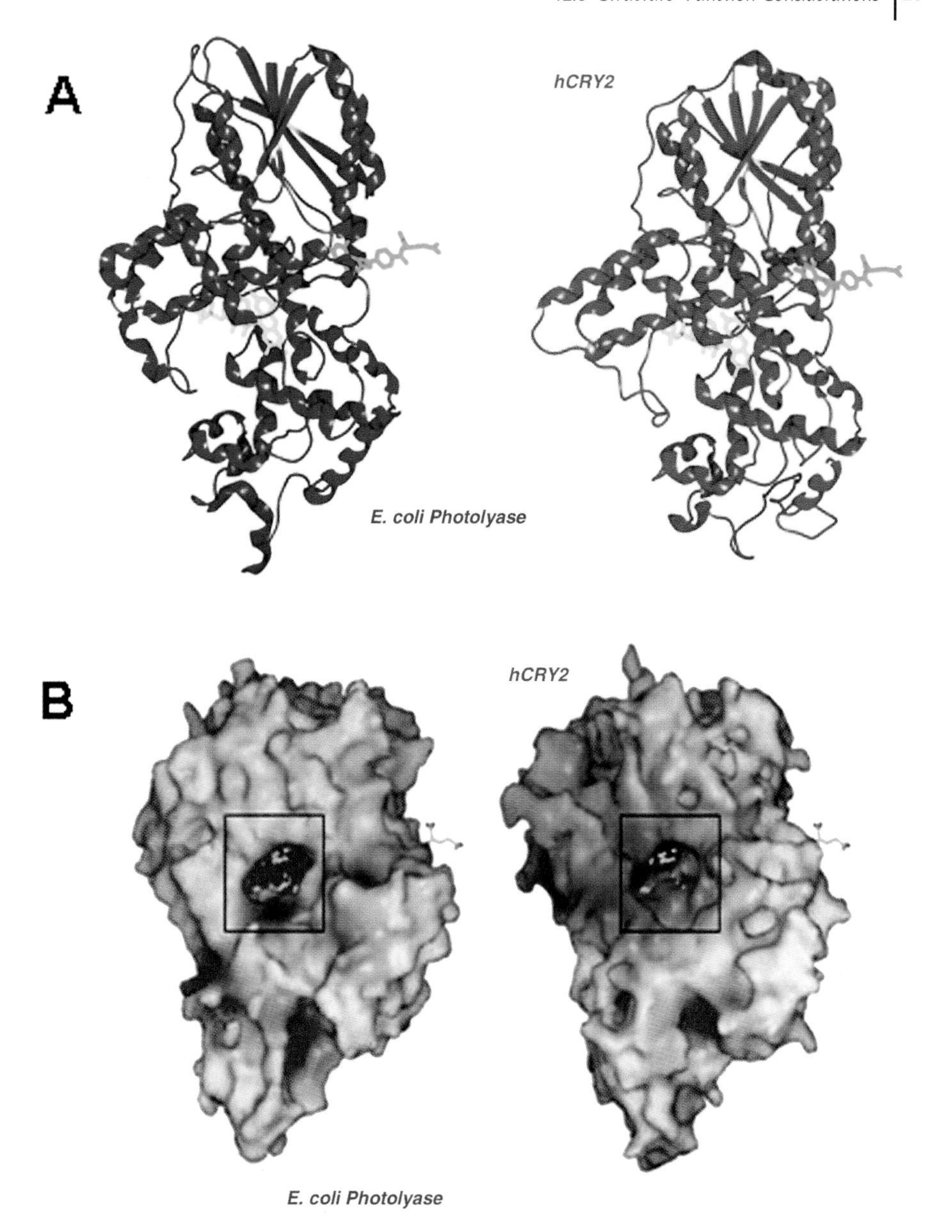

E. coli Photolyase

Figure 12.1 Structures of E. coli photolyase and human cryptochrome 2. The photolyase structure was obtained by X-ray crystallography. The cryptochrome structure was generated by computational methods using photolyase as a template. Both ribbon diagrams and surface potential representation of the molecules are shown. In the ribbon diagrams yellow indicates FAD and green indicates MTHF. In the surface potential representation blue, red, and white indicate basic, acidic, and hydrophobic residues, respectively. Note that only the photolyase homology region of hCRY2 is shown; The C-terminal 103 amino acids were not included in the modeling. From Park et al., (1995) and Özgür and Sancar (2003).

a C-terminal α-helical domain (residues 204–471), that are connected with a long loop that wraps around the α/β domain. The methenyltetrahydorfolate (MTHF) photoantenna is bound in a shallow cleft between the two domains and the flavin adenine dinucleotide (FAD) cofactor is deeply buried within the α-helical domain. A surface-potential representation of the molecule reveals a positively charged groove running the length of the molecule. A hole in the middle of this groove leads to the flavin in the core of the α-helical domain. The enzyme binds DNA around the cyclobutane dimer in a light-independent reaction and "flips out" the dimer dinucleotide into the active site cavity in contact with flavin to produce a stable enzyme-substrate complex. In the absence of light this complex eventually dissociates without repair. Exposure of the E-S complex to light initiates catalysis: MTHF absorbs a photon, transfers the excitation energy by Förster dipole–dipole resonance electronic energy-transfer mechanism to the flavin; then the excited flavin [$^1(FADH^-)^*$] transfers an electron to the pyrimidine dimer. The resulting pyrimidine-dimer radical undergoes bond rearrangements to generate two pyrimidines, and the flavin neutral radical generated during the reaction is restored to the catalytically competent $FADH^-$ by back electron transfer (Sancar, 2003); finally the enzyme and product dissociate from one another.

An inspection of cryptochrome structures reveal such a high similarity to that of photolyase that raises the question as to why cryptochromes cannot repair DNA. Currently there is no obvious answer to this question.

In considering the photolyase reaction mechanism as a model for the mode of action of cryptochrome the following points must be kept in mind. First, the active form of flavin in photolyase is the two-electron reduced and deprotonated flavin adenine dinucleotide ($FADH^-$). This cofactor is readily oxidized in vitro to either the one-electron reduced flavin blue neutral radical ($FADH^•$) or to the two electron oxidized FAD_{ox} and both forms are catalytically inert. Second, of the two chromophores in photolyase only $FADH^-$ is indispensable. Enzyme without the second chromophore MTHF [or 8-hydroxy-5-deazariboflavin (8-HDF) in certain photolyases] is still active (albeit with much lower efficiency, under limiting light conditions, because of the low extinction coefficient of $FADH^-$). Under physiological conditions when the enzyme contains both chromophores and is exposed to sunlight, >90% of the photons that initiate catalysis are absorbed by MTHF because of its higher extinction coefficient and longer-wavelength absorption maximum (ε~25 000 M^{-1} ^{cm}–1 at λ_{max}~380–420 nm) compared to FADH (ε~6000 M^{-1} ^{cm}–1 at λ_{max}~360 nm). Third, for catalysis to occur, the photon must be absorbed by the enzyme in the enzyme-substrate complex. A photon absorbed by free enzyme generates an excited state that decays by fluorescence or nonradiative mechanisms within 1–2 nanosecond with no lasting effect on the binding to or subsequent activity of photolyase on its substrate. Finally, of all of the photolyase/cryptochrome family blue-light photoreceptors purified only the *S. cerevisiae* photolyase (Sancar et al., 1987) and the *V. cholerae* cryptochrome 1 retain stoichiometric amounts of both chromophores and the flavin cofactor in the active $FADH^-$ form after purification under aerobic conditions (Worthington et al., 2003).

Heterologously expressed human cryptochromes have been purified and shown to contain both FAD and MTHF by fluorescence spectroscopy. Both cofactors are at grossly sub-stoichiometric amounts; these cryptochromes exhibit a 420 nm maximum which is due mostly to FAD_{ox} absorption and cannot be considered the absorption maximum of the native cryptochromes (Hsu et al., 1996; Özgür and Sancar, 2003). The oxidation of the flavin or the loss of a fraction and of all of both chromophores are not problems unique to cryptochromes as this has been observed with some photolyases as well. For example, Drosophila (6–4) photolyase expressed in *E. coli* contains only about 5% flavin in the inactive FAD_{ox} form and even less folate (Zhao et al., 1997). Therefore, the affinities of cryptochromes to their cofactors are not necessarily weaker than those of photolyases to their cofactors. Nevertheless, cofactor oxidation and loss during purification of cryptochromes has been a serious impediment in biochemical characterization of these photoreceptors. Most of our understanding of cryptochromes is based on genetic data and some cell-based in vivo biochemical experiments. Even though the role of cryptochrome in the circadian clock was first defined in mammals (Hsu et al., 1996; Miyamoto and Sancar, 1998; Thresher et al., 1998) currently the *Drosophila* cryptochrome is better characterized and therefore will be presented first.

12.4
Drosophila melanogaster Cryptochrome

Drosphila cryptochrome was discovered in a genetic screen for mutants affecting circadian rhythms (Stanewsky et al., 1998) shortly after the report directly implicating mammalian cryptochrome in circadian regulation (Miyamoto and Sancar, 1998). Flies, like most metazoans, have an internal timekeeping mechanism that controls the timing of locomotor activity of the animal. Even when kept in complete darkness, flies maintain a nearly-24-h rhythm behavior. The genetic basis of this phenomenon has been extensively studied (see Van Gelder, 2003; Van Gelder et al., 2003). However, the fly clock, like all circadian clocks, does not keep exact 24-hour time; it needs to be continuously synchronized to the external world in order to maintain a stable phase relationship with local time. External light–dark cycles are the dominant synchronizing signal for circadian clocks in all species.

While the most parsimonious model for light entrainment of rhythms would suggest that the visual system provides the necessary input to the brain centers containing the circadian clock, it had been recognized since at least 1971 that this model was oversimplified or possibly incorrect. All visual photopigments are opsin-based and this is dependent on vitamin A for chromophore generation. When Zimmerman and Goldsmith (1971) grew flies on vitamin A-depleted media, they found that the visual system of these flies lost 3 log units of sensitivity (as measured by electroretinogram); however, the photic sensitivity of the flies for circadian entrainment was unchanged. These results suggested the existence of a non-opsin photopigment underlying circadian entrainment.

In screening for EMS-generated mutants in circadian rhythmicity, Stanewsky and colleagues discovered a novel, recessive gene that resulted in apparently arrhythmic flies in constant conditions (Stanewsky et al., 1998). However, further study demonstrated that these flies were capable of maintaining free-running circadian rhythmicity under constant conditions, provided the flies were first synchronized to an external temperature cycle. This result strongly suggested that the phenotype of these flies was not caused by a defect in the circadian rhythm system *per se*, but by an inability to synchronize their clocks to external light–dark cycles. This hypothesis was confirmed when the effects of short pulses of light on phase-shifting of the circadian rhythms of flies were studied: *mutant*^{b} flies were essentially incapable of shifting their rhythms to short pulses of light. Molecular cloning of this locus demonstrated that the mutation mapped to the *Drosophila* homolog of cryptochrome. The mutation (an Asp→Asn substitution at amino acid 410) appeared to be loss of function or genetically null [although it must be noted that this change in *E. coli* photolyase does not cause a null phenotype (Sancar, 2003)]. The mutant allele was named *cryptochrome*^{baby}, abbreviated *cry*^{b}. *Cry*^{b} mutant flies were found to have only partial responses to lighting cycles. It had been known for many years that the eyes of the fly are not necessary for the circadian clock to entrain to light–dark cycles; mutants lacking eyes, oscelli, and other photoreceptors are still capable of normal entrainment (Wheeler et al., 1993) . While the initial report of the *cry*^{b} mutation noted that it had an added, epistatic effect on the *norpA* visual-system mutant (suggesting that the *cryptochrome*-dependent and classical visual pathways were somewhat independent), these flies still maintained some ability to sense light and entrain their circadian rhythms to light–dark cycles. However, *norpA* mutants likely do not have a complete visual loss phenotype, and likely have preservation of photoreceptor function of the minor Hofbauer-Buchner eyelet photoreceptors (located internal to the compound eyes) . Generation of double mutants of the pan-photoreceptor mutant *glass* and *cry*^{b} resulted in flies that could not entrain their rhythms to external light–dark cycles (Helfrich-Forster et al., 2001). As the *cry*^{b} mutation was essentially epistatic to *glass*, this strongly suggested parallel roles for *Drosophila* cryptochrome and classical photoreception in entrainment of the *Drosophila* circadian clock to light–dark cycles.

A more direct role for cryptochromes in nonvisual photoreception in *Drosophila* was suggested by studies on the cell-autonomy of cryptochrome function. Many *Drosophila* tissues will continue to show circadian rhythms of gene expression even when cultured in vitro (Plautz et al., 1997). These rhythms – even in tissues like the sensory bristles of the leg – could be independently entrained by light–dark cycles, demonstrating cell-autonomy of both the circadian timekeeping mechanism and its photoreceptive system. Similar results have also been seen in the isolated Malpighian tubules of the fly (Giebultowicz and Hege 1997; Hege et al., 1997). By constructing flies that expressed cryptochrome only in the circadian pacemaker neurons for locomotion (by effectively driving cryptochrome expression under the *pigment dispersing factor* enhancer/promoter (Renn et al., 1999)), Emery et al. (Emery et al., 2000) were able to rescue the circadian photoentrainment in *cry*^{b} mutant flies. Thus, cryptochrome – acting cell-specifically within the pacemaker cells – is capable of rescuing photosensitivity. The primacy of cryptochrome for circadian photoreception was further suggested by testing the effect of constant lighting on *Drosophila* circadian rhythms. Flies kept under continuous lighting (LL) will normally become behaviorally arrhythmic within several days. However, *cry*^{b} flies kept under these conditions showed persistent free-running rhythms, suggesting that the flies were oblivious to the presence of constant light (Emery et al., 2000). From these studies, the authors concluded that *cry* is an essential component of a deep-brain photoreceptor in the fly.

While these genetically-based experiments demonstrated the necessity of cryptochrome function for normal light-entrainment of the circadian clock, they could not demonstrate the sufficiency of cryptochrome protein as a photopigment. Ceriani et al. expressed *Drosophila* cryptochrome in a yeast two-hybrid experiment, and performed directed interaction experiments with known components of the *Drosophila* circadian pathway (Ceriani et al., 1999). They found a specific interaction between cryptochrome and the protein encoded by the *Timeless* gene. This interaction is biologically plausible, as light is known to cause a rapid degradation of timeless protein by targeting the protein to the proteosome (Hunter-Ensor et al., 1996; Yang et al., 1998; Naidoo et al., 1999). The remarkable finding in *Drosophila* is that cryptochrome and timeless proteins interacted only when the yeast were kept in the light – the interaction was not seen when the yeast were kept in the dark. As yeast contains no known photosensory system, this result suggests that heterologously expressed cryptochrome is sufficient to form a functional photopigment that undergoes some light-dependent conformational change or post-translational modification. Expressed cry^b protein was not capable of interacting with timeless protein under light or dark conditions. To date, this is the best evidence that animal cryptochromes form functional photopigments.

Genetically, it appears that *cry* and the clock gene *period* (*per*) also interact; Rosato et al. (Rosato et al., 2001) noted a temperature-dependent difference in the number of short-period double $cry^b;per^s$ flies, with increasing proportion of short-period animals under higher temperature conditions. These authors then tested for direct interactions between the Period and Cryptochrome using the yeast two-hybrid system, and found that a specific fragment of Period protein (amino acids 233–685) interacted with Cry only in light, and further that this interaction was temperature-dependent. Deletion analysis revealed that the light-dependency of the interaction was dependent on the C-terminus of the Cryptochrome protein.

Several lines of evidence also suggest that *cryptochrome* acts in a light-dependent manner in vivo. A critical component of the circadian oscillator in *Drosophila* is the inhibition of the complex of the Cycle:Clock protein complex on the E-box-containing promoter of the *period* and *timeless* genes by the Timeless:Period complex (Ceriani et al., 1999). Cryptochrome's function on this portion of the oscillator mechanism appears to be to de-activate the repressor (at least in part by causing light-dependent degradation of Timeless). Transfection of a *Drosophila* embryonic cell line with all components of this pathway showed light-dependent activation by Clock (i.e. derepression of Period: Timeless). This activation may have been mediated by another light-dependent mechanism than that elucidated in the yeast two-hybrid system, however, as coimmunoprecipitation experiments showed Timeless-Cryptochrome interactions in both dark-cultured and light-pulsed cells. Some of the light-dependence of this system may be due to light-dependent nuclear localization of the Cryptochrome-Timeless-Period complex; the authors noted nuclear localization of this complex in about 10% in dark-reared cells and 40% in cells subjected to a 3-h light pulse.

Some pleiotropy in cryptochrome function is suggested by studies of circadian rhythms other than locomotion in fly. The antennae of the fly have a circadian sensitivity to chemical stimulants, which persists even in antennae cultured ex vivo (Krishnan et al., 1999) . In analyzing this rhythm, Krishnan et al. (Krishnan et al., 2001) noted that free-running rhythmicity in antennal rhythms is lost in the *cry^b* mutant (not just entrainability to light). Thus, cryptochrome appears to play an essential role in the generation of a peripheral circadian rhythm in the fly, and is therefore pleiotropic in its functions. A similar theme of pleiotropic activity in generation of circadian rhythmicity and mediation of light entrainment will be seen in analysis of mammalian cryptochrome function.

12.5 Mammalian Cryptochromes, Circadian Rhythmicity, and Nonvisual Photoreception

Murine circadian rhythms have a free-running period of about 23.5 h, and thus require daily synchronization to external time cues – predominantly the light–dark cycle – to maintain 24-hour rhythmicity. The murine circadian clock pacemaker has an anatomical locus in the suprachiasmatic nucleus (SCN) of the hypothalamus; ablation of this nucleus results in mice with behavioral arrhythmicity (Stephan and Zucker 1972; Weaver 1998). Unlike *Drosophila*, however, the murine circadian oscillator does not have a cell-autonomous photoreceptor, as isolated SCN tissue in culture cannot be entrained to external lighting cycles (Herzog and Huckfeldt 2003). Enucleated mice, or mice genetically lacking optic nerves, cannot entrain their circadian clocks to external lighting cycles, demonstrating that the circadian photoreceptor in the mouse is contained in the eyes (Freedman et al., 1999; Wee et al., 2002). However, the simple hypothesis that the classical photoreceptors – the rods and cones – mediate light information reaching the suprachiasmatic nucles and controlling the timing of behavior is incorrect: mice lacking all classical photoreceptors (i.e., the *cl;rdta* mouse that expresses diphtheria toxin in the rods and cones) still entrain their rhythms normally (Ebihara and Tsuji 1980; Freedman et al., 1999). The predicted inner retinal photoreceptive cell was discovered in 2002 by Berson and colleagues (Berson et al., 2002), and is found in a subset of retinal ganglion cells – the cells whose axons make up the optic nerve. The intrinsically photoreceptive retinal ganglion cells (ipRGCs) are those cells whose axons project specifically to brain centers subserving nonvisual, light sensitive processes, such as circadian rhythm entrainment and pupillary light responsiveness (Hattar et al., 2002).

Both murine cryptochromes (abbreviated *mCry1* and *mCry2*) are expressed in the retina (Miyamoto and Sancar 1998; Miyamoto and Sancar 1999). In situ hybridization suggests that the cryptochromes are predominantly expressed in the inner retina, most notably in the ganglion cell layer (which includes the ganglion cells and amacrine cells), but to a lesser extent in the inner nuclear layer and the outer retina. To date, antibody staining of murine tissue has not yielded definitive immunohistochemistry. However, reliable staining for cryptochrome 2 has been achieved in human ocular samples (Thompson et al., 2003). Cryptochrome 2 appears to be expressed in about 70% of the retinal ganglion cells (Figure 12.2). The overlap of these cells with the intrinsically photosensitive retinal ganglion cells is unknown at present. In addition to ocular expression, mammalian cryptochromes are widely expressed in a number of tissues, including the suprachiasmatic nucleus. This immediately suggests that cryptochrome expression is not sufficient to confer photoresponsiveness to a tissue, as the SCN is not directly photoresponsive.

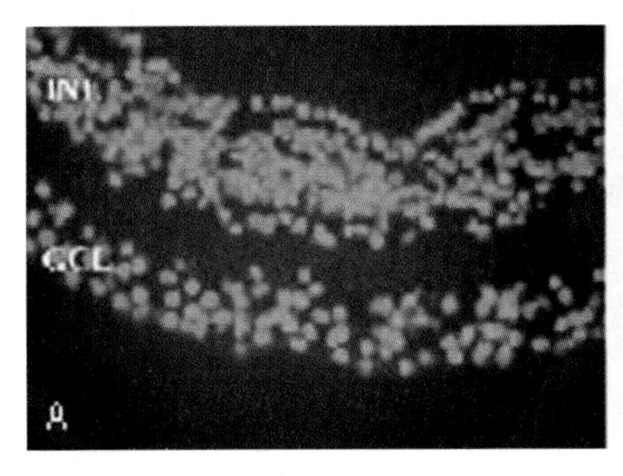

Figure 12.2 Expression of cryptochrome 2 in the human retina. Double label histochemistry showing hCRY2 expression compared to overall cellular density as revealed by staining with the nuclear marker DAPI in the macula (A) DAPI staining. (B) hCRY2 staining. GCL, ganglion cell layer; INL, inner nuclear layer. From Thompson et al., (2003).

Genetic "knockouts" of the cryptochromes have been produced by two laboratories (Thresher et al., 1998; van der Horst et al., 1999; Vitaterna et al., 1999). Assaying the circadian phenotypes of these mice revealed unexpected findings. While in *Drosophila*, loss of cryptochrome function results in animals with intact free running circadian rhythms, but with abnormal entrainment phenotypes, the phenotype of the compound *mCry1*$^{-/-}$;*mCry2*$^{-/-}$ mutant in mice showed arrhythmicity in free-running conditions (Figure 12.3). However, in light–dark cycles, the mice showed rhythmic behavior. This behavior likely represents "masking", the phenomenon by which light directly controls behavior (Mrosovsky 1999), rather than entrainment of the circadian clock.

The individual cryptochrome mutations by themselves show alterations in the free-running period of circadian rhythmicity (Thresher et al., 1998; van der Horst et al., 1999; Vitaterna et al., 1999; Van Gelder et al., 2002). Mice lacking both copies of *mCry1* show substantial shortening of their circadian rhythms, while mice lacking both copies of *mCry2* have long period rhythms. While the two genes are largely able to substitute for each other with respect to the generation of circadian rhythmicity, they rescue rhythmicity with differential efficacy. In a *mCry1*$^{-/-}$ background, two copies of *mCry2* are necessary to rescue rhythmicity, while only a single copy of *mCry1* is generally necessary to rescue rhythmicity in a *mCry2*$^{-/-}$ background (Van Gelder et al., 2002).

Is the arrhythmic phenotype of *mCry1*$^{-/-}$;*mCry2*$^{-/-}$ mice fundamentally different from the arrhythmic phenotype initially seen in *Drosophila*? Perhaps these mice are

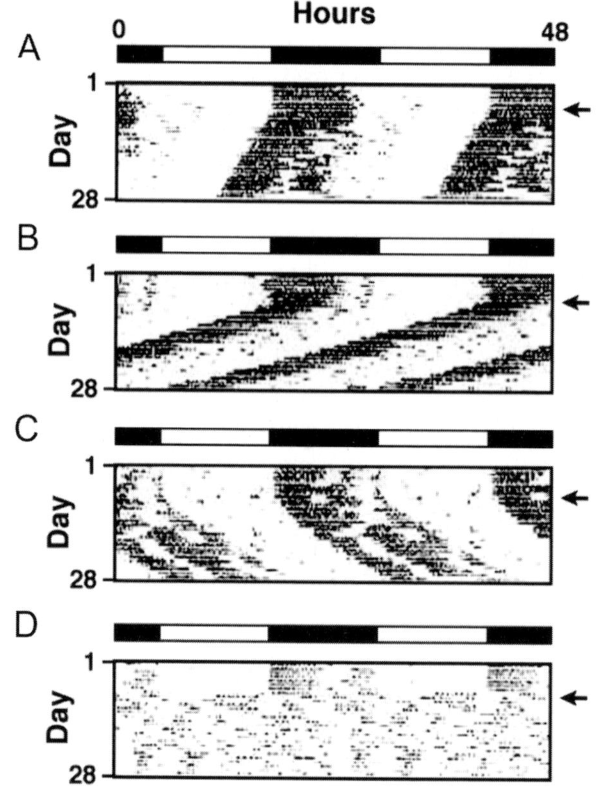

Figure 12.3 Effect of cryptochrome mutations on the mouse circadian clock. These are actograms of mouse wheel running activity as a function of 12 h light and 12 h clock (LD 12:12) cycle. The bar at the top shows light and dark phases. The *y* axis indicates activity (wheel rpm) for 28 days and the *x* axis indicates time of day. The activity profile is double plotted: first line shows activity for the first day on the left side and the second day on the right; the second line shows activity for the second day on the left and third day on the right and so on. Mice were switched from LD 12:12 into constant darkness DD on the say indicated by arrows. Under DD, free-running period lengths of mutant mice are altered: (a) wt, 23.7 h; (b) *Cry1*$^{-/-}$ 22.7 h; (c) *Cry2*$^{-/-}$, 24.7 h; (d) *Cry1*$^{-/-}$ *Cry2*$^{-/-}$, arrhythmic. From Vitaterna *et al* (1999).

merely desynchronized, but the (presumably) cell-autonomous oscillators of the SCN are each still able to keep time? Recent imaging experiments in which circadian rhythmicity can be followed by measuring reporter gene activity of single cells strongly suggests that this is not the case; individual *mCry-* cells appear to be arrhythmic (Yamaguchi et al., 2003). In cell culture-based assays, cryptochrome expression can potently repress transcription of the circadian clock gene *period* by interfering with activation of the "E-box-" containing promoter elements by the clock genes Bmal1 and Clock (Griffin et al., 1999; Kume et al., 1999). Cryptochrome appears to form a complex with mPer and directly inhibit activation of the E-box by the positive transcription factors (Lee et al., 2001; Reppert and Weaver 2001). This contrasts with the role of cryptochrome in *Drosophila,* where its primary function appears to cause light-dependent degradation of Timeless (and thus functionally activate transcription from Clock:Cycle, rather than repress Clock:Bmal1 as is the case in the mammal). Furthermore, while mammalian cryptochromes can form yeast two-hybrid associations with multiple circadian clock gene products (including Clock, Bmal1, Per1, Per2, and Timeless), unlike the case with *Drosophila* cryptochrome these interactions appear to be light-independent in yeast (Griffin et al., 1999).

Does this mean that cryptochrome has no role in phototransduction in the mouse? In *Drosophila,* cryptochrome mutations had a partial phenotype for circadian entrainment, which was made complete by compounding the cryptochrome mutation on a genetic background that lacked visual photoreceptor function (*glass*). To test whether a similar functional redundancy between classical photoreceptor pathways and cryptochromes exists, Selby et al. generated triply mutant *mCry1*$^{-/-}$;*mCry2*$^{-/-}$; *rd/rd* mice and tested these animals for circadian entrainment (Figure 12.3) (Selby et al., 2000). While both *rd/rd* and *mCry1*$^{-/-}$;*mCry2*$^{-/-}$ mice showed fairly normal rhythmicity in light–dark conditions, the compounded triple mutants were largely arrhythmic (Selby et al., 2000; Van Gelder et al., 2002). However, a few animals still showed substantial diurnal rhythmicity. To determine whether this arrhythmicity correlated with loss of signaling from the eye to the SCN, Selby et al. also examined the induction of the immediate-early gene *c-fos* by light in the SCN (Figure 12.4). While *rd/rd* animals show super-normal levels of *c-fos* induction by light during the subjective night (i.e, those times when the time of the animals' circadian clocks correspond to the dark period), the *mCry1*$^{-/-}$;*mCry2*$^{-/-}$ animals showed 10–20-fold reduction and the *rd/rd;mCry1*$^{-/-}$;*mCry2*$^{-/-}$ animals showed about 10,000-fold reduction in photosensitivity for *c-fos* induction under limiting light and the triply mutant mice exhibited reductions of *c-fos* induction to ~17% of control levels even at the brightest lights tested (Selby et al., 2000; Thompson et al., 2004). These results suggest pleiotropic actions for mCry in both phototransduction and light entrainment.

One other notable observation in the *mCry1*$^{-/-}$;*mCry2*$^{-/-}$ animals is a reversal of lighting preference – a subset of *mCry1*$^{-/-}$;*mCry2*$^{-/-}$ animals consolidate their activity into the light period, and thus transform from nocturnal to diurnal behavior (Van Gelder et al., 2002). It is difficult to reconcile this finding with a pure circadian phenotype, as lighting preference reversal has not been seen in other genetically arrhythmic mouse lines, such as *mPer1*$^{-/-}$;*mPer2*$^{-/-}$ or *Bmal*$^{-/-}$ (Zheng et al., 1999; Albrecht et al., 2001; Bae et al., 2001; Bunger et al., 2000).

Interpretation of the results of the retinal degenerate animals lacking cryptochromes has been problematic, however. Since the circadian clock is clearly dysfunctional in cryptochrome-deficient animals, how does one interpret reduced amplitude of *c-fos* induction in the suprachiasmatic nucleus? Behaviorally, one cannot claim that mammalian cryptochromes are influencing circadian entrainment (as they clearly do in *Drosophila*), as it is meaningless to talk about the entrainment of a clock-less animal. Thus, it has remained ambiguous whether cryptochromes are truly participating in "nonvisual phototransduction" (Van Gelder 2001). A more compelling case for cryptochromes' roles in these processes could be made if these proteins influenced another nonvisual photoreceptive process. In 1927, Keeler (discov-

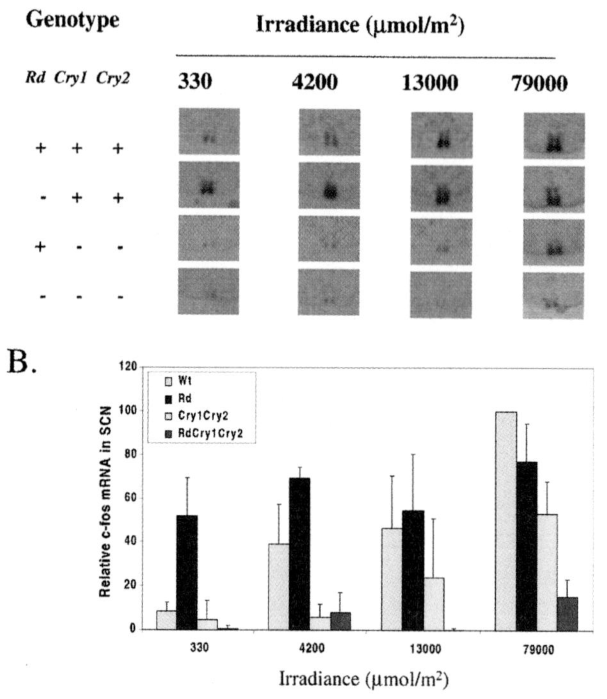

Figure 12.4 Functional redundancy of cryptochromes and visual opsins in retinohypothalamic photoreception/phototransduction. Acute induction of *c-fos* in the SCN was measured by *in situ* hybridization. Mice of the indicated genotypes were exposed to four doses of white light and examined for *c-fos* induction in the SCN after 30 min. (A) Representative SCN images from peak section. (B) Quantitative analysis *c-fos* induction in SCN as a function of light dose. The induction levels are expressed relative to the wild type at the highest dose used, which is taken as 100%. The bars indicate standard deviation ($n = 3–9$). *From Selby et al.*, (2000).

erer of the *rd/rd* mouse) noted that the pupillary light response of these mutant animals remained remarkably intact (Keeler 1927). Lucas et al. more recently reinvestigated this phenomenon and determined that the pupillary light response remains intact in mice lacking both rods and cones, and has a blue-light action spectrum centering around 480 nm (Lucas et al., 2001). This therefore would appear to be a second "nonvisual" photoresponse. (Photic suppression of pineal melatonin production likely constitutes a third such phenomenon (Lucas et al., 1999))

To assess the role of cryptochromes in this phenomenon, Van Gelder et al. (Van Gelder et al., 2003) examined the pupillary light response of *rd/rd* mice, cryptochrome-deficient mice, and *rd/rd;mCry1*$^{-/-}$*;mCry2*$^{-/-}$ mice. They found that, consistent with the earlier reports, the pupillary light response to mid-blue light (470 nm) of *rd/rd* mice is about 1 log reduced in sensitivity from that of wild-type animals, while the pupillary light response of *mCry1*$^{-/-}$*;mCry2*$^{-/-}$ animals was not affected.

However, the conjoint *rd/rd;mCry1*$^{-/-}$*;mCry2*$^{-/-}$ mouse had pupillary light responses that were less than 10% of *rd/rd* animals, and less than 1% of wild-type (Figure 12.5). At very bright light intensities, however, the *rd/rd;mCry1*$^{-/-}$*;mCry2*$^{-/-}$ mice still showed substantial photoresponses, demonstrating that cryptochrome is not absolutely required for pupillary light responses. These results largely parallel the findings in circadian entrainment; however, as the pupillary light response appears to be largely independent of the circadian rhythm response, these results strongly implicate cryptochromes as participating in the nonvisual phototransducive process.

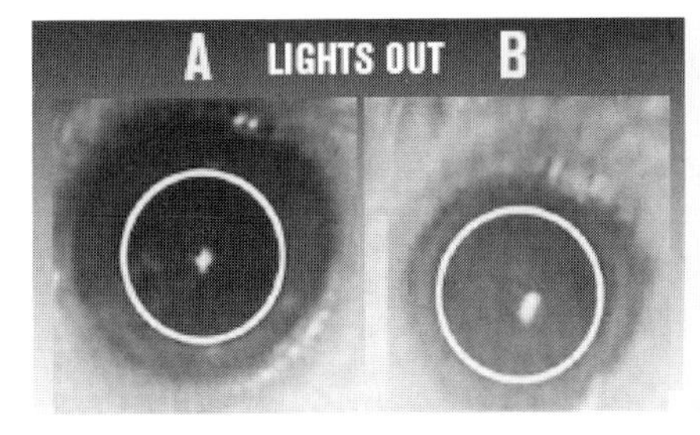

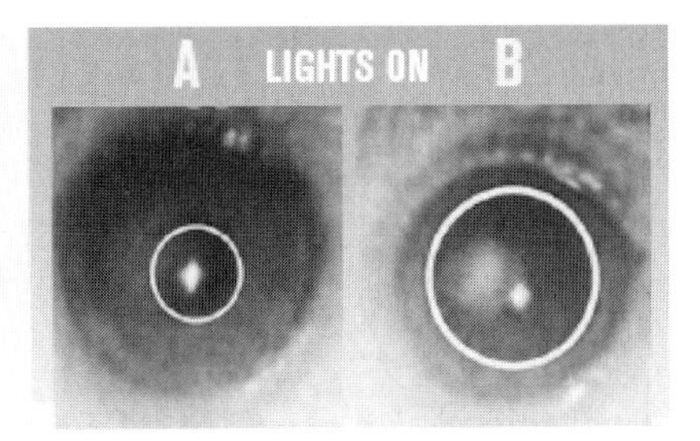

Figure 12.5 Effect of cryptochrome mutations on pupillary light response of *rd/rd* mouse. Still frames of digital infrared videography of mouse pupils are shown after 30 min dark adaptation (left, "Lights Out") and 30 s following illumination with 1×10^{13} photons cm^{-2} s^{-1} of 470 nm narrow bandpass filtered light. Pupillary margin has been manually highlighted in each case; white spot cornea is reflection of infrared light source. A = rd/rd; B = rd/rd; mCry1$^{-/-}$; mCry2$^{-/-}$.

If cryptochromes are not necessary for either c-fos induction in the SCN or the pupillary light response of *rd/rd* animals under bright light conditions, what is? A second candidate photopigment expressed in the inner retina, melanopsin, is the leading candidate. Melanopsin was originally identified in frog dermal melanophores as an opsin-like molecule (Provencio et al., 1998). Provencio et al. cloned a mammalian homolog and demonstrated that it was expressed nearly exclusively a subset of retinal ganglion cells (Provencio et al., 2000). Subsequent work has shown that the menalopsin-expressing cells have substantial overlap with the intrinsically photosensitive retinal ganglion cells (Hattar et al., 2002; Provencio 2002).

Several groups successfully knocked out melanopsin (Panda et al., 2002; Ruby *et al.*, 2002; Lucas et al., 2003). Interestingly, the melanopsin knockout had a minimal phenotype, showing mildly reduced amplitudes of circadian phase shifts following short light exposures (Panda et al., 2002; Ruby et al., 2002), and showing a slight reduction in pupillary light responsiveness at high irradiance levels (Lucas et al., 2003). Immediate-early *c-fos* induction in the SCN was not attenuated in these animals (Ruby et al., 2002). When the melanopsin knockout alleles were compounded with mutations in the classical photoreceptors, however, very strong phenotypes emerged. Compounding the melanopsin$^{-/-}$ allele with either *rd/rd* (Panda et al., 2003) or with independent loss of rod and cone function genes (*Cgna*$^{-/-}$ *and Pde*$^{-/-}$) (Hattar et al., 2003) resulted in mice with intact circadian clocks that were incapable of entrainment or masking, and that showed no pupillary light responsiveness.

The likely explanation for the additivity seen between outer retinal mutations and either melanopsin or cryptochrome knockout alleles is probably anatomically based. The intrinsically photosensitive retinal ganglion cells have dendritic arbors in the inner plexiform layer of the retina, and thus receive input from the outer retinal receptors. In the absence of either melanopsin or cryptochrome function, these cells can still serve as "conduits" for light information from the outer retina to reach the SCN and the centers for pupillary light responses (particularly the olivary pretectal nucleus [OPN]). Conversely, in the absence of outer retinal photoreceptors, intrinsically photosensitive retinal ganglion cells are sufficient – in a melanopsin- and partially cryptochrome-dependent manner – for phototransduction to these brain regions. Only when both systems are incapacitated is a strong "circadianly-blind" phenotype seen. This model would predict that ablation of the intrinsically photosensitive retinal ganglion cells themselves would have a strong phenotype.

Taken together, the genetic approach to understanding nonvisual photoreception would suggest the following:

1. Either outer retinal photoreception (mediated by rods and/or cones) or inner retinal photoresponses (mediated by the intrinsically photosensitive retinal ganglion cells) are sufficient for photons reaching the eye to be transduced to signals reaching the nonvisual centers of the brain controlling circadian rhythms and pupillary light reception.
2. Melanopsin is required for inner retinal photoreception, while cryptochrome is necessary only for full amplitude of the response
3. Cryptochrome function is not sufficient to rescue photic responses in mice with both outer retinal degeneration and loss of melanopsin function.

A parsimonious model explaining this might posit that melanopsin is the primary photopigment in the intrinsically photosensitive retinal ganglion cells, while cryptochromes act downstream to amplify the melanopsin signal. However, several predictions made by this model do not appear to be correct. If melanopsin is a photopigment, its chromophore should be a retinal, such as 11-*cis* retinaldehyde. One would predict that this photopigment would require dietary vitamin A for its synthesis, and therefore that dietary depletion of vitamin A should lead to loss of its function. While it is impossible to deplete a mouse of vitamin A completely by diet (due to hepatic stores), mobilization of hepatic retinol to the eye can be blocked by inactivation of the serum retinol-binding protein gene (*rbp*). *rbp*$^{-/-}$ mice that are deprived of dietary vitamin A lose outer retinal function (as measured by electroretinography) within three months (Quadro et al., 1999). After ten months of depletion, most mice have no detectable retinol in the outer retina. However, light-induced immediate-early gene induction in the SCN of these mice remains intact (Thompson et al., 2001). Similarly, reconstitution of 11-*cis* retinal (the chromophore for all known vertebrate opsins) requires the activity of the lecithin-retinol acyl transferase gene (LRAT). Mice lacking this gene activity also lose outer retinal function as measured by electroretinography, but these animals retain pupillary light responses to very bright lights, as do mice mutant in the RPE65 gene (which is also required for normal retinaldehyde reisomerization) (Batten et al., 2004). A further inconsistency in this model relates to the matching of action spectrum and absorption spectrum. Both circadian phase shifting in *rd/d;cl* mice and pupillary light responses have similar action spectra (Hattar et al., 2003), (Lucas et al., 2001), with peaks at ~480 nm, and a shape that can be fitted with an opsin template. However, heterologously expressed melanopsin has an absorption spectrum centered at 424 nm (Newman et al., 2003). Whether this reflects a fundamental difficulty with the hypothesis that melanopsin serves as primary photoreceptor, or with misfolding or absence of correct post-translational modifications in heterologously-expressed melanopsin is presently not known. It is important to remember that genetic experiments cannot prove that a particular protein is serving as photopigment in a system; one can only say that melanopsin apoprotein is necessary for inner retinal photoreception in outer retinal degenerate mice.

12.6 Cryptochromes of Other Animals

Like *Drosophila*, zebrafish (*Danio rerio*) have cell-autonomous circadian oscillators. Circadian rhythms of individual zebrafish organs can be entrained by light–dark cycles (Whitmore et al., 2000), and at least one embryonic cell primary cell line also shows light entrainable rhythms in vitro (Pando et al., 2001). Zebrafish have seven cryptochrome family members (in addition to *bona fide* photolyase and 6–4 photolyase) (Kobayashi et al., 2000). Interestingly, three of the zebrafish cryptochromes strongly resemble the *Drosophila* cryptochromes, while the other three resemble the mammalian cryptochromes. These latter cryptochromes appear to be able to repress Clock:Bmal-mediated transcription (akin to their mammalian homologs) (Kobayashi et al., 2000). The seventh zebrafish cryptochrome is similar to the *Arabidopsis* Cry-DASH (Cry3) and some bacterial Crys (Brudler et al., 2003).

The action spectrum for light-induction of the zPer2 gene in the Z3 embryonic cell line has been determined (Cermakian et al., 2002). The spectrum was found to be maximally sensitive to blue and ultraviolet light, with a peak sensitivity at ~380 nm. The shape of the curve was not consistent with an opsin chromophore. Additionally, HPLC measurement of 11-cis and all-trans retinal failed to detect these chromophores in the Z3 cell line. Although zebrafish do express a melanopsin homolog, its expression appears to be limited to the eye (Bellingham et al., 2002). The Z3 cells do express numerous cryptochrome homologs; in the absence of loss-of-function data, however, it is impossible to ascribe this response to cryptochrome action.

Xenopus laevis expresses at least four cryptochrome genes (Zhu and Green 2001), one homologous to *mCry1* and two homologous to *mCry2*. These genes are expressed rhythmically in the frog's eyes, and are capable of suppressing Clock:Bmal1-driven gene expression in a light-independent manner in cell culture (Zhu and Green 2001). To date, potential photoreceptive roles for these molecules have not been assessed. Some of the biochemistry of their transcriptional repressor activity has been assayed, however, and reveals differences in the necessity of the flavin (with xCry1 still repressing transcription despite mutation of most of the conserved flavin-binding residues, but xCry2s sensitive to any change in flavin binding residues) (Zhu and Green 2001). Interestingly, the C-terminus of cryptochrome doesn't seem to be important for actual repressive activity in *Xenopus,* but contains an essential nuclear localization domain.

12.7. Conclusions and Future Directions

The animal cryptochrome was first described as an EST photolyase ortholog in humans (Adams et al., 1995). Shortly afterwards, Hsu *et al* (1996) demonstrated that neither this nor a second photolyase-like gene they discovered in humans encodes photolyases, named these genes *hCRY1* and *hCRY2* and proposed that the human cryptochromes may function as circadian photoreceptors. This was the first suggestion implicating cryptochrome in the circadian clock in any organism. Subsequent research has confirmed that cryptochromes are in fact very important clock proteins. Ironically, however, while current evidence has unambiguously established cryptochromes as core clock proteins their photoreceptive function is a subject of considerable debate. This is mainly because the photochemical reaction carried out by cryptochrome is not known for mammalian cryptochromes or for cryptochromes from any other organism for that matter. Identification of the primary photochemical step carried out by cryptochrome would be necessary to prove formally that cryptochrome is a photosensory pigment in animals or even in plants where cryptochrome is taken for granted to be a photosensory pigment.

References

Adams, M.D., Kerlavage, A.R., Fleischmann, R.D., Fulder, R.A., Bult, C.S., et al. (1995) *Nature* **377**, 3–171.

Ahmad, M. and A. R. Cashmore (1993) *Nature* **366**(6451): 162–6.

Albrecht, U., B. Zheng, D. Larkin, Z. S. Sun and C. C. Lee (2001) *J. Biol. Rhythms* **16**(2): 100–4.

Bae, K., X. Jin, E. S. Maywood, M. H. Hastings, S. M. Reppert and D. R. Weaver (2001) *Neuron* **30**(2): 525–36.

Batten, M. L., Y. Imanishi, T. Maeda, D. C. Tu, A. R. Moise, D. Bronson, D. Possin, R. N. Van Gelder, W. Baehr and K. Palczewski (2004) *J Biol Chem* **279**(11): 10422–32.

Bellingham, J., D. Whitmore, A. R. Philp, D. J. Wells and R. G. Foster (2002) *Brain Res Mol Brain Res* **107**(2): 128–36.

Berson, D. M., F. A. Dunn and M. Takao (2002) *Science* **295**(5557): 1070–3.

Brautigam, C., and Deisenhofer, J. (2004) in press.

Briggs, W.R., and Christe, J.M. (2002) *Trends Plant Sci.* **7**, 204–210.

Brudler, R., Hitomi, K., Daiyasu, H., Toh, H., Kucho, K., Ishiura, M., Kanehisa, M., Roberts, V.A., Todo, T., Tainer, J.A., and Getzoff, E.D. (2003) *Mol. Cell.* **11**, 59–67.

Bunger, M.K., Wilsbacher, L.D., Moran, S.M., Clendenin, C., Radcliffe, L.A., Hogenesch, J.B., Simon, M.C., Takahashi, J.S., and Bradfield, C.A. (2000) *Cell* **103**, 1009–1017.

Ceriani, M. F., T. K. Darlington, D. Staknis, P. Mas, A. A. Petti, C. J. Weitz and S. A. Kay (1999) *Science* **285**(5427): 553–6.

Cermakian, N., M. P. Pando, C. L. Thompson, A. B. Pinchak, C. P. Selby, L. Gutierrez,

D. E. Wells, G. M. Cahill, A. Sancar and P. Sassone-Corsi (2002) *Curr Biol* **12**(10): 844–8.

Ebihara, S. and K. Tsuji (1980) *Physiol Behav* **24**(3): 523–7.

Emery, P., R. Stanewsky, J. C. Hall and M. Rosbash (2000) *Nature* **404**(6777): 456–7.

Emery, P., R. Stanewsky, C. Helfrich-Forster, M. Emery-Le, J. C. Hall and M. Rosbash (2000) *Neuron* **26**(2): 493–504.

Freedman, M. S., R. J. Lucas, B. Soni, M. von Schantz, M. Munoz, Z. David-Gray and R. Foster (1999) *Science* **284**(5413): 502–4.

Froy, O., D. C. Chang and S. M. Reppert (2002) *Curr Biol* **12**(2): 147–52.

Giebultowicz, J. M. and D. M. Hege (1997) *Nature* **386**(6626): 664.

Griffin, E. A., Jr., D. Staknis and C. J. Weitz (1999) *Science* **286**(5440): 768–71.

Hattar, S., H. W. Liao, M. Takao, D. M. Berson and K. W. Yau (2002) *Science* **295**(5557): 1065–70.

Hattar, S., R. J. Lucas, N. Mrosovsky, S. Thompson, R. H. Douglas, M. W. Hankins, J. Lem, M. Biel, F. Hofmann, R. G. Foster and K. W. Yau (2003) *Nature* **424**(6944): 75–81.

Hege, D. M., R. Stanewsky, J. C. Hall and J. M. Giebultowicz (1997) *J. Biol. Rhythms* **12**(4): 300–8.

Helfrich-Forster, C., C. Winter, A. Hofbauer, J. C. Hall and R. Stanewsky (2001) *Neuron* **30**(1): 249–61.

Herzog, E. D. and R. M. Huckfeldt (2003) *J Neurophysiol* **90**(2): 763–70.

Hoffman, P. D., A. Batschauer and J. B. Hays (1996) *Mol Gen Genet* **253**(1–2): 259–65.

Hsu, D. S., X. Zhao, S. Zhao, A. Kazantsev, R. P. Wang, T. Todo, Y. F. Wei and A. Sancar (1996) *Biochemistry* **35**(44): 13871–7.

Hunter-Ensor, M., A. Ousley and A. Sehgal (1996) *Cell* **84**(5): 677–85.

Keeler, C. E. (1927) *Amer J Physiol* **81**: 107–112.

Kim, S. T., K. Malhotra, H. Ryo, A. Sancar and T. Todo (1996) *Mutation Res* **363**(2): 97–104.

Kobayashi, Y., T. Ishikawa, J. Hirayama, H. Daiyasu, S. Kanai, H. Toh, I. Fukuda, T. Tsujimura, N. Terada, Y. Kamei, S. Yuba, S. Iwai and T. Todo (2000) *Genes Cells* **5**(9): 725–38.

Krishnan, B., S. E. Dryer and P. E. Hardin (1999) *Nature* **400**(6742): 375–8.

Krishnan, B., J. D. Levine, M. K. Lynch, H. B. Dowse, P. Funes, J. C. Hall, P. E. Hardin and S. E. Dryer (2001) *Nature* **411**(6835): 313–7.

Kume, K., M. J. Zylka, S. Sriram, L. P. Shearman, D. R. Weaver, X. Jin, E. S. Maywood, M. H. Hastings and S. M. Reppert (1999) *Cell* **98**(2): 193–205.

Lee, C., J. P. Etchegaray, F. R. Cagampang, A. S. Loudon and S. M. Reppert (2001) *Cell* **107**(7): 855–67.

Li, Y.F., Kim, S.T., and Sancar, A. (1993) *Proc. Natl. Acad. Sci. USA* **90**, 4389–4393.

Lin, C., Robertson, D.E., Ahmad, M., Raibekas, A.A., Jorns, M.S., Dutton, P.L., and Cashmore, A.R. (1995) *Science* **269**, 968–970.

Lucas, R. J., R. H. Douglas and R. G. Foster (2001) *Nature Neuroscience* **4**(6): 621–6.

Lucas, R. J., M. S. Freedman, M. Munoz, J. M. Garcia-Fernandez and R. G. Foster (1999) *Science* **284**(5413): 505–7.

Lucas, R. J., S. Hattar, M. Takao, D. M. Berson, R. G. Foster and K. W. Yau (2003) *Science* **299**(5604): 245–7.

Miyamoto, Y. and A. Sancar (1998) *Proc Natl Acad Sci U S A* **95**(11): 6097–102.

Miyamoto, Y. and A. Sancar (1999) *Brain Res Mol Brain Res* **71**(2): 238–43.

Mrosovsky, N. (1999) *Chronobiol Int* **16**(4): 415–29.

Naidoo, N., W. Song, M. Hunter-Ensor and A. Sehgal (1999) *Science* **285**(5434): 1737–41.

Nakamura, T. J., K. Shinohara, T. Funabashi, D. Mitsushima and F. Kimura (2001) *Neurosci Res* **41**(1): 25–32.

Newman, L. A., M. T. Walker, R. L. Brown, T. W. Cronin and P. R. Robinson (2003) *Biochemistry* **42**(44): 12734–8.

Özgür, S. and Sancar, A. (2003) *Biochemistry* **42**, 2926–2932.

Panda, S., I. Provencio, D. C. Tu, S. S. Pires, M. D. Rollag, A. M. Castrucci, M. T. Pletcher, T. K. Sato, T. Wiltshire, M. Andahazy, S. A. Kay, R. N. Van Gelder and J. B. Hogenesch (2003) *Science* **301**(5632): 525–7.

Panda, S., T. K. Sato, A. M. Castrucci, M. D. Rollag, W. J. DeGrip, J. B. Hogenesch, I. Provencio and S. A. Kay (2002) *Science* **298**(5601): 2213–6.

Pando, M. P., A. B. Pinchak, N. Cermakian and P. Sassone-Corsi (2001) *Proc. Natl. Acad. Sci. USA* **98**(18): 10178–83.

Park, H.W., Kim, S.T., Sancar, A., and Deisenhofer, J. (1995) Crystal structure of DNA photolyase from *Escherichia coli*. *Science* **268**, 1866–1872.

Plautz, J. D., M. Kaneko, J. C. Hall and S. A. Kay (1997) *Science* **278**(5343): 1632–5.
Provencio, I. (2002) *Nature* **415**(6871): 493.
Provencio, I., G. Jiang, W. J. De Grip, W. P. Hayes and M. D. Rollag (1998) *Proc. Natl. Acad. Sci. USA* **95**(1): 340–5.
Provencio, I., I. R. Rodriguez, G. Jiang, W. P. Hayes, E. F. Moreira and M. D. Rollag (2000) *J Neurosci* **20**(2): 600–5.
Quadro, L., W. S. Blaner, D. J. Salchow, S. Vogel, R. Piantedosi, P. Gouras, S. Freeman, M. P. Cosma, V. Colantuoni and M. E. Gottesman (1999) *EMBO J* **18**(17): 4633–44.
Renn, S. C., J. H. Park, M. Rosbash, J. C. Hall and P. H. Taghert (1999) [erratum appears in Cell 2000 Mar 31;101(1):following 113.]. *Cell* **99**(7): 791–802.
Reppert, S. M. and D. R. Weaver (2001) *Annu Rev Physiol* **63**: 647–76.
Rosato, E., V. Codd, G. Mazzotta, A. Piccin, M. Zordan, R. Costa and C. P. Kyriacou (2001) *Curr Biol* **11**(12): 909–17.
Ruby, N. F., T. J. Brennan, X. Xie, V. Cao, P. Franken, H. C. Heller and B. F. O'Hara (2002) *Science* **298**(5601): 2211–3.
Sancar, A. (2000) *Annu Rev Biochem* **69**: 31–67.
Sancar, A. (2003) *Chem Rev* **103**(6): 2203–37.
Sancar, G.B., Smith, F.W., and Heelis, P.F. (1987) *J. Biol. Chem.* **282**, 15457–15465.
Selby, C. P. and A. Sancar (1999) *Photochemistry and Photobiology* **69**(1): 105–7.
Selby, C. P., C. Thompson, T. M. Schmitz, R. N. Van Gelder and A. Sancar (2000) *Proc. Natl. Acad. Sci. U S A* **97**(26): 14697–702.
Stanewsky, R., M. Kaneko, P. Emery, B. Beretta, K. Wagersmith, S. A. Kay, M. Rosbash and J. C. Hall (1998) *Cell* **95**(5): 681–692.
Stephan, F. K. and I. Zucker (1972) *Proc. Natl. Acad. Sci. USA* **69**(6): 1583–6.
Thompson, C. L., W. S. Blaner, R. N. Van Gelder, K. Lai, L. Quadro, V. Colantuoni, M. E. Gottesman and A. Sancar (2001) *Proc Natl Acad Sci U S A* **98**: 11708–11713.
Thompson, C. L., C. B. Rickman, S. J. Shaw, J. N. Ebright, U. Kelly, A. Sancar and D. W. Rickman (2003) *Invest Ophthalmol Vis Sci* **44**(10): 4515–21.
Thresher, R. J., M. H. Vitaterna, Y. Miyamoto, A. Kazantsev, D. S. Hsu, C. Petit, C. P. Selby, L. Dawut, O. Smithies, J. S. Takahashi and A. Sancar (1998) *Science* **282**(5393): 1490–4.
van der Horst, G. T., M. Muijtjens, K. Kobayashi, R. Takano, S. Kanno, M. Takao, J. de Wit, A. Verkerk, A. P. Eker, D. van Leenen, R. Buijs, D. Bootsma, J. H. Hoeijmakers and A. Yasui (1999) *Nature* **398**(6728): 627–30.
Van Gelder, R. N. (2001) *Ophthalmic Genetics* **22**(4): 195–205.
Van Gelder, R. N. (2003) *Sci STKE* **2003**(209): tr6.
Van Gelder, R. N., T. M. Gibler, D. Tu, K. Embry, C. P. Selby, C. L. Thompson and A. Sancar (2002) *J Neurogen* **16**, 181–203.
Van Gelder, R. N., E. D. Herzog, W. J. Schwartz and P. H. Taghert (2003) *Science* **300**(5625): 1534–5.
Van Gelder, R. N., R. Wee, J. A. Lee and D. C. Tu (2003) *Science* **299**(5604): 222.
Vitaterna, M. H., C. P. Selby, T. Todo, H. Niwa, C. Thompson, E. M. Fruechte, K. Hitomi, R. J. Thresher, T. Ishikawa, J. Miyazaki, J. S. Takahashi and A. Sancar (1999) *Proc Natl Acad Sci U S A* **96**(21): 12114–9.
Wang, H., L. G. Ma, J. M. Li, H. Y. Zhao and X. W. Deng (2001) *Science* **294**(5540): 154–8.
Weaver, D. R. (1998) *J. Biol. Rhythms* **13**(2): 100–12.
Wee, R., A. M. Castrucci, I. Provencio, L. Gan and R. N. Van Gelder (2002) *J Neurosci* **22**: 10427–10433.
Wheeler, D. A., M. J. Hamblen-Coyle, M. S. Dushay and J. C. Hall (1993) *J. Biol. Rhythms* **8**(1): 67–94.
Whitmore, D., N. S. Foulkes and P. Sassone-Corsi (2000) *Nature* **404**(6773): 87–91.
Worthington, E.N., Kavakli, I.H., Berrocal-Tito, G., Bondo, B.E., and Sancar, A. (2003) *J. Biol. Chem.* **278**, 39143–39154.
Yamaguchi, S., H. Isejima, T. Matsuo, R. Okura, K. Yagita, M. Kobayashi and H. Okamura (2003) *Science* **302**(5649): 1408–12.
Yamamoto, K., T. Okano and Y. Fukada (2001 *Neurosci Lett* **313**(1–2): 13–6.
Yang, Z., M. Emerson, H. S. Su and A. Sehgal (1998) *Neuron* **21**(1): 215–23.
Zhao, X., Liu, J., Hsu, D.S., Taylor, J.S., and Sancar, A. (1997) *J. Biol. Chem.* **272**, 28971–28979.
Zheng, B., D. W. Larkin, U. Albrecht, Z. S. Sun, M. Sage, G. Eichele, C. C. Lee and A. Bradley (1999 *Nature* **400**(6740): 169–73.
Zhu, H. and C. B. Green (2001) *Curr Biol* **11**: 1945–1949.
Zhu, H. and C. B. Green (2001) *Mol Vis* **7**: 210–5.
Zimmerman, W. F. and T. H. Goldsmith (1971) *Science* **171**(976): 1167–9.

13
Blue Light Sensing and Signaling by the Phototropins

John M. Christie and Winslow R. Briggs

13.1
Introduction

Plants are dependent on sunlight for photosynthesis. As a consequence, plants have the ability to sample the surrounding light environment and use this information to optimize their growth and development. In particular, wavelengths in the blue/UV-A region of the electromagnetic spectrum (from 320 to 500 nm) act to control a wide range of plant responses (Briggs and Huala, 1999; Lin, 2002). Some of these serve to maximize photosynthetic potential in weak light, while others act to prevent damage to the photosynthetic apparatus in excess light. These responses include phototropism, light-induced stomatal opening, and chloroplast movement in response to changes in light intensity (Christie and Briggs, 2001; Briggs and Christie, 2002; Kagawa and Wada, 2002; Liscum, 2002; Kagawa, 2003; Wada et al., 2003).

The phototropins (phot1 and phot2) are flavoprotein photoreceptors that mediate the three above-mentioned responses in *Arabidopsis thaliana*. In addition, the phototropins have been shown to control other blue light-activated processes, including leaf expansion (Sakamoto and Briggs, 2002) and the rapid inhibition of hypocotyl elongation in dark-grown seedlings (Folta and Spalding, 2001). Hence, phot1 and phot2 function to regulate a number of photoresponses in plants besides phototropism, after which they were originally named (Christie et al., 1999; Briggs et al., 2001a).

Our knowledge of the phototropins and how these photoreceptors function has increased dramatically since the isolation of the first phototropin gene back in 1997 (Huala et al., 1997). In this chapter, we focus on some of the recent advances relating to phototropins with respect to their structural and biochemical properties. We also describe our current understanding of the signaling events that follow phototropin excitation and how these events may be linked to particular phototropin-mediated responses. The biophysical properties of the phototropins and their chromophore-binding domains are covered in more detail elsewhere in this book (Swartz and Bogomolni, Chapter 14), as are structural studies of these domains (Crosson, Chapter 15).

Handbook of Photosensory Receptors. Edited by W. R. Briggs, J. L. Spudich

ISBN 3-527-31019-3

13.2 Phototropin Structure and Function

13.2.1 Discovery of Phototropin

Much of our understanding of phototropin photoreceptors has come from genetic analysis using the model plant *Arabidopsis thaliana*. *Arabidopsis*, or thale cress as it is more commonly known, is not the most exciting plant to look at, but its small size and short life cycle combined with its plentiful seed production make it an ideal genetic tool for laboratory work. More importantly, *Arabidopsis* can be easily manipulated to generate mutants that show altered characteristics. And it was the isolation of *Arabidopsis* mutants altered in phototropism that eventually led to the cloning and characterization of the first phototropin gene.

The ***n***on-***p***hototropic ***h***ypocotyl (*nph*) mutants of *Arabidopsis* show impaired stem (or hypocotyl) phototropism to low intensities of unilateral blue light (Liscum and Briggs, 1995). In particular, one class of *nph* mutants, the *nph1* mutant, was found to lack the activity of a plasma membrane-associated protein that becomes phosphorylated upon irradiation with blue light (Figure 13.1). The encoded protein, originally designated NPH1, was therefore hypothesized to represent a phototropic receptor that undergoes autophosphorylation in response to blue light (Liscum and Briggs, 1995). Later experiments confirmed this hypothesis and the NPH1 protein was renamed phototropin 1 (phot1) after its functional role in phototropism (Christie et al., 1999).

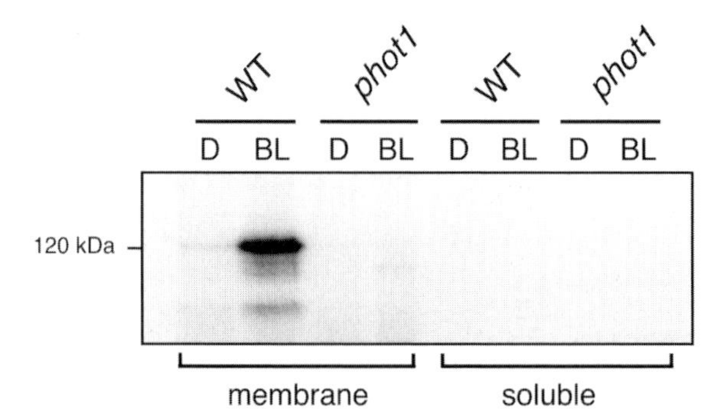

Figure 13.1 Blue light-induced kinase activity of phot1 in *Arabidopsis* membranes. Autoradiograph showing autophosphorylation of phot1 in membrane protein extracts prepared from 3-day-old dark-grown *Arabidopsis* seedlings. Protein extracts were prepared under a red safe light and given a mock irradiation (D for dark) or a pulse of blue light (BL) prior to the addition of radio-labeled ATP. In membrane extracts prepared from wild-type seedlings (WT), phot1 undergoes autophosphorylation in response to blue light. This response is lacking in the *phot1* null mutant (previously *nph1*). No phot1 kinase activity is detectable in soluble protein extracts from wild-type seedlings, indicating that phot1 is membrane associated.

13.2.2
Phot1: a Blue Light-activated Receptor Kinase

The *NPH1* gene, or *PHOT1* gene as it will now be called, encodes a serine/threonine protein kinase (Figure 13.2 A). The kinase domain of phot1 is located at its C terminus and contains all 11 conserved subdomains typical of protein kinases (Hanks and Hunter, 1995). When expressed in insect cells, phot1 undergoes autophosphorylation in response to blue light irradiation (Figure 13.2 B) implying that the heterologously expressed protein is a functional photoreceptor kinase (Christie et al., 1998). Indeed, mutation of an essential aspartate residue within subdomain VII of the phot1 kinase domain results in a loss of light-dependent autophosphorylation in the insect cell system, demonstrating that light-dependent phosphorylation of phot1 is mediated by phot1 itself and not some other kinase present in insect cells (Christie et al., 2002). Furthermore, recombinant phot1 binds the vitamin-B-related chromophore flavin

A

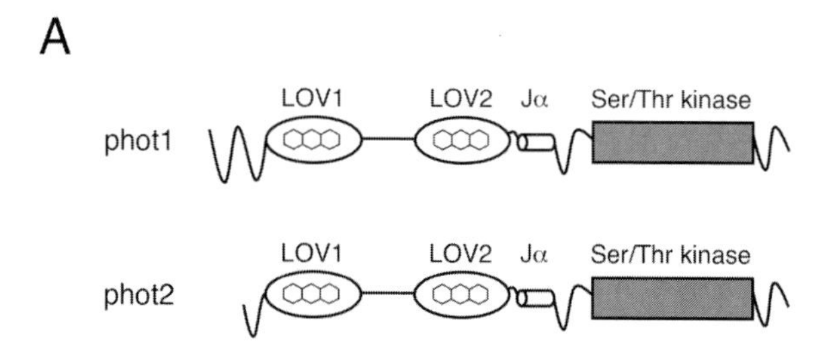

B

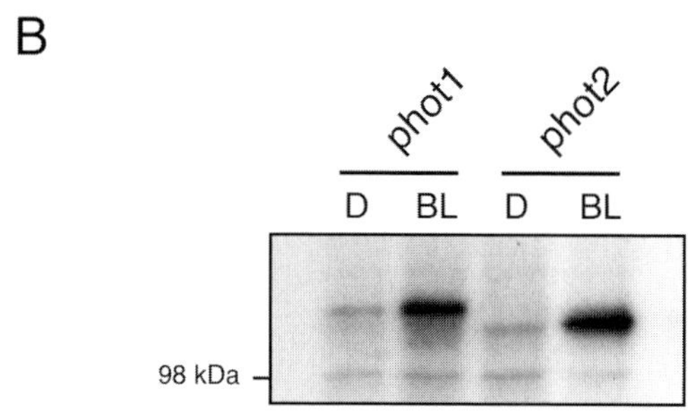

Figure 13.2 Protein structures and kinase activites of *Arabidopsis* phot1 and phot2. (A) Schematic drawing of *Arabidopsis* phot1 and phot2 (996 and 915 amino acids respectively). The light-sensing LOV domains, each of which binds FMN are indicated. The serine/threonine kinase domains are shown in grey. A conserved α-helix at the C-terminal position of LOV2 (Jα) is also shown. Displacement of this helix in response LOV2 photoexcitation has been proposed to result in activation of the C-terminal kinase domain (Harper et al., 2003). (B) Blue light-induced autophosphorylation activity of phot1 and phot2 expressed in insect cells. Insect cells expressing either phot1 or phot2 were grown in complete darkness for 3 days. Soluble protein extracts were isolated under a red safe light. Protein samples were given either a mock irradiation (D) or irradiated with blue light (BL) prior to the addition of radio-labeled ATP. Samples were resolved by protein gel electrophoresis and exposed to autoradiography.

monocleotide (FMN) and displays spectral characteristics consistent with the properties of a photoreceptor for phototropism (Christie et al., 1998). Further details of the biochemical and photophysiological properties of the light-activated kinase reaction have been presented elsewhere (Briggs et al., 2001b).

The N-terminal region of phot1 contains a repeated motif of 110 amino acids that belongs to the large and diverse superfamily of **P**ER/**A**RNT/**S**IM (PAS) domains (Taylor and Zhulin, 1999). PAS domains are found in proteins throughout nature and are often associated with cofactor binding. However, the PAS domains of phot1 are most closely related to a subset of proteins within the PAS-domain superfamily that are regulated by external signals such as **l**ight, **o**xygen or **v**oltage. Thus, the two PAS domains of phot1, with approximately 40% amino acid identity, were assigned the acronym LOV and named LOV1 and LOV2 respectively (Huala et al., 1997). As will be discussed below, both LOV domains of phot1 and phot2 serve as binding sites for the chromophore FMN and are directly involved in light sensing (Christie et al., 1999; Salomon et al., 2000).

13.2.3
Phot2: a Second Phototropic Receptor

Homologs of phot1 have been identified in several plant species including oat, maize, rice, pea, and ferns (for the latter, see Nozue et al., 2000; Kagawa et al., 2004), and more recently the unicellular green alga *Chlamydomonas rheinhardti* (Briggs et al., 2001a; Huang et al., 2002; Kasahara et al., 2002b). A second member of the *Arabidopsis* phototropin family, phot2, shows considerable homology to phot1 (Figure 13.2 A) (Jarillo et al., 1998). Like phot1, phot2 contains two FMN-binding LOV domains and a kinase domain, and undergoes autophosphorylation in response to blue light (Figure 13.2 B; Sakai et al., 2001; Christie et al., 2002). Both phot1 and phot2 contain a conserved α-helix, designated Jα, located at the C-terminal region of LOV2 (Figure 13.2 A). This helical region has been proposed to couple photoexcitation of LOV2 to activation of the C-terminal kinase domain (Harper et al., 2003) and will be discussed in more detail later (Section 13.4.1).

While *phot1* mutants show impaired hypocotyl phototropism to low intensities of blue light (Liscum and Briggs, 1995), they retain phototropic responsiveness to high-intensity blue light (Sakai et al., 2000; Sakai et al., 2001), demonstrating the presence of a second phototropic photoreceptor in *Arabidopsis*. The curvature response to high light intensities is severely impaired in the *phot1phot2* double mutant (Sakai et al., 2001). Thus, both phot1 and phot2 mediate phototropism, but phot2 functions only under high light intensities. The functional activity of phot2 under high light intensities most likely requires differential gene expression; *PHOT2* gene expression in dark-grown seedlings is induced by light (Jarillo et al., 2001b; Kagawa et al., 2001) and is mediated by the red/far-red light receptor phytochrome A (phyA; Tepperman et al., 2001). In contrast to phot2, phot1 appears to be constitutively expressed and functions as the primary phototropic receptor in etiolated *Arabidopsis* because *phot2* single mutants exhibit normal phototropic curvature to both low and high intensities of blue light (Sakai et al., 2001), indicating that phot1 functions over a broad range of light conditions.

13.2.4
Phototropins: Photoreceptors for Movement and More

Phototropism is not the only form of photomovement found in plants. Photomovement can also occur at the cellular level. The opening of stomata (pores in the epidermis) in response to blue light allows the plant to regulate CO_2 uptake for photosynthesis. Recent genetic studies have shown that this response is controlled redundantly by phot1 and phot2 (Kinoshita et al., 2001). But unlike the case for phototropism, phot1 and phot2 contribute equally to stomatal opening by working across the same light-intensity range.

In addition to the cellular level, photomovement in plants can also occur at the subcellular level. For instance, chloroplasts move in response to changes in blue light intensity (Kagawa, 2003; Wada et al., 2003). Like phototropism and stomatal opening, chloroplast movement is a process that helps to regulate the photosynthetic efficiency of a plant. At low light intensities, chloroplasts redistribute themselves within the cell to maximize light capture for photosynthesis. This process is often referred to as the low-light accumulation response. On the contrary, chloroplasts under high-light conditions reorganize themselves to avoid the potentially damaging effects of excess light (Kasahara et al., 2002a) by a process known as the high-light avoidance response. Genetic analysis with phototropin-deficient mutants has shown that phot1 and phot2 overlap in function to control the low-light accumulation response (Sakai et al., 2001). Yet, the high light avoidance response is controlled exclusively by phot2 (Jarillo et al., 2001b; Kagawa et al., 2001).

It is now apparent through the study of phototropin-deficient mutants that the phototropins are not only involved in controlling phototropism, stomatal opening and chloroplast movement in *Arabidopsis*. These photoreceptors are now associated with controlling other growth responses to blue light as well as phototropism, including leaf expansion (Sakamoto and Briggs, 2002) which is partially induced by blue light (van Volkenburgh, 1999). The leaves of some plant species can follow the position of the sun throughout the day, in a process known as solar tracking. Solar tracking is regulated by blue light (Yin, 1938) and it is postulated that the phototropins may also function to control this photomovement response (Briggs and Christie, 2002).

It is worth noting that phot1 and phot2 not only overlap in function, but also exhibit distinct functional roles: phot2 acts as the sole photoreceptor for the chloroplast high-light avoidance response (Jarillo et al. 2001b; Kagawa *et al.* 2001), and phot1 alone has been shown to mediate the rapid inhibition of hypocotyl elongation by blue light (Folta and Spalding, 2001). More recently, phot1 activity has been reported to control the destabilization of specific nuclear and chloroplast transcripts in *Arabidopsis*. Both *Lhcb* and *rbcL* transcripts, although transcribed in different subcellular compartments, are destabilized by high-intensity blue light (Folta and Kaufman, 2003). The blue light-dependent destabilization of these transcripts is impaired in *phot1* mutants, demonstrating a role for phot1 in regulating the level of chloroplast and nuclear-encoded transcripts during plant growth and development. A summary of all known phototropin-mediated responses identified to date is shown in Figure 13.3.

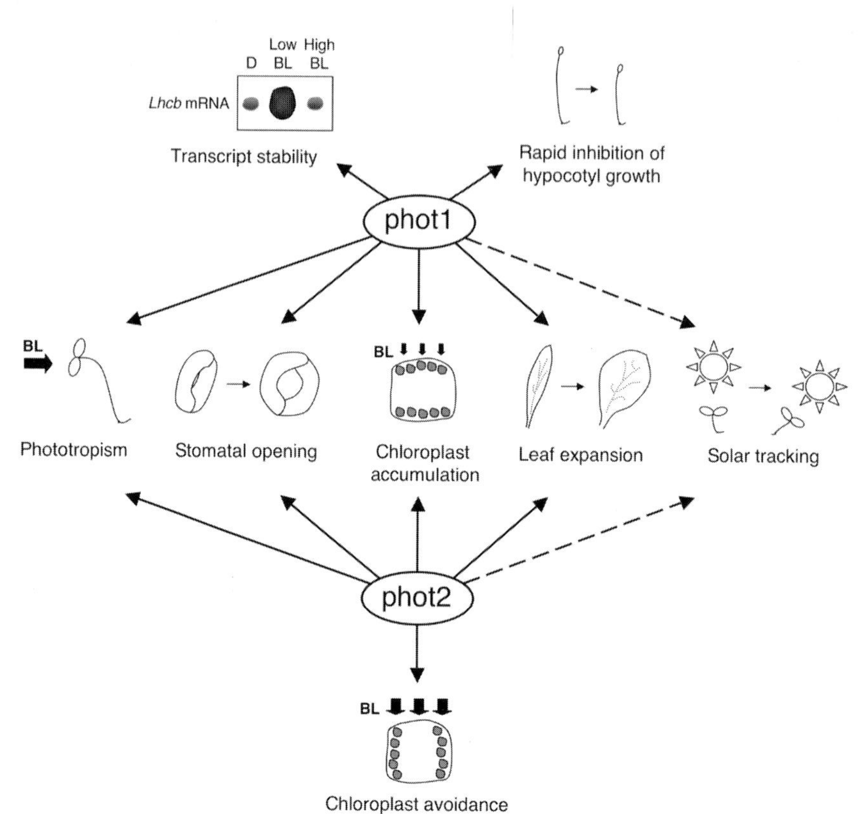

Figure 13.3 Range of physiological and biological responses mediated by phot1 and phot2. When phot1 and phot2 are activated by blue light (BL) they overlap in function to mediate several responses. These are shown in the center and include, from left to right: phototropism, stomatal opening, chloroplast accumulation movement, leaf expansion and possibly solar tracking. By contrast, the chloroplast avoidance movement is only controlled by phot2. Similarly, phot1 alone plays a role in mediating the rapid inhibition of hypocotyl elongation by blue light and the destabilization of *Lhcb* (and *rbcS*) transcripts in response to high intensity blue light treatment.

A major task for researchers in the future will be to unravel the signaling events that couple photoreceptor activation to specific blue light-induced responses, such as phototropism and chloroplast movement. Yet significant progress in this area has already begun. Before we address what is known regarding these downstream events, we briefly review our current knowledge of the photochemical and biochemical mechanisms associated with phototropin receptor activation by blue light.

13.2.5 Overview of Phototropin Activation

Both phot1 and phot2 are hydrophilic proteins but have been shown to localize to, and co-purify with the plasma membrane in *Arabidopsis* (Sakamoto & Briggs, 2002; Harada et al., 2003) and other plant species (Briggs et al., 2001b). Although the nature of their association with the plasma membrane is unknown, it is likely that phot1 and phot2 undergo post-translational modification and/or bind a protein cofactor to facilitate membrane interaction.

For convenience, phototropin activation can be viewed as a series of events beginning with the absorption of blue light by the LOV domains (Figure 13.4 A). In the dark or ground state, the phototropin receptor remains unphosphorylated and inactive. Upon illumination, light sensing by LOV2 is considered to result in a protein conformational change that involves a highly conserved α-helical region, designated Jα (Figure 13.4 B). Evidence for this light-induced protein structural change will be discussed later (Section 13.4.1). Although the exact role of LOV1 in regulating phototropin activity is not fully understood (see Section 13.3.3), the displacement of Jα in response to LOV2 photoexcitation is hypothesized to lead to activation of the C-terminal kinase domain, which in turn results in autophosphorylation of the photoreceptor protein (Figure 13.4 C). At least for phot1, autophosphorylation has been shown to occur on multiple serine residues (Palmer et al., 1993; Short et al., 1994; Salomon et al., 1996).

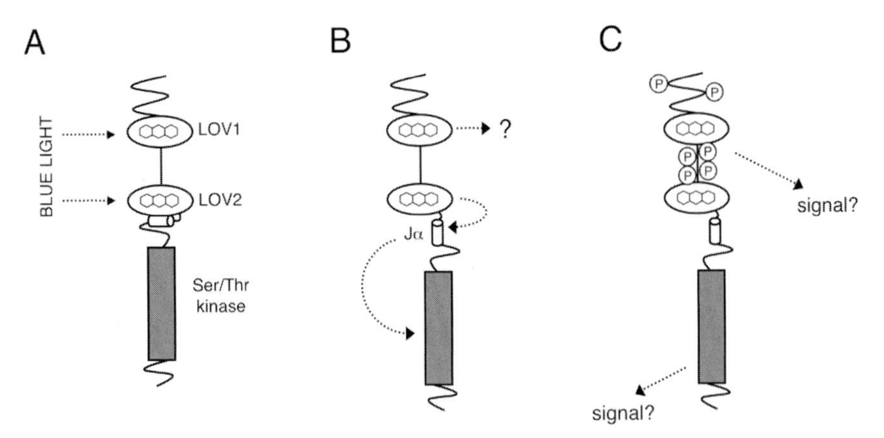

Figure 13.4 Overview of phototropin activation by blue light. (A) In the dark or ground state, the phototropin receptor is unphosphorylated and inactive. Upon irradiation, the LOV domains detect blue light. (B) Photoexcitation of LOV2 results in a protein conformational change that involves the displacement of a highly conserved α-helix from the surface of LOV2 (Jα). This protein structural change is hypothesized to lead to activation of the C-terminal kinase domain. The function of LOV1 is presently unknown. (C) Activation of the kinase domain consequently leads to autophosphorylation of the photoreceptor protein. It is unknown whether autophosphorylation is involved in receptor signaling or whether phototropin initiates signaling via substrate phosphorylation.

The above description outlining the photochemical and biochemical mechanisms associated with phototropin receptor activation is clearly oversimplified. In the following sections, we review in more detail what is known regarding each of these reaction steps, but it is important to note that there are many aspects associated with this light-driven molecular switch that remain to be addressed. For example, it is presently unknown whether light-mediated autophosphorylation of the phototropins plays a role in receptor signaling or is involved in some other function, say receptor desensitization, or both. It is also unknown whether phot1 and phot2 can phosphorylate an interacting substrate. Nevertheless, mutant alleles of *PHOT1* and *PHOT2* carrying single amino acid substitutions in the kinase domain have been identified indicating that kinase activity is essential for signaling (Huala et al., 1997; Kagawa et al., 2001). We will discuss the subject of phototropin kinase activation in more detail after addressing the photochemical mechanisms associated with light sensing by the phototropin LOV domains.

13.3
LOV Domain Structure and Function

13.3.1
Light Sensing by the LOV Domains

A detailed account of the photochemical properties of the phototropin LOV domains is described elsewhere in this book (Swartz and Bogomolni, Chapter 14) as is a detailed account of LOV-domain structure (Crosson, Chapter 15). Therefore, related aspects of the phototropin LOV domains that have been covered already will be mentioned only briefly.

The LOV1 and LOV2 domains from several phototropin proteins have been produced, either singly or in tandem, in *Escherichia coli* (Christie et al., 1999; Kasahara et al., 2002b). LOV-domain fusion proteins purified from *E. coli* are highly fluorescent and function as binding sites for the chromophore FMN (Christie et al., 1999). FMN binding is non-covalent and each LOV domain binds one molecule of FMN. The spectral properties of recombinant LOV1 and LOV2 fusion proteins are similar to those of the full-length photoreceptor protein expressed in insect cells (Christie et al., 1998; Kasahara et al., 2002b), showing absorption in the blue/UV-A regions of the spectrum that closely matches the action spectrum for phototropism (Baskin and Iino, 1987).

Salomon et al. (2000) were the first to show that LOV domains expressed and purified from *E. coli* function as light sensors. When irradiated with blue light, both LOV1 and LOV2 undergo a complex spectral change (Figure 13.5), characteristic of that known to accompany the formation of a stable bond between the C(4a) carbon of the flavin isoalloxazine ring and a cysteine residue within the flavoprotein (Miller et al., 1990). These light-induced spectral changes are fully reversible in darkness, indicating that the photoproduct generated has the ability to regenerate back to the initial ground state.

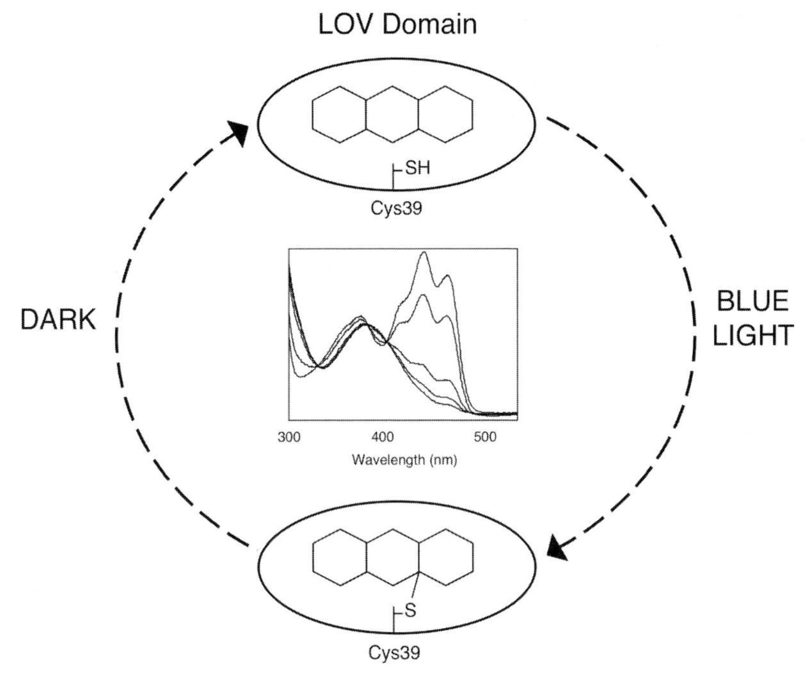

Figure 13.5 Schematic representation of LOV-domain photochemistry. Blue light drives the formation of a covalent bond between the FMN chromophore and a conserved cysteine residue within the LOV domain (designated Cys39). This process is self-contained and is fully reversible in darkness. The graph (centre) illustrates the light-induced absorbance changes observed for the LOV2 domain of phot1 over several seconds. The uppermost spectrum at 450 nm represents the protein sample in the dark state. The sample is then irradiated with blue light and the absorption spectra recorded at 1-s intervals. The light-induced absorbance changes detected are characteristic of the formation of a flavin-cysteinyl adduct. Adapted from Salomon et al. (2000).

All phototropin LOV domains identified to date contain a cysteine residue situated within a highly conserved motif, GRNCRFLQ (Crosson et al., 2003). This cysteine corresponds to residue 39 within the 110 amino-acid stretch of LOV1 and LOV2, and has been designated Cys39 for convenience (Salomon et al., 2000; Swartz et al., 2001; Kasahara et al., 2002b; Christie et al., 2002). Replacement of Cys39 with either alanine or serine results in a loss of photochemical reactivity of LOV1 and LOV2 (Salomon et al., 2000), implying that the photochemical reaction involved requires the formation of a covalent bond between the FMN chromophore and the side chain of Cys39 (Figure 13.5). Formation of this bond, known as a flavin-cysteinyl adduct, has subsequently been confirmed by a plethora of biophysical studies (Salomon et al., 2001; Swartz et al., 2001; Holzer et al., 2002; Iwata et al., 2002; Ataka et al., 2003; Kottke et al., 2003; Kennis et al., 2003) and protein crystallography (Crosson and Moffat, 2001, 2002; Federov et al., 2003). Indeed, the crystal structures of LOV1 and LOV2 provide a snapshot of the domains at the atomic level. Both domains are very similar in structure and resemble a molecular hand, holding the flavin chromophore tightly within its grasp with the sulfhydryl group of Cys39 located close to the C(4a) position of the FMN isoalloxazine ring. (For simplicity, the cysteine sulfur in Figure 13.5 is represented as a sulfhydryl group but there is the possibility that it may exist as a thiolate anion as well. Swartz and Bogomolni discuss this matter in detail in Chapter 14)

13.3.2
LOV is all Around

Crosson and Moffat (2001) were the first to obtain the crystal structure of a phototropin LOV domain. For their crystallographic studies, they used the LOV2 domain from a novel photoreceptor from the fern *Adiantum capillus-veneris*, designated phytochrome 3 (phy3), which is a chimera of the red/far-red light photoreceptor phytochrome and phototropin (Nozue et al., 1998; Nozue et al., 2000). Phy3 has been shown to be required for red light-activated phototropism and chloroplast relocation in *Adiantum*, both of which are regulated by red and blue light in this organism (Kawai et al., 2003).

The phy3 LOV2-domain structure closely resembles that of other PAS domains, including PYP, HERG, and FixL (Crosson and Moffat, 2001). Its protein fold comprises five anti-parallel β-sheets and four α-helices that form a central pocket to accommodate the FMN chromophore. The crystal structure of *Adiantum* phy3 LOV2 is very similar to that recently obtained for the LOV1 domain of *Chlamydomonas* phot (Federov et al., 2003). However, the side chain of Cys39 in *Chlamydomonas* phot LOV1 appears to exist in two conformations instead of one as was found in *Adiantum* phy3 LOV2. It should be noted that the LOV1 domain structure of *Chlamydomonas* phot was solved under liquid nitrogen temperatures. In contrast, the structure of *Adiantum* phy3 LOV2 was solved at room temperature, which may account for the difference observed. A more detailed description of LOV domain structure can be found elsewhere in this book (Crosson, Chapter 15).

From the structure of *Adiantum* phy3 LOV2, Crosson and Moffat (2001; 2002) were able to pinpoint 10 amino acid residues, in addition to Cys39, that come into contact with the FMN chromophore via hydrogen bonding or Van der Waals forces. Three other proteins containing a single LOV domain, with these exact same residues, have been found in *Arabidopsis* (Crosson et al., 2003): ZTL/ADO (Somers et al., 2000; Jarillo et al., 2001a), FKF1 (Nelson et al., 2000), and LKP2 (Schultz et al., 2001) all of which are associated with a regulation of the circadian clock. Mutants at the *ZTL/ADO* locus show a much-lengthened period for their circadian rhythms (Somers et al., 2000; Jarillo et al., 2001a), whereas overexpression of *LKP2* results in arrhythmic phenotypes for several circadian responses and causes a loss of the photoperiodic control of flowering (Schultz et al., 2001). Similarly, an absence of *FKF1* results in a late-flowering phenotype and alters the expression of clock-regulated genes (Nelson et al., 2000). Although none of these findings establish these encoded proteins as photoreceptors, the presence of a canonical, phototropin-like LOV domain in each of these proteins and their close interaction with both circadian responses and photoperiodism support this possibility. In fact, Imaizumi et al. (2003) have shown that the LOV domains of ZTL/ADO, FKF1 and LKP2 all bind FMN and undergo a blue light-activated photochemical reaction identical to that of the phototropin LOV domains (Salomon et al., 2000). Curiously, all three LOV domains fail to revert back to the dark state, in dramatic contrast to the phototropin LOV domains. Whether this failure to recover to the dark state is functionally relevant or physiologically meaningful remains to be determined, but demonstration of photochemical reactivity leading to the formation of a flavin-cysteinyl adduct provides strong support for the hypothesis that these proteins function as blue-light receptors in *Arabidopsis*. Schultz discusses these putative photoreceptors in the context of circadian rhythms and day-length responses in detail in Chapter 16.

Beside the LOV domain, ZTL/ADO, FKF1 and LKP2 share no structural homology with the phototropins. Rather they have an F-box (related to targeting other proteins for degradation) located at the C-terminal region of the protein followed by 6 kelch-domain repeats that form a propeller-like structure thought to be involved in protein-protein interactions. Indeed, ZTL has been reported to modulate circadian clock function by targeting TOC1, a key component of the circadian oscillator, for degradation (Mas et al., 2003). Interestingly, the LOV domains of *Arabidopsis* ZTL/ADO, FKF1, and LKP2, unlike the phototropin LOV domains, contain an additional amino acid insert of 9–11 residues hypothesized to accommodate a larger flavin cofactor such as FAD rather than FMN (Crosson et al., 2003). Nonetheless, as mentioned above, the LOV domains of ZTL/ADO, FKF1 and LKP2 have been reported to bind FMN (Imaizumi et al., 2003), at least when expressed in *E. coli*.

In addition to plants, single LOV-domain containing proteins have also been identified in fungi (Crosson et al., 2003). The LOV domain of WC-1, a photoreceptor for many but not all blue light responses in the filamentous fungus *Neurospora crassa* (Froehlich et al., 2002; He et al., 2002), binds flavin adenine dinucleotide (FAD). A second LOV-domain containing protein in *Neurospora*, VIVID (VVD), serves as the photoreceptor that allows the fungus to adapt to changes in light intensity (Schwerdtfeger and Linden, 2003). Like the LOV domains of ZTL/ADO, FKF1, and LKP2, the LOV domains of WC-1 and VVD contain an 11-amino acid insert that is not found in phototropin LOV domains. Other than the LOV domain, WC-1 and VVD do not share any sequence homology either with each other or with the phototropins. WC-1 is a zinc-finger protein (Ballario et al., 1996), whereas VVD is a relatively small cytosolic protein with no homology to proteins of known function (Heinzten et al., 2001).

When expressed and purified from *E. coli*, VVD binds a flavin chromophore that forms a flavin-cysteinyl adduct when irradiated with blue light (Schwerdtfeger and Linden, 2003). Like the phototropin LOV domains, the conserved photoactive cysteine is essential for the photochemical reactivity and function of VVD in *Neurospora*. Intriguingly, VVD is able to bind both FAD and FMN when expressed in *E. coli*. Whether VVD exhibits this flexibility in chromophore binding in *Neurospora* has yet to be determined. Nevertheless, Cheng *et al.* (2003) have demonstrated that the LOV domain of VVD can partially replace the function of the WC-1 LOV domain, suggesting that these domains are, at least in part, functionally interchangeable. Further information regarding these and other fungal photoreceptors is discussed elsewhere in this book (Dunlap and Loros, Chapter 18).

Single LOV domain-containing proteins have even been identified in bacteria (Crosson et al., 2003). These proteins typically contain a single LOV domain coupled to a specific output domain, such as a histidine kinase, a phosphodiesterase, or a STAS domain (Crosson et al., 2003). STAS domains are generally found in bacterial **s**ulfate **t**ransporters and **anti****s**igma factor antagonists (Aravind and Koonin, 2000). YtvA, a LOV-STAS protein from *Bacillus subtilis* has been shown to bind FMN and undergoe a blue light-activated photocycle analogous to that of the phototropin LOV domains (Losi et al., 2002). Though the presence of a LOV domain with its characteristic photocycle strongly implies that YtvA functions as a photoreceptor, the effects of blue light on *Bacillus subtilis*, a non-photosynthetic soil organism, are unknown. Nonetheless, the presence of LOV domain-containing proteins throughout various kingdoms of life indicate that this functional light-sensing mechanism is not just restricted to plants but has been conserved throughout evolution.

13.3.3
Are Two LOVs Better than One?

Although single LOV domain-containing proteins have been identified in plants, fungi and bacteria, the phototropins are the only proteins identified to date that possess two LOV domains (Briggs et al., 2001a). So why do phototropins possess two LOV domains? Are two LOV domains better than one? At present, the significance of two chromophore-binding sites within the phototropin molecule is not fully understood. Energy transfer occurring between LOV1 and LOV2 seems unlikely given that their respective chromophores are tightly bound within the LOV domain apoprotein (Crosson and Moffat, 2001; 2002; Federov et al., 2003). Similarly, the absorption maxima of LOV1 and LOV2 are only a few nm apart (Salomon et al., 2000), an arrangement that would not favour unidirectional energy transfer. Recent studies, however, have uncovered some insights into the functions of LOV1 and LOV2.

Christie et al. (2002) have used the Cys39Ala mutation, which blocks LOV domain photochemistry, to investigate the individual roles of LOV1 and LOV2 in regulating phototropin function in *Arabidopsis*. In brief, the photochemical activity of LOV2 is required to mediate phot1 kinase activity and to elicit phot1-mediated hypocotyl phototropism in response to low intensities of unilateral blue light. LOV1 photochemistry, on the other hand, plays at most a minor light-sensing role in regulating the photochemical reactivity of phot1 and is not sufficient to elicit phot1-mediated hypocotyl phototropism under low light conditions or phot1 kinase activity. Thus, at least for phototropism, LOV2 is essential for phot1 function in *Arabidopsis*. Similarly, initial studies suggest that phot2, like phot1, operates through a mechanism by which LOV2 acts as the principal light-sensing domain (Christie et al., 2002). Indeed, Kagawa et al. (2004) have recently shown that the LOV2 domain of phot2, in the absence of LOV1, is sufficient to mediate the chloroplast avoidance movement in *Adiantum*. It is worth noting, however, that the LOV1 domain of phot2 is still able to mediate a small degree of light-activated autophosphorylation (Christie et al., 2002). This situation is not apparent for phot1. It will now be important to establish whether the LOV1 and LOV2 domains of phot2 mediate autophosphorylation on the same or different amino acid residues.

Although the above findings demonstrate that LOV2 plays an important role in regulating phototropin activity, the exact role of LOV1 remains unclear. The LOV1 domain of oat phot1a has been reported to self-dimerize, whereas the LOV2 domain does not (Briggs et al., 2001b; Salomon et al., 2004). Likewise, the LOV domain of WC-1 from *Neurospora* has been shown to homodimerize *in vitro* (Ballario et al., 1998). LOV1 may therefore play a role in receptor dimerization. If so, receptor dimerization may be affected by light, and in turn, control the sensitivity of a phototropin-associated signaling complex (Liscum and Stowe-Evans, 2000). Alternatively, LOV1 may be involved in regulating phototropin-activated processes other than phot1-mediated phototropism and the phot2-induced chloroplast avoidance movement. Evidently, further work is needed to understand the exact role of LOV1 in controlling phototropin function and elucidate whether two LOVs are indeed better than one.

LOV1 and LOV2 have also been shown to exhibit different photosensitivities and reaction kinetics (Salomon et al., 2000; Kasahara et al., 2002b), consistent with their apparently distinct light-sensing roles in regulating photoreceptor function (Christie et al., 2002). Though studies of individual LOV domains have been instrumental in uncovering the primary mechanisms associated with phototropin photochemistry (Briggs and Christie, 2002; see Swartz and Bogomolni, Chapter 14), photochemical studies have shown that bacterially expressed fusion proteins containing both LOV domains (designated LOV1+LOV2) possess photochemical properties that more closely resemble those of full-length phot1 and phot2 expressed in insect cells (Kasahara et al., 2002b). Hence, truncated LOV1+LOV2 fusion proteins represent the more appropriate model system to study phototropin photochemistry in relation to the full-length photoreceptor proteins.

While the LOV1+LOV2 fusion proteins of phot1 and phot2 exhibit similar approximate quantum efficiencies, their times for dark regeneration differ significantly (Kasahara et al., 2002b). Dark-regeneration kinetics for phot1 are far slower than those observed for phot2. The reason for this discrepancy is unclear. However, given that the full-length photoreceptor proteins expressed in insects also show this difference in dark regeneration kinetics (Kasahara et al., 2002b), it seems likely that this phenomenon has some functional significance. Notably, autophosphorylation of phot1 *in vitro* has been shown to possess a memory for a light pulse when subsequently transferred to darkness prior to the addition of radio-labeled ATP (Short et al., 1992, Palmer et al., 1993; Salomon et al., 1996; Christie et al., 1998). This memory capability of phot1 for a light pulse is only lost in darkness after 10 minutes or more, which closely corresponds to the kinetics of dark regeneration observed for phot1 photochemistry (Kasahara et al., 2002b) and may reflect a return of the photoactivated system to its initial ground state. On the other hand, the rapid recovery observed for phot2 would be expected to yield steady-state levels of photoproduct much lower than those of phot1 at a given fluence rate of blue light. As a result, higher light intensities would be required to drive phot2 to the same photostationary equilibrium as phot1.

Whether the difference in dark regeneration kinetics observed for phot1 and phot2 relates to their physiological functions remains to be determined. Phot1 and phot2 have been reported to exhibit different photosensitivities in activating chloroplast accumulation movement in response to low light intensities, with phot2 requiring a higher light threshold for activity than phot1 (Kagawa and Wada, 2000; Sakai et al., 2001). Still, the mechanisms associated with phototropin recovery *in vivo* are undoubtedly more complex than a simple reversal of LOV-domain photochemistry. As discussed below, phototropin recovery for light-activated phosphorylation *in vivo* involves dephosphorylation of the photoreceptor protein by an as yet unidentified protein phosphatase rather than degradation and resynthesis of the photoreceptor protein.

13.4 From Light Sensing to Receptor Activation

13.4.1 LOV Connection

It is generally accepted that light sensing by the LOV domains results in a protein conformational change that somehow leads to activation of the C-terminal kinase domain (Crosson et al., 2003). How then does light absorption and subsequent flavin-cysteinyl adduct formation lead to such a structural change(s) and kinase activation? The photoexcited crystal structure of the *Adiantum* phy3 LOV2, compared to that of the ground state shows only minor, light-induced protein changes, all within the vicinity of the FMN chromophore (Crosson and Moffat, 2002). This is not too surprising given the probable structural constraints of the crystal lattice. In contrast, solution difference infrared absorbance spectroscopy (FTIR) indicates that photoactivation of purified LOV2 is accompanied by changes in the LOV domain apoprotein (Swartz et al., 2002; Iwata et al., 2003; see Swartz and Bogomolni, Chapter 14; Crosson, Chapter 15). In addition, circular dichroism measurements using purified LOV2 from oat phot1a suggest that light-induced flavin-cysteinyl adduct formation is followed by a loss of α-helicity (Corchnoy et al., 2003). But the greatest detail so far regarding light-induced protein structural changes associated with LOV2 has come from solution nuclear magnetic resonance (NMR) spectroscopy. Using NMR, Harper et al. (2003) have identified an amphipathic α-helix located just C-terminal of LOV2 whose structure is altered in response to light (Figure 13.4). This 20-amino acid region, designated Jα, associates with the surface of LOV2 in the dark. The interaction between the Jα helix and LOV2 is disrupted following illumination and flavin-cysteinyl adduct formation. This structural mechanism has been proposed to couple photoexcitation of LOV2 to activation of the C-terminal kinase domain (Harper et al., 2003).

The structural consequences of LOV2 photoexcitation may serve as a light-induced activator of the C-terminal kinase domain. Alternatively, LOV2 may serve as an autoinhibitor, repressing kinase activity in the dark whereupon photoexcitation would function to relieve this repression. A similar PAS/kinase domain interaction mechanism has been proposed for regulating the activities of the bacterial oxygen sensor, FixL (Gong et al., 1998), and the novel eukaryotic protein kinase, PAS kinase (Rutter et al., 2001). Intriguingly, collective alignment of LOV1 or LOV2 domains from all phototropins identified to date reveals that peptide sequences which can form the Jα-helix are only found associated with LOV2 and not LOV1 (K. Gardner, personal communication). Thus, the structural consequences accompanying LOV1 photoexcitation are likely to differ from those following photactivation of LOV2, consistent with the apparent distinct functionality observed for these two LOV domains (Christie et al., 2002).

13.4.2
Phototropin Autophosphorylation

While a specific role for LOV1 remains to be elucidated, light activation of the phototropins ultimately results in autophosphorylation of the photoreceptor protein. The kinase domain of plants phototropins belongs to the family of cAMP-dependent protein kinases (Hanks and Hunter, 1995), although it seems unlikely that light-activated phototropin autophosphorylation requires cAMP. In particular, the phototropin kinase domain is a member of the AGC-VIII subfamily of protein kinases (Watson, 2000). There are two differences between the AGC-VIII subfamily of kinases and the majority of other protein kinases (Watson, 2000). First, the AGC-VIII family contains a DFD amino-acid motif instead of a DFG motif in subdomain VII. The Asp of the DFG motif is required for chelating Mg^{2+}, an ion necessary for phosphate transfer (Hanks and Hunter, 1995) and is essential for phototropin kinase activity (Christie et al., 2002). Second, there is an additional peptide sequence between subdomains VII and VIII in the phototropin kinase domains. The nature of this region varies depending on the kinase, but subdomain VIII is typically involved in the recognition of peptide substrates (Hanks and Hunter, 1995).

As mentioned previously, autophosphorylation, at least for phot1, has been shown to occur on multiple serine residues (Palmer et al., 1993; Short et al., 1994; Salomon et al., 1996). Consistent with phosphorylation on multiple sites, phot1 from several plant species has been reported to show reduced electrophoretic mobility after blue light irradiation (Short et al., 1993; Liscum and Briggs, 1995; Knieb et al., 2004). A recent study by Salomon et al. (2003) has identified eight serine residues within oat phot1a that become phosphorylated upon illumination. Two of these sites (Ser27, Ser30) are located upstream of LOV1, near the N terminus of the protein. The remaining six sites (Ser274, Ser300, Ser317, Ser325, Ser332, and Ser349) are located in the peptide region between LOV1 and LOV2. Salomon et al. (2003) also demonstrated that phot1 autophosphorylation *in vivo* is fluence dependent: the two serine residues situated at the N terminus are phosphorylated in response to low fluences of blue light, whereas the other sites are phosphorylated either at intermediate or high fluences. This hierarchical pattern of phosphorylation is only observed when blue light irradiation is performed *in vivo*. No such fluence discrimination is detected when the blue light irradiation is given *in vitro*, suggesting that the fluence dependency for phot1 autophosphorylation *in vivo* depends on the presence of an additional factor(s).

At least *in vivo*, the biochemical consequences of low-fluence phosphorylation could be quite different from those arising from high-fluence phosphorylation. Salomon et al. (2003) have proposed, as Briggs (1996) had earlier, that low-fluence phosphorylation of phot1 might initiate signaling by modifying the interaction status between the receptor and a specific signaling partner. In fact, phot1a from *Vicia faba* (broad bean) has recently been reported to interact with a 14-3-3 protein whose binding is dependent on the phosphorylation of a particular serine residue within the protein (Kinoshita et al., 2003). This interaction will be discussed in more detail below (Section 13.5.3). The high-fluence-activated phosphorylation sites, on the contrary, have been proposed to play some other role e.g. receptor desensitization (Briggs, 1996; Christie and Briggs, 2001; Briggs and Christie, 2002; Liscum, 2002; Salomon et al., 2003). Site-directed mutagenesis of these sites and their effect on photoreceptor function will help to provide information on the role of autophosphorylation in phototropin activation and signaling.

So far, there is no evidence to indicate that phototropins initiate signaling through activation of a phosphorylation cascade. However, a truncated version of phot2 comprising only the LOV2 domain and the C-terminal kinase domain is able to complement the chloroplast high-light avoidance response in a *phot2* mutant of *Adiantum* (Kagawa et al., 2004). Given that the sites of phototropin autophosphorylation are located before LOV2 (Salomon et al., 2003), it is tempting to speculate that phot2 uses some means other than autophosphorylation to bring about this response. Could this signaling mechanism involve the phosphorylation of an as yet unidentified interacting partner? Further studies are now required to test such a hypothesis.

13.4.3
Phototropin Recovery

As mentioned earlier, it is generally viewed that the phototropins are unphosphorylated in the dark or ground state and that light activation results in autophoshorylation of the photoreceptor protein. Autophosphorylation of phot1 *in vivo* has been shown to return to its unphosphorylated or inactive state in darkness following a saturating pulse of blue light (Short and Briggs 1990; Hager and Brich, 1993; Salomon et al., 1997a; Kinoshita et al., 2003). Moreover, the recovered photoreceptor system can be rephosphorylated in response to a second blue light pulse (Hager et al., 1993; Salomon et al., 1997a; Kinoshita et al., 2003). These findings therefore demonstrate that the phot1, and most likely phot2, have the ability to regenerate back to the non-phosphorylated form.

How do the phosphorylated receptors regenerate back to the dark or inactive state? One possibility is that the phosphorylated form of the receptor is degraded and is, at the same time, accompanied by *de novo* synthesis of the unphosphorylated form. Indeed, prolonged exposure of phot1 to blue light has been shown to result in a gradual decrease in phot1 protein levels in dark-grown *Arabidopsis* seedlings (Sakamoto and Briggs, 2002). However, this decrease in phot1 protein levels in response to continuous blue light irradiation is apparent only after many hours. In contrast, the recovery of phot1 to its non-phosphorylated form *in vivo* occurs 20–90 minutes after a blue light pulse (Short and Briggs 1990; Hager and Brich, 1993; Salomon et al., 1997a; Kinoshita et al., 2003). Thus, the reduction of phot1 protein levels in response to a prolonged light exposure most likely represents a long-term adaptation process.

Further studies now indicate that phototropin recovery *in vivo* involves dephosphorylation of the photoreceptor system in addition to a dark reversal of LOV-domain photochemistry as discussed previously. Using a specific antibody to *Arabidopsis* phot1, Knieb et al. (2004) have confirmed that phot1 from several plants species exhibits reduced electrophoretic mobility after blue light irradiation, first noted by Short et al. (1993), and commonly associated with protein phosphorylation (Beebe and Corbin, 1986). The mobility shift observed for phot1 corresponds to an increase size of 2–3 kDa. This size increase is transient and gradually reverts back to the original size in darkness within 60–90 min, consistent with the protein being dephosphorylated. Similarly, Salomon et al. (2003) have shown that the recovery of unphosphorylated oat phot1a *in vivo* occurs in a fluence-dependent manner; sites that are phosphorylated by high fluences are dephosphorylated first followed by those that are phosphorylated at intermediate and low fluences respectively. This fluence-dependent decline in phosphorylation most likely reflects dephosphorylation of phot1 by an as yet unidentified protein phosphatase rather than a degradation and resynthesis of the photoreceptor protein. Whether dephosphorylation of phot1 represents the rate-limiting step in the recovery for light-activated autophosphorylation *in vivo* remains unknown.

Blue light irradiation has been reported to result in a rapid reduction (within minutes) in the association of phot1 with the plasma membrane (Sakamoto and Briggs, 2002; Knieb et al., 2004). The significance of this partial redistribution of phot1 is currently unknown. Whether it represents the dissociation of a phototropin-associated signaling complex (Liscum and Stowe-Evans, 2000) or is somehow involved in photoreceptor signaling requires further investigation. Resolving the relationships between receptor dephosphorylation and photoproduct decay within the LOV domains and how these events are linked to intracellular movements like those observed for phot1 upon excitation will help to improve our understanding of the short-term processes associated with phototropin recovery. Incidentally, the decrease in phot1 protein levels mentioned above in response to continuous blue light irradiation is reported to coincide with a reduction in *PHOT1* transcript levels (Kanagae et al., 2000; Sakamoto and Briggs, 2002; Elliot et al., 2004). This light-induced decrease in *PHOT1* mRNA levels is dependent on the photoactivation of phytochrome (Elliot et al., 2004).

13.5
Phototropin Signaling

13.5.1
Beyond Photoreceptor Activation

While much progress has been made in elucidating the photochemical and biochemical properties of the phototropins, less information is known regarding the downstream signaling events that follow phototropin activation. The ultimate challenge for researchers will be to identify these components and determine how these are connected to particular phototropin-mediated responses. Exciting progress in this area has already begun. In the following section, we briefly discuss recent insights that have been made regarding phototropin signaling and how these relate to photoreceptor localization and certain phototropin-mediated responses.

13.5.2
Phototropism

Phototropism is important for germinating seedlings, whereby the emerging shoot must grow towards the light in order to survive. Generally, shoots show positive phototropism; movement towards a light source, whereas roots exhibit negative phototropic movement. Positive phototropism is mediated by an increase in growth on the shaded side of the stem resulting from an accumulation of the growth hormone auxin (Iino, 2001).

At present, very little is known regarding how phototropin activation by blue light leads to an accumulation of auxin in the shaded side of the stem to bring about a curvature response. The model currently favored begins with an establishment of a light gradient across the hypocotyl in response to a directional light stimulus (Iino, 2001). This results in an unequal activation of photoreceptors across the phototropically stimulated organ. Indeed, Salomon et al. (1997b; 1997c) have shown that unilateral irradiation induces a gradient in phot1 autophosphorylation across oat coleoptiles, with a higher level of phosphorylation on the irradiated side than on the shaded side. The biochemical gradient produced somehow leads to a lateral accumulation of auxin on the shaded side of the stem (Iino, 2001). Although Cholodny and Went first introduced this hypothesis over 70 years ago with respect to the oat coleoptile (Iino, 2001), it has proved markedly difficult to demonstrate as a general model. Recent support for this mechanism has come from the use of an auxin-sensitive gene reporter, designated *DR5::GUS*, whose activity correlates with auxin measurements. This reporter has been used successfully to demonstrate the establishment of an auxin gradient across a phototropically stimulated *Arabidopsis* hypocotyl (Friml et al., 2002).

Further support for the Cholodny-Went model of lateral auxin movement in dicots has recently come from the identification of a putative auxin transporter in *Arabidopsis* that is localized to the plasma membrane at the outer lateral side of hypocotyl endodermal cells (Friml et al., 2002). PIN3 is a member of the PIN-FORMED or PIN family of putative auxin efflux carriers of which there are eight members in *Arabidopsis* (Friml and Palme, 2002). Mutants lacking PIN3 exhibit reduced phototropism, suggesting that PIN3 (in conjunction with some other PIN family member) may function to bring about a lateral movement of auxin to the shaded side of the stem. How the activity of such a transporter is controlled by the phototropins is not known. Yet, a lateral relocalization of PIN3 has been shown to occur in roots following a gravity stimulus (Friml et al., 2002), raising the possibility that phototropin activation could modulate PIN3 distribution in such a way as to bring about a lateral asymmetry of PIN3 across the stem in response to a phototropic stimulus.

Given the known role of auxin in phototropism (Koller, 2000; Iino, 2001), it is perhaps not surprising that phot1 is found in cells associated with polar auxin transport (Sakamoto and Briggs, 2002). The main mode of active auxin transport in stems is polar or unidirectional, a movement from cell to cell from the tip of the stem (where it is predominantly synthesized) to the base (Iino, 2001). Polar auxin transport, unlike lateral auxin movements, can be measured readily and has been attributed to the action of specific auxin influx and efflux carriers (Friml, 2003). Members of the AUX subfamily of amino-acid permeases have been identified as candidate auxin-influx carriers (Swarup et al., 2000). AUX1 is localized to the apical surface of root cells, consistent with its role in controlling auxin influx (Swarup et al., 2001). PIN1, on the other hand, a member of the PIN family, is localized to the basal side of the cell in roots and shoots consistent with its role in auxin efflux (Gälweiler et al., 1998). The asymmetric distribution of these two types of putative transporters is thought to determine the polarity of auxin flow.

Sakamoto and Briggs (2002) have examined the intracellular distribution of phot1 in *Arabidopsis* by fusing the photoreceptor to green fluorescent protein (GFP). The native *PHOT1* promoter was used to drive expression of phot1-GFP in the transgenic lines obtained. In cells associated with polar auxin transport, phot1 is more strongly localized to the plasma membrane adjacent to the apical and basal walls, rather than the side walls, placing the receptor in an ideal location to influence the activity or distribution of auxin influx and efflux carriers. How this mode of transport is connected to the proposed lateral accumulation of auxin on the shaded side of a phototropically stimulated stem is not known. However, inhibitors of polar auxin transport have been shown to block phototropism (Friml et al., 2002) suggesting that this mode of transport is required to generate the differential auxin distribution associated with phototropism. Support for this conclusion has come from recent experiments demonstrating that the amplitude of phototropic curvature in *Arabidopsis* is influenced by changes in the basal distribution of PIN1 (Noh et al., 2003). This alteration in PIN1 distribution is reported to involve AtMDR1, a member of a different putative transporter family associated with polar auxin transport (Noh et al., 2001; Noh et al., 2003). However, the role of AtMDR1 in regulating phototropism and auxin transport is poorly understood at present and requires further investigation.

To date, only two proteins have been shown to interact with phototropin (phot1). One is the scaffold-type protein, NPH3 (Motchoulski and Liscum, 1999). NPH3 is a novel protein containing several protein-protein interacting motifs and has been shown to interact with phot1 *in vitro*. The second is a 14-3-3 protein, which will be discussed in the following section. NPH3 was first identified through the isolation of phototropism mutants of *Arabidopsis* (Liscum and Briggs, 1995; 1996). Indeed, NPH3 appears to be essential for phototropism since *nph3* mutants exhibit impaired hypocotyl phototropism to both low and high intensities of unilateral blue light without affecting phot1 autophosphorylation (Liscum and Briggs, 1995; 1996; Motchoulski and Liscum, 1999; Sakai et al., 2000). Like phot1 and phot2, NPH3 is associated with the plasma membrane and most likely serves as a bridge to bring components of a phot1 photoreceptor complex together (Liscum and Stowe-Evans, 2000). A protein closely related to NPH3, designated root phototropism 2 (RPT2), has been isolated from a separate genetic screen (Sakai et al., 2000). In contrast to *NPH3*, *RPT2* gene expression is enhanced at increased light intensities similar to the situation found for *PHOT2* gene expression (Jarillo et al., 2001b; Kagawa et al., 2001; Tepperman et al., 2001). Thus, RPT2 in conjunction with phot2 seems to play a role in mediating phototropic curvature under high-light conditions.

13.5.3 Stomatal Opening

Phot1 has been shown, by means of GFP fluorescence, to localize at or near the plasma membrane of stomatal guard cells (Sakamoto and Briggs, 2002), consistent with its role in blue light-induced stomatal opening. While epidermal strips from *phot1* and *phot2* single mutants show blue light-induced stomatal opening (Kinoshita et al., 2001), the *phot1phot2* double mutant fails to exhibit this response, demonstrating that both phototropins redundantly mediate blue light-induced stomatal opening in *Arabidopsis* (Kinoshita et al., 2001). In addition, stomatal guards cells of the *phot1phot2* double mutant fail to extrude protons in response to blue light treatment (Kinoshita et al., 2001). Proton extrusion is essential for stomatal opening and is known to result from activation of the plasma membrane proton-ATPase (Schroeder et al., 2001). In brief, proton extrusion by guards cells creates an electrochemical gradient that drives K^+ and Cl^- uptake which in turn causes a rise in osmotic potential, resulting in water influx, swelling of guard cells, and consequent stomatal opening. In fact, blue light activates the guard cell proton-ATPase via phosphorylation of its C terminus (Kinoshita and Shimizaki, 1999), providing a tentative connection between blue light-induced proton extrusion and phototropin activation. Could the proton-ATPase be a potential substrate for phototropin kinase activity? This vital question still remains to be addressed.

Recent biochemical studies provide a further connection between phototropin activation and the plasma membrane proton-ATPase. Kinoshita and Shimizaki (2002) have shown that the guard-cell proton-ATPases, VHA1 and VHA2 from *Vicia faba* (broad bean) bind a 14-3-3 protein at their C terminus when phosphorylated in response to blue light. 14-3-3 proteins belong to a highly conserved protein family that generally bind to phosphorylated target proteins and play a central role in regulating signaling in eukaryotic cells (Sehnke et al., 2002). While the role of 14-3-3 binding in the activation of the proton-ATPase is not clear, Kinoshita et al. (2003) have gone on to show that phot1a from broad bean also binds a 14-3-3 protein after undergoing autophosphorylation by blue light. This binding is reversible and like phot1 autophosphorylation recovers to a basal level in darkness approximately 15 minutes following a blue light pulse, indicating that 14-3-3 binding to phot1 is dependent on autophosphorylation.

The binding of a 14-3-3 protein to phot1a from broad bean requires phosphorylation of Ser358 situated between LOV1 and LOV2 (Kinoshita et al., 2003), equivalent to Ser325 in oat phot1a, a residue that is phosphorylated in response to intermediate fluences of blue light (Salomon et al., 2003). Therefore, consistent with the mechanism proposed by Briggs (1996) and Salomon et al. (2003), autophosphorylation of phot1 in response to low and/or intermediate fluence rates of blue light may initiate signaling by binding a 14-3-3 protein. 14-3-3 binding may in turn allow the photoreceptor system to associate with or modify the activity of the guard cell proton-ATPase. Studies are now required to clarify the significance and functional consequences of 14-3-3 binding to light-activated phot1 and determine its role in activation of the plasma membrane proton-ATPase. For instance, it is unknown whether the 14-3-3 protein that binds phosphorylated phot1 is the same one that binds to the phosphorylated proton-ATPase. Interestingly, Kinoshita et al. (2003) observed that blue light-dependent 14-3-3 binding to phot1 is not restricted to stomatal guard cells. The authors found that light-induced 14-3-3 binding to phot1 was also detectable in dark-grown seedlings and mature leaves of broad bean and other plant species, including *Arabidopsis,* indicating that 14-3-3 binding may represent a common event associated with phototropin signaling.

13.5.4 **Chloroplast Movement**

The location of phot1 close to the plasma membrane of *Arabidopsis* mesophyll cells (Sakamoto and Briggs, 2002) is consistent with its role in mediating blue light-activated chloroplast movement (Kagawa and Wada 2000, Kagawa et al., 2001, Kagawa, 2003; Wada et al., 2003). Similarly, phot2 has been reported to co-purify with the plasma membrane from *Arabidopsis* leaves (Harada et al., 2003). However, the mechanisms coupling phototropin activation at the plasma membrane to chloroplast movements are poorly understood. Similarly, it is not yet known whether chloroplast accumulation and chloroplast-avoidance movements share the same signals. All the same, recent genetic studies indicate that both chloroplast accumulation and chloroplast-avoidance movements in *Arabidopsis* are mediated through changes in the cytoskeleton.

Many experiments involving the use of cystoskeletal inhibitors have implicated a role for actin filaments in mediating chloroplast movement (Wada et al., 2003). In addition, the isolation of mutants impaired in their chloroplast avoidance response has led to the identification of a novel actin-binding protein, designated chloroplast unusual positioning 1 (CHUP1). CHUP1 is required for the positioning of chloroplasts to both low and high light intensities in *Arabidopsis* (Kasahara et al., 2002a; Oiwaka et al. 2003). Mutants lacking CHUP1 exhibit aberrant chloroplast positioning and light-induced movement compared to wild-type plants, in that chloroplasts are constantly gathered at the bottom of palisade cells. CHUP1 contains multiple functional domains some of which are involved in mediating protein-protein interactions. One domain located in the N-terminal half of the protein is an actin-binding motif and has been shown to bind F-actin *in vitro* (Oikawa et al., 2003). Therefore, CHUP1 most likely interacts with the actin-based cytoskeleton *in vivo*. The extreme N-terminal region of CHUP1 contains a hydrophobic segment that on its own confers the ability to target GFP into the chloroplast envelope (Oikawa et al., 2003), suggesting that CHUP1 could function at the periphery of the chloroplast outer membrane. In contrast to *Arabidopsis* phot1 and phot2 mutants, which are specifically defective in chloroplast accumulation movement and/or chloroplast avoidance movement, *chup1* mutants are altered in both the positioning and movement of chloroplasts. Hence, CHUP1 most likely represents an essential component of the machinery required for chloroplast positioning and movement, rather than being directly associated with phototropin signaling.

A possible signal connecting photoreceptor activation at the plasma membrane to the chloroplasts is the versatile intracellular messenger, calcium (Wada et al., 2003). Several approaches have been used to demonstrate that blue light irradiation leads to an increase in cytosolic calcium concentrations $[Ca^{2+}]_{cyt}$, in *Arabidopsis*. Baum et al. (1999) were the first to report a blue light-induced elevation of $[Ca^{2+}]_{cyt}$ in *Arabidopsis* seedlings transformed with the gene encoding the Ca^{2+}-luminescent protein, aequorin. This response is both rapid and transient (rises and falls over a period of 80 seconds) and is severely attenuated in a null mutant of phot1. Harada et al. (2003) have recently extended these aequorin measurements and demonstrated that phot2, in addition to phot1, mediates a rapid blue light-dependent increase in $[Ca^{2+}]_{cyt}$ in *Arabidopsis* leaves. Moreover, Stoelzle et al. (2003) have used patch-clamping techniques to identify a phototropin-activated plasma membrane Ca^{2+} channel in *Arabidopsis* mesophyll cells.

Using various pharmacological agents, Harada et al. (2003) were able to show that phot1 and phot2 can mediate an influx of Ca^{2+} from the apoplast through the activation of a Ca^{2+} channel(s) at the plasma membrane, whereas phot2 alone can induce a release of Ca^{2+} from intracellular stores via phospholipase C-mediated phosphoinositide signaling. A separate study involving microelectrode impalement also suggests that phot2 mediates Ca^{2+} movement from intracellular stores whereas phot1 activation exclusively results in an influx of Ca^{2+} from the apoplast (Babourina et al., 2002). It is worth noting that phot1 and phot2 induce an increase in $[Ca^{2+}]_{cyt}$ in *Arabidopsis* leaves with different photosensitivities (Harada et al., 2003) similar to that found for the activation of chloroplast accumulation movement, in which phot2 requires a higher light threshold for activity than phot1 (Kagawa and Wada, 2000; Sakai et al., 2001). These observations thus provide a tenuous link between calcium and phototropin-mediated chloroplast movements in *Arabidopsis*.

Although studies from several plant species provide evidence for an involvement of Ca^{2+} in eliciting chloroplast movements (Wada et al., 2003), none of the above findings conclusively demonstrate a role for Ca^{2+} in phototropin-induced chloroplast movement in *Arabidopsis*. Indeed, such blue light-induced Ca^{2+} increases are not only restricted to leaves but are also found in hypocotyls (Babourina et al., 2002; Stoelzle et al., 2003), indicating that Ca^{2+} may act as an intracellular messenger in other phototropin-mediated responses. Research is now required to establish how the blue light-induced Ca^{2+} fluxes described above are connected to specific phototropin-mediated responses. Nonetheless, a recent study by Folta et al. (2003) indicates that the early transient rise in $[Ca^{2+}]_{cyt}$ is associated with at least one, but not all phot1-mediated responses in *Arabidopsis*. These studies are described in the following section.

13.5.5
Rapid Inhibition of Hypocotyl Growth by Blue Light

Transfer of dark-grown seedlings to blue light causes an inhibition of hypocotyl growth. This growth inhibition response can be separated into two distinct phases: a rapid, transient response occurring within a few minutes after the onset of blue light treatment and a slow response that continues for many hours (Spalding, 2000; Parks et al., 2001). The *phot1* single mutant has been shown to lack the rapid growth-inhibition response (Folta and Spalding, 2001). Evidently, phot2 plays no role in the rapid inhibition of hypocotyl growth as *phot1* mutants completely lack this response (Folta and Spalding, 2001). These findings are consistent with the expression pattern of *PHOT2*, whereby transcripts are not abundant in dark-grown seedlings but increase upon illumination (Jarillo et al., 2001b; Kagawa et al., 2001; Tepperman et al., 2001). By contrast, the slow growth inhibition response is mediated by the cryptochromes, a second family of blue light receptors in plants (Lin and Shalitin, 2003; see Batschauer, Chapter 10; Cashmore, Chapter 11).

Using the Ca^{2+}-specific chelator BAPTA Folta et al. (2003) were able to demonstrate that the rapid increase in $[Ca^{2+}]_{cyt}$ levels in response to blue light treatment is associated with phot1-mediated inhibition of hypocotyl growth. Equivalent concentrations of BAPTA were found to be effective in preventing both the rise in $[Ca^{2+}]_{cyt}$ and phot1-mediated hypocotyl growth inhibition by blue light. More importantly, the same chelator treatment did not impair phot1-mediated phototropism in *Arabidopsis*, indicating that the rise in $[Ca^{2+}]_{cyt}$ is a signaling step in the transduction process linking phot1 activation to hypocotyl growth inhibition and not phototropism. The signaling pathway for phot1-mediated inhibition of hypocotyl growth also differs from that for phot1-mediated phototropism in that it does not require the phot1-interacting protein, NPH3 (Folta and Spalding, 2001). Once again, these findings suggest that the signal transduction pathways for these two phototropin-mediated responses are distinct. NPH3, however, is not only involved in phototropism as this protein, in addition to phot1, is required for the blue light-dependent destabilization of *Lhcb* and *rcbS* transcripts in *Arabidopsis* (Folta and Kaufman, 2003). It now remains to be seen whether other phototropin-mediated responses such as light-induced chloroplast movement have a strong dependence on rapid changes in $[Ca^{2+}]_{cyt}$.

13.6
Future Prospects

Although exciting progress has been made in the last few years since the discovery of the phototropins, most of this progress has concerned the photochemical and structural properties of the phototropin LOV domains. Much work remains to be done to improve our understanding of how these photoreceptors function to bring about a diverse range of responses. More information is required to understand how these blue light-activated kinases initiate signaling and to identify the cellular events downstream of phototropin activation. At present, it is still unknown why the phototropins contain two LOV domains with different photochemical properties, and how these differences relate to their physiological functions. Moreover, in contrast to phot1, very little is known regarding the subcellular distribution of phot2. Clearly, there are many issues still to be addressed.

Even so, it is now apparent through the study of phototropin-deficient mutants that these blue light receptors not only function to regulate phototropism, after which they were first named, but activate a wide range of blue light responses. More extraordinary, and by some weird twist of fate, is the recent discovery that the LOV domains are actually involved in love; phototropin from *Chlamydomonas* has been shown to serve as the photoreceptor mediating various steps in the cycle for sexual reproduction (Huang and Beck, 2003). Perhaps further analysis of phototropin-deficient mutants will uncover other exciting roles for these blue light receptors. Mutants lacking phot1 and phot2 have been reported to exhibit residual phototropism under high light intensities of unilateral blue light (Sakai et al., 2001). Similarly, Talbott et al. (2003) have recently shown that the *phot1phot2* double mutant still exhibits blue light-induced stomatal opening under certain light conditions. Whether these residual photoresponses can be attributed to a leaky *phot2* mutation or to cryptochrome or phytochrome action or some other novel photoreceptor system awaits further investigation.

References

yAtaka K., Hegemann P. and Heberle J. (2003) *Biophy. J.* 84: 466–474.

Aravind L. and Koonin E. V. (2000) *Curr. Biol.* 10: R53–55.

Babourina O., Newman I. and Shabala S. (2002) *Proc. Natl. Acad. Sci. USA* 99: 2433–2438.

Ballario P., Talora C., Galli D., Linden H. and Macino G. (1998) *Mol. Microbiol.* 29: 719–729.

Ballario P., Vittorioso P., Magrelli A., Talora C., Cabibbo A. and Macino G. (1996) *EMBO J.* 15: 1650–11657.

Baum G., Long J. C., Jenkins G. I. and Trewavas A. J. (1999) *Proc. Natl. Acad. Sci. USA* 96: 13554–13559.

Baskin T. I. and Iino M. (1987) *Photochem. Photobiol.* 46: 127–136.

Beebe S. J. and Corbin J. D. (1986) In: *The Enzymes, Ed. 3, Vol. XVII, Part A,* pp. 44–100, Boyer P. D. and Krebs E. G. (eds.) Academic Press, New York.

Briggs W. R. (1996) In: *UV/Blue Light: Perception and Responses in Plant,* meeting held in Marburg, Germany, August, 1996, Abstracts p. 49.

Briggs W. R., Beck C. F., Cashmore A. R., Christie J. M., Hughes J., Jarillo J., Kagawa T., Kanegae H., Liscum E., Nagatani A., Okada, K. Salomon M., Rüdiger W., Sakai T. Takano M., Wada M. and Watson J. C. (2001a) *Plant Cell* 13, 993–997.

Briggs W. R. and Christie J. M. (2002) *Trends Plant Sci.* 7: 204–210.

Briggs W. R., Christie J. M. and Salomon M. (2001b) *Antiox. Redox Signaling* 3: 775–788.

Briggs W. R. and Huala E. (1999) *Annu. Rev. Cell Dev. Biol.* 15: 33–62.

Cheng P., He Q., Yang Y., Wang L. and Liu Y. (2003) *Proc. Natl. Acad. Sci. USA* 100: 5938–5943.

Christie J. M. and Briggs W. R. (2001) *J. Biol. Chem.* 276: 11457–11460.

Christie J. M., Reymond P., Powell G., Bernasconi P., Reibekas A. A., Liscum E. and Briggs W. R. (1998) *Science* 282: 1698–1701.

Christie J. M., Salomon M., Nozue K., Wada M. and Briggs W. R. (1999) *Proc. Natl. Acad. Sci. USA* 96: 8779–8783.

Christie J. M., Swartz T. E., Bogomolni R. and Briggs W. R. (2002) *Plant J.* 32: 205–219.

Corchnoy S. B., Swartz T. E., Lewis J. W., Szundi I., Briggs W. R. and Bogomolni R. A. (2003) *J. Biol. Chem.* 278: 724–731.

Crosson S., and Moffat. K. (2001) *Proc. Natl. Acad. Sci. USA* 98: 2995–3000.

Crosson S. and Moffat K. (2002) *Plant Cell* 14: 1067–1075.

Crosson S., Rajagopal S. and Moffat K. (2003) *Biochemistry* 42: 2–10.

Elliot R. C., Platten D., Watson J. C. and Reid J. B. (2004) *J. Plant Physiol.*, 161: 265–270.

Federov R., Schlichting I., Hartmann E., Domratcheva T., Fuhrmann M. and Hegemann P. (2003) *Biophys. J.* 84: 2474–2482.

Folta, K. M. and Kaufman L. S. (2003) *Plant Mol Biol.* 51(4): 609–618.

Folta K. M., Lieg E. J., Durham T. and Spalding E. P. (2003) *Plant Physiol.* 133: 1464–1470.

Folta K. M. and Spalding E. P. (2001) *Plant J.* 26: 471–478.

Friml J. (2003) *Curr. Opin. Plant Biol.* 6: 7–12.

Friml J. and Palme K. (2002) *Plant Mol. Biol.* 49: 273–84.

Friml J., Wisniewska J., Benkova E., Mendgen K. and Palme K. (2002) *Nature* 415: 806–809.

Froehlich A., Liu Y., Loros J. J. and Dunlap J. C. (2002) *Science* 297: 815–819.

Gälweiler L., Guan C., Müller A., Wisman E., Mendgen K., Yephremov, A. and Palme K. (1998) *Science* 282: 2226–2230.

Gong W., Hao B., Mansy S. S., Gonzalez G., Gilles-Gonzalez M. A. and Chan M. K. (1998) *Proc. Natl. Acad. Sci. USA* 95: 15177–15182.

Hager A. and Brich M. (1993) *Planta* 189: 567–576.

Hager A., Brich M. and Balzen I. (1993) *Planta* 190: 120–126.

Hanks, S.K. and Hunter, T. (1995) *FASEB J.* 9: 576–610.

Harada A., Sakai T. and Okada K. (2003) *Proc. Natl. Acad. Sci. USA* 100: 8583–8588.

Harper S. M., Neil L. C. and Gardner K. H. (2003) *Science* 301: 11541–1544.

He Q., Cheng P., Yang Y., Wang L., Gardner K. H. and Liu Y. (2002) *Science* 297: 840–843.

Heintzen C., Loros J. J. and Dunlap J. C. (2001) *Cell* 2001 104: 453–464.

Holzer W., Penzkofer A., Fuhrmann M. and Hegemann P. (2002) *Photochem. Photobiol.* 75: 479–487.

Huala E., Oeller P. W. Liscum E., Han I.-S., Larsen E. and Briggs W. R. (1997) *Science* 278: 2121–2123.

Huang K. and Beck C. F. (2003) *Proc. Natl. Acad. Sci. USA* 100: 6269–6274.

Huang K., Merkle T. and Beck C. F. (2002) *Physiol. Plant.* 114: 613–622.

Iino M. (2001) In: *Photomovement*, Häder, D.-P and Lebert M. eds, pp. 659–811.

Imaizumi T., Tran H. G., Swartz T. E., Briggs, W. R. and Kay S. A. (2003) *Nature* 426: 302–306.

Iwata T., Nozaki D., Tokutomi S., Kagawa T, Wada M. and Kandori H. (2003) *Biochemistry* 42: 8183–8191.

Iwata T., Tokutomi S. and Kandori H. (2002). *J. Am. Chem. Soc.* 124: 11840–11841.

Jarillo J. A., Ahmad M. and Cashmore A. R. (1998) NPL1 (Accession No. AF053941): *Plant Physiol.* 117: 719.

Jarillo J. A., Capel J., Tang R.-H., Yang H.-Q., Alonso J. M. Ecker J. R. and Cashmore A. R. (2001a) *Nature* 410: 487–490.

Jarillo J. A., Gabrys H., Capel J., Alonso J. M., Ecker J. R. and Cashmore A. R. (2001b) *Nature* 410: 592–594.

Kagawa T. (2003) *J. Plant Res.* 116: 77–82.

Kagawa T., Kasahara M., Abe T., Yoshida S. and Wada M. (2004) *Plant Cell Physiol.*, in press.

Kagawa T., Sakai T., Suetsugu N., Oikawa K., Ishiguro S., Kato T., Tabata S., Okada K. and Wada M. (2001) *Science* 291: 2138–2141.

Kagawa T. and Wada M. (2000) *Plant Cell Physiol.* 41: 84–93.

Kagawa T. and Wada M. (2002) *Plant Cell Physiol.* 43: 367–371.

Kanegae H., Tahir M., Savazzini F., Yamamoto K., Yano M., Sasaki T., Kanegae T., Wada M. and Takano M. (2000) *Plant Cell Physiol.* 4: 415–423.

Kasahara M., Kagawa T., Oikawa K., Suetsugu N., Miyao M. and Wada M. (2002a) *Nature* 420: 829–832.

Kasahara M., Swartz T. E., Olney M. O., Onodera A., Mochizuki N., Fukuzawa H., Asamizu E., Tabata S., Kanegae H., Takano M., Christie J. M., Nagatani A. and Briggs W. R. (2002b) *Plant Physiol.* 129: 762–773.

Kawai H., Kanegae T., Christensen S., Kiyosue T., Sato Y., Imaizumi T., Kadota A. and Wada M. (2003) *Nature* 421: 287–290.

Kennis J. T., Crosson S., Gauden M., van Stokkum I. H., Moffat K. and van Grondelle R. (2003) *Biochemistry* 42: 3385–3392.

Kinoshita T. and Shimazaki K. (1999) *EMBO J.* 18: 5548–55558.

Kinoshita T. and Shimazaki K. (2002) *Plant Cell Physiol.* 43: 1359–1365.

Kinoshita T., Emi T., Tominaga M., Sakamoto K., Shigenaga A., Doi M. and Shimazaki K. (2003) *Plant Physiol.* 133:1453–1463.

Kinoshita T., Doi M., Suetsugu N., Kagawa T., Wada M. and Shimizaki K.-I. (2001) *Nature* 414: 656–660.

Knieb E., Salomon M. and Rudiger W. (2004) *Planta* 218: 843–851.

Koller. D. (2000) *Advances Bot. Res.* 33: 35–131.

Kottke T., Heberle J., Hehn D., Dick B. and Hegemann P. (2003) *Biophys. J.* 84: 1192–2001.

Lin C. (2002) Blue light receptors and signal transduction. *Plant Cell Supplement*: S207–S225.

Lin C. and Shalitin D. (2003) *Annu. Rev. Plant Biol.* 54: 469–496.

Liscum E. (2002) In C. R. Somerville and E. M. Meyerowitz, eds, The Arabidopsis Book. American Society of Plant Biologists, Rockville, MD, doi/10.1199/tab.0074/, http://www/aspb.org.org/publciations/arabidopsis/

Liscum E. and Briggs W. R. (1995) *Plant Cell* 7: 473–485.

Liscum E. and Briggs W. R. (1996) *Plant Physiol.* 112: 291–296.

Liscum E. and Stowe-Evans E. L. (2000) *Photochem. Photobiol.* 72: 273–282.

Losi A., Polverini E., Quest B. and Gärtner W. (2002) *Biophys. J.* 82: 2627–2634.

M s P., Kim W.-I., Somers D. E. and Kay S. A. (2003) *Nature* 426; 567–570.

Miller S. M., Massey V., Ballou D., Williams C. H. Jr., Distefano M. D., Moore M. J. and Walsh C. T. (1990) *Biochemistry* 29: 2831–2841.

Motchoulski A. and Liscum E. (1999) *Science* 286: 961–964.

Nelson D. C., Lasswell J., Rogg L. E., Cohen M. A. and Bartel B. (2000) *Cell* 101: 331–340.

Noh B., Murphy A. S. and Spalding E. P. (2001) *Plant Cell* 13: 2441–2454.

Noh B., Bandyopadhyay A., Peer W. A., Spalding E. P. and Murphy A. S. (2003) *Nature* 424: 999–1002.

Nozue K., Kanegae T., Imaizumi T., Fukada S., Okamoto H., Yeh K. C., Lagarias J. C. and Wada M. (1998) *Proc. Natl. Acad. Sci. USA* 95: 15826–15830.

Nozue K., Christie J. M., Kiyosue T., Briggs W. R. and Wada M. (2000) *Plant Physiol.* 122: 1457.

Oikawa K., Kasahara M., Kiyosue T., Kagawa T., Suetsugu N., Takahashi F., Kanegae T., Niwa Y., Kadota A. and Wada M. (2003) *Plant Cell* 15: 2805–2815.

Palmer J. M., Short T. W., Gallagher S. and Briggs W. R. (1993) *Plant Physiol.* 102: 1211–1218.

Parks B. M., Folta K. M. and Spalding E.P. (2001) *Curr. Opin. Plant Biol.* 4: 436–440.

Rutter J., Michnoff C. H., Harper S. M., Gardner K.H. and McKnight S. L. (2001) *Proc. Natl. Acad. Sci. USA* 98: 8991–8996.

Sakai, T., Kagawa, T., Kasahara, M., Swartz, T.E., Christie, J.M., Briggs, W.R., Wada, M. and Okada, K. (2001) *Proc. Natl Acad. Sci. USA* 98: 6969–6974.

Sakai T., Wada T., Ishiguro S. and Okada K. (2000) *Plant Cell* 12: 225–236.

Sakamoto K. and Briggs W. R. (2002) *Plant Cell* 14: 1723–1735.

Salomon M., Christie J. M., Knieb E., Lempert U. and Briggs W. R. (2000) *Biochemistry* 39: 9401–9410.

Salomon M., Eisenreich W., Dürr H., Schleicher E., Knieb E., Massey V., Rüdiger W., Müller F., Bacher A. and Richter G. (2001) *Proc. Natl. Acad. Sci. USA* 98: 12357–12361.

Salomon M., Knieb E., von Zeppelin T. and Rüdiger W. (2003) *Biochemistry* 42: 4217–4225.

Salomon M., Lempert U. and Rüdiger W. (2004) *FEBS Lett.* 572: 8–10.

Salomon M., Zacherl M., Luff L. and Rüdiger W. (1997a) *Plant Physiol.* 115: 493–500.

Salomon M., Zacherl M. and Rüdiger W. (1997b) *Bot. Acta* 110: 214–216.

Salomon M., Zacherl M. and Rüdiger W. (1997c) *Plant Physiol.* 115: 485–491.

Salomon M., Zacherl M. and Rüdiger W. (1996) *Planta* 199: 336–342.

Schroeder J. I., Allen G. J., Hugouvieux V., Kwak J. M. and Waner D. (2001) *Annu. Rev. Plant Physiol. Plant Mol. Biol.* 52: 627–658.

Schultz, T. F., Kiyosue T. Yanofsky M., Wada M., and Kay S. A. (2001) *Plant Cell* 13: 2659–2670.

Schwerdtfeger C. and Linden H. (2003) *EMBO J.* 22: 4846–4855.

Sehnke P. C., DeLille J. M. and Ferl R. J. (2002) *Plant Cell Supplement* 14: S339–S354.

Short T. W. and Briggs W. R. (1990) *Plant Physiol.* 92: 179–185.

Short T. W., Porst M. and Briggs W. R. (1992) *Photochem. Photobiol.* 55: 773–781.

Short T. W., Porst M., Palmer J. M., Fernbach E. and Briggs W. R. (1994) *Plant Physiol.* 104: 1317–1324.

Short T. W., Reymond P. and Briggs W. R. (1993) *Plant Physiol.* 101: 647–655.

Somers D. E., Schultz T. F., Milnamow M., and Kay S. (2000) *Cell* 101: 319–329.

Spalding E. P. (2000) *Plant Cell Environ.* 23: 665–674.

Stoelzle S., Kagawa T., Wada M., Hedrich R. and Dietrich P. (2003) *Proc. Natl. Acad. Sci. USA* 100: 1456–1461.

Swartz T. E., Corchnoy S. B., Christie J. M., Lewis J. W., Szundi I., Briggs W. R. and Bogomolni R. A. (2001) *J. Biol. Chem.* 276: 36493–36500.

Swartz T. E., Wenzel P. J., Corchnoy S. B., Briggs W. R. and Bogomolni R. A. (2002) *Biochemistry* 41: 7182–7189.

Swarup R., Marchant A. and Bennett M. J. (2000) *Biochem Soc. Trans.* 28: 481–485.

Swarup R., Friml J., Marchant A., Ljung K., Sandberg G., Palme K. and Bennett M. (2001) *Genes Dev.* 15: 2648–2653.

Talbott L. D., Shmayevich I. J., Chung Y., Hammad J. W. and Zeiger E. (2003) *Plant Physiol.* 133: 1522–1529.

Taylor, B. L. and Zhulin, I. B. (1999) *Microbiol. Mol. Biol. Rev.* 63: 479–506.

Tepperman J. M., Zhu T., Chang H. S., Wang X. and Quail P. H. (2001) *Proc. Natl. Acad. Sci. USA* 98: 9437–9442.

Van Volkenburgh E. (1999) *Plant Cell Environ.* 22: 1463–1473.

Wada M., Kagawa T. and Sato Y. (2003) *Annu. Rev. Plant Biol.* 54: 455–468.

Watson J. C. (2000) *Advances Bot. Res.* 32: 149–184.

Yin. H. C. (1938) *Amer. J. Botan.* 25: 1–6.

14
LOV-domain Photochemistry

Trevor E. Swartz and Roberto A. Bogomolni

14.1
Introduction

The LOV domains are the light-sensing modules of the phototropins and related chromoproteins. Previous chapters have introduced the physiology mediated by the phototropins in plants and have discussed in general terms the LOV domain-mediated signal transduction process that results in activation of their kinase domains (Christie and Briggs, Chapter 13). Briefly, in this process, light absorbed by the flavin chromophore results in the transient formation of a covalent flavoprotein adduct and a localized protein structural perturbation (Briggs et al. 2002). Intramolecular propagation of this conformational change results in activation of the kinase moiety and hierarchical autophosphorylation at several sites located in the protein region between the two LOV domains and near the N terminus. Thus, adduct formation is viewed as a light-driven molecular switch that activates the subsequent molecular events.

Since the discovery of the LOV domains *in planta*, a variety of LOV domain-containing proteins have been identified in prokaryotes and eukaryotes (Crosson et al. 2003; Crosson, Chapter 15). The LOV domains are versatile blue-light sensors and are found coupled to a diverse list of signaling elements. In all cases, it appears that the LOV domain is the primary sensory module that conveys a signal to protein domains with functions as diverse as regulation of gene expression, regulation of protein catabolism, or serine/threonine kinase activation in eukaryotes and histidine kinases in prokaryotes. The LOV domains utilize photochemistry unlike that in any other known light sensors, such as the rhodopsins, phytochromes, xanthopsins, cryptochromes and BLUF proteins (Van der Horst et al. 2004). In addition, the LOV domains are versatile light sensors in that they are soluble proteins, absorb maximally in the blue and UV-A, do not require synthesis of a special chromophore but utilize a ubiquitous cellular component, and are able to adjust their photocycle kinetics and quantum yield depending presumably on the physiology they are mediating.

The initial work on the phototropin LOV domains demonstrated that they bind an FMN chromophore and exhibit reversible photochemistry (Salomon et al. 2000). The

Handbook of Photosensory Receptors. Edited by W. R. Briggs, J. L. Spudich

ISBN 3-527-31019-3

spectral similarities between the light-induced absorbance changes observed in isolated LOV domains and those associated to redox reactions in mercuric ion reductase upon formation of a cysteinyl-flavin adduct (Miller et al. 1990) strongly suggested such reaction occurring in the LOV domains as a result of photoactivation, with transient formation of a protein-FMN bond (cysteinyl adduct). As expected, these spectral changes are absent in a mutant in which the reactive cysteine is replaced by alanine (C39A) (Salomon et al. 2000). Time-resolved spectroscopy showed that the LOV domains undergo a photocycle, which is characterized by a series of transient intermediates (Salomon et al. 2000; Swartz et al. 2001; Kottke et al. 2002) with a spontaneous return to the ground-state of the protein in the dark. In addition, these studies showed that the C39A mutant has in fact a truncated photoreaction in which the system returns thermally from an excited state to its initial ground state (Swartz et al. 2001). We present in this chapter these photochemical processes and the accompanying molecular structural perturbations. We first discuss the protein ground-state structure and spectroscopy and then the photocycle kinetics, structure, and spectral properties of the intermediates and the mechanism of transitioning between photointermediates.

14.2 The Chromoprotein Ground State Structure and Spectroscopy

14.2.1 Structure of the Chromoprotein and its Chromophore Environment

The first crystal structure of a closely related phototropin LOV domain (*Adiantum capillus-veneris* phy3 LOV2) was solved to 2.7 Å (Crosson et al. 2001) and described elsewhere (Crosson, Chapter 15). Here, we mention some of the key points relevant to the photochemistry discussion. The LOV domain structure clearly demonstrated the very specific and tight binding of FMN within a protein pocket (Crosson et al. 2001; Fedorov et al. 2003). The FMN is held in place non-covalently through interaction with a hydrogen-bonding network on the pyrimidime side of the isoalloxazine ring and with hydrophobic residues on the dimethylbenzene moiety (Crosson et al. 2001). Specifically, the FMN N3, O2, and O4 all hydrogen-bond to protein side chains; in addition, the phosphate group on the FMN ribityl chain interacts with the guanidinium groups of two arginine residues to form salt bridges. The hydroxl groups of the FMN ribityl side chain form hydrogen bonds to the LOV-domain protein. There are two water molecules in close proximity to the chromophore. Both are within hydrogen-bonding distance of the hydroxl groups of the FMN ribityl chain.

14.2.2
FMN Electrostatic Environment within the Protein

14.2.2.1 Chromophore Absorption

The proximity of the FMN molecule to the protein is apparent in the spectroscopic properties (absorption, fluorescence, and circular dichroism) of the LOV domains as compared to that of free FMN. Figure 14.1 shows both the absorption spectrum of *Avena sativa* (oat) phot1-LOV2 and free FMN in H_2O. Typical of the LOV domains (Figure 14.1), the ground form has major absorption bands around 370 nm and 450 nm, with vibronic structure at around 360 nm and 380 nm, and 425 nm and 475 nm, respectively (Salomon et al. 2000; Swartz et al. 2001). This vibronic structure, which is absent in the absorption of free FMN in H_2O at room temperature (Figure 14.1), is the consequence of a reduction of inhomogeneous broadening of the flavin chromphore within the protein binding pocket as compared to that of free FMN (Schuttrigkeit et al. 2003) and is a reflection of the tight binding of FMN held in place by both hydrogen bonding and hydrophobic interactions. The reported extinction coefficients for the two domains of oat phot1 are for LOV1, $\varepsilon 449$ = 12 200 mol^{-1} cm^{-1} and $\varepsilon 370$ = 10 000 mol^{-1} cm^{-1}; for LOV2, $\varepsilon 447$ = 13 800 mol^{-1} cm^{-1} and $\varepsilon 378$ = 8700 mol^{-1} cm^{-1} (Salomon et al. 2000). The main absorption band around 450 nm is due to the first singlet state transition $S_0 \rightarrow S_1$, whereas the 360 nm band is associated with the $S_0 \rightarrow S_2$ transition (Müller 1991). Stark Spectroscopy provides information on changes in charge distribution in molecules upon optical excitation by measuring the changes in permanent dipole moment (permanent polarity) and in molecular polarizability (inducible polarity) associated with a transition. Work on model flavins (Stanley 2001) has shown that both transitions, S_1 and S_2, are accompanied by small changes in the permanent excited state dipole moment relative to the ground state. This slight increase in polarity in the excited states (less than 10%) implies a small change in the electrostatic interaction of the FMN with the immediate environment upon excitation.

In contrast to the minor alterations in permanent dipole moment, the change in polarizability associated with the $S_0 \rightarrow S_2$ transition (370 nm band) is much larger than that of the $S_0 \rightarrow S_1$ transition (450 nm band), making its absorption spectrum sensitive to the dielectric environment around the flavin-binding sites in proteins (Stanley 2001). The vibronic components of the absorption spectrum of the LOV domains in this 370 nm region differs among different domains, showing changes in both the wavelengths of maximal absorption and the relative intensities. In oat phot1-LOV2, the 360 nm peak is considerably lower than the 380 nm peak (dome shape), whereas in oat phot1-LOV1 domain, the peaks are slightly shifted and of nearly equal height (flat top shape) (Salomon et al. 2000). These features are very similar to those observed between N3 methyl flavins (Stanley 2001) (flat top) and free FMN (dome top) at low temperatures, where all flavins show these vibronic bands.

Within a given LOV domain, the vibronic structure and peak frequencies of the 370 nm band are particularly sensitive to replacement of the cys39 with an alanine. This sensitivity strongly suggests the presence of electrostatic interactions between the sulfur of cys39 and the FMN chromophore (Swartz et al. 2001). Comparison of

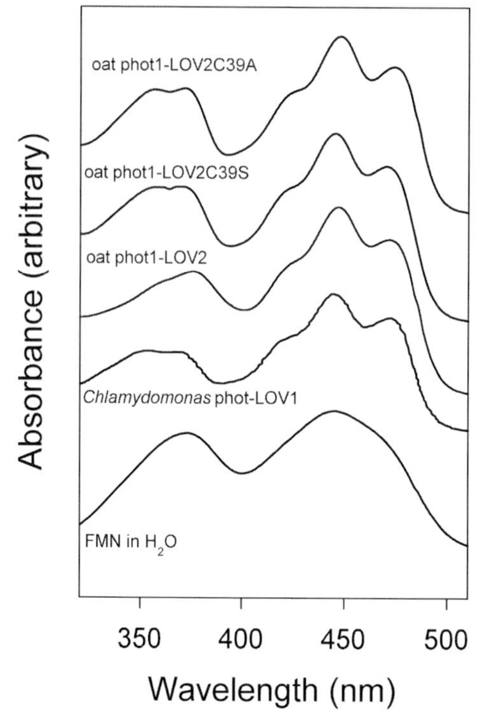

Figure 14.1 FMN vibronic structure evident for chromophore in LOV-domain binding pocket. UV-A absorption band of LOV domains is sensitive to the position, ionization state, and/or residue at position 39. Also, note difference in 370 nm absorption band between *Chlamydomonas* phot-LOV1 and wild-type oat phot1-LOV2.

the absorption spectra of various LOV domains (*At*-phot1-LOV1, LOV2 and *At*-phot2-LOV1, LOV2) shows that the major difference in absorption is the change in shape of the 360 nm peak, and a slight ~2 nm shift of the 450 nm peak (Sakai et al. 2001). *At*-Phot1-LOV2 and phy3-LOV2 show the typical dome shape, whereas *Clamydomonas* phot-LOV1 (Kottke et al. 2002), *At*-Phot1-LOV1, *At*-Phot2-LOV2, and the oat phot1-LOV2-C39A mutant show the flat top feature. In oat phot1-LOV2, the flatness of the spectrum seems to reflect a decrease in the environment polarity (C39 is more polar than C39A, Figure 14.1) (Corchnoy et al. 2003). Replacement of a cysteine with a more polar serine (C39S) has negligible effect on the UV-A absorption band, but it decreases fluorescence quenching (see below). This difference suggests a stronger polar interaction by cysteine as compared to serine. The stronger polar interaction by the cysteine residue could be explained by differences in the position of the side chains in the mutants, or alternatively by the presence of the net charge of a thiolate, S^-, a partially ionized thiol or perhaps involvement of a charge transfer complex. Although these perturbations of the short wavelength band must reflect effects of local fields on the chromophore electronic polarizability, the available structural data for phy3-LOV2 and *Chlamydomonas* phot-LOV1 domains provides no information on groups involved in these specific interactions.

At room temperature, differences in the vibronic structure of the 370 nm peak of *Chlamydomonas* phot-LOV1 and phy3-LOV2 are evident [phy3-LOV2 absorption is similar to that of oat phot1-LOV2 (Kennis et al. 2003) (Salomon et al. 2000) (Figure 14.1)]. This could reflect differences in the location of the sulfur of cys39 with respect to the FMN chromophore and/or differences in the binding-site electrostatics. The X-ray structure of the *Chlamydomonas* Phot-LOV1 was solved to 1.9 Å (Fedorov et al. 2003). In comparisons of the X-ray crystal structures of *Chlamydomonas* phot-LOV1 (Fedorov et al. 2003) and phy3-LOV2 (Crosson et al. 2001), the major discrepancy is in the position of the sulfur group of cys39. In *Chlamydomonas* phot-LOV1, the cysteine appears in two conformations, whereas in phy3-LOV2, the cysteine appears in a single conformation. However, the structure of *Chlamydomonas* phot-LOV1 was solved at liquid nitrogen temperatures, whereas that of phy3-LOV2 was solved at room temperature. It should be noted that phy3-LOV2 also displayed the cysteine in two conformations when solved at 77 K (Crosson, Chapter 15). The presence of these conformers alone does not explain the subtle, but significant, spectral differences that must originate from different chromophore environments. These different environments could account for their different reactivities, quantum yields, and cycling times (Kasahara et al. 2002). Therefore, it may not be appropriate to generalize the photochemical mechanisms proposed for a particular LOV domain to all LOV domains.

14.2.2.2 **Chromophore Fluorescence and Raman Scattering**

Free FMN has a relatively high fluorescence quantum yield and low efficiency for photochemical reactions in solution (Weber 1950). In an efficient system, photochemistry has to compete efficiently with both radiative (fluorescence) and thermal relaxation (internal conversion, IC) processes. Fluorescence efficiency of free FMN approaches 26% (Weber 1950), whereas in the protein binding pocket, it is around 13% (Kennis et al. 2003). Although protein–chromophore interactions listed above and binding-site electrostatics may have an effect on fluorescence yield, a major contribution to the decrease in fluorescence of FMN in the chromoprotein seems to be the competing process of intersystem crossing (ISC) into the flavin triplet state caused by the presence of the cysteine sulfur atom in the vicinity of the flavin chromophore [see below, (Holzer et al. 2002; Kennis et al. 2003)]. The fluorescence quantum yield of phy3 LOV2 is 0.13 (Kennis et al. 2003) and that of *Chlamydomonas* phot-LOV1 is 0.17 (Kennis et al. 2003). The LOV domains' fluorescence quantum yield is sensitive to both the chromophore binding-site electrostatic environment and the presence of the sulfur of cys39 (Salomon et al. 2000; Swartz et al. 2001).

The influence of binding-site electrostatics is evident from the relative marked increase in fluorescence yield upon substitution of the cys39 for a serine (0.13 to about 0.19) and alanine (0.13 to about 0.24) (Swartz et al. 2001). The increased spin orbit coupling in presence of sulfur is expected to enhance the $S_1 \rightarrow T_1$ transition by the ISC process, decreasing the fluorescence yield. The triplet yield depends on both the ISC rate and the internal conversion (IC) rate. The ISC rate increase for either oat phot1-

LOV2 vs. free FMN (Kennis et al. 2003) or oat phot1-LOV2 vs. oat phot1-LOV2C39A (Schuttrigkeit et al. 2003) has been reported to be about 2.4. There is disagreement in the reported values for triplet yield in the presence and absence of sulfur. The quantum yield of triplet formation (Φ_T) for free FMN has been reported to be about 0.6 (Kennis et al. 2003; Schuttrigkeit et al. 2003). There was no measurable increase in triplet formation in phy3-LOV2 ($\Phi_T = 0.6$) (Kennis et al. 2003). However, in oat phot1-LOV2, Φ_T was reported as 0.83 [consistent with an earlier report of 0.88 (Swartz et al. 2001)], an increase from 0.68 for oat phot1-LOV2C39A (Schuttrigkeit et al. 2003). The reported discrepancy in Φ_T between oat phot1-LOV2 and phy3-LOV2 originates from different values assigned to the IC rates. A low triplet yield of 0.26 reported for the *Chlamydomonas* phot-LOV1 domain is in sharp contrast to these high values in oat phot1-LOV2 (Islam et al. 2003). These discrepancies may reflect variability in the sample preparation and/or differences in fusion proteins used [calmodulin binding protein (CBP), maltose binding protein (MBP), and histidine-tagged protein] or the different LOV domains studied.

Both the absorption spectrum of the ground state (see above) and fluorescence data suggest that differences not apparent in the structural data must exist in the environment directly interacting with the FMN chromophore of various LOV domains studied to date. It is arguable from these data that there may be subtle positional differences in the cys39 sulfur in various LOV domains. An additional factor not addressed is that FMN in the LOV domains is restricted in a rigid binding pocket where conformational strain may affect the evolution of the system in the excited state, modifying the various relaxation channels, including fluorescence. The LOV domains' fluorescence emission shows vibronic structure that is absent in the free FMN in solution (Kennis et al. 2003). The vibronic structures of the fluorescence emission and excitation spectra reflect vibrational progressions in the ground and excited states, respectively. In general, identical progressions (mirror images) are indicative of small or no changes in the molecular coordinates of the atomic nuclei between ground and excited state. In oat phot1-LOV2, the progressions for fluorescence emission and excitation are about 1070 cm^{-1} and 1200 cm^{-1}, respectively. This difference is indicative of changes in the nuclear configurations in the excited state (Schuttrigkeit et al. 2003).

Fluorescence yield is affected by electrostatic interactions in the excited state. Ionic electrostatic potential provides the strongest and longest range interaction. Therefore, both location and ionization state of the sulfur of cys39 could be expected to influence the flavin fluorescence yield. In mercuric ion reductase, a cysteine sulfur is located slightly closer to isoalloxazine C(4a) (Müller 1991) than in phy3-LOV2 (Crosson et al. 2001) (2.7 Å vs. 4.2 Å,) and its ionization state is the dominant factor in controlling the fluorescence yield of the chromophore at low pH (Miller et al. 1990). It was first suggested by fluorescence pH titrations that the oat phot1-LOV2 sulfur of cys39 existed predominantly as a thiolate (S^-) at neutral pH (Swartz et al. 2001). The FMN fluorescence in oat phot1-LOV2 is independent of pH in the range 4–10. At high pH, there is nearly complete fluorescence quenching upon titration of the flavin N3 with an apparent pK around 10.5. At low pHs (3–4) there is a marked increase in the fluorescence of oat phot1-LOV2 (yield increases 0.13 to 0.20). As in the mercuric ion reductase system (Miller et al. 1990), this result was interpreted as a reduction in fluorescence quenching of the FMN upon protonation of the sulfur thiolate. It was suggested that cys39 was in a proton equilibrium with a putative base with an apparent pK around 4–4.5. This equilibrium would yield a ground state containing a Cys39 thiolate as the dominant species at neutral pH. In the reductase system, the fluorescence of FAD is fully quenched by the neighboring thiolate anion, while in the LOV2 domain, because of the longer distance, a thiolate effect would not expected to be as large.

This interpretation is not consistent with the reported disappearance of an S-H vibration (negative difference band) at around 2500 cm^{-1} upon formation of the cysteinyl adduct in phy3-LOV2 (Iwata et al. 2002) and *Chlamydomonas* phot-LOV1 (Ataka et al. 2003). Loss of an S–H bond is expected when a protonated ground state sulfur group (thiol) forms an adduct with the FMN. Band assignment was confirmed by the expected shift to lower frequencies caused by H/D isotopic replacement when measurements were carried out in D_2O. The amplitude of the negative SH vibration band would be expected to be pH dependent in a thiol/thiolate equilibrium in which the thiolate is the reactive species, because the thiol would ionize as thiolate is consumed in the reaction. The lack of pH dependence of the FTIR S–H band between pH 7 and pH 10 (77–250 K) argues against such a pK equilibrium in phy3-LOV2 (Iwata et al. 2003). Based on the apparent pH independence of the S–H difference band amplitude, it was suggested that the phy3-LOV2 C39 thiol has a pKa higher than 10, inferring that it is buried in a hydrophobic environment but able to equilibrate with bulk deuterons (Iwata et al. 2003).

The fusion protein used for this study (Iwata et al. 2003) is larger than that used for the X-ray crystallography work, with 20 and 40 additional amino acids on the N terminus and C terminus, respectively, and a CBP tag. It should be noted that LOV2 fusion proteins used in different laboratories contain varying amounts of native sequence and specific tags that may or may not have been cleaved, in addition to the domain sequence used for two available X-ray crystallographic studies of LOV domains (Crosson et al. 2001; Fedorov et al. 2003). These additional components may influence the solution structure, molecular folding, solvent accessibility of specific residues, and more important the stabilization of conformations favoring different positions of the reactive cysteine. Adduct formation occurs at cryogenic temperatures [phy3 LOV2, (Iwata et al. 2003); oat phot1-LOV2, (T.E. Swartz and R.A. Bogomolni unpublished; M. Gauden, S. Crosson, I. Van Stokkum, R. van Grondelle, K. Moffat and J. Kennis unpublished)]. At these low temperatures, only 36% and 64% of the ground state converts into the adduct at 77 K and 100 K, respectively (Iwata et al. 2003), and a temperature-dependent amplitude of the S–H band is attributed to an increase in the extinction coefficient for that absorption band at lower temperatures (Iwata et al. 2003). These observations point to the importance of molecular dynamics and other processes in the reaction mechanism.

Pre-resonance Raman of the *Chlamydomonas* phot-LOV1 (Ataka et al. 2003) and oat phot1-LOV2 (J. Ciorciari and R.A.Bogomolni unpublished) ground states obtained with 752 nm and 785 nm Raman probe beams, respectively, show spectra very similar to that of free FMN, consistent with non-covalent binding of the FMN within the

LOV domain pocket. Hydrogen bonding of FMN in the protein causes slight shifts in some of the bands, confirming spectroscopically hydrogen bonds seen in the crystal structure (Ataka et al. 2003). The pre-resonance Raman spectrum of the ground state is dominated by chromophore vibrational bands but shows some contribution of the protein amide vibrations in the 1600–1700 cm^{-1} range that overlap with chromophore carbonyl bands (Ataka et al., 2003; J. Ciorciari and R.A. Bogomolni unpublished).

14.2.2.3 **Chromo-peptide Optical Activity**

The circular dichroism (CD) spectrum of the oat phot1-LOV2 ground state in the near UV-visible regions shows negative bands that coincide with absorption bands of the chromophore (Salomon et al. 2000; Corchnoy et al. 2003). This optical activity is not present in the ground state spectrum of free FMN in aqueous solution (S. Corchnoy and R. A. Bogomolni unpublished), suggesting that FMN in LOV2 is in a protein-asymmetric environment. In the mid-UV region, there are CD bands in the aromatic region, perhaps contributed by tryptophan and high-energy chromophore bands. The far UV CD spectrum shows a positive band around 193 nm and negative bands at 208 nm and 224 nm, indicative of the presence of alpha-helical secondary structure, and the overall spectrum is consistent with the secondary-structure content shown in the X-ray crystal structure (Corchnoy et al. 2003).

14.3 Photochemistry

14.3.1 Photocycle Kinetics and Structure of its Intermediates

To date, almost a dozen LOV domains have been shown to undergo similar chromophore structural changes during their photocycles. However each has distinct kinetic characteristics and quantum yields and some have truncated photocycles (Salomon et al. 2000; Kasahara et al. 2002; Kottke et al. 2002; Losi et al. 2002; Imaizumi et al. 2003; Schwerdtfeger et al. 2003). Early studies of the photochemistry were carried out in the *E coli*-expressed oat phot1-LOV2 domain (Salomon et al. 2000; Swartz et al. 2001). Photoreactions of other FMN-binding LOV domains studied afterwards were shown to be qualitatively similar. Following light absorption, the LOV domains undergo a photocycle characterized by a series of transient photo-intermediates (Swartz et al. 2001; Kottke et al. 2002; Losi et al. 2002). Presently, two excited states and one metastable thermal intermediate have been kinetically resolved in the LOV domain photocycle. Briefly, at room temperature and slightly alkaline pH, light absorption at around 450 nm promotes the $S_0 \rightarrow S_1$ transition to the FMN singlet excited state that decays with a half-time around 2 ns into a red-absorbing species that absorbs maximally at 660 nm (Kennis et al. 2003). This species, labeled $LOV2^{T}_{660}$, has been shown to be an FMN triplet state (Figure 14.2 A) (Swartz et al. 2001; Kottke et al. 2002; Losi et al. 2002; Kennis et al. 2003). $LOV2^{T}_{660}$ decays in microseconds into the second metastable intermediate, which absorbs maximally at 390 nm and is labeled

LOV^S_{390} (Swartz et al. 2001). The LOV^S_{390} intermediate thermally relaxes back to the ground state, LOV^D_{447} (Figure 14.2 B).

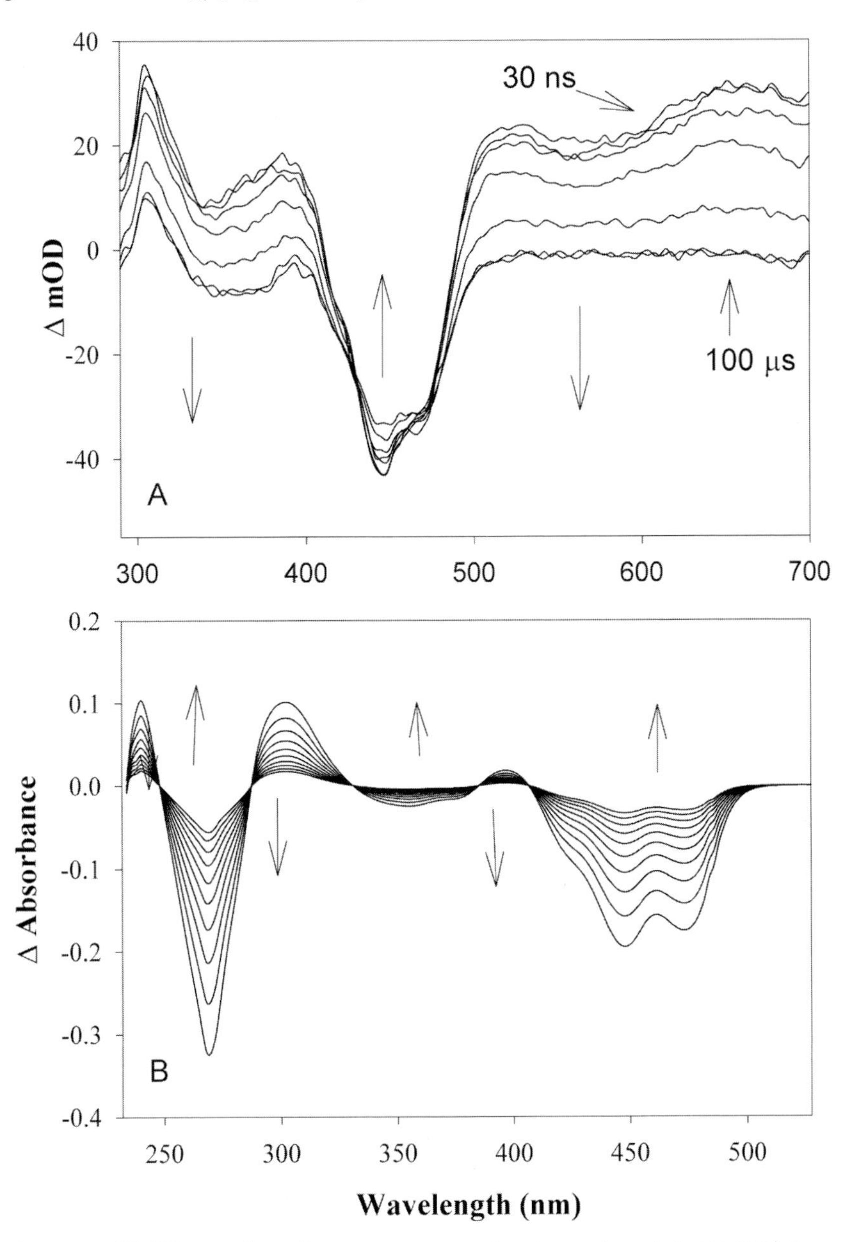

Figure 14.2 (A) difference absorption spectra of oat phot1-LOV2 after excitation with a 477-nm laser pulse. Spectra collected at 0.03, 0.13, 0.33, 1.0, 3, 10, and 100 μs after light excitation. (B) Time-resolved light-induced absorption changes for oat phot1-LOV2 between 1 and 100 s after light excitation. Spectra are taken every 15 s. Arrows indicate spectral changes with time.

Global kinetic analysis of laser flash-induced absorption changes of oat phot1-LOV2 (actinic wavelength 477 nm, 4-ns pulse) measured from 30 ns to 100 μs gives a single decay time constant of 2 μs in this time window, and isosbestic points indicate a single resolvable, spectroscopically identifiable species (Swartz et al. 2001). The data were globally fit by assuming a linear kinetic scheme with two intermediates ($LOV2^T_{660}$ and LOV^S_{390}) and the spectra of both species was calculated (Swartz et al. 2001) (Figure 14.3). Because the measured amount of the initial bleach of the ground state is half that of the amount of LOV^S_{390} formed, a simultaneous 1:1 split decay of $LOV2^T_{660}$ back to the ground state $LOV2^D_{447}$ and forward to LOV^S_{390} was included in the kinetic scheme, resulting in a calculated time constant of 4 μs for each decay direction (Swartz et al. 2001). The calculated $LOV2^T_{660}$ spectrum closely matches that of the triplet state of FMN (Figure 14.3) (Swartz et al. 2001). Measurements on the femtosecond timescale show direct conversion of the FMN triplet state from the singlet excited state in about 2–3 ns (Kennis et al. 2003). Since no other intermediates are seen on this time scale, the triplet state forms directly from the excited singlet state via intersystem crossing (ISC) (Kennis et al. 2003; Schuttrigkeit et al. 2003). Further analysis of the triplet-state spectrum suggests a mixture of species in which the FMN N5 exists as an equilibrium of both protonated and unprotonated forms (Kennis et al. 2003).

The LOV^S_{390} species that is formed from the triplet state decays spontaneously in the dark back to the ground state. The rate of return of the ground state varies from a few seconds to many minutes (Salomon et al. 2000; Kasahara et al. 2002; Losi et al.

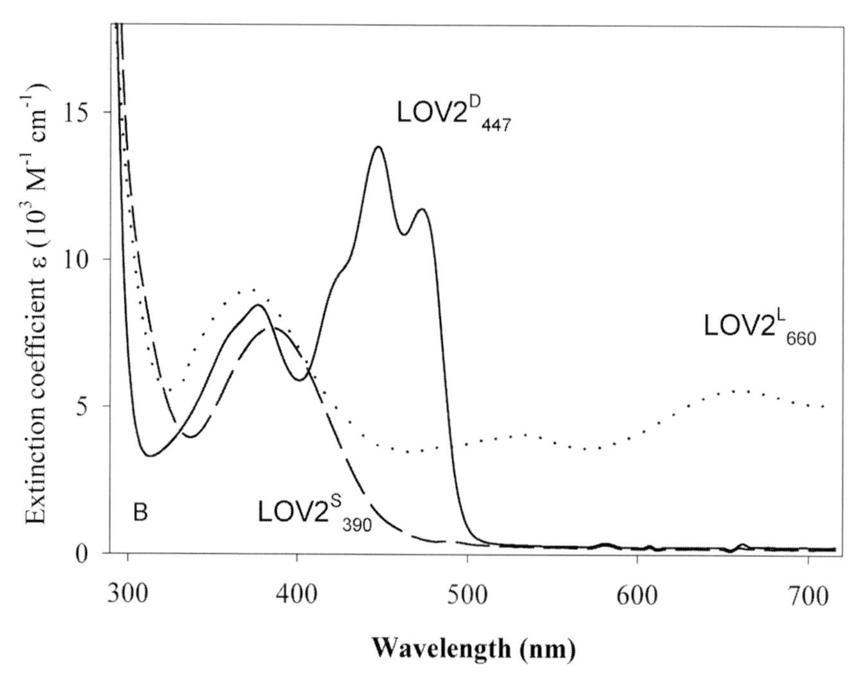

Figure 14.3 Ground-state oat phot1-LOV2 spectrum and calculated intermediate spectra. LOV^L_{660} is the triplet state. The LOV^S_{390} state contains the cysteinyl adduct.

2002; Schwerdtfeger et al. 2003), with some LOV domains not returning to the ground state at all (Imaizumi et al. 2003). This long-lived metastable intermediate involves the formation of a protein-FMN covalent bond. Specifically an S–C bond is formed between the sulfur of cys39 and the C(4a) carbon of FMN (Figure 14.4). Because this bond forms in microseconds and decays in seconds, continuous illumination of a LOV2 sample with blue light converts most of the sample into LOV^S_{390}; this property has conveniently allowed for structural and spectroscopic studies of the LOV^S_{390} intermediate. Conformation of the C–S bond in LOV^S_{390} was obtained in this way by X-ray crystallography, and NMR (Salomon et al. 2001; Crosson et al. 2002; see Crosson, chapter 15). Formation of the cysteine-C(4a) bond during the photocycle was also inferred from difference infrared-absorbance spectroscopy (FTIR). Three vibrational bands occurring at 1580 cm^{-1}, 1550 cm^{-1}, and 1350 cm^{-1}, that are strongly coupled to the stretching of the C(4a)–N5 double bond, disappear upon formation of $LOV2^S_{390}$ (Swartz et al. 2002; Ataka et al. 2003), and two new bands form at 1516 cm^{-1} and 1536 cm^{-1} (Swartz et al. 2002). Normal mode calculations on a simplified model C(4a) thio-adduct predict the disappearance of the former bands and the appearance of these two bands on adduct formation (Swartz et al. 2002). They are associated with symmetric and antisymmetric stretching modes involving all three rings of an isoalloxazine containing an sp^3 hybridized carbon at position C(4a). Thus, FTIR results also show formation of a bond at the C(4a) position (Swartz et al. 2002; Ataka et al. 2003). The light-induced CD changes in the 250–500 nm range show the appearance of a large positive band at 270–290 nm upon formation of $LOV2^S_{390}$. This band coincides with a higher energy FMN absorption band around 270 nm. The large optical

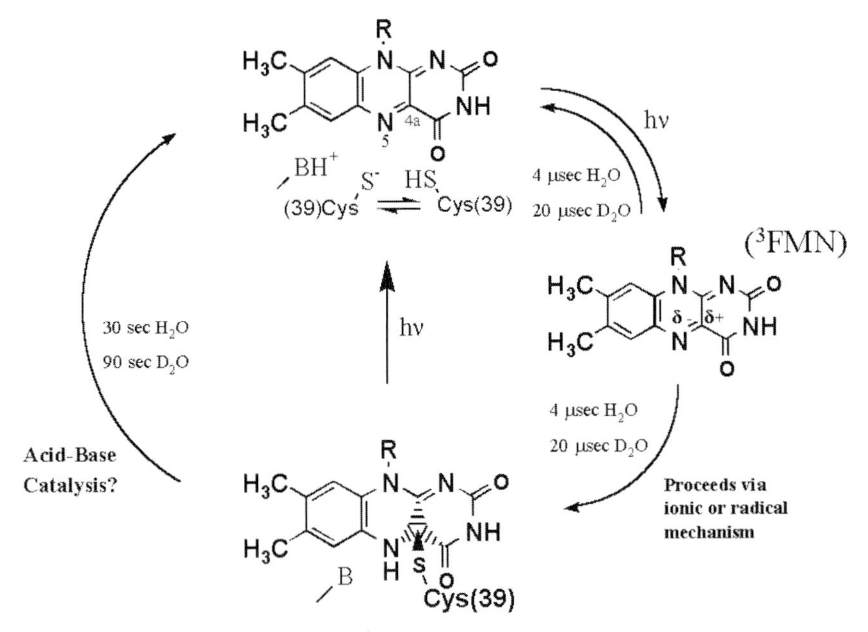

Figure 14.4 Oat phot1–LOV2 reaction scheme.

activity associated with this band is consistent with the formation of an sp^3 chiral C(4a) center in the $LOV2^S_{390}$ adduct (Salomon et al. 2000; Corchnoy et al. 2003).

Both circular dichroism (CD) and difference FTIR have demonstrated protein conformational perturbations associated with the photocycle (Swartz et al. 2002; Ataka et al. 2003; Corchnoy et al. 2003; Iwata et al. 2003). Numerous positive and negative FTIR bands are observed in the protein amide regions (Swartz et al. 2002; Ataka et al. 2003; Iwata et al. 2003), and CD difference spectra show transient loss in alpha helicity (Corchnoy et al. 2003). The magnitude and character of these protein conformational changes depend on the size and sequence of the specific LOV constructs used for the studies. In particular, these structural changes become more apparent in larger constructs containing additional protein segments on the C-terminal end of the LOV2 domain. In a larger construct, difference 3D NMR has revealed protein perturbations of an amphipathic alpha-helix C-terminal from the LOV2 domain (Harper et al. 2003; see Crosson, Chapter 15), the direction in which signal propagation is expected in the full-length chromoprotein. However, the physiological significance of this structural change is not clear because these are truncated phototropin molecules. In the full construct, the relevant structural change may propagate further without major changes in such an intervening segment.

14.3.2
Photo-backreaction

It was inferred that in *Chlamydomonas* phot-LOV1 the inability to fully bleach the ground state was due to a photo-backreaction from LOV^S_{390} to LOV^D_{447} (Kottke et al. 2002). This photo-backreaction was directly measured in phy3-LOV2 and occurs in 100 ps (Kennis et al. 2004), with quantum yield estimated to be between 0.2 and 0.3 (Kennis et al. 2004). It is unknown if this UV-A induced back-reaction is physiologically relevant.

14.4
Reaction Mechanisms

14.4.1
Adduct Formation

14.4.1.1 Proton Transfer Reaction Initiates Triplet State Decay

The rate of adduct formation is 5 times slower in D_2O than in H_2O, confirming that the rate-limiting step between triplet state and adduct formation is a proton-transfer reaction (Corchnoy et al. 2003). The triplet state (LOV^T_{660}) in wild-type oat phot1-LOV2, decays simultaneously back to LOV^D_{447}, the ground state, and LOV^S_{390} in both D_2O and H_2O. The observed forward decay of LOV^T_{660}, to LOV^S_{390} is five times slower in D_2O (Corchnoy et al. 2003). Analysis of the data shows that the forwards/backward 1:1 split ratio for its decay is unchanged in D_2O and that the backward decay to the ground state is also slowed down 5-fold. This analysis suggests that both decay reactions of the triplet (forward to the adduct state and back to the ground state) are rate limited by a single proton-transfer reaction. The magnitude of the deuterium result is suggestive of a primary isotope effect (Melander et al. 1980). It is reasonable that this primary isotope effect is protonation of N5 (see mechanism below); however, at this time there is no explicit data to support this hypothesis.

Mutation of cysteine 39 to an alanine (LOV2C39A) results in a truncated photocycle as compared to the wild type. Because these proteins are missing the reactive cysteine, they do not form the LOV^{S}_{390} intermediate; however, after a short laser flash, these proteins do form the LOV^{T}_{660} intermediate. This intermediate $LOV2C39A^{T}{}_{660}$ decays directly back to the ground state ($LOV2C39A^{D}_{447}$) in 72 μs, one order of magnitude slower than wild type (Swartz et al. 2001). The presence of a sulfur atom in the WT oat phot1-LOV2 domain may contribute to the triplet decay presumably by increasing the rate of spin flipping due to spin orbit coupling (Swartz et al. 2001). Not unexpectedly, the decay of $LOV2C39A^{T}{}_{660}$ back to the ground state, $LOV2C39A^{D}_{447}$, does not show a deuterium effect (Corchnoy et al. 2003).

Several alternate reaction pathways for the formation of the adduct following initial activation to the triplet state have been proposed. They include direct transfer of a proton from the cysteine thiol to N5 of the triplet state followed by formation of the S-C bond (Crosson et al. 2001), excited-state proton transfer to N5 preceding triplet formation followed by reaction of a thiolate with FMN C(4a) (Kennis et al. 2003), involvement of a flavosemi-quinone free radical and reaction with a sulfur radical (Bittl et al. 2003; Kay et al. 2003; Kottke et al. 2003), and proton transfer to the triplet FMN followed by reaction of a C(4a) carbo-cation with a cys39 thiolate (Swartz et al. 2001). Available evidence is insufficient to establish the correct mechanism. Common to all mechanisms is the redistribution of charge around the FMN N5-C(4a) double bond that occurs in the FMN triplet state, resulting in an increase of basicity of N5 (Song 1968; Fedorov et al. 2003; Neiss et al. 2003). In the ionic recombination mechanism, protonation of N5 either from the C39 thiol (Crosson et al. 2001; Kennis et al. 2003) or from an as yet unidentified acid group (Swartz et al. 2001), draws electronic density from the N5-C(4a) double bond, leaving C(4a) as a reactive carbo-cation that is attacked by the ionized sulfur of cysteine 39, forming the flavin-cysteinyl adduct.

Kennis et al. (2003) proposed that proton transfer from the C39 thiol occurs in the excited state with transient thiolate formation, whereas Swartz et al. (2001) have invoked the existence of a ground-state thiolate and a putative proton-acceptor residue that donates a proton to N5, initiating the split decay of the triplet state to either adduct or ground state. Vibrational-spectroscopy evidence that the sulfur of cysteine exists as a thiol has led to the argument that the proton originates directly from the cysteine SH. Although the proton could come from another group, no other acid groups have been identified in the vicinity of the chromophore. The crystal structure, however, shows two water molecules that could conduct a proton from a residue that is not directly interacting with the chromophore. The C39A mutants do produce a neutral semiquinone radical under high light intensities (see below) (Kay et al. 2003). Formation of the neutral semquinone requires donation of a hydrogen atom from an as yet unidentified group.

It is generally accepted that in thiol proteases, the cysteine in the catalytic site which forms a thiolate-imidazolium ion pair is separated by about 4 Å from the imidazole ring of a histidine (Bergmann et al. 1997). The pKa values of imidazole and thiol in solution are around 6 and 8.5 respectively. In the thiol proteases, these pKa values shift to 8 and 4.5, respectively, with the thiol becoming a stronger acid than the imidazole (McGrath 1999). The pKa of FMN N5 in water is ~0 in the ground state and jumps to about pKa ~5 in the triplet state. In phy3-LOV2, the sulfur of cys 39 is 3.7 Å from N5 and 4.2 Å from C(4a) (Crosson et al. 2001). It would not be surprising that they would interact electrostatically and modulate each other's pKa values in both the ground state and triplet state. The expectation is that the N5 pKa is increased and the thiol pKa is decreased due to the interaction. It should also be noted that hydrogen bonding has been shown to modulate the pKa of N5 in the excited state (Yagi et al. 1980). Presumably, the pKa values in the triplet state would approach values consistent with a spontaneous proton transfer.

The radical-pair mechanism involves formation of a flavin semiquinone from the FMN triplet state in a one-electron photoreduction that may be followed by a proton transfer to give an anionic or neutral flavin radical (Figure 14.5). In this scenario, the sulfur donates either an electron (and then the proton) or a hydrogen atom resulting in formation of a sulfur radical and neutral semiquinione radical. The adduct would form by recombination of the two radicals. The oat phot1-LOV2 triplet state absorption spectrum shows no evidence of a semiquinone (Swartz et al. 2001). It is possible that the triplet state is the rate limiting step, and that the rapid rate of disappearance of the semiquinone would not allow its transient accumulation and its detection in the absorption changes. Kay et al. (Kay et al. 2003) point out that protonation of the FMN triplet state is not necessarily the rate-limiting step in a radical-pair mechanism, whereas protonation of the triplet state should be a rate-limiting step in an ionic mechanism. The ionic mechanism would allow adduct formation in the triplet state, whereas the reaction-pair mechanism would require the reaction in a singlet state (antiparallel spins recombine to form a bond) (Kay et al. 2003).

The radical recombination mechanism proposed for adduct formation was based on the detection of a flavine-semiquinone radical in a C39A mutant of oat phot1-LOV2 [labeled C450A in that work (Kay et al. 2003)]. In the absence of sulfur, the photoreaction produces significant amounts of the neutral flavo-semiquinone that absorb maximally around 600 nm. The mechanism proposed for this reaction involves formation of the triplet state, followed by proton and electron transfers from unidentified donor(s). In wild type oat phot1-LOV2, the proposed mechanism conserves these features and simply substitutes the thiol for the unidentified donor (Kay et al. 2003). At physiological light intensities, the formation of a neutral flavosemiquinone was not observed in LOV2C39A (Swartz et al. 2001), but it was observed at very high light intensities (T. E. Swartz and R.A. Bogomolni unpublished). At similar high light intensities, a small fraction of wild-type oat phot1-LOV2 also forms a neutral semiquinone (T.E. Swartz and R.A. Bogomolni unpublished). At these light intensities, most of the sample forms the adduct and undergoes a fully reversible photocycle returning to the ground state in tens of seconds. The long-lived semiquinone forms via a branched reaction pathway. It returns to the ground state in tens of minutes to hours, presumably re-oxidized by atmospheric oxygen (T.E Swartz and R.A. Bogomolni unpublished). Figure 14.5 summarizes the various reaction mechanisms proposed for adduct formation.

It should be mentioned that although the *Chlamydomonas* phot-LOV1 photocycle contains intermediates similar to those of oat phot1-LOV2, it involves a more complex kinetic scheme. The *Chlamydomonas* phot-LOV1 triplet state decays forward with two time constants, one of 800 ns and one of 4 μs; it has been suggested that the triplet state exists as a mixture of two species with different decay times (Kottke et al. 2002). However, because their spectral properties are nearly identical, it is not possible to discriminate between the two (Kottke et al. 2002). In contrast to oat phot1-LOV2, *Chlamydomonas* phot-LOV1 does not show an appreciable return to the ground state from the triplet state. The isotope deuterium effect has not been reported on the *Chlamydomonas* phot-LOV1 adduct formation or decay.

14.4.2 Adduct Decay

The mechanism of dark recovery of the ground state $LOV2^{D}_{447}$ from LOV^{S}_{390}, which requires breakage of a carbon sulfur-bond, remains a mystery. Synthetic C4a-sulfur adduct model compounds have been used for structural studies (Müller 1991), and the C(4a)-Sulfur covalent bond is strong (typically greater than 200 kJ mol^{-1}) and stable in aqueous media. Breakage of this bond in the protein environment must involve a catalytic process provided by specific protein residues in the FMN binding pocket. In oat phot1-LOV2, the rate of adduct decay shows only a slight decrease at lower pH [apparent pK around 6.5 (Corchnoy et al. 2003)], but in *Chlamydomonas* phot-LOV1 there is a marked pH dependence in which the decay rate increases several fold between pH 8 and pH 3 with an apparent pK around 5–6 (Kottke et al. 2002), indicating that it is base-catalyzed. These different pH sensitivities demonstrate another example of the differences in chemical behavior between LOV domains. Abstraction of the N(5) proton (acid–base catalysis) could be the event that initiates the back reaction. The dark recovery rate in oat phot1-LOV2 is 3 times slower in D_2O than in H_2O, suggesting that a proton-transfer reaction is the rate-limiting step (Swartz et al. 2001).

Because the deuterium effect is only a factor of three, the rate-limiting step in the back reaction is perhaps not a primary isotope effect and could reflect breaking of a hydrogen bond rather than a proton transfer. Because the pKa of N5 is modulated by hydrogen bonding on the flavin (Yagi et al. 1980), perhaps the variability of relaxation in different LOV domains is a measure of this hydrogen bonding. Kinetics of CD changes in the far UV indicate that the protein returns to the ground state with the same kinetics as the chromophore and with identical kinetic deuterium isotope effects suggesting concerted events and a common rate-limiting proton-associated event (Corchnoy et al. 2003). The X-ray structure fails to suggest a suitable catalytic basic group in the vicinity of N5. The back reaction has been measured in almost a dozen different LOV domains, with rates varying from seconds to minutes to no return of the ground state (Salomon et al. 2000; Kasahara et al. 2002; Kottke et al. 2002; Losi et al. 2002; Imaizumi et al. 2003; Schwerdtfeger et al. 2003). The sequences of

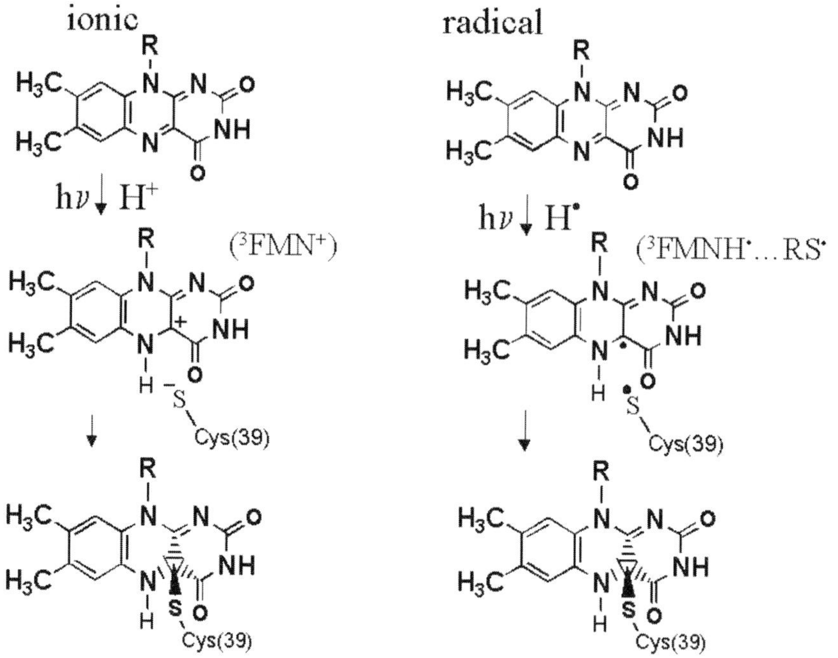

Figure 14.5 Proposed reaction mechanisms for adduct formation.

these various LOV domains present no clear insight into the varying relaxation rates. Difference FTIR spectra show hydrogen-bonding perturbations associated with the two buried waters that are in close proximity to the chromophore (Iwata et al. 2003), suggesting that these waters are possibly involved in the photochemistry. They may form a proton-conducting channel to a basic group not in the immediate vicinity of the chromophore, a mechanism that has been found in other proteins. It has been suggested that mediation of the structure of a conserved salt bridge occurs during the photocycle and results in a transmitted signal (Crosson et al. 2003).

14.4 Future Perspectives

The LOV domains are the photosensitive module of larger proteins that control varying physiological responses. The studies reviewed here have been carried out in isolated fusion chromoproteins expressed in heterologous systems (*E. coli*). Therefore the photochemical behavior and molecular changes observed in these systems do not provide information on the important mechanism of intramolecular signal transduction in the native chromoproteins. Neither the structure nor the aggregation state of full-length phototropin is known. Measurements on the aggregation state of LOV1 (M. Salomon, personal communication) suggest phot1 forms a dimer raising the possibility of cis vs. trans signal transduction. In addition, the native photoreceptor is membrane bound. It is unknown if the protein is post-translationally modified with a membrane anchor or binds a membrane protein and if these components affect photochemistry. The photocycle kinetics of full-length phototropin expressed in insect cells (Kasahara et al. 2002) differ from those of isolated LOV domains but resemble those of peptides containing both LOV1 and LOV2 domains (Kasahara et al. 2002). The effect of phosphorylation on photochemistry is also unknown. These questions will only be answered when native protein is isolated from plants.

References

Ataka, K., P. Hegemann, et al. (2003). *Biophys. J.* 84(1): 466–474.

Bergmann, E. M., S. C. Mosimann, et al. (1997). *J. Virol.* 71(3): 2436–2448.

Bittl, R., C. Kay, et al. (2003). *Biochemistry* 42(28): 8506–8512.

Briggs, W. and J. Christie (2002). *Trends Plant Sci.* 7(5): 204–210.

Corchnoy, S., T. Swartz, et al. (2003). *J. Biol. Chem.* 278(2): 724–731.

Crosson, S. and K. Moffat (2001). *Proc. Natl. Acad. Sci. USA* 98(6): 2995–3000.

Crosson, S. and K. Moffat (2002). *Plant Cell* 14(5): 1067–1075.

Crosson, S., S. Rajagopal, et al. (2003). *Biochemistry* 42(1): 2–10.

Fedorov, R., I. Schlichting, et al. (2003). *Biophys. J.* 84(4): 2474–2482.

Harper, S., L. Neil, et al. (2003). *Science* 301(5639): 1541–1544.

Holzer, W., A. Penzkofer, et al. (2002). *Photochem. Photobiol.* 75(5): 479–487.

Imaizumi, T., H. Tran, et al. (2003). *Nature* 426(6964): 302–306.

Islam, S., A. Penzkofer, et al. (2003). *Chem. Phys.* 291(1): 97–114.

Iwata, T., D. Nozaki, et al. (2003). *Biochemistry* 42(27): 8183–8191.

Iwata, T., S. Tokutomi, et al. (2002). *J. Am. Chem. Soc.* 124(40): 11840–11841.

Kasahara, M., T. Swartz, et al. (2002). *Plant Physiol.* 129(2): 762–773.

Kay, C., E. Schleicher, et al. (2003). *J. Biol. Chem.* 278(13): 10973–10982.

Kennis, J., S. Crosson, et al. (2003). *Biochemistry* 42(12): 3385–3392.

Kennis, J. T. M., I. H. M. van Stokkum, et al. (2004). *J. Am. Chem. Soc.* 126(14): 1412–1413.

Kottke, T., B. Dick, et al. (2003). *Biochemistry* 42(33): 9854–9862.

Kottke, T., J. Heberle, et al. (2002). *Biophys. J.* 84(2): 1192–1201.

Losi, A., E. Polverini, et al. (2002). *Biophys. J.* 82(5): 2627–2634.

McGrath, M. E. (1999). *Annu. Rev. Biophys. Biomolec. Struct.* 28: 181–204.

Melander, L. and W. Saunders (1980). *Reaction rates of isotopic molecules.* New York, Wiley.

Miller, S. M., V. Massey, et al. (1990). *Biochemistry* 29(11): 2831–41.

Müller, F. (1991). *Chemistry and biochemistry of flavoenzymes.* Boca Raton, CRC Press.

Neiss, C. and P. Saalfrank (2003). *Photochem. Photobiol.* 77(1): 101–109.

Sakai, T., T. Kagawa, et al. (2001). 98(12): 6969–74.

Salomon, M., J. Christie, et al. (2000). *Biochemistry* 39(31): 9401–9410.

Salomon, M., W. Eisenreich, et al. (2001). *Proc. Natl. Acad. Sci. U. S. A.* 98(22): 12357–12361.

Schuttrigkeit, T., C. Kompa, et al. (2003). *Chem. Phys.* 294(3): 501–508.

Schwerdtfeger, C. and H. Linden (2003). *Embo J.* 22(18): 4846–4855.

Song, P. S. (1968). *Photochem. Photobiol.* 7(3): 311–3.

Stanley, R. J. (2001). *Antioxid. Redox Signal.* 3(5): 847–866.

Swartz, T., S. Corchnoy, et al. (2001). *J. Biol. Chem.* 276(39): 36493–36500.

Swartz, T., P. Wenzel, et al. (2002). *Biochemistry* 41(23): 7183–7189.

Van der Horst, M. A. and K. J. Hellingwerf (2004). *Accounts Chem. Res.* 37(1): 13–20.

Weber, G. (1950). *Biochem. J.* 47: 114–121.

Yagi, K., N. Ohishi, et al. (1980). *Biochemistry* 19(8): 1553–7.

15
LOV-Domain Structure, Dynamics, and Diversity

Sean Crosson

15.1
Overview

Light, oxygen, or voltage (LOV) domains, a subset of the PER-ARNT-SIM (PAS) superfamily (Taylor and Zhulin, 1999), were originally identified as the loci for blue-light absorption in the plant photoreceptor kinases known as phototropins (Christie et al., 1999). These domains have since been shown to act as photosensory modules in DNA-binding and F-box proteins that regulate circadian rhythms in fungi (Cheng et al., 2003) and higher plants (Imaizumi et al., 2003), and have been identified in the genomes of a variety of other organisms including several species of photosynthetic and non-photosynthetic eubacteria (Crosson et al., 2003; Losi et al., 2002). With the exception of the phototropins, which possess two tandem LOV domains adjacent to a carboxy-terminal serine/threonine kinase (Huala et al., 1997), these photosensory modules are typically found as a single copy at the amino-terminal region of a structurally and functionally diverse set of proteins (Crosson et al., 2003). LOV domains contain approximately 110 amino acids, bind a single molecule of flavin, and undergo a unique photochemical transformation in response to blue-light absorption in which a conserved cysteine residue forms a covalent bond with the 4a carbon of the flavin cofactor (Crosson and Moffat, 2002; Salomon et al., 2001). In the phototropins, light-driven formation of this cysteinyl-C(4a) adduct leads to upregulation of the carboxy-terminal serine/threonine kinase (Briggs and Christie, 2002). The function of LOV domains in proteins other than phototropin is not as well understood.

The molecular and structural basis of how LOV domains regulate the activity of the kinase domain of phototropins, or indeed any other domain to which they are attached, is an active area of investigation. This chapter focuses on recent work describing the structure and dynamics of LOV domains in their dark and photoexcited states. Included is a comparative analysis of LOV-domain structure that discusses the relation to other PAS domains and the structurally similar GAF domain family (Aravind and Ponting, 1997). In addition, an updated analysis of proteins containing LOV domains outlining the diversity in their domain structure and taxonomic distribution is included.

Handbook of Photosensory Receptors. Edited by W. R. Briggs, J. L. Spudich

ISBN 3-527-31019-3

15.2
LOV Domain Architecture and Chromophore Environment

X-ray crystallographic studies of *Chlamydomonas* phototropin LOV1 (Fedorov et al., 2003) and maidenhair fern phy3 LOV2 (Crosson and Moffat, 2001) reveal that LOV domains exhibit a prototypical PAS fold (Pellequer et al., 1998) consisting of a five-stranded antiparallel β-sheet flanked by the helix-turn-helix motif αA/αB, a single 3_{10} helical turn α'A, and the 15-residue connector helix αC (Figure 15.1 A). A least-squares fit of the refined coordinates of *Chlamydomonas* phototropin LOV1 and fern phy3 LOV2 confirms that the secondary structural elements and tertiary architecture of LOV1 and LOV2 are nearly identical (main chain root-mean-square deviation = 0.73 Å). Embedded in the PAS fold is a single molecule of flavin mononucleotide (FMN) that, in the dark state, is non-covalently bound to the core of the protein by a series of polar interactions with the pyrimidine moiety and nonpolar interactions with the dimethylbenzene moiety of the isoalloxazine ring. Additional hydrogen-bond and charge-charge interactions stabilize the ribityl side chain and terminal phosphate within the LOV fold (Figure 15.1 B). To date, all LOV domains have been shown to bind FMN with the exception of the LOV domains of the *Neurospora* circadian regulators, White Collar 1 (Wc-1) and VIVID (VVD). Wc-1 binds flavin adenine dinucleotide (FAD) and requires this cofactor for its function (He et al., 2002), while VVD can bind both FAD and FMN when expressed in *E. coli*. No structural informa-

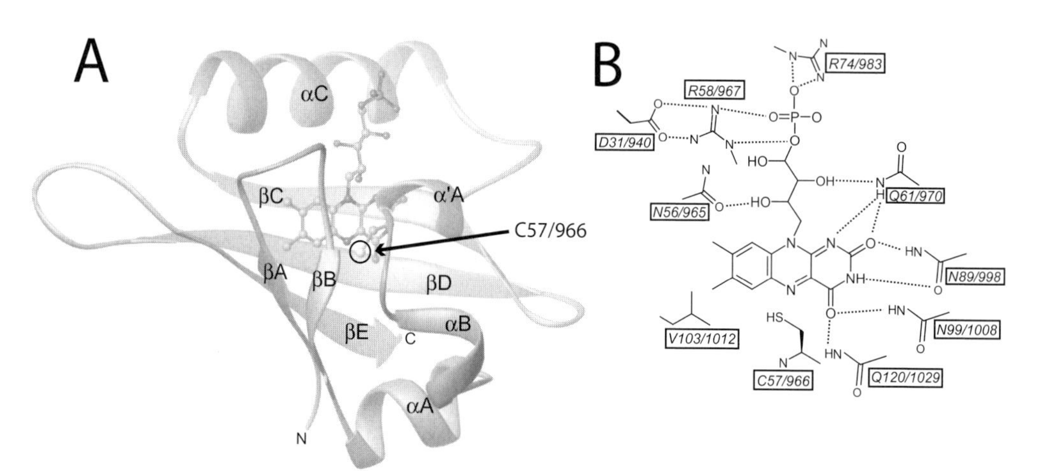

Figure 15.1 LOV domain architecture and cofactor-binding environment. (A) Ribbon diagram of maidenhair fern phy3 LOV2 (PDB accession number 1G28). Secondary structure elements are marked on the diagram; nomenclature follows reference (Crosson and Moffat, 2001). The conserved cysteine side chain is shown attached to the amino terminal end of helix α'A and is labeled with a circle and an arrow. The FMN cofactor is shown in the core of the domain. (B) Hydrogen bond network between the LOV domain polypeptide and the FMN cofactor (dashed lines). The 1.9 Å structure of *Chlamydomonas* phototropin LOV1 (PDB accession number 19NL) was used to determine hydrogen bonds. Residues forming bonds are labeled with the first number corresponding to the *Chlamydomonas* phototropin LOV1 structure and the second corresponding to the phy3 LOV2.

tion is available on how the LOV domain accommodates the adenine nucleotide moiety of FAD, which is bound to the ribityl phosphate and is thus predicted to be positioned outside of the flavin-binding core of the LOV fold.

A conserved cysteine residue, corresponding to C966 in fern phy3 LOV2 and C57 in *Chlamydomonas* phototropin LOV1, flanks the *si* face of the flavin isoalloxazine ring. This residue is in nearly identical positions in the LOV1 and LOV2 crystal structures (Figure 15.2). In the 2.7 Å dark-state structure of phy3 LOV2 solved at room temperature (Crosson and Moffat, 2001), the S_γ of C966 is 4.2 Å from C(4a). Thus, facile rotation about the C_α–Cα bond combined with small movements in either the protein main chain or in the flavin cofactor is sufficient to bring this conserved cysteine within covalent bonding distance of C(4a). The 1.9 Å dark-state structure of *Chlamydomonas* phot LOV1 solved at cryogenic temperatures (Fedorov et al., 2003) reveals two conformers of the cysteine side chain. The predominant conformer (at 70 ± 10% occupancy) is similar to the refined position of C966 in phy3 LOV2. The second conformer is rotated ~80° about the C_α–Cβ bond, bringing the S_γ within 3.5 Å of C(4a) (Figure 15.2). Evidence for two conformers of this conserved cysteine in the dark-state structure of phy3 LOV2 is present in electron density maps calculated from a partially refined structure determined at cryogenic temperatures (S. Crosson and K. Moffat, unpublished results). It is uncertain what role temperature has on the formation of this minor conformer and whether this structure is important for LOV domain photochemistry and function under biological conditions. Regardless, the conserved cysteine is closely positioned to flavin atom C(4a) in the crystal structures of LOV1 and LOV2, allowing for adduct formation without a large conformational change in the protein.

15.3
Photoexcited-State Structural Dynamics of LOV Domains

While X-ray crystallographic and NMR structural analysis of proteins can provide a great deal of insight into function, traditional versions of these techniques provide only a static, time-averaged structure with little or no information on molecular motion. Understanding small- and large-scale protein motions is necessary to decipher processes ranging from enzyme-substrate specificity to allostery. LOV domains provide an excellent model system to understand the dynamics of protein structure because they are relatively small and exhibit a self-contained photocycle (Salomon et al., 2000). Thus illumination with blue light can initiate structural changes in the protein that can be probed using both steady-state and time-resolved crystallography and NMR. Recently, several investigators have taken advantage of this property of LOV domains and solved steady-state photoexcited structures of phototropin LOV1 and LOV2 domains in both crystals and solution. These experiments are beginning to provide a picture of LOV-domain signaling with downstream partners.

Changes in LOV domain structure upon blue light illumination were first documented in a set of one-dimensional ^{13}C, ^{15}N, and ^{31}P NMR experiments using oat phototropin LOV2 containing an isotopically-enriched FMN chromophore (Salomon

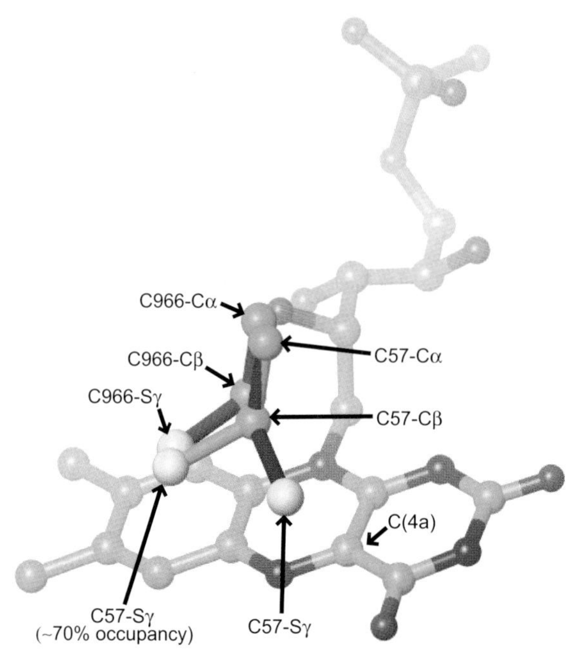

Figure 15.2 Dark-state conformers of the conserved cysteine side chain in LOV1 and LOV2. The dark-state structure of phy3 LOV2 (1G28) reveals one conformer of the conserved cysteine (labeled C966 with bonds colored red). The dark-state structure of *Chlamydomonas* phototropin LOV1 (19NL) has two conformers of the cysteine side chain (labeled C57 with bonds colored blue and green); the predominant conformer in LOV1 (at ~70%) is in a similar position to the cysteine side chain position in the dark-state structure of phy3 LOV2. The 4a carbon of the isoalloxazine ring of FMN is labeled. Atoms are colored by elements: carbon, green; nitrogen, blue; oxygen, red; phosphorus, pink; sulfur, yellow.

et al., 2001). These experiments demonstrated that the photoproduct was indeed a cysteinyl-C(4a) adduct and revealed changes in the chemical environments of the polypeptide backbone, the terminal phosphate, and the ribityl chain of FMN in response to photon absorption. The three-dimensional structure of a LOV domain in a photoexcited state was later solved by collecting X-ray diffraction data on a single crystal of fern phy3 LOV2 under continuous illumination at room temperature (Crosson and Moffat, 2002). While this steady state structure allowed the direct observation of a covalent bond between the conserved cysteine (C966) and the FMN C(4a) carbon, there were surprisingly very few changes in the overall structure of the protein as evidenced by difference Fourier maps calculated against the dark-state structure factor amplitudes (Crosson and Moffat, 2001) (Figure 15.3 A). These data demonstrate that, in the context of the 104 amino acids visible in the phy3 LOV2 electron density maps, protein motion in response to photon absorption is small and is concentrated around the isoalloxazine ring of the FMN cofactor. The largest motion between the dark- and illuminated-state structures is in the conserved cysteine side chain, which undergoes a simple chi1 rotation to bring the S_γ sulfur within an appropriate distance to form the covalent adduct with C(4a). Additionally, the flavin ring rotates ~8° so that residues with hydrogen bonds to the isoalloxazine moiety can maintain these interactions (Figure 15.3 B). Similar studies conducted on the light state of the *Chlamydomonas* phototropin LOV1 domain at cryogenic temperatures yielded similar results (Fedorov et al., 2003). Namely, electron density for the covalent cysteinyl-C(4a) adduct is clearly present with protein structural changes confined to the areas around the flavin ring. However, this illuminated-state structure of LOV1 shows slightly different geometry at the conserved cysteine (C57) as well as some small movement in the region of the terminal phosphate of FMN that is not evident in phy3 LOV2 structure.

While these crystallographic experiments provide valuable information on the structure of the cysteinyl adduct state, the limited light-induced conformational changes seen in these structures fail to explain how adduct formation can signal through the surrounding LOV domain to result in kinase activation. A possible model for this process has been provided by recent solution NMR experiments on oat phototropin LOV2 (Harper et al., 2003). These multidimensional NMR experiments were carried out on a construct of LOV2 containing an additional 40 amino acids after the last C-terminal residues evident in the crystal structure of phy3 LOV2. This C-terminal extension is conserved in all phototropin LOV2 domains, including phy3 LOV2. Intriguingly, roughly 20 of these residues form an amphipathic α-helix that docks against the five-stranded antiparallel β-sheet of the core LOV fold (Figure 15.4 A). Pulsed illumination of this LOV2 construct causes dramatic changes in $^{15}N/^{1}H$ HSQC and other spectra, indicating extensive blue light-induced structural changes. Identical experiments conducted on mutant LOV2 protein lacking the conserved cys-

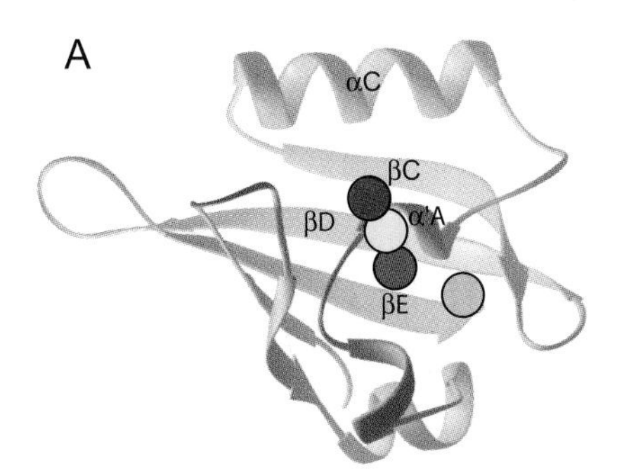

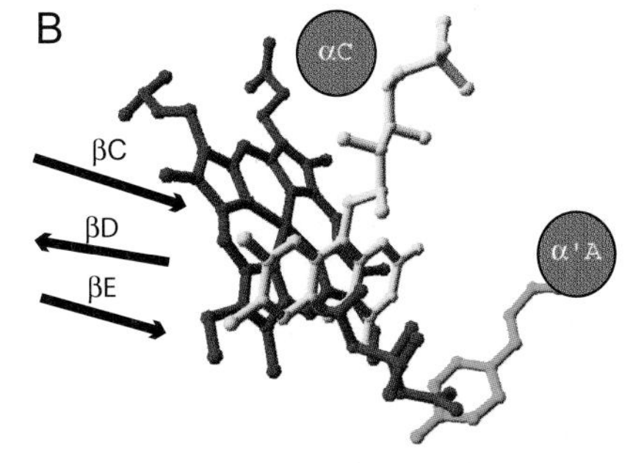

Figure 15.3 Variable ligand binding properties of the PAS domain family. A least squares superposition of the structural coordinates of fern phy3 LOV2 (Crosson and Moffat, 2001), *Bradyrhizobium japonicum* FixL-PAS (Gong et al., 1998), *Halorhodospira halophila* PYP (Borgstahl et al., 1995),and *Klebsiella pneumoniae* CitA-PAS (Reinelt et al., 2003) shows the ligand that binds each of these PAS domains to occupy a similar region of the structure. (A) Colored circles (FMN, yellow; heme, red; para-hydroxycinnamic, green; citrate, purple) are drawn onto the polypeptide structure of phy3 LOV2. (B) Ball-and-stick diagram using the same color scheme shows, in greater detail, how FMN, heme, para-hydroxycinnamic acid, and citrate are positioned relative to each other. Secondary structure elements are marked with arrows (beta sheet) and circles (helices). Panel B represents the LOV fold rotated counter-clockwise 45° across the plane of the page with respect to the orientation of phy3 LOV2 in panel A.

teine reveal no spectral changes in the lit state, suggesting that changes in the wild-type LOV2 domain are a direct result of cysteinyl-C(4a) adduct formation.

A comparison of a variety of parameters, including NMR chemical shifts, ^{2}H exchange protection factors, as well as increased susceptibility to proteolysis, demonstrates that while the core LOV fold is slightly destabilized in the adduct state, it retains the same overall secondary and tertiary structure. However, the C-terminal amphipathic helix, which is not present in the LOV2 crystal structures, is displaced from the β-sheet and unfolds in response to illumination (Figure 15.4 B). This adduct-induced change in secondary and tertiary structure likely serves as an allosteric switch controlling the activity of the C-terminal kinase domain, as mutations in the conserved cysteine which prevent the light-induced structural changes also render kinase activity completely insensitive to illumination. Helix undocking and unfolding from the β-sheet of the PAS/LOV fold also occurs in the photoexcited state of PYP (Hoff et al., 1999). Indeed, protein-protein interactions through the β-sheet may be a general feature of several PAS-mediated signaling processes (Erbel et al., 2003). Additional experiments on mutant LOV domains and larger LOV constructs are necessary to understand how adduct formation results in helix undocking from the β-sheet and how this undocking subsequently regulates the activity of the C-terminal kinase.

15.4
Comparative Structural Analysis of LOV Domains

As mentioned earlier, LOV domains form a subset of the PAS structural superfamily. A previous comparative structural analysis of phy3 LOV2 to the PAS domains of *Bradyrhizobium japonicum* FixL (Gong et al., 1998), *Halorhodospira halophila* photoactive yellow protein (PYP) (Borgstahl et al., 1995), and the human ERG potassium channel (Cabral et al., 1998) revealed a high degree of structural homology even though sequence homology between these domains is low (Crosson and Moffat, 2001). Lack of sequence homology between the PAS domains has created difficulties in annotation of these ubiquitous protein modules. Indeed, two of the five PAS structures deposited in the Protein Data Bank (the periplasmic domain of *Klebsiella pneumoniae* CitA (Reinelt et al., 2003) and the N-terminal domain of the human ERG potassium channel (Cabral et al., 1998)) were not known to be PAS domains until they were solved. This combination of sequence plasticity and conservation of structure suggests that the PAS fold can be utilized in many different ways.

An example of how PAS architecture has been adapted for different functions is evident in the variable ligand-binding properties of PAS domains. Many PAS proteins function as sensors that relay cellular signals in direct response to ligand binding and/or physicochemical changes at the ligand-binding site. Among the small molecular ligands that have been shown to bind PAS domains are: 1) iron protoporphyrin IX in the rhizobial oxygen-sensor kinase, FixL (Gilles-Gonzalez et al., 1991); 2) para-hydroxycinnamic acid in the bacterial photosensor, PYP (Baca et al., 1994; Hoff et al., 1994); 3) flavin adenine dinucleotide (FAD) in the aerotaxis sensor of *E. coli* (Bibikov et al., 2000), the NifL redox sensor of *Azotobacter vinelandii* (Hill et al., 1996), and the

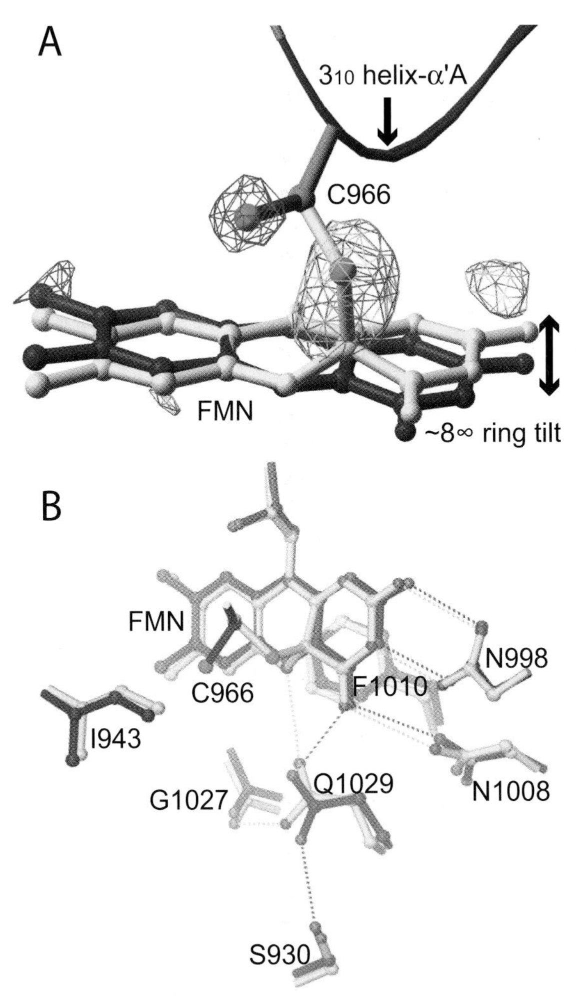

Figure 15.4 Conformational change in the FMN binding pocket of photoexcited LOV2. (A) Fourfold noncrystallographic symmetry-averaged light-minus-dark difference Fourier map contoured at $\pm 4\sigma$ in which σ is the root-mean-square value of the electron density. The conserved cysteine and the flavin ring for the dark (blue) and the photoexcited (yellow) structures are shown. Negative difference density (blue) and positive density (yellow) indicate Cys and ring motion upon illumination. (B) Side chains exhibiting significant displacements between the dark (blue) and photoexcited (yellow) structures in response to cysteinyl-flavin C(4a) adduct formation. Hydrogen bonds between the protein and FMN cofactor in the dark and photoexcited structures are indicated by blue and yellow dotted lines, respectively. A 2.6 to 3.5 Å range for hydrogen bonding was used. Atoms are colored by elements: nitrogen, light blue; oxygen, red; sulfur, green. Atoms colored blue in the dark structure and yellow in the photoexcited structure are carbon. In addition to the residues exhibiting motion in phy3 LOV2, the cryo-illuminated structure of *Chlamydomonas* phototropin LOV1 (Fedorov et al., 2003) shows additional displacements in Asn56, which forms a hydrogen bond with the ribityl chain, in the phosphate of FMN, and in Arg58, which forms a salt bridge with the phosphate. Figure from reference (Crosson and Moffat, 2002); reprinted with permission of the American Society of Plant Biologists.

photosensory LOV domain of *Neurospora* Wc-1 (He et al., 2002); 4) flavin mononucleotide (FMN) in the photosensory LOV domains of *Arabidopsis* and other phototropins, ZTL, FKF1, and LKP2 (Imaizumi et al., 2003); 5) small aromatic compounds in mammalian PAS kinase (Amezcua et al., 2002); and 6) citrate in the CitA receptor histidine kinase of *Klebsiella* (Reinelt et al., 2003). Thus, PAS domains are able to bind a chemically and structurally distinct range of ligands and communicate information from the ligand binding site to intra- and intermolecular partners. A discussion of the structural basis of this phenomenon in LOV domains will follow in a later section.

A least-squares comparison of the four types of ligand-binding PAS domain structures available in the Protein Data Bank, including PYP, the LOV domain of phototropins, the FixL heme-binding domain, and the periplasmic citrate-binding domain of CitA, shows that all exhibit high structural homology at the level of the protein main chain (less than 3.0 Å root mean square deviation). The most structurally conserved region of all PAS structures is the five-stranded antiparallel β-sheet. This element of secondary/tertiary structure exhibits remarkable conservation, with a root mean square deviation of approximately 1 Å between structures [see Figure 15.3 A of (Crosson and Moffat, 2001)]. All four of the ligand-binding PAS domains position their ligands in a similar region within the core of the fold suggesting a common mode of communication from the ligand-binding core to the surface of the protein (Figure 15.5).

Bioinformatic analysis of sequence from the phytochrome family of red-light photoreceptors has identified additional domains that are homologous to PAS/LOV domains. Namely, the bilin-binding GAF and PHY domains of this versatile photoreceptor family, which are positioned between several predicted PAS domains in the phytochrome polypeptide, are predicted to exhibit a PAS-like fold (Montgomery and Lagarias, 2002). Indeed, crystallographic analysis of a yeast GAF protein confirms the structural similarity between PAS and GAF domains (Ho et al., 2000). Thus phytochome is likely constructed of multiple repeating units of a structurally similar fold (Montgomery and Lagarias, 2002). How these PAS/GAF/PHY structural units come together to form the tertiary/quaternary structure of the holophytochrome dimer is not known. Nevertheless, it is notable that both the phototropin and phytochrome families of plant photoreceptors utilize very similar structural modules for signaling.

15.5 LOV-Domain Diversity

Genetic, biochemical, and biophysical studies have revealed a great deal about the structure and function of LOV domains from the phototropin family. Notably, a BLAST search of GenBank using a LOV flavin-binding consensus sequence (Crosson and Moffat, 2001) reveals numerous non-phototropin LOV proteins in plants, fungi and eubacteria that are predicted to bind flavin and exhibit a canonical LOV photocycle (Crosson et al., 2003). While flavin binding and photochemistry has only been confirmed in a handful of these proteins (Imaizumi et al., 2003; Losi et al.,

2002; Salomon et al., 2000), conservation of the flavin-binding consensus is a strong predictor that they contain *bona fide* LOV domains. Besides the ZTL (Somers et al., 2000), FKF1 (Nelson et al., 2000) and LKP2 (Schultz et al., 2001) circadian regulators

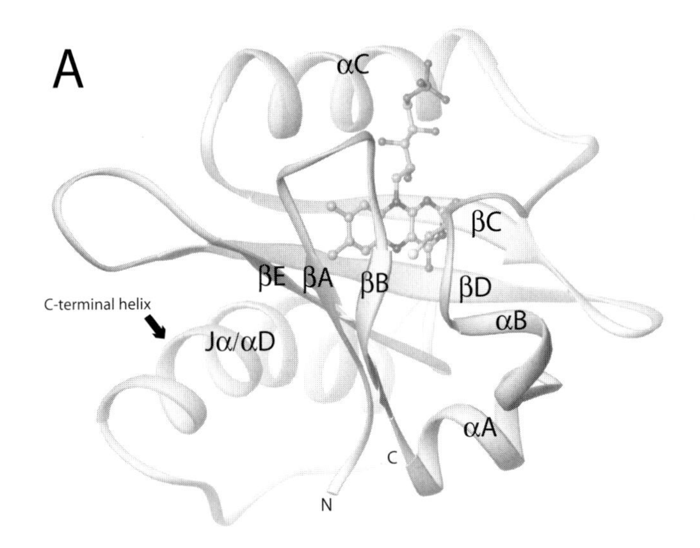

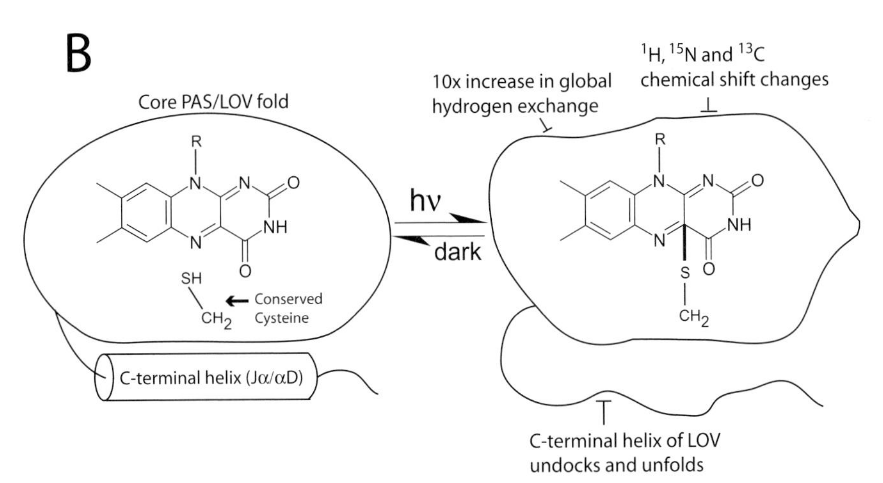

Figure 15.5 Light induced tertiary structure changes in oat phototropin LOV2. (A) Ribbon diagram of oat phototropin1 LOV2 solved by multidimensional NMR (structural coordinates courtesy of the Gardner Lab, University of Texas Southwestern Medical Center at Dallas). Diagram shows the C-terminal helical extension (Jα/αD) that is not present in the crystal structures of phy3 LOV2 and phototropin LOV1. This amphipathic helix docks against the *β*-sheet of the PAS/LOV fold. The flavin cofactor and conserved cysteine are shown in the core of the fold. (B) Cartoon summary of oat phototropin1 LOV2 NMR data shows C-terminal helix undocking and unfolding, an increase in global hydrogen/deuterium exchange in the core PAS/LOV fold, widespread changes in ^{1}H, ^{13}C, and ^{15}N chemical shifts, and increased proteolysis in response to adduct formation. Panel B adapted from Figure S3 of reference (Harper et al., 2003).

of *Arabidopsis* and the Wc-1 (Crosthwaite et al., 1997) and VVD (Heintzen et al., 2001) circadian regulators of *Neurospora*, little is known about the function of these putative photoreceptors. Notably, LOV domains are fused to a broad array of signal-output domains and range in size from the very small *Neurospora* VVD, containing little more than a single LOV domain, to hybrid photoreceptor kinases in cyanobacteria containing upwards of 1800 amino acids (Crosson et al., 2003).

LOV proteins can be assigned to six functional categories based on their Pfam/Smart/COG annotation in the NCBI Conserved Domain Database (Marchler-Bauer et al., 2003): 1) phototropins; 2) two-component signaling proteins; 3) STAS/sulfate transport or sigma factor regulators; 4) GGDEF/EAL-phosphodiesterase/cyclases; 5) G protein regulators (RGS proteins); and 6) proteins regulating circadian rhythm (Table 15.1 and Figure 15.6). With few exceptions, members of these categories have very similar domain construction, with a LOV domain(s) in the amino-terminal region of the protein followed by some type of signaling output (e.g. kinase, STAS, GGDEF) or DNA binding domain (e.g. GATA-type zinc finger) (Figure 15.6). With the exception of the hypothetical RGS protein of *Magnaporthe grisea*, LOV proteins in eukaryotes fall into either the phototropin or circadian regulator categories.

Bacterial LOV domains are fused to common prokaryotic signal output domains such as histidine kinases, response regulators, and GGDEF/EAL domains. Little is known about the function of these bacterial proteins and how light may affect their activity. Indeed, the antisigma factor antagonist, YtvA, of *Bacillus subtilis* is the only bacterial LOV protein with a known function (Akbar et al., 2001). However, no light-dependent activity *in vivo* or in vitro is known for YtvA. Understanding the biological role of bacterial LOV proteins is one of the frontiers of blue light photobiology. Surprisingly, many of these putative photoreceptors are present in organisms that are both non-photosynthetic and have no known behavioral response to light. It will be interesting to see what role LOV proteins play in the biology of such species (e.g. *Caulobacter crescentus, Bacillus subtilis, Listeria monocytogenes,* and *Magnetospirillum magnetotacticum*). Importantly, the possibility that some prokaryotic LOV proteins are not directly responding to light, but are sensing some other intra- or extracellular signal cannot be ruled out.

How PAS domains transmit a signal to inter- and intramolecular partner domains is a long-standing question in the study of PAS proteins. Recently, a picture of a conserved signaling pathway across the PAS β-sheet has begun to emerge (Erbel et al., 2003) that is supported by solution NMR work on oat phototropin1 LOV2 showing helix undocking from the β-sheet in response to adduct formation (Harper et al., 2003). Future experiments that compare the biochemistry, structure, and dynamics of full-length prokaryotic and eukaryotic LOV photoreceptors will allow us to identify how LOV domains signal to their structurally- and functionally-variable partner domains and determine if the mode of signaling has been conserved over time.

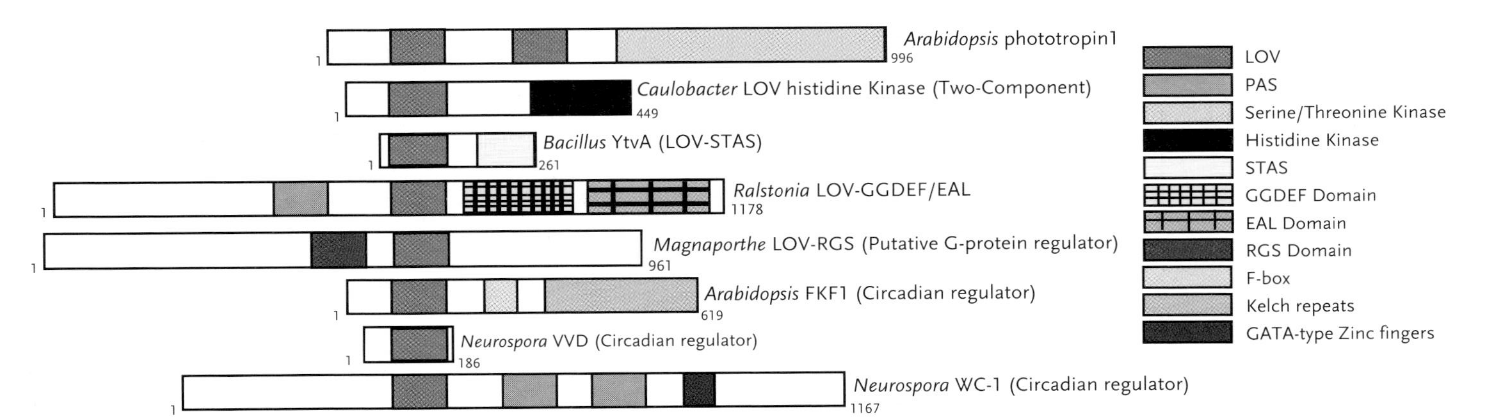

Figure 15.6 Categories of proteins containing LOV domains. Examples of the six major subsets of LOV proteins are represented: (1) phototropins, (2) two-component proteins, (3) LOV-STAS proteins, (4) LOV-GGDEF/EAL proteins, (5) LOV-RGS proteins, (6) LOV circadian regulators. Domains were annotated using the NCBI Conserved Domain Database (cutoff score $<e^{-6}$). A brief description of the putative biological/enzymatic activity of each of these domains is included in the text.

Table 15.1 LOV proteins present in GenBank as of January 2004. Hypothetical proteins are followed by their accession numbers.

Group	LOV protein
1) Phototropin/phy3	*Arabidopsis thaliana* phot1 and phot2
	Oryza sativa phot1 and phot2
	Avena sativa phot1
	Zea mays phot1
	Adiantum capillus-veneris phot
	Pisum sativum phot1
	Chlamydomonas phot
	Spinacia oleracea phot
	Vicia faba phot
	Adiantum capillus-veneris phy3
	Onoclea sensibilis phy3
	Hypolepis punctata phy3
	Dryopteris filix-mas phy3
2) LOV Histidine Kinases and Hybrid Histidine Kinase/Response Regulators	*Xanthomonas axonopodis* AAM37406
	Xanthomonas campestris AAM41699
	Pseudomonas syringae AAO56389
	Caulobacter crescentus AAK22272
	Brucella melitensis AAL53921
	Brucella suis AAN33777
	M. magnetotacticum MAGN4020
	M. magnetotacticum MAGN5031
	N. aromaticivorans SARO2721
	Pirellula sp. 1 RB4511
	Thermosynechococcus elongatus LL1282
	Noctoc PCC7120 ALL2875
	Nostoc punctiforme NPUN0349
	Anabaena PCC7120 BAB74574
3) LOV Response Regulator	*N. aromaticivorans* SARO0132
4) LOV STAS Proteins	*Bacillus subtilis* YtvA
	Oceanobacillus iheyensis BAC12544
	Listeria monocytogenes LMO0799
	Listeria innocua LIN0792
5) LOV GGDEF/EAL Proteins	*Noctoc punctiforme* NPUN5680
	Nostoc PCC7120 ALR3170
	Pseudomonas syringae PSYR0372
	Synechocystis P6803 SLR0359
	Ralstonia solanacearum RSP0254
	Anabaena PCC7120 BAB74869
6) Putative G Protein regulator	*Magnaporthe grisea* MG08735
7) LOV Circadian Regulators	*Neurospora crassa* VVD
	Neurospora crassa Wc-1
	Arabidopsis thaliana ZTL
	Arabidopsis thaliana FKF1
	Arabidopsis thaliana LKP2
8) Other LOV Proteins	*Arabidopsis thaliana* AAC05351
	Magnaporthe grisea MG07517
	Pseudomonas putida PP4629
	Chloroflexus aurantiacus CHLO3495
	Rhodobacter sphaeroides RSPH2966

Acknowledgements

This work has benefited from countless discussions with Keith Moffat, Spencer Anderson, Jason Key, and Sudar Rajagopal regarding protein structure, function, and dynamics. My work on LOV domains would not have been possible without early help from Winslow Briggs and John Christie. I also thank Kevin Gardner, John Kennis, and Trevor Swartz for many insightful discussions during the course of this work. Comments from Kevin Gardner and Shannon Harper greatly improved this manuscript.

References

Akbar, S., Gaidenko, T. A., Kang, C. M., O'Reilly, M., Devine, K. M., and Price, C. W. (2001) *J Bacteriol. 183*, 1329–1338.

Amezcua, C. A., Harper, S. M., Rutter, J., and Gardner, K. H. (2002) *Structure 10*, 1349–1361.

Aravind, L., and Ponting, C. P. (1997) *Trends Biochem. Sci. 22*, 458–459.

Baca, M., Borgstahl, G. E. O., Boissinot, M., Burke, P. M., Williams, D. R., Slater, K. A., and Getzoff, E. D. (1994) *Biochemistry 33*, 14369–14377.

Bibikov, S. I., Barnes, L. A., Gitin, Y., and Parkinson, J. S. (2000) *Proc. Natl. Acad. Sci. USA 97*, 5830–5835.

Borgstahl, G. E. O., Williams, D. R., and Getzoff, E. D. (1995) *Biochemistry 34*, 6278–6287.

Briggs, W. R., and Christie, J. M. (2002) *Trends Plant Sci. 7*, 204–210.

Cabral, J. H. M., Lee, A., Cohen, S. L., Chait, B. T., Li, M., and Mackinnon, R. (1998) *Cell 95*, 649–655.

Cheng, P., He, Q. Y., Yang, Y. H., Wang, L. X., and Liu, Y. (2003) *Proc. Natl. Acad. Sci. USA 100*, 5938–5943.

Christie, J. M., Salomon, M., Nozue, K., Wada, M., and Briggs, W. R. (1999) *Proc. Natl. Acad. Sci. USA 96*, 8779–8783.

Crosson, S., and Moffat, K. (2001) *Proc. Natl. Acad. Sci. USA 98*, 2995–3000.

Crosson, S., and Moffat, K. (2002) *Plant Cell 14*, 1067–1075.

Crosson, S., Rajagopal, S., and Moffat, K. (2003) *Biochemistry 42*, 2–10.

Crosthwaite, S. K., Dunlap, J. C., and Loros, J. J. (1997) *Science 276*, 763–769.

Erbel, P. J. A., Card, P. B., Karakuzu, O., Bruick, R. K., and Gardner, K. H. (2003) *Proc. Natl. Acad. Sci. USA 100*, 15504–15509.

Fedorov, R., Schlichting, I., Hartmann, E., Domratcheva, T., Fuhrmann, M., and Hegemann, P. (2003) *Biophys. J. 84*, 2474–2482.

Gilles-Gonzalez, M. A., Ditta, G. S., and Helinski, D. R. (1991) *Nature 350*, 170–172.

Gong, W. M., Hao, B., Mansy, S. S., Gonzalez, G., Gilles-Gonzalez, M. A., and Chan, M. K. (1998) *Proc. Natl. Acad. Sci. USA 95*, 15177–15182.

Harper, S. M., Neil, L. C., and Gardner, K. H. (2003) *Science 301*, 1541–1544.

He, Q. Y., Cheng, P., Yang, Y. H., Wang, L. X., Gardner, K. H., and Liu, Y. (2002) *Science 297*, 840–843.

Heintzen, C., Loros, J. J., and Dunlap, J. C. (2001) *Cell 104*, 453–464.

Hill, S., Austin, S., Eydmann, T., Jones, T., and Dixon, R. (1996) *Proc. Natl. Acad. Sci. USA 93*, 2143–2148.

Ho, Y. S. J., Burden, L. M., and Hurley, J. H. (2000) *EMBO J. 19*, 5288–5299.

Hoff, W. D., Dux, P., Hard, K., Devreese, B., Nugterenroodzant, I. M., Crielaard, W., Boelens, R., Kaptein, R., Vanbeeumen, J., and Hellingwerf, K. J. (1994) *Biochemistry 33*, 13959–13962.

Hoff, W. D., Xie, A., Van Stokkum, I. H. M., Tang, X. J., Gural, J., Kroon, A. R., and Hellingwerf, K. J. (1999) *Biochemistry 38*, 1009–1017.

Huala, E., Oeller, P. W., Liscum, E., Han, I. S., Larsen, E., and Briggs, W. R. (1997) *Science 278*, 2120–2123.

Imaizumi, T., Tran, H. G., Swartz, T. E., Briggs, W. R., and Kay, S. A. (2003) *Nature 426*, 302–306.

Losi, A., Polverini, E., Quest, B., and Gartner, W. (2002) *Biophys. J. 82*, 2627–2634.

Marchler-Bauer, A., Anderson, J. B., DeWeese-Scott, C., Fedorova, N. D., Geer, L. Y., He, S. Q., Hurwitz, D. I., Jackson, J. D., Jacobs, A. R., Lanczycki, C. J., Liebert, C. A., Liu, C. L., Madej, T., Marchler, G. H., Mazumder, R., Nikolskaya, A. N., Panchenko, A. R., Rao, B. S., Shoemaker, B. A., Simonyan, V., Song, J. S., Thiessen, P. A., Vasudevan, S., Wang, Y. L., Yamashita, R. A., Yin, J. J., and Bryant, S. H. (2003) *Nucleic Acids Res. 31*, 383–387.

Montgomery, B. L., and Lagarias, J. C. (2002) *Trends Plant Sci. 7*, 357–366.

Nelson, D. C., Lasswell, J., Rogg, L. E., Cohen, M. A., and Bartel, B. (2000) *Cell 101*, 331–340.

Pellequer, J. L., Wager-Smith, K. A., Kay, S. A., and Getzoff, E. D. (1998) *Proc. Natl. Acad. Sci. USA 95*, 5884–5890.

Reinelt, S., Hofmann, E., Gerharz, T., Bott, M., and Madden, D. R. (2003) *J. Biol. Chem. 278*, 39189–39196.

Salomon, M., Christie, J. M., Knieb, E., Lempert, U., and Briggs, W. R. (2000) *Biochemistry 39*, 9401–9410.
Salomon, M., Eisenreich, W., Durr, H., Schleicher, E., Knieb, E., Massey, V., Rudiger, W., Muller, F., Bacher, A., and Richter, G. (2001) *Proc. Natl. Acad. Sci. USA 98*, 12357–12361.
Schultz, T. F., Kiyosue, T., Yanovsky, M., Wada, M., and Kay, S. A. (2001) *Plant Cell 13*, 2659–2670.
Somers, D. E., Schultz, T. F., Milnamow, M., and Kay, S. A. (2000) *Cell 101*, 319–329.
Taylor, B. L., and Zhulin, I. B. (1999) *Microbiol. Mol. Biol. Rev. 63*, 479–506.

16
The ZEITLUPE Family of Putative Photoreceptors

Thomas F. Schultz

16.1
Introduction

Genetic analysis of the circadian system in *Arabidosis* has led to the identification of a novel family of photoreceptors, the ZTL family. Molecular genetic analysis of this family of proteins indicates that its members play key roles within the circadian system. Members of this family appear to be unique in possessing protein motifs that suggest they perceive light and feed light signals into the circadian clock via light-dependent changes in protein ubiquitination. The goal is this chapter is to summarize the data on the ZTL family members and to integrate these data into a model for their mode of action within the circadian system of *Arabidopsis*.

16.2
Circadian Clocks

Due to the rotation of the earth, organisms have evolved under continuous fluctuations in light and temperature environments. These fluctuations occur with a 24-hour rhythmicity and organisms spanning all major kingdoms have evolved a circadian clock allowing them to anticipate these regular and constant changes in their environment. The hallmark of processes under circadian control is that they continue to exhibit rhythmic behavior with an approximate 24-hour periodicity even when transferred to constant conditions. The rhythmic control of these processes has fascinated biologists for more than 100 years since their first descriptions. The circadian system consists of three broad domains: input or entrainment pathways, the central oscillator which generates the overt rhythms, and processes that are under circadian control, also known as outputs. Recent advances in the field of circadian biology has led to molecular insights into circadian mechanisms including the identification of the *ZTL* gene family.

In order to understand the functions of the ZTL family fully, a brief description of the circadian clock is necessary. Current models of circadian oscillators in all organ-

Handbook of Photosensory Receptors. Edited by W. R. Briggs, J. L. Spudich

ISBN 3-527-31019-3

isms consist of feedback loops based on transcription and translation (Young et al., 2001). Positive factors drive expression of a repressor at a constant rate. The repressor accumulates within the cell until it reaches a threshold level. Once this threshold is reached, the repressor feeds back and suppresses its own expression. Once its protein levels fall below the threshold, the positive factors enhance its accumulation and start the cycle again. This simplified feedback model affords multiple control points that are able to affect the pace at which the clock progresses through its cycle. In Drosophila, PER and TIM form the negative arm of the feedback loop and function to repress their own promoters. Their activities are regulated by entry into the nucleus, phosphorylation, and protein stability (Panda et al., 2002), and alterations in any of these control points causes changes in the pace at which the clock progresses. In *Arabidopsis*, a feedback loop has been described between two Myb-domain transcription factors (the repressors CCA1 and LHY) and the TOC1 protein (Figure 16.1), a pseudo-response regulator protein (Alabadi et al., 2001). Phosphorylation of CCA1 has been shown to regulate the pace at which the clock runs and alterations in protein stability has been described for both LHY and TOC1 (Kim et al., 2003; Mas et al., 2003a; Sugano et al., 1999). Multiple control points exist in all circadian systems analyzed and these control points function to maintain a 24-hour periodicity and are utilized to make the oscillator entrainable to environmental stimuli such as light and temperature.

Entrainment pathways function to sense environmental signals such as changes in light and temperature and feed these signals into the circadian oscillator. These pathways are not considered part of the oscillator itself but are crucial for the oscillator to maintain a stable relationship between itself and its environment. The mechanisms by which temperature entrainment occurs are largely unknown, while light entrainment has been at least partially elucidated (Devlin and Kay, 2001). In mammalian systems, clock entrainment appears to occur via ocular light perception. Removal of the

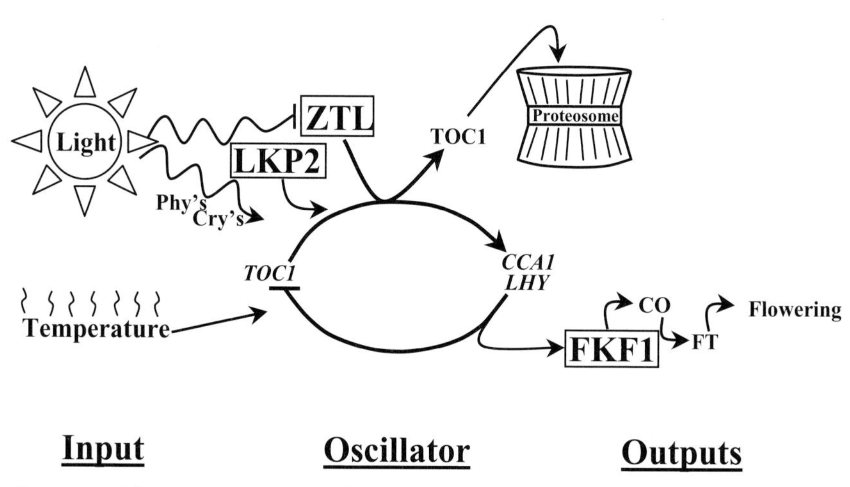

Figure 16.1 Schematic representation showing the positions occupied by ZTL family members within the circadian system.

eyes results in a loss of the ability to entrain the circadian clock (Foster et al., 1991). However, classical ocular photoreceptors, the rhodopsins, do not appear to be the only photoreceptors mediating entrainment. Mutations that abolish light perception via rhodopsins resulting in complete blindness (*rd* mutants) do not abolish the ability to entrain the circadian clock (Foster et al., 1991). Recently a novel gene related to opsins was identified and named melanopsin (Provencio et al., 2000). This photoreceptor exhibits proper cell-specific expression profiles and appears to function in light input to the clock. However, loss of melanopsin alone does not appear to abolish circadian photoperception either, indicating that multiple photoreceptors constitute this function (Panda et al., 2003). Melanopsin in combination with the rhodopsins appear to form the basis for circadian photoperception in mammalian systems, although more circadian photoreceptors may continue to be discovered. The mammalian cryptochromes have also been described as candidate circadian photoreceptors and they may play a role in phototransduction to the mammalian circadian clock, although a direct role in light perception has yet to be established (see Van Gelder and Sancar, Chapter 12).

Unlike mammalian systems, plants do not exhibit tissue-specific light perception and all plants cells appear to possess the ability to detect light. However, as with mammalian systems, plants also utilize multiple photoreceptors for light input to the clock. Known plant photoreceptors, the phytochromes and cryptochromes, appear to play key roles in circadian photoperception (Devlin and Kay, 2000; Somers et al., 1998). Phytochromes are the classical red-light photoreceptors in plants and consist of multigene families. There are five phytochromes present in the *Arabidopsis* genome. The two most prominent and well-studied are phyA and phyB. Loss of either of these phytochromes results in period lengthening under constant red-light. phyB appears to be the predominant red-light photoreceptor at higher fluences and phyA appears to be predominant at lower fluences. Having multiple red-light photoreceptors allows the plant to respond to a much wider range of fluences, a strategy that is also utilized for blue-light photoperception.

In *Arabidopsis*, the cry1 and cry2 genes constitute the cryptochrome gene family and function in blue-light perception. Again the presence of two blue-light photoreceptors appears to allow the plant to respond to a much wider range of fluences. The similarities between the plant and mammalian entrainment systems are worth noting. Neither system uses a single circadian photoreceptor and both systems employ multiple receptors allowing responses to various light qualities and quantities. The evolution of such an elaborate and redundant system of perception suggests a crucial role for clock entrainment in the fitness of the organism. As further analysis of clock entrainment proceeds, novel mechanisms of photoperception will certainly continue to be discovered.

16.3 SCF Ubiquitin Ligases

Ubiquitination has emerged as key mechanism for regulating numerous cellular processes such as transcription, signal transduction, cell division, and development. Ubiquitin molecules are transferred onto target proteins via a cascade of steps from E1 to E2 to E3 ubiquitin ligases (Hershko and Ciechanover, 1998). The E3 ubiquitin ligases confer specificity onto the system via protein: protein interactions with target molecules. E3 ubiquitin ligases represent an incredibly diverse family of proteins and have been categorized into five general classes (Vierstra, 2003): HECT-domain proteins, SCF protein complexes, RING/U-box, APC, and VBC-Cul2 proteins. Analysis of the *Arabidopsis* genome has demonstrated the importance of SCF (Skp, Cullin, F-box) complexes in plant biology. The yeast Cyclin F protein was the founding member of the F-box motif. The function of F-box proteins is to confer specificity onto SCF complexes by interacting with target proteins. Nearly 700 putative F-box proteins are encoded in the *Arabidopsis* genome (Gagne et al., 2002). This number represents nearly 2.5% of the genome and contains twice as many as the RING-finger class of E3 ubiquitin ligases, the second largest group in *Arabidopsis*. The staggering number of these proteins indicates the importance plants have placed on this family of ubiquitin ligases.

The ZTL family of proteins belongs to the SCF class of E3 ubiquitin ligases. Members of the ZTL family contain an F-box domain and six kelch repeats at their C-termini (Figure 16.2) (Nelson et al., 2000; Schultz et al., 2001; Somers et al., 2000). The kelch domain forms a β-propeller structure, a structurally conserved motif found in a large number of proteins (Adams et al., 2000). Each kelch repeat forms a single blade of the propeller-like structure and the motif is thought to mediate protein:protein interactions. Out of the nearly 700 *Arabidopsis* F-box proteins, over 90 have kelch repeats at their C-termini indicating that F-box/kelch proteins are fairly common. What makes the ZTL family members unique is the presence of a PAS/LOV domain at their N-termini (Figure 16.2), with only three proteins in the *Arabidopsis* genome exhibiting this unique architecture. In addition, phylogenetic analysis of all known Arabidopsis F-box proteins groups this family together in a single clade with no additional members (Gagne et al., 2002). PAS/LOV domains are remarkably diverse and are found in organisms spanning prokaryotes and eukaryotes. Their functions can be broadly categorized into two groups, they either mediate small molecule binding or participate in protein:protein interactions (Gu et al., 2000) or possibly both. The PAS/LOV domains most closely related to the ZTL family of proteins belongs to the phototropin1/2 (phto1/2) and WHITE COLLAR-1 (WC-1) proteins (Figure 16.2), both of which are known photoreceptors (see Christie and Briggs, Chapter 13, Swartz and Bogomolni, Chapter 14). The biochemical and physiological analysis of the ZTL family members supports their role in photoperception within the circadian system.

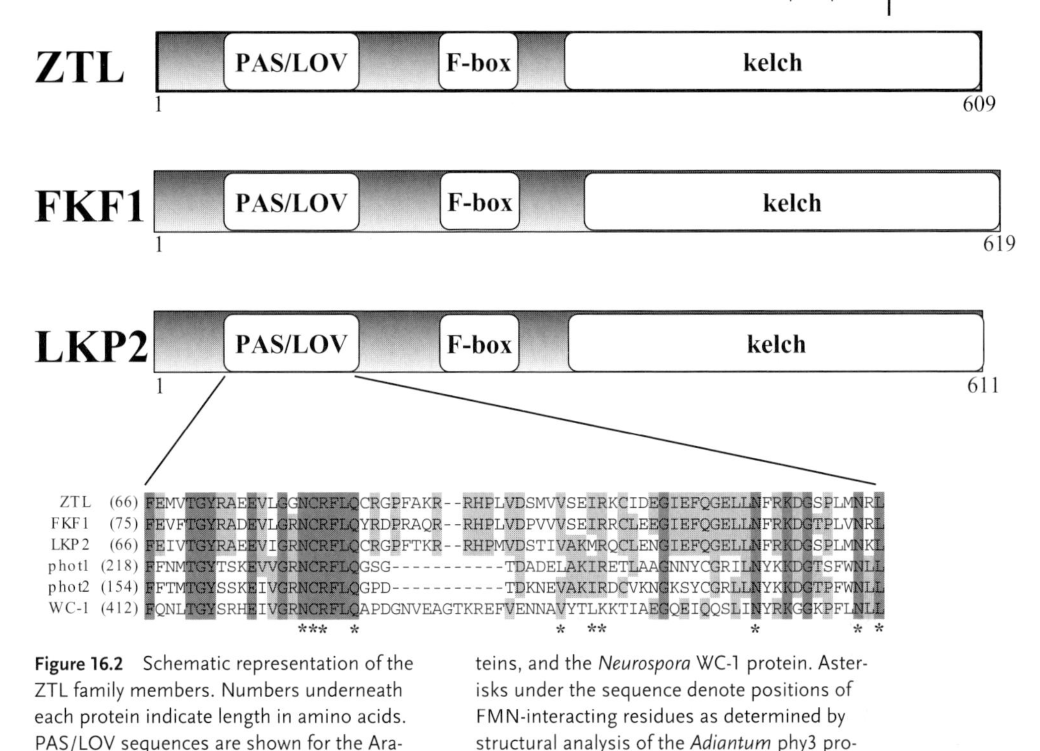

Figure 16.2 Schematic representation of the ZTL family members. Numbers underneath each protein indicate length in amino acids. PAS/LOV sequences are shown for the Arabidopsis ZTL family members, phot1/2 proteins, and the *Neurospora* WC-1 protein. Asterisks under the sequence denote positions of FMN-interacting residues as determined by structural analysis of the *Adiantum* phy3 protein (Crosson and Moffat, 2001).

16.4 Photoperception

As described above, the most striking and unique feature of the ZTL family of proteins is the presence of a PAS/LOV domain at the N terminus of an F-box protein. This unusual combination suggests a role for these proteins in light-dependent protein ubiquitination and the physiological and biochemical analysis of this gene family support this conclusion. The PAS/LOV domains most closely related to the ZTL family of proteins belong to the phot1/2 and WC-1/2 proteins, two photoreceptor gene families involved in *Arabidopsis* phototropism (Christie et al., 1998) and the *Neurospora* circadian clock (Froehlich et al., 2002). Extensive biochemical and photochemical analysis of phot1/2 LOV domains has shown that these domains bind flavin mononucleotide (FMN), a blue light-absorbing chromophore, in a non-covalent manner. Upon absorption of light, a covalent bond is formed between FMN and a conserved Cys residue within the LOV domain (see Christie and Briggs, Chapter 13 and Crosson, Chapter 15 for further details). Furthermore, this covalent bond is labile and spontaneously reverts to a non-activated state very rapidly in the photoreceptors mentioned above. These results indicate that the LOV domains of phot1/2 perform a photocycle and are able to switch into an active state rapidly in response to light and return to the inactive state in the dark.

Recently, it's been shown that the LOV domains of all three ZTL family members exhibit similar although not identical photochemical properties (Imaizumi et al., 2003). FMN co-purifies with these fusion proteins when expressed in bacteria, and these fusion proteins exhibit photochemical activity. In response to flashes of blue light, the proteins exhibit a characteristic shift in absorption spectra consistent with the formation of a covalent cysteinyl bond similar to that observed for the LOV domains of phot1/2. Interestingly, while the LOV domains from phot1/2 spontaneously revert to a resting state quite rapidly in the dark, the covalent bond formed by the ZTL family LOV domains appears to be stable, lasting for hours. This result would suggest that either the ZTL family of proteins do not photocycle and are competent to absorb light only once, or in contrast to phot1/2, require an accessory protein(s) for their reversion to a non-photoactivated state. Another possibility is that the lack of reversion is a property of bacterially-expressed fusion protein that is not exhibited by full-length endogenous protein *in planta*.

16.5
The ZTL Gene Family

Members of this gene family (*ZTL/FKF1/LKP2*, Figure 16.2) were identified using two basic approaches, forward genetics and bioinformatics. The founding member was obtained from a forward genetic screen for circadian clock mutants. Mutations in the *ZEITLUPE* (*ZTL*) gene result in a reduction in the pace at which the clock runs ("zeitlupe" is German and roughly translates to "slow motion") (Somers et al., 2000). The second member, identified concurrently with *ZTL*, was isolated as a late flowering mutant and was named *FKF1* (***F**lavin-binding, **K**elch repeat, **F**-box*) (Nelson et al., 2000). Finally, the *LKP2* (***LO**V, **K**elch, **P**rotein2*) gene was identified in a search of the *Arabidopsis* genome for putative novel photoreceptors containing a PAS/LOV domain (Schultz et al., 2001). The combination of a PAS/LOV domain with an F-box makes the members of this gene family unique and stimulated strong interest in their potential roles as novel photoreceptors. Consequently, multiple groups identified these proteins independently leading to a number of names being published for each member. The *ZTL* gene was identified multiple times and descriptions exist under *LKP1* (Kiyosue and Wada, 2000), *FKF1-like protein 2* (Nelson et al., 2000), and *ADAGIO1* (*ADO1*) (Jarillo et al., 2001). *FKF1* is also annotated as *ADO3* and *LKP2* is annotated as *ADO2* (Jarillo et al., 2001). For the purposes of this review, each gene will be referred to as the name with which it first appeared in publication. It should be noted that naming all three based on presumed circadian function could be misleading since *FKF1* does not appear to alter the pace at which the clock runs but rather more specifically regulates flowering time in response to day length.

16.5.1
ZTL

The *ZTL* gene was first identified in a screen for circadian clock mutants in *Arabidopsis* using luciferase imaging of whole seedlings (Millar et al., 1995). The *ztl* mutants isolated from this type of screen (*ztl-1* and *ztl-2*) both exhibited long-period circadian phenotypes (Somers et al., 2000). Under free-running conditions (constant light and temperature), *CAB2* expression oscillates with an approximate 27-hour periodicity in *ztl* mutants compared to 24 hours for wild-type plants. The long-period phenotype was observed for multiple clock outputs (CCR2 expression and leaf movement rhythms) suggesting a pervasive effect of the mutation on the circadian clock. Although both mutants exhibited semi-dominant phenotypes in segregating populations, the mutants appear to be loss-of-function and an insertion line with no detectable mRNA was extensively characterized and exhibited identical circadian phenotypes (Jarillo et al., 2001). In contrast to the loss-of-function phenotypes, over-expression of *ZTL* (gain-of-function) resulted in either arrhythmic or short period phenotypes (Somers et al., 2004). The severity of the clock phenotype showed a strong correlation with ZTL expression levels. Clock phenotypes ranged from short period to arrhythmicity as ZTL expression levels increased. In addition to circadian phenotypes, both *ztl* mutants and *ZTL* over-expressors also exhibited short- and long-hypocotyl phenotypes (respectively), and a slight early flowering phenotype under short day growth conditions (*ztl* mutants) or late flowering phenotype under long days (ZTL over-expressors). Both hypocotyl-length and flowering-time phenotypes are typically observed in clock mutants.

One of the key observations for the *ztl-1* and *ztl-2* point mutations came from the analysis of fluence-rate response curves for period length (Somers et al., 2000). The long period clock phenotype was dependent on the fluence rate of light. At lower fluences the differential between mutant and wild type period lengths were greater than at higher fluences. This observation argued strongly for a light-dependent role of ZTL in the circadian clock. Recently it was shown that loss-of-function *ztl* mutants exhibit a long period phenotype under constant darkness, a result indicating a more central role in the circadian system beyond light input.

The ZTL protein possesses an F-box domain, suggesting that it functions as an E3 ubiquitin ligase and targets other proteins for ubiquitination. In searching for candidate targets of ZTL, an inverse relationship was observed between *ztl* mutants and another key clock gene, *TOC1* (Mas et al., 2003b). Reduction in the levels of TOC1 result in short period phenotypes under constant light (Strayer et al., 2000) and increases in TOC1 result long period phenotypes (Mas et al., 2003b); these phenotypes mirror the *ztl* mutant phenotypes leading to the hypothesis that ZTL may target TOC1 for degradation.

This hypothesis was recently confirmed by demonstrating a direct physical interaction between these two proteins both in yeast-interaction screens and in plants using co-immunoprecipitation assays (Mas et al., 2003a). Furthermore, TOC1 levels are greatly increased in *ztl* mutants and the *in vitro* degradation of TOC1 was much slower in extracts isolated from *ztl* mutants compared to wild-type extracts. These results support the conclusion that ZTL mediates the ubiquitination and degradation of TOC1 protein. Furthermore, TOC1 protein is much more stable in the light than in the dark, suggesting that light inhibits ZTL activity. These results support the model that ZTL is a circadian photoreceptor that targets a core clock component (TOC1) for degradation in the dark (Figure 16.1). This model is consistent with the fluence rate-dependent period-length phenotypes observed for *ztl* mutants (Somers et al., 2000). At higher fluences ZTL activity in wild-type seedlings is inhibited and TOC1 is more stable such that the differential between mutant and wild-type period lengths is smaller. At lower fluences, wild-type ZTL activity is higher and TOC1 is degraded more rapidly. The period-length difference between wild type and *ztl* mutants is greatest under conditions where ZTL activity is the highest and this appears to be at lower fluence rates. These results indicate that ZTL functions as a clock parameter within the circadian system. However it does not participate directly in the oscillator itself but rather alters the stability of a clock component (TOC1) resulting in alterations in the rate at which the clock progresses through its cycle.

16.5.2 FKF1

Arabidopsis is a facultative long-day plant flowering more rapidly under long-day growth conditions (16 h light:8 h dark) than under short-day conditions (8 h light:16 h dark). Day-length dependent processes are also known as photoperiodic responses. Classic experiments showed that plants, as well as other organisms, utilize their circadian clocks to sense changes in day-length and the pathway between the circadian clock and the induction of flowering has been amenable to genetic analysis (Yanovsky and Kay, 2003). Two key proteins have been identified which are central to photoperiodic control of flowering, CONSTANS (CO) and FLOWERING LOCUS-T (FT). Expression of *FT* is a key regulator of the developmental transition to flowering and is the last committed step before the transition occurs (Kardailsky et al., 1999). CO is a putative transcription factor that directly regulates *FT*, and the combination of light at the end of the day and CO expression is required for the induction of FT (Suarez-Lopez et al., 2001; Yanovsky and Kay, 2002). The circadian expression pattern of *CO* is complex and is crucial for its regulation of flowering time. Under long days, *CO* expression exhibits a bi-phasic pattern with a peak occurring at approximately 10 hours after lights on and a second peak occurring approximately 4 hours after lights off (Suarez-Lopez et al., 2001). The first peak of expression appears to be crucial for the photoperiodic induction of flowering. When this peak occurs during the light period (i.e. in long days), CO is able to induce expression of *FT* which in turn induces the transition to flowering. When this peak occurs after lights off (i.e. in short days), CO is unable to induce expression of FT and subsequently flowering.

The *fkf1* mutant was isolated as a large deletion on the bottom of chromosome 1 that resulted in a late-flowering phenotype under long day conditions (Nelson et al., 2000). This late-flowering phenotype was limited to long days and was not observed in short days indicating a loss of photoperiodic control over time to flowering. Loss of photoperiodic control of flowering is common to many clock mutants. In contrast to *ZTL*, *fkf1* mutants do not appear to have strong circadian phenotypes but do have a much stronger flowering-time phenotype (Nelson et al., 2000). The expression of FKF1 is regulated by the circadian clock with peak expression occurring late in the day and its expression may be a direct target of the core clock (Imaizumi et al., 2003). The observations that *fkf1* mutants result in a loss of photoperiodic control over flowering but do not appear to affect the circadian clock itself, and that FKF1 is clock-regulated, place FKF1 on the output side of the clock on a pathway mediating photoperiodic control over time to flowering.

Recent results indicate that FKF1 appears to be required for the first peak in *CO* expression under long days (Imaizumi et al., 2003). In *fkf1* mutants, this peak is absent and CO expression is confined to the night. This key result explains the flowering time phenotype of *fkf1* mutants at the molecular level and place FKF1 on the regulatory pathway between the circadian clock and photoperiodic control over flowering. Furthermore, plants entrained in short days respond to extended light during the first cycle after transfer to long days and this response requires FKF1. This result suggests that FKF1 is required for sensing light and shifting the pattern of CO expression to a long day mode. Thus the phase of FKF1 expression in the presence of light appears to be crucial for perception of day length and its subsequent regulation of time to flowering. FKF1 may be the photoreceptor that mediates the plant's photoperiodic response. However, further biochemical analysis of its mode of action is still needed.

16.5.3 LKP2

The third member of this gene family appears to function within the circadian system in much the same way as ZTL, although less data are available. Over-expression of LKP2 results in aberrant clock-regulated expression of *CAB2*, *CCR2*, *CCA1*, *LHY*, and *TOC1* under constant light (Schultz et al., 2001). The pervasive effects on multiple clock outputs suggest a central role for LKP2 in the circadian system under constant light. The expression of *CCR2* in the *LKP2* over-expresser was assayed under constant darkness and also found to be arrhythmic, indicating a light-independent role for LKP2 in the clock. Over-expression also resulted in long-hypocotyl phenotypes and late flowering in short days, phenotypes similar to those observed for ZTL. These phenotypes are all limited to gain-of-function (over-expression) studies and loss-of-function alleles have not been described either from forward genetic screens or from insertion lines. These results suggest that ZTL may be the primary circadian photoreceptor, and while over-expression of *LKP2* is able to affect the clock, it may have a distinct and undiscovered role within the cell. Further analysis of LKP2, particularly of loss-of-function alleles, is needed to ascertain its function within the *Arabidopsis* circadian system.

16.6
Summary

The ZTL photoreceptors belong to a small gene family consisting of three members. These genes are unique and possess a photoactive PAS/LOV domain at the N-terminus of an F-box protein. Out of nearly 700 F-box proteins in *Arabidopsis*, only the ZTL family members contain a PAS/LOV motif. Genetic and physiological analysis of ZTL indicates that it plays a central role in the circadian system, and molecular analysis suggests that it targets a core clock protein (TOC1) for degradation in a light-dependent manner (Figure 16.1). The LKP2 protein appears to play a similar role but further analysis of this family member is needed to determine its mode of action. The FKF1 protein appears to function on the output pathway leading from the circadian clock to photoperiodic control of flowering time. FKF1 is required for proper timing of *CO* expression, a key flowering gene, under long days although its mechanism of action remains unknown. The PAS/LOV domains of each of these proteins bind FMN and exhibits photochemistry consistent with their roles as photoreceptors that mediate light-dependent ubiquitination.

References

Adams, J., Kelso, R., and Cooley, L. (2000) *Trends Cell Biol.* **10**:17–24.

Alabadi, D., Oyama, T., Yanovsky, M. J., Harmon, F. G., Mas, P., and Kay, S. A. (2001) *Science* **293**:880–883.

Christie, J. M., Reymond, P., Powell, G. K., Bernasconi, P., Raibekas, A. A., Liscum, E., and Briggs, W. R. (1998) *Science* **282**:1698–1701.

Crosson, S. and Moffat, K. (2001) *Proc. Natl. Acad. Sci. USA* **98**:2995–3000.

Devlin, P. F. and Kay, S. A. (2000) *Plant Cell* **12**:2499–2510.

Devlin, P. F. and Kay, S. A. (2001) *Annu. Rev. Physiol.* **63**:677–694.

Foster, R. G., Provencio, I., Hudson, D., Fiske, S., De Grip, W., and Menaker, M. (1991) *J.Comp. Physiol. [A]* **169**:39–50.

Froehlich, A. C., Liu, Y., Loros, J. J., and Dunlap, J. C. (2002) *Science* **297**:815–819.

Gagne, J. M., Downes, B. P., Shiu, S. H., Durski, A. M., and Vierstra, R. D. (2002) *Proc. Natl. Acad. Sci. USA* **99**:11519–11524.

Gu, Y. Z., Hogenesch, J. B., and Bradfield, C. A. (2000) *Annu. Rev. Pharmacol. Toxicol.* **40**:519–561.

Hershko, A. and Ciechanover, A. (1998) *Annu. Rev. Biochem.* **67**:425–479.

Imaizumi, T., Tran, H. G., Swartz, T. E., Briggs, W. R., and Kay, S. A. (2003) *Nature* **426**:302–306.

Jarillo, J. A., Capel, J., Tang, R. H., Yang, H. Q., Alonso, J. M., Ecker, J. R., and Cashmore, A. R. (2001) *Nature* **410**:487–490.

Kardailsky, I., Shukla, V. K., Ahn, J. H., Dagenais, N., Christensen, S. K., Nguyen, J. T., Chory, J., Harrison, M. J., and Weigel, D. (1999) *Science* **286**:1962–1965.

Kim, J. Y., Song, H. R., Taylor, B. L., and Carre, I. A. (2003) *EMBO J.* **22**:935–944.

Kiyosue, T. and Wada, M. (2000) *Plant J.* **23**:807–815.

Mas, P., Kim, W. Y., Somers, D. E., and Kay, S. A. (2003a) *Nature* **426**:567–570.

Mas, P., Alabadi, D., Yanovsky, M. J., Oyama, T., and Kay, S. A. (2003b) *Plant Cell* **15**:223–236.

Millar, A. J., Carre, I. A., Strayer, C. A., Chua, N. H., and Kay, S. A. (1995) *Science* **267**:1161–1163.

Nelson, D. C., Lasswell, J., Rogg, L. E., Cohen, M. A., and Bartel, B. (2000) *Cell* **101**:331–340.

Panda, S., Hogenesch, J. B., and Kay, S. A. (2002) *Nature* **417**:329–335.

Panda, S., Provencio, I., Tu, D. C., Pires, S. S., Rollag, M. D., Castrucci, A. M., Pletcher, M. T., Sato, T. K., Wiltshire, T., Andahazy, M., Kay, S. A., Van Gelder, R. N., and Hogenesch, J. B. (2003) *Science* **301**:525–527.

Provencio, I., Rodriguez, I. R., Jiang, G., Hayes, W. P., Moreira, E. F., and Rollag, M. D. (2000) *J. Neurosci.* **20**:600–605.

Schultz, T. F., Kiyosue, T., Yanovsky, M., Wada, M., and Kay, S. A. (2001) *Plant Cell* **13**:2659–2670.

Somers, D. E., Devlin, P. F., and Kay, S. A. (1998) *Science* **282**:1488–1490.

Somers, D. E., Kim, W. Y., and Geng, R. (2004) *Plant Cell* **16**:769–782.

Somers, D. E., Schultz, T. F., Milnamow, M., and Kay, S. A. (2000) *Cell* **101**:319–329.

Strayer, C., Oyama, T., Schultz, T. F., Raman, R., Somers, D. E., Mas, P., Panda, S., Kreps, J. A., and Kay, S. A. (2000) *Science* **289**:768–771.

Suarez-Lopez, P., Wheatley, K., Robson, F., Onouchi, H., Valverde, F., and Coupland, G. (2001) *Nature* **410**:1116–1120.

Sugano, S., Andronis, C., Ong, M. S., Green, R. M., and Tobin, E. M. (1999) *Proc. Natl. Acad. Sci. USA* **96**:12362–12366.

Vierstra, R. D. (2003) *Trends Plant Sci.* **8**:135–142.

Yanovsky, M. J. and Kay, S. A. (2002) *Nature* **419**:308–312.

Yanovsky, M. J. and Kay, S. A. (2003) *Nat. Rev. Mol. Cell Biol.* **4**:265–276.

Young, M. W. and Kay, S. A. (2001) *Nat. Rev. Genet.* **2**:702–715.

17
Photoreceptor Gene Families in Lower Plants

Noriyuki Suetsugu and Masamitsu Wada

17.1
Introduction

Both blue and red light mediate many aspects of growth and development in plants. Through extensive research using a flowering plant *Arabidopsis thaliana*, it was shown that most developmental and physiological phenomena in response to blue and red light were mediated by three kinds of blue light receptors: cryptochrome (cry), phototropin (phot) and members of the ZTL/FKF/LKP family which were very recently shown likely to be blue-light receptors (Imaizumi et al., 2003), and by the phytochrome family, respectively (Figure 17.1).

Two *A. thaliana* cryptochrome genes (*CRY1* and *CRY2*) regulate de-etiolation, gene expression, flowering, entrainment of the circadian clock, and so on (Lin and Shalitin, 2003). Cryptochromes have amino(N)-terminal light-sensing domains similar to the microbial type-I photolyases (Figure 17.1) that repair cyclobutane pyrimidine dimers. However, cryptochromes have no photolyase activity. The carboxy(C)-terminal extension of cryptochromes is thought to be a blue-light signal-transducing domain (Figure 17.1), since overexpression of the C-terminal extension of CRY1 or CRY2 alone confers a constitutive de-etiolation phenotype in *A. thaliana* (Yang et al., 2000). Not only plants, but also fly, mouse, and zebrafish have cryptochrome gene(s) which regulate circadian clock (Sancar, 2000), although the origin of animal cryptochromes is thought to be different from that of plants (Cashmore et al., 1999; see Van Gelder and Sancar, Chapter 12).

Phototropin consists of two LOV (Light, Oxygen and Voltage) domains at the N terminus and a serine/threonine (Ser/Thr) kinase domain at the C terminus (Figure 17.1) (Briggs et al., 2001; Briggs and Christie, 2002). A LOV domain binds one flavin mononucleotide (FMN) and absorbs blue light. Phototropins are plasma membrane-localized proteins (Briggs et al., 2001; Sakamoto and Briggs, 2002) and shows blue light-regulated autophosphorylation in vitro and in vivo. The *A. thaliana PHOTOTROPIN1* gene (*PHOT1*) was identified through analysis of a non-phototropic mutant *nph1* (Huala et al., 1997). Subsequently, it was shown that *PHOT2* gene mediates the chloroplast avoidance movement under high fluence-rate blue light (Kagawa et

Handbook of Photosensory Receptors. Edited by W. R. Briggs, J. L. Spudich

ISBN 3-527-31019-3

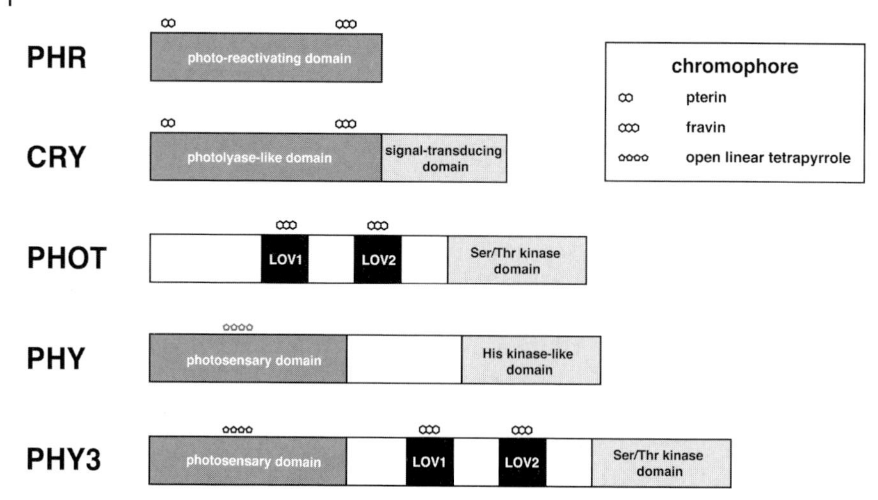

Figure 17.1 Plant photoreceptors. Photolyase (or photo-reactivating enzyme: PHR) repairs UV-induced DNA lesions by absorbing blue light through a pterin and a flavin adenine dinucleotide. Cryptochrome (cry) consists of a photolyase-like N-terminal domain (photolyase-like domain) and a C-terminal domain necessary for cryptochrome functions (signal-transducing domain). Cryptochrome utilizes a pterin and a flavin adenine dinucleotide as chromophore. Phototropin (phot) has two LOV domains (LOV1/2) and a serine/threonine kinase (Ser/Thr kinase domain). A LOV domain binds one flavin mononucleotide (FMN). Phytochrome (phy) consists of a photosensory domain at the N terminus (photosensary domain), which uses an open linear tetrapyrrole as chromophore, and C-terminal histidine kinase-like domain (His kinase-like domain). Fern phytochrome 3 (phy3) is a chimeric photoreceptor having a phytochrome photosensory domain and a complete phototropin domain (two LOV domains and a serine/threonine kinase).

al., 2001; Jarillo et al., 2001). The two phototropins redundantly regulate phototropism, chloroplast accumulation movement (Sakai et al., 2001), stomatal opening (Kinoshita et al., 2001) and leaf expansion (Sakai et al., 2001; Sakamoto and Briggs, 2002) in *A. thaliana*.

Phytochrome is a red/far-red (R/FR) light photoreceptor bearing an open linear tetrapyrrole chromophore (Figure 17.1), and is photoreversibly converted between the R light-absorbing Pr form, and FR light-absorbing Pfr form. The latter is the biologically active form. Therefore, if a response is R/FR reversible, the response must be regulated by the phytochrome system (Nagy and Schäfer, 2002; Quail, 2002). Various responses are mediated by phytochrome(s) in plants such as germination of seeds or fern spores, inhibition of stem elongation, flowering, and gene expression. Molecular genetic analyses of phytochrome responses in *A. thaliana* that has five phytochrome genes (*PHYA-E*) showed that the phytochrome genes may have specific or overlapping functions to regulate various responses. These studies have identified many of the downstream components of phytochrome signal transduction (Nagy and Schäfer, 2002; Quail, 2002). Among five phytochromes in *A. thaliana*, phyB predominantly mediates R/FR-photoreversible responses, the so-called low-fluence response (LFR). phyA mediates the very low-fluence responses (VLFR) and the FR-high irradiance response (FR-HIR). Nuclear localization of phytochromes was found to be essential for many physiological responses in *A. thaliana* (Matsushita et al., 2003; Nagy and Schäfer, 2002; Quail, 2002). It is thought that phytochromes directly modulate gene expression through interaction with transcription factors in the nucleus (Martínez-García et al., 2000; Quail, 2002).

Many physiological and developmental responses regulated by both blue and red light have been extensively studied in ferns, mosses, and green alga, due to their simple architecture at the cell level (Wada and Kadota, 1989). Gametophytic cells of ferns and mosses offer particularly suitable systems for studying photobiology (Suetsugu and Wada, 2003). They have a simple organization such as a single cell, linearly chained cells, or a two-dimensional sheet of a single layer of cells (Figure 17.2). Photoresponses at cell level such as cell division, cell elongation, phototropism, and chloroplast movement can be observed under a microscope. Furthermore, the localization of photoreceptor molecules can be deduced by irradiation with a microbeam or polarized light. But the difficulties in biochemical and genetic analyses have prevented the identification of phytochrome and blue-light photoreceptor molecules for these photoresponses. *Chlamydomonas reinhardtii* is a unicellular green alga and is the only model system capable of being analyzed by molecular genetics in lower plants. Phototaxis and sex determination are blue light-induced responses extensively studied photobiologically and genetically in this alga (Hegemann et al., 2001). Here we review photoreceptor genes (*CRY, PHOT,* and *PHY*) from a fern *Adiantum capillus-veneris,* a moss *Physcomitrella patens* (Figure 17.2), and a green alga *C. reinhardtii.* These plant species are the only examples in which these photoreceptor genes have been cloned and characterized.

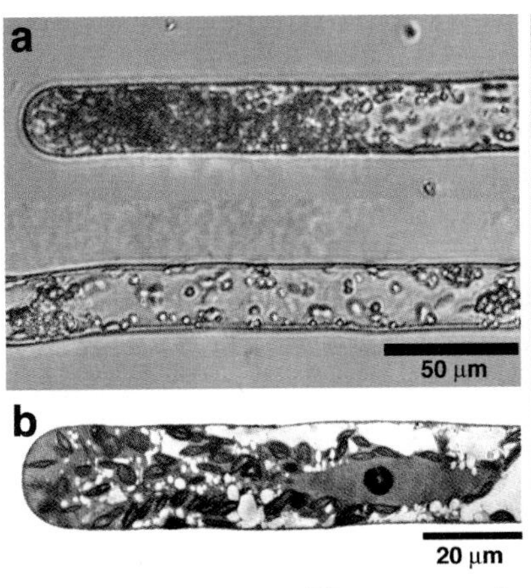

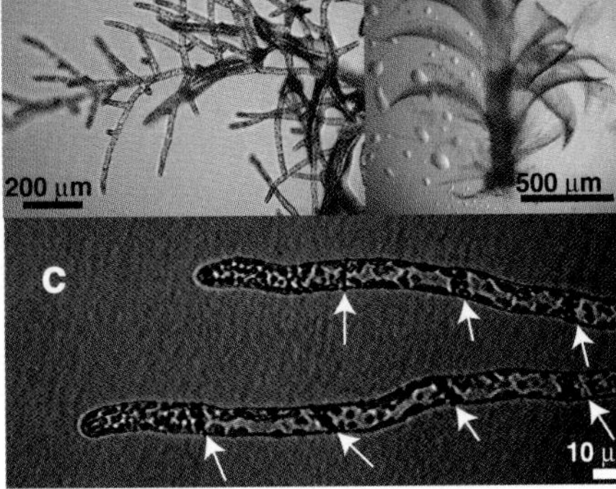

Figure 17.2 Photographs of fern and moss species mentioned in this chapter. (A) The fern *Adiantum capillus-veneris.* (a) A red light-grown protonemal cell. Apical and basal parts of the protonemata are shown. (b) A longitudinal section of the apical part of a protonema. (B) The moss *Physcomitrella patens.* (a) Protonemata. (b) A gametophore. (c) The apical cells of red light-grown protonemata. The position of septums is indicated with arrows.

17.2 Cryptochromes

17.2.1 *Adiantum capillus-veneris*

Five *CRY* genes (*CRY1-5*) were cloned from *A. capillus-veneris* (Kanegae and Wada, 1998; Imaizumi et al., 2000). Their N-terminal photolyase-like domains show high similarity to each other and to other plant CRY proteins. None of the five cryptochrome proteins showed photolyase activity (Imaizumi et al., 2000). The Trp-277 in the *Escherichia coli* PHR protein, which is important for photolyase function, is replaced by Leu in CRY1 and CRY5, and by Phe in CRY2, CRY3 and CRY4. Phylogenetic analysis indicated that *Adiantum CRY1* and *CRY2* are in one cluster, and *CRY3* and *CRY4* are in another cluster. *CRY5* is closer to *CRY1/CRY2* cluster but is not included in either group. Therefore, gene duplications might have occurred independently three times. The C-terminal extensions of *Adiantum* CRYs have less similarity to other proteins, but contains a DAS (DQXVP-Acidic-STAES) domain conserved among plant cryptochromes (Figure 17.3; Lin and Shalitin, 2003), except for CRY5, which has no C-terminal extension. *Adiantum* CRY1 has DQXVP motif and an acidic region, but does not have STAES motif (Imaizumi et al., 2000). Through genetic analyses in *A. thaliana*, it was demonstrated that a DAS domain is important in the biological function of a cryptochrome (Lin and Shalitin, 2003). Given that *Chlamydmonas reinhardtii* CPH1 has no recognizable DAS domain in its C-terminal extension and that *Sinapis alba* SA-PHR1 as well as *Adiantum* CRY5 have no C-terminal extension, unidentified domain(s) other than the DAS domain may also mediate cryptochrome function.

All of the *A. capillus-veneris CRY* genes were expressed in both the sporophytic (dark- or light-grown leaves) and gametophytic stages (dark-imbibed spores, light-grown protonemata and prothallia) with different expression patterns (Imaizumi et al., 2000). The amounts of *CRY1* and *CRY2* mRNA slightly increased after spore germination and stayed at the same level during gametophytic and sporophytic stages. *CRY3* mRNAs accumulated in spores and prothallia slightly less than in other tissues. The accumulation of *CRY4* mRNA occurred mainly in dark-grown spores and leaves, but not in light-grown tissues, suggesting that the expression or accumulation of *CRY4* mRNAs were repressed by light. The amount of *CRY5* mRNA was highly concentrated in sporophytic tissues. The amount of all five *CRY* mRNAs increased one day after the start of spore imbibition in darkness. Although the mRNA levels of *CRY1-4* remained at the same level during the following 6 days of dark-imbibition, the level of *CRY5* decreased to the level before imbibition.

In *A. capillus-veneris*, red light induces spore germination whereas blue light inhibits it (Furuya et al., 1997). When spores were irradiated continuously with red or blue light after 4 days in darkness, the amounts of *CRY1-3* mRNA did not vary much during light irradiation. Both red and blue light were effective in decreasing the amounts of *CRY4* mRNA. Red light reduced the level 50-fold and blue light reduced it 5-fold. Unexpectedly, the accumulation of *CRY5* mRNA was induced by both red

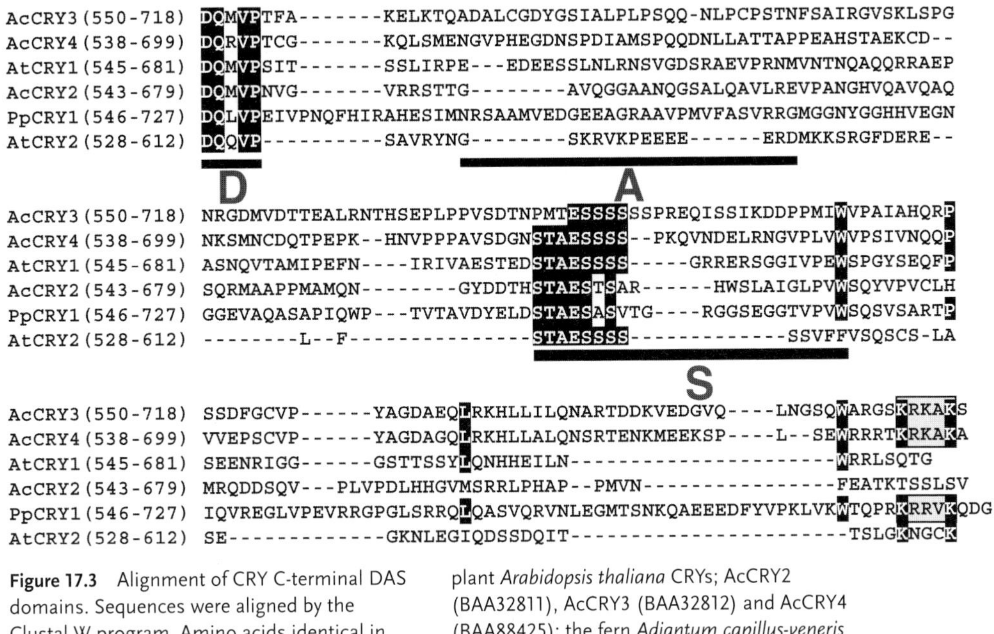

Figure 17.3 Alignment of CRY C-terminal DAS domains. Sequences were aligned by the Clustal W program. Amino acids identical in four among six CRYs are black-boxed. The DAS sub-domains are underlined. The putative nuclear localization signals in AcCRY3, AcCRY4 and PpCRY1 are gray-boxed. AtCRY1 (Q43125) and AtCRY2 (Q96524): the flowering plant *Arabidopsis thaliana* CRYs; AcCRY2 (BAA32811), AcCRY3 (BAA32812) and AcCRY4 (BAA88425): the fern *Adiantum capillus-veneris* CRYs; PpCRY1 (BAA83338 for PpCRY1a and BAB70665 for PpCRY1b): the moss *Physcomitrella patens* CRY. Amino acid sequences of the C-terminal domains of PpCRY1a and PpCRY1b are identical.

and blue light, with 300- to 400-fold increase compared to the dark level. In experiments using pulse irradiation in combination of red, blue, or far-red light, the increase of *CRY5* mRNAs by both red and blue light showed far-red light reversibility, a diagnostic characteristic of phytochrome regulation. Red/far-red reversibility was observed in the reduction of *CRY4* mRNAs, but the effect of blue light was not canceled by irradiation with far-red light, indicating that the regulation of *CRY4* mRNA accumulation was mediated by both phytochrome(s) and blue light photoreceptor(s). Importantly, the changes of the amounts of *CRY4* or *CRY5* mRNA preceded spore germination, suggesting that this regulation of mRNA level of *CRY4* or *CRY5* may be prerequisite for spore germination. In *A. thaliana*, blue light-dependent phosphorylation and subsequent degradation (for cry2) is likely to be crucial for cryptochrome function rather than the regulation of mRNA accumulation (Shalitin et al., 2002, 2003). Whether *A. capillus-veneris* cryptochromes can be phosphorylated by blue light remained to be determined.

Arabidopsis cry1 and cry2 are nuclear proteins. cry2 localizes constitutively in nucleus regardless of light condition (Guo et al., 1999; Kleiner et al., 1999), whereas cry1 accumulates in the nucleus in darkness and is exported to the cytosol under blue light (Yang et al., 2000). β-glucuronidase (GUS) genes fused to each *Adiantum CRY* cDNA were transiently expressed in fern gametophytes, and the localization patterns were analyzed after incubating in darkness, or under red or blue light (Imaizumi et al., 2000). GUS-CRY3 and GUS-CRY4 were localized to the nucleus, but GUS-CRY1, GUS-CRY2 and GUS-CRY5 were not. This localization pattern also occurred when GUS-CRY fusion proteins were expressed in onion epidermal cells. CRY3 and CRY4 have a monopartite nuclear-localization signal at the C-terminal extension (Figure 17.3). As expected, the C-terminal extension of CRY3 and CRY4 was sufficient for nuclear localization. GUS-CRY1, GUS-CRY2 and GUS-CRY5 showed diffuse staining pattern regardless of light conditions, suggesting that these CRYs were cytosolic proteins. GUS-CRY3 showed nucleo-cytoplasmic partitioning according to light condition; nuclear-localized in darkness or under red light, and exported to cytosol under blue light. This localization pattern is similar to that of *Arabidopsis* cry1 (Yang et al., 2000). Like *Arabidopsis* cry2 (Guo et al., 1999; Kleiner et al., 1999), GUS-CRY4 accumulated in the nucleus under all three light conditions.

From experiments with microbeams or polarized-light irradiation, the subcellular localization of blue light photoreceptor molecules for various photoresponses in gametophytes of *A. capillus-veneris* has been deduced (Wada and Furuya, 1978; Yatsuhashi et al., 1987; Hayami et al., 1992; Furuya et al., 1997). The blue-light receptor molecules that are involved in the inhibition of spore germination and promotion of cell division in protonemal cells in *A. capillus-veneris* are likely to be localized in or close to the nuclear compartment in darkness or red light before blue light irradiation, respectively (Furuya et al., 1997; Wada and Furuya, 1978). cry3 and cry4 are possible candidates for the blue-light receptor for suppression of spore germination and the induction of cell division, since their mRNAs increased in the dark, and GUS-CRY3 was localized in the nucleus in darkness and under red light, and GUS-CRY4 was also in the nucleus under the all light conditions. But blue light has a negative effect on the accumulation of cry3 or cry4 in the nucleus; that is cry3 was exported to the cytosol and the amount of *CRY4* mRNA was reduced under blue light. In *A. thaliana*, blue light induces cry2 phosphorylation and subsequent degradation (Shalitin et al., 2002). This blue light-induced cry2 degradation is thought to be a mechanism to regulate its activity in blue light and to sensitize cry2 in the absence of light. In *A. capillus-veneris*, cry3 and cry4 may mediate spore germination and cell division through sophisticated autoregulation at the mRNA or protein level in the nucleus.

17.2.2
Physcomitrella patens

Two *CRY* genes (*CRY1a* and *CRY1b*) were identified in the moss *P. patens* (Imaizumi et al., 2002). Their cDNA sequences are almost identical except for the 5' or 3' untranslated region, and the encoded proteins differ by one single amino acid at the position 80. Furthermore, genomic DNA sequences also shared almost identical nucleotide sequences overall. DNA gel blot analysis indicated the existence of two *CRY* genes in the genome of *P. patens*. These *CRY* genes are likely to be very recently duplicated. Moss CRYs showed the highest homology with *A. capillus-veneris* CRYs and have a C-terminal extension with a DAS domain and a monopartite nuclear localization signal as is the case for AcCRY3 and AcCRY4 (Figure 17.3; Imaizumi et al.,

2000). When GUS-CRY1a or GUS–CRY1b were expressed in the *P. patens* protoplasts, these fusion proteins were localized in the nucleus regardless of light condition as is the case for Accry4. Also, different light conditions did not change *CRY* mRNA levels in moss protonemata.

In *Physcomitrella patens*, knockout lines by gene targeting via homologous recombination can be generated (Schaefer, 2002). Since blue-light responses in this moss had been poorly investigated, analysis of cryptochrome disruptants offered not only information on cryptochrome function, but also information on unidentified blue-light responses. Single (*cry1a* or *cry1b*) and double (*cry1a cry1b*) *CRY* disruptants were generated and were analyzed for growth and development under the different light conditions (white, blue or red light) (Imaizumi et al., 2002). Under blue light but not white or red light, protonemal growth of *CRY* disruptants was different from wild type; the colony diameters were greater and the colony density was thinner. This colony morphology of *CRY* disruptants under blue light resulted from the defects in side-branch formation and in the growth of side-branch initials. Although less side-branches were formed in *CRY* disruptants, the number of gametophore buds differentiated from side-branch initials was higher in *CRY* disruptants than in wild type. Therefore, moss *CRY* genes controls gametophore induction by promoting the formation and the growth of side-branch initials and by repressing the subsequent differentiation from side-branch initials to buds. Gametophore growth was also regulated by cryptochrome-mediated blue-light signaling. Under white or blue light but not red light, stems of *CRY* disruptants were longer than those of wild type. The size of leaves in *CRY* disruptants under blue light was smaller than in wild type. Even under white or red light, the leaf size of *cry1a cry1b* double disruptants was smaller than that of wild type whereas that of *cry1a* or *cry1b* and wild type was the same. Together, it was demonstrated that cryptochrome-mediated blue-light signals inhibit stem elongation and promote leaf expansion in *P. patens*.

Since the developmental responses shown to be regulated by cryptochromes are also controlled by auxin in *P. patens*, the interaction between light and auxin in the developmental responses was examined in wild type and *CRY* disruptants (Imaizumi et al., 2002). Auxin treatment resulted in protonemal colonies that had a greater diameter and thinner density in a concentration-dependent manner under all three light conditions. *CRY* disruptants were hypersensitive to auxin under white and blue light, but not red light. Auxin-treated *cry1a* or *cry1b* colonies had similar morphology to *cry1a cry1b* colonies cultured on medium without auxin, suggesting that *cry1a cry1b* was hypersensitive to endogenous auxin. When auxin sensitivity of wild type or a *cry* disruptant was analyzed by transient assay at which a *GUS* gene fused with auxin-responsive soybean *GH3* promoter were expressed in the moss protoplasts, *cry1a cry1b* was hypersensitive to auxin at all of the auxin doses tested under white and blue light compared to wild type. *cry1a* or *cry1b* were hypersensitive only under blue light. Furthermore, the transcripts of native auxin-inducible genes, the *P. patens GH3-like protein 1* (*PpGH3L1*) gene and the *P. patens indole-3-acetic acid 1* (*PpIAA1*) gene, accumulated faster in *cry1a cry1b* than in wild type. Even without auxin application, the amounts of *PpGH3L1* and *PpIAA1* mRNAs in *cry1a cry1b* were higher than that in wild type.

Thus, blue-light signal transduction by cryptochromes mediates developmental responses to suppress auxin signaling in *P. patens*. Since moss crys are constitutively localized in nucleus, the suppression of auxin signaling by cryptochromes must be a nuclear event, possibly in auxin-inducible gene transcription. In *A. thaliana*, the auxin-mediated degradation of IAA proteins by the ubiquitin pathway is crucial for auxin responses and gene expression (Dharmasiri and Estelle, 2002). Involvement of cryptochromes in ubiquitin-mediated protein degradation is suggested from the recent discovery that *Arabidopsis* cry1 and cry2 interacts to COP1, which is negative regulator of photomorphogenesis and has a RING-finger domain which is found in some E3 ubiquitin ligase proteins (Wang et al., 2001; Yang et al., 2001). Moss COP1 may promote auxin-induced IAA protein degradation, but under blue light active crys may inactivate COP1, resulting in suppression of IAA protein degradation. Further analyses of moss crys will clarify the mechanism of suppression of auxin signaling by blue light.

17.2.3 *Chlamydomonas reinhardtii*

The *CPH1* (*Chlamydomonas* photolyase homolog 1) gene encodes a type-I CPD photolyase-like protein (Small et al., 1995). The encoded protein (CPH1) has an N-terminal photolyase-like domain that is more closely related to plant cryptochromes than to type-I photolyases, with 49% identity to *Arabidopsis* CRY1. When expressed in a photolyase-deficient *E. coli* mutant strain, CPH1 did not complement the UV-sensitive phenotype of a mutant *E. coli* strain indicating that CPH1 has no photolyase activity, as the case with plant cryptochromes (Peterson et al., 1999). Although Small et al (1995) first reported that *CPH1* gene encodes a polypeptide of 867 amino acids with a predicted molecular mass of 91 kDa, their new work showed that the previous cDNA sequence is incorrect and that CPH1 protein consists of 1007 amino acids with a predicted molecular mass of about 105 kDa (Reisdorph and Small, 2004).

CPH1 has the longest C-terminal extension among all known cryptochromes. However, its extension has no recognizable DAS domain and instead is glycine and alanine rich. Therefore, CPH1 may regulate its downstream component by some novel mechanism. Western-blot analysis indicated that there are two sizes of CPH1 protein (126 and 143 kDa) in vivo. This is different from that of recombinant CPH1 protein expressed in *E. coli* (110 kDa), suggesting that CPH1 proteins are likely to be subjected to post-translational modification (Reisdorph and Small, 2004). CPH1 accumulates only in darkness, but is rapidly degraded in response to blue light (Reisdorph and Small, 2004), as is the case of *Arabidopsis* cry2 (Shaltin et al., 2002). Interestingly, red light is also effective in the degradation of CPH1. Light-induced CPH1 degradation is inhibited by the proteasome inhibitor MG132 but not by cycloheximide. Thus, it is likely that light activates the proteasome pathway to degrade CPH1 protein rather than reducing CPH1 protein synthesis (Reisdorph and Small, 2004). An in vivo function for the protein encoded by the *CPH1* gene remains to be determined. Overexpression of CPH1 gene in *C. reinhardtii* resulted in no recognizable phenotype (Hegemann et al., 2001). Analyses of *CPH1* knockout strains are needed to reveal CPH1 function.

17.3 Phototropins

17.3.1 *Adiantum capillus-veneris*

In gametophytic cells of *A. capillus-veneris*, phototropism and chloroplast photorelocation movement are regulated by both red and blue light (Kadota et al., 1982; Hayami et al., 1986; Yatsuhashi et al., 1985; Kagawa and Wada, 1994). Since the effect of red light but not blue light is completely canceled by far-red light, both phytochrome(s) and blue light receptor(s) must mediate these responses. Dichroism of these responses in polarized light suggests that the photoreceptor molecules are membrane-localized or close to the plasma membrane (Yatsuhashi et al., 1987; Hayami et al., 1992). phot1 and phot2 redundantly regulate the chloroplast accumulation response and phototropism in *A. thaliana* (Sakai et al., 2001), and phot2 alone mediates the chloroplast avoidance response under high fluence-rate blue light (Kagawa et al., 2001; Jarillo et al., 2001). Since phototropins are plasma membrane-localized proteins (Briggs et al., 2001; Sakamoto and Briggs, 2002), a phototropin is a strong candidate for the blue light receptor involved in phototropism and chloroplast movement in *A. capillus-veneris*.

Two *A. capillus-veneris* phototropin genes, *AcPHOT1* and *AcPHOT2,* were isolated by RT-PCR using a degenerate primer pair (Nozue et al., 2000; Kagawa et al., 2004). Recombinant LOV domains (LOV1, LOV2 or LOV1+LOV2) of AcPHOT1 or AcPHOT2 showed photochemical properties similar to those of phototropins from *A. thaliana, Oryza sativa,* and *C. reinhardtii* (Kasahara et al., 2002), indicating that the two Acphot chromoproteins function as blue-light receptors (Kagawa et al., 2004). Both phototropins were expressed not only in gametophytes, but also in sporophytes where blue light-induced chloroplast movement and phototropism occurs (Kawai et al., 2003; Wada and Sei, 1994). Very recently, we isolated mutants deficient in **b**lue **h**igh light-dependent **c**hloroplast movement, *bhc* (Kagawa et al., 2004). Among eleven mutants isolated, *bhc*-07 and *bhc*-08 showed strong phenotypes. Under microbeam irradiation, their chloroplasts accumulated in the beam spot even with blue light of more than 100 W m^{-2}. Therefore, *bhc*-07 and *bhc*-08 are defective in the avoidance response but not in the accumulation response. This phenotype resembles that in *phot2* mutants of *A. thaliana* (Kagawa et al., 2001; Jarillo et al., 2001), suggesting that these two *bhc* mutants must have lesions in the *AcPHOT2* gene. Sequencing the *AcPHOT2* gene in these mutants revealed that both *bhc*-07 and *bhc*-08 have deletions in this gene. Furthermore, transient over-expression of *AcPHOT2* cDNA in two *bhc* mutants rescued the defect in the avoidance response, confirming that *AcPHOT2* is essential for the chloroplast avoidance response in *A. capillus-veneris* (Kagawa et al., 2004).

Kagawa et al. (2004) investigated the functional domains of AcPHOT2 essential for the avoidance response by transient-expression analysis. Results showed that the LOV2 domain but not the LOV1 domain is necessary for the avoidance response. Conversion of cysteine 529 to alanine in the LOV2 domain of AcPHOT2, a mutation that disrupts cysteinyl-adduct formation with FMN (Salomon et al., 2000; Crosson and Moffat, 2002; Kasahara et al., 2002), made Acphot2 inactive. These results are consistent with those of Christie et al. (2002), who demonstrated that cysteinyl-adduct formation in LOV2 domain of AtPHOT1 is essential for phototropism and blue light-activated autophosphorylation in vivo. However, when the lifetime of the signal for the chloroplast avoidance movement was estimated as a period between the time when blue light irradiation are stopped and the cessation of movement or the beginning of a subsequent accumulation response, the shortest signal half-lives are 1 to 2.5 min, similar to those of the activated Acphot2-LOV1+LOV2 domain (about 2 min) but not the LOV2 domain alone (about 17 s), both of which were determined spectroscopically (Kagawa et al., 2004). Therefore, although the LOV1 domain of AcPHOT2 is not essential as a photoreceptor, LOV1 may modulate the lifetime of the LOV2 activation to allow the chloroplast avoidance response to function under physiological conditions.

The deletion of the last 40 amino acids C-terminal of the Ser/Thr kinase domain disrupted Acphot2 function, but the deletion of the last 9 or 20 amino acids did not. This C-terminal extension is well conserved among *Arabidopsis* AGC protein kinases (cAMP-dependent protein kinase **A**, cGMP-dependent protein kinase **G** and phospholipids-dependent protein kinase **C**) including the phototropin family (Kagawa et al., 2004; Bögre et al., 2003). Bögre et al. (2003) report that some AGC kinases have PDK1 (3-Phosphoinositide-Dependent protein Kinase 1)-interacting fragments in the last 4~6 residues and that the interaction of AGC kinases with PDK1 and their phosphorylation by PDK1 may be necessary for AGC kinase function. However, phototropins do not have this motif (Bögre et al., 2003). This absence is consistent with the results of Kagawa et al (2004), in which the last 20 amino acids are dispensable for Acphot2 function. Thus, the C-terminal portion of Acphot2 (amino acid 978~997) must play an important role in mediating the chloroplast avoidance response without functioning as a PDK1-interacting motif. This portion has two proline residues that are conserved in most of AGC kinases except for that of *Adiantum* PHY3 (Nozue et al., 1998b). These two proline residues may have important function in regulating the AGC kinase function of the phototropins (Kagawa et al., 2004).

17.3.2
Physcomitrella patens

Four *PHOT* genes were identified in *P. patens* by cDNA library screening and the functions of these genes were analyzed (Kasahara et al., 2004). A phylogenic tree drawn using the combined sequences of LOV1, LOV2, and Ser/Thr kinase domains showed that PpPHOTs clustered in a new group outside the PHOT1 or PHOT2 clusters for seed plants (Figure 17.4). Therefore, the four *P. patens* phototropin genes were named *PpPHOTA1, A2, B1,* and *B2.* PpPHOTA1 and PpPHOTA2, PpPHOTB1 and PpPHOTB2, respectively, are highly similar to each other. The two LOV domains and the Ser/Thr kinase domain of these PpPHOTs are well conserved and similar to those of other phototropins, but the PpPHOT N-terminal stretches upstream of the LOV1 domains (250~350 amino acids) are considerably longer than those of other plant phototropins (100~200 amino acids) (Figure 17.5). *Adiantum* PHOT1 also has an N-terminal stretch of about 250 amino acids (Nozue et al., 2000; Kagawa et al., 2004). However, the significance of this difference between PpPHOTs and other phototropins is unknown, since the function of the N-terminal stretch remains to be determined. *PpPHOTA1* mRNA accumulated under red, blue and white light conditions and was much reduced in darkness, whereas *PpPHOTA2, PpPHOTB1* and *PpPHOTB2* genes were constitutively expressed regardless of light conditions (Kasahara et al., 2004).

In *P. patens,* chloroplast movement is induced by blue light as well as red light (Kadota et al., 2000; Sato et al., 2001). To investigate whether Ppphots mediates these responses, each *PpPHOT* gene was disrupted by homologous recombination (*photA1, photA2, photB1,* and *photB2*), and further double (*photA1photA2* and *photB1photB2*) and triple *PpPHOT* disruptants (*photA2photB1photB2*) were generated (Kasahara et al., 2004). Blue light-induced chloroplast movement in basal cells of protonemata of *photA1, photB1* and *photB2* single disruptants behaved as did wild type protonemata, whose chloroplasts accumulated under weak blue light (0.02~50 W m^{-2}) and escaped from strong blue light (more than 50 W m^{-2}). However, in *photA2* disruptants, the accumulation response but not the avoidance response was induced regardless of the intensity of blue light, indicating that PpphotA2 functions as the blue light receptor for the avoidance response just as phot2 in *A. thaliana* and *A. capillus-veneris* (Kagawa et al., 2001, 2004; Jarillo et al., 2001) does.

Whereas the *photA1photA2* double disruptant showed a phenotype similar to that of the *photA2* single disruptant, the sensitivity of *photB1photB2* to blue light inducing the avoidance response was lower than that of wild type cells, so that the avoidance response did not occur up to 200 W m^{-2} of blue light. Given that *photA2* (and also *photA1photA2*) lacks the avoidance response and that *photB1* and *photB2* show a wild type phenotype, it is suggested that PpphotB1 and PpphotB2 redundantly regulate the avoidance response but are not essential (Kasahara et al., 2004). Although the single or double *PpPHOT* disruptants retained the normal accumulation response, *photA2photB1photB2* triple disruptant showed lower sensitivity to blue light. The accumulation response was not found at 0.02~50 W m^{-2}, and was weak at 100 W m^{-2} in the triple disruptant. These results indicate that *PpPHOTA2, PpPHOTB1* and *PpPHOTB2* redundantly mediate the chloroplast accumulation response and the residual response under a high intensity of blue light must be mediated by *PpPHOTA1* (Kasahara et al., 2004). Generation and characterization of quadruple *PpPHOT* diruptants are awaited. Interestingly, the effect of *PpPHOT* disruption on the avoidance response in the tip cells is different from that in the basal cells, where *photA1, photA2,* and *photA2photB1photB2* showed the avoidance response at 200 W m^{-2} blue light like wild type. However, *photA1photA2* lacked the avoidance response in the tip cells, indicating that both *PpPHOTA1* and *PpPHOTA2* redundantly mediate the avoidance response in the tip cells (Kasahara et al., 2004).

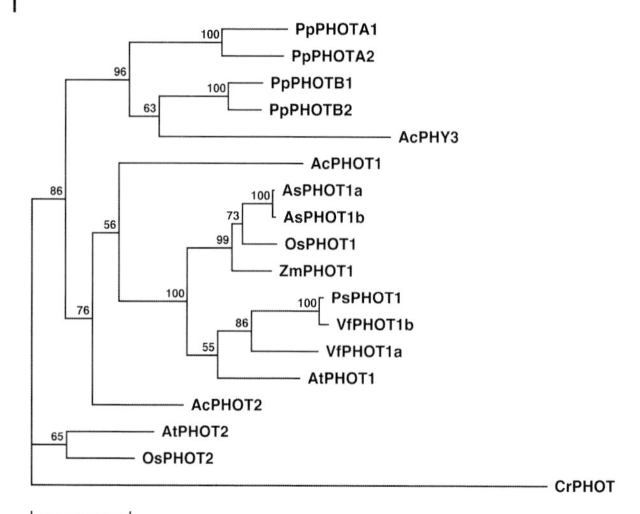

Figure 17.4 Phylogenetic tree and protein structure of phototropins. Combined amino acid sequences of LOV1, LOV2 and serine/threonine kinase domain were aligned using the Clustal W program. The phylogenetic tree was generated using the neighbor-joining analysis software MOLPHY (ftp://ftp.ism.ac.jp/pub/ISMLIB/MOLPHY/). Numbers indicate local bootstrap probabilities. Scale bar indicates the evolutionary distance. Accession numbers are AcPHOT1 (BAA95669, *Adiantum capillus-veneris* PHOT1); AcPHOT2 (AB115545, *A. capillus-veneris* PHOT2); AcPHY3 (BAA36192, *A. capillus-veneris* PHY3); AsPHOT1a (AAC05083, *Avena sativa* PHOT1a); AsPHOT1b (AAC05084, *A. sativa* PHOT1b); AtPHOT1 (AAC01753, *Arabidopsis thaliana* PHOT1); AtPHOT2 (AAC272293, *A. thaliana* PHOT2); CrPHOT1 (CAC94941, *Chlamydomonas reinhardtii* PHOT); OsPHOT1 (BAA84780, *Oryza sativa* PHOT1); OsPHOT2 (BAA84779, *O. sativa* PHOT2); PpPHOTA1 (AB163420, *Physcomitrella patens* PHOTA1); PpPHOTA2 (AB163421, *P. patens* PHOTA2); PpPHOTB1 (AB163422, *P. patens* PHOTB1); PpPHOTB2 (AB163423, *P. patens* PHOTB2); PsPHOT1 (AAM15725, *Pisum sativa* PHOT1); VfPHOT1a (BAC23098, *Vicia fava* PHOT1a); VfPHOT1b (BAC23099, *V. fava* PHOT1b); ZmPHOT1 (AAB88817, *Zea mays* PHOT1).

Phytochrome-mediated chloroplast movement occurs in *P. patens* (Kadota et al., 2000; Sato et al., 2001). Surprisingly, 1 W m^{-2} of red light cannot induce any chloroplast movement in *photA2photB1photB2* and 30 or 400 W m^{-2} of red light induced the weak accumulation response but not the avoidance response (Kasahara et al., 2004). Since phototropins do not absorb red light (Christie et al., 1999; Kasahara et al., 2002; Kagawa et al., 2004) and the existence of *PHY3*-like photoreceptor gene (Nozue et al., 1998b; Kawai et al., 2003) in *P. patens* has not been reported, phot-regulated signal transduction may function as downstream of phytochromes. Sato et al. (2001) previously demonstrated that the motility system (i.e. cytoskeleton) is differently affected by red light and by blue light: microtubule-dependent under red light, and actin- and/or microtubule-dependent under blue light. Possibly, red light-activated phytochrome may utilize the phot-dependent microtubule system.

Arabidopsis phototropins mediate a blue light-induced calcium influx into the cytoplasm (Baum et al., 1999; Babourina et al., 2002). At least in mesophyll cells, this

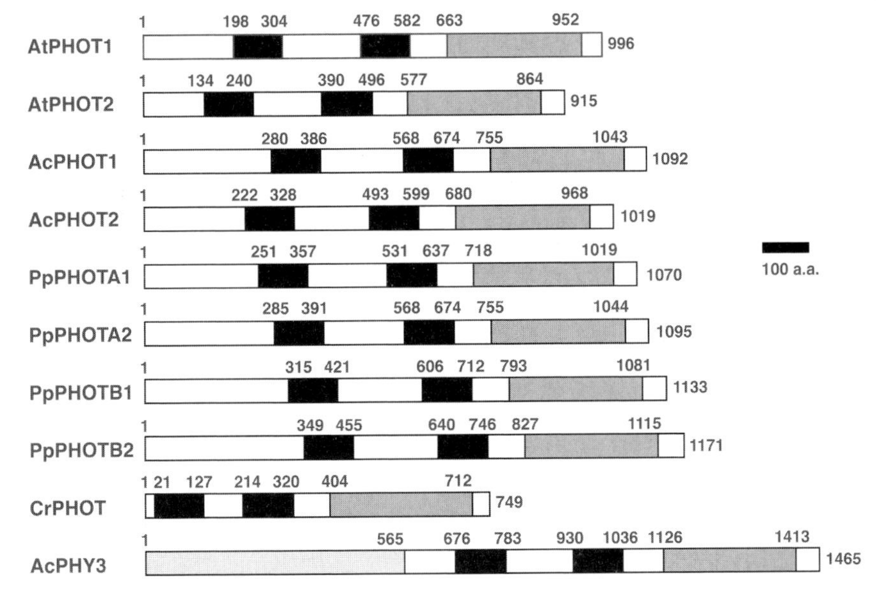

Figure 17.5 Domain organization of phototropins from *A. thaliana, A. capillus-veneris, P. patens* and *C. reinhardtii*. The two LOV domains are indicated as black rectangles. A serine/threonine kinase domain is indicated as a gray rectangle. The phytochrome photosensory domain in PHY3 is indicated as a light gray rectangle. The number "1" is the first methionine. The beginning and the last amino acid numbers of each domain are indicated.

calcium influx is dependent on calcium-permeable channels on the plasma membrane (Stoelzle et al., 2003). Previous work using apoaequorin transformants indicated that blue light but not red light increases the concentration of cytosolic calcium in *P. patens* (Russell et al., 1998). As with the phototropin-mediated calcium influx (Baum et al., 1999; Stoelzle et al., 2003), a blue light-induced calcium influx is inhibited by application of the calcium channel blocker lanthanum (La^{3+}) (Russell et al., 1998). Note that La^{3+} is completely ineffective in inhibiting the light-induced chloroplast movement in *P. patens* protonemal cells (Sato et al., 2003). Therefore, it is likely that the phot-mediated calcium influx into cytoplasm is not required for phot-mediated chloroplast movement.

17.3.3 *Chlamydomonas reinhardtii*

Although many land plants such as *A. capillus-veneris, Oryza sativa* and *A. thaliana* have two or more *PHOT* genes, *C. reinhardtii* has only one *PHOT* gene, *CrPHOT* (Holzer et al., 2002; Kasahara et al., 2002; Huang et al., 2002). CrPHOT has a very short amino acid stretch (ca. 20 amino acids) in front of the LOV1 domain whereas other plant phototropins have longer N-terminal stretches (ca. 100~350 amino acids) (Figure 17.5). The spectroscopic characteristics of the LOV domain from CrPHOT are similar to those of land-plant phototropins (Holzer et al., 2002; Kasahara et al., 2002). Crphot was shown to be a membrane-associated protein in vivo, and Crphot–GFP was localized to endogenous membranes when transiently expressed in tobacco protoplasts (Huang et al., 2002). Together, it was shown that Crphot is a bona-fide phototropin.

However, existence of a *PHOT* gene in *C. reinhardtii* is surprising, since this alga does not show phototropism, chloroplast movement, or stomatal opening. Blue light mediates phototactic movements and the progression of the sexual life cycle in this alga (Sineshchekov and Govorunova, 1999; Beck and Haring, 1996). Although phototactic movement is known to be mediated by retinal-based opsin-like photoreceptor (Sineshchekov and Govorunova, 1999; Ebrey, 2002; see Sineshchekov and Spudich, Chapter 2), the photoreceptor(s) mediating the sexual life cycle was suggested to be a flavin-based photoreceptor molecule (Weissig and Beck, 1991; Beck and Haring, 1996). During the sexual life cycle of *C. reinhardtii*, three steps are regulated by blue light: gamete formation from pregamete, the maintenance of mating competence in gametes, and zygote germination (Beck and Haring, 1996). Recently, it was demonstrated that *CrPHOT* mediates these three steps (Huang and Beck, 2003). When the PHOT protein level was reduced by RNA interference (RNAi) and the resulting phot-deficient strains were analyzed, these strains showed greatly impaired gamete formation, reactivation of mating competence of dark-inactivated gametes, and zygote germination. Blue light-induced gene expression in the late phase of gametogenesis was also impaired in phot-depleted strains. Given that Crphot is a membrane-associated protein (Huang et al., 2002), it is interesting that phot mediates blue light-induced gene expression during the sexual life cycle in *C. reinhardtii*. Although phot-depleted strains still had some responsiveness to blue light for mediating the sexual life cycle, another blue-light receptors is not likely to be involved in these responses. First, *C. reinhardtii* cryptochrome CPH1 is light-labile protein (Reisdorph and Small, 2004). Therefore, this receptor protein likely does not mediate the conversion of pregametes to gametes and zygote germination because both processes require the illumination with blue light for at least 1 hour. Second, southern blot analysis showed that a second phototropin gene was not likely in the *C. reinhardtii* genome (Huang et al., 2002). It may simply be that knockdown of *PHOT* gene expression by RNAi is incomplete. Analyses of other *PHOT*-knockout strains will clarify this question.

17.4 Phytochromes in Lower Plants

17.4.1 Conventional Phytochromes

Phytochrome responses have been extensively studied in lower-plant species. Chloroplast movement in the green algae *Mougeotia* and *Mesotaenium* is one of the best-known phytochrome responses (Haupt and Scheuerlein, 1990). The phytochrome system also mediates chloroplast movement in some ferns such as *A. capillus-veneris* and the moss *P. patens*. But the regulation of chloroplast movement by phytochrome is likely to be an exceptional case, because red light is ineffective in inducing chloroplast movement in most higher-plant species (Wada et al., 2003). Various physiological responses in lower-plant species are influenced by phytochrome; spore germination, protonemal growth, cell cycle, side-branching, rhizoid formation and membrane potential etc. (references in Wada and Kadota, 1989). Unlike seed plants, VLFR or FR-HIR responses have not yet been found in lower-plant species. Some full-length phytochrome gene sequences (genome or cDNA) are available for various lower plant species (Wada et al., 1997): the green algae *Mougeotia scalaris* (Winands and Wagner, 1996) and *Mesotaenium caldariorum* (Lagarias et al., 1995), the liverwort *Marchantia paleacea* (Suzuki et al., 2001), the mosses *P. patens* (Kolukisaoglu et al., 1993) and *Ceratodon purpureus* (Thümmler et al., 1992; Hughes et al., 1996; Pasentsis et al., 1998), and the ferns *Selaginella martensii* (Hanelt et al., 1992) and *A. capillus-veneris* (Okamoto et al., 1993; Nozue et al, 1998a, b). Although most of these phytochromes have a domain structure similar to that of conventional phytochromes from seed plants, *C. purpureus* PHYCER and *A. capillus-veneris* PHY3 are peculiar types of phytochrome. Instead of having a histidine kinase-like domain at the C terminus, PHYCER and PHY3 have a tyrosine kinase-like domain or a complete phototropin domain, respectively (Figures 17.1, 17.5; Thümmler et al., 1992; Nozue et al., 1998b).

Because of the difficulties of transformation and a genetic approach, the in vivo function of these phytochromes remains to be determined. Overexpression of *M. caldariorum PHY1b* or *A. capillus-veneris PHY1* in *A. thaliana* exhibited a dominant negative effect on inhibition of hypocotyl elongation under red and white or far-red light, respectively (Wu and Lagarias, 1997; Okamoto et al., 1997). Therefore, these lower plant phytochromes are likely to function improperly in *A. thaliana*.

Mutational analysis of phytochrome-mediated phototropism was performed in *C. purpureus* (Lamparter et al., 1996, 1997; Esch and Lamparter, 1998; Esch et al., 1999). Two classes of *C. purpureus* aphototropic mutants (*ptr*) were isolated. Class 1 *ptr* mutants are deficient not only in phototropism but also in chlorophyll accumulation under red light and contain less spectrally active phytochrome than wild type. These phenotypes were rescued by feeding the tetrapyrrole biliverdin or phycocyanobilin but not protoporphylin or heme, suggesting that class 1 *ptr* mutants were impaired in the biosynthesis of the phytochrome chromophore, probably in the conversion of heme into biliverdin (Lamparter et al., 1996, 1997; Esch and Lamparter, 1998). Confirming these results, microinjection of heme oxygenase genes of rat or *A. thaliana* rescued class 1 *ptr* mutants (Brücker et al., 2000). Class 2 *ptr* mutants are aphototropic but do not show the defect in chlorophyll accumulation and have spectrally active phytochrome at a level similar to that of wild type (Lamparter et al., 1996; Esch et al., 1999).

17.4.2
Phytochrome 3 in Polypodiaceous Ferns

The *phytochrome 3* gene (*PHY3*) was isolated by genomic-library screening with the *Adiantum PHY1* cDNA chromophore-binding region as a probe (Nozue et al., 1998b). *PHY3* is a single-copy gene in *A. capillus-veneris*, has no intron, and encodes a 1465 amino acid protein. *PHY3* mRNA is expressed in both gametophytic and sporophytic tissues. The 565 amino acid N-terminal chromophore-binding domain is similar to that of other phytochromes, but PHY3 C-terminal domain is not typical among any other phytochromes, and surprisingly contains a complete phototropin domain (Figure 17.1, 17.5). The two LOV domains and Ser/Thr kinase domain of the PHY3 protein show high similarity to the corresponding domains of the phototropins. When recombinant PHY3 proteins were reconstituted with the chromophore precursor phycocyanobilin *in vitro*, the absorption-difference spectrum of the phy3 holoprotein was very similar to those of other recombinant phytochromes (Nozue et al., 1998b). Furthermore, recombinant LOV domains bind flavin mononucleotide (FMN) at the equimolar ratio, and have absorption and fluorescence excitation spectra similar to those of the recombinant LOV domains from *A. thariana* or *Avena sativa* phototropins (Christie et al., 1999). Therefore, it seemed possible that phy3 could mediate not only red/far-red light responses but also blue light responses. Given that phototropism and chloroplast movement in *A. capillus-veneris* was likely to be regulated by membrane-associated phytochrome(s) and blue light photoreceptor(s), phy3 was a likely candidate for the photoreceptor for these responses.

We isolated red light-aphototropic (*rap*) mutants from *A. capillus-veneris* (Kadota and Wada, 1999). *rap* mutants lack red light-induced chloroplast movement (both the accumulation and avoidance responses) as well as phototropism (Figure 17.5), but retain normal blue light-induced phototropism and chloroplast movement. Importantly, other phytochrome-dependent responses (such as spore germination, tip growth, and cell division) and all other examined blue light responses in *A. capillus-veneris* were normal in *rap* mutants. *rap* mutants were defective in phototropism and chloroplast movement in sporophytic tissues as well as in gametophytic tissues (Figure 17.5; Kawai et al., 2003). Recently, it was shown that the *rap* mutants had their mutations in the *PHY3* gene (Kawai et al., 2003). Transient expression of the *PHY3* gene in *rap* mutant cells rescued the red light-induced chloroplast movement, confirming that phy3 is the red light photoreceptor for chloroplast movement. Besides *A. capillus-veneris*, many ferns such as *Dryopteris filix-mas, Onoclea sensibilis* and others show red light-induced phototropism and chloroplast movement and some polypodiaceous ferns including *D. filix-mas, O. sensibilis* and *Hypolepis punctata* have *PHY3*–homologous sequences. More primitive ferns such as *Osmunda japonica* (Osmundaceae) and

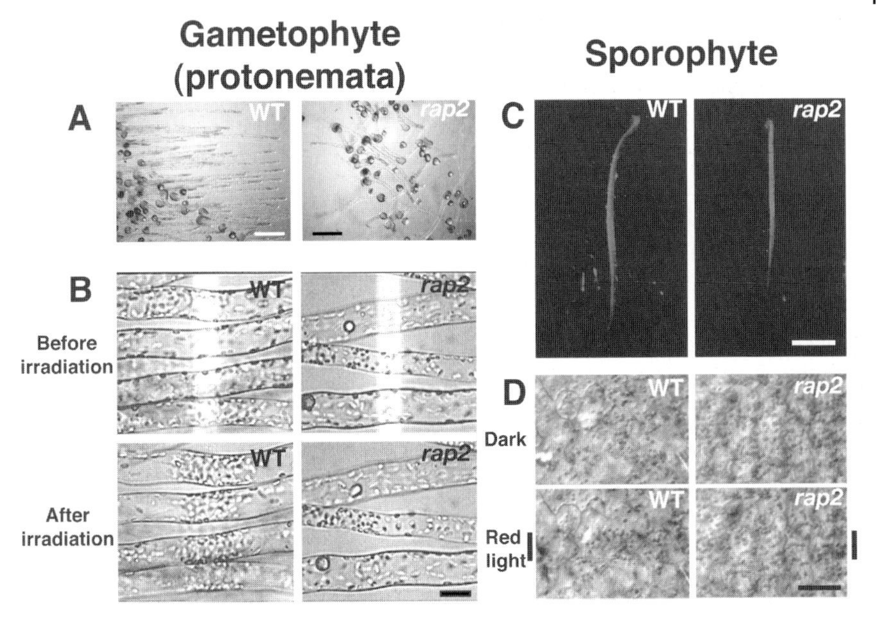

Figure 17.6 *rap* mutant phenotypes in gametophytic and sporophytic tissues of *A. capillus-veneris*. The *rap* mutants are deficient in phototropism and chloroplast movement both in gametophyte and in sporophyte. (A) Red light-induced phototropism in protonemata of wild type (left) and *rap2* (right). When irradiated with red light (0.5 W m^{-2}) from the right-hand side, wild type protonemata elongated towards the light source. However, *rap2* protonemata randomly elongated regardless of the direction of light. The scale bar represents 200 μm. (B) Red light-induced chloroplast movement in protonemata of wild type (left) and *rap2* (right). When irradiated for 3 h with a 20 μm beam of red light of 1.5 W m^{-2}, the accumulation of chloroplasts in the irradiated area is induced in wild type but not in *rap2* protonemata. The scale bar represents 20 μm. (C) Red light-induced phototropism in sporophytic leaves of wild type (left) and *rap2* (right). Dark-grown leaves were irradiated with red light (0.22 μmol m^{-2} s^{-1}) from the right-hand side. Whereas the wild-type leaf bends towards the light source, the *rap2* mutant leaf does not. The scale bar represents 5 mm. (D) Red light-induced chloroplast movement in a sporophytic leaf of wild type (left) and *rap2* (right). Dark-adapted leaf (Dark) was irradiated with red light (0.55 μmol m^{-2} s^{-1}) in a microbeam 20 μm in width (indicated by the black bars) for 3 h. A green band appeared at the irradiated area in wild type, but not in the *rap2* mutant. The scale bar represents 50 μm.

Lygodium japonicum (Schizaeaceae) do not have *PHY3* gene and also do not show phototropism and chloroplast movement in response to red light (Kawai et al., 2003). Thus, both red light-induced phototropism and chloroplast movement in higher ferns result from the gain of a *PHY3* gene.

What is the ecological and evolutional significance of the gain of a *PHY3* gene in polypodiaceous ferns? When the phototropic responses were analyzed, sporophytic leaves of *rap* mutants were less sensitive to white light at a low fluence rate than those of wild type (Kawai et al., 2003). Whereas phototropic curvature was saturated under a white light fluence rate of above about 10^{-3} μmol m^{-2} s^{-1}, *rap* mutants showed no apparent phototropic responses under the same light conditions.

Although *AcPHOT1* and *AcPHOT2* are expressed in sporophytic tissues, the lack of a *PHY3* gene has a detrimental effect on white-light sensitivity. Given that phy3-deficient mutants also lack red light-induced phototropism and chloroplast movement in sporophytic tissues, *Adiantum* phy3 is important for capturing low light in the sporophyte generation. Fossil records for most of the polypodiaceous ferns were found in a stratum dating back between the Upper Cretaceous and the Tertiary (Taylor and Taylor, 1993; Tidwell and Ash, 1994), when the inland forests dominated by gymnosperms were expanding (Taylor and Taylor, 1993; Willis and McElwain, 2002). Very recently, Schneider et al (2004) showed that the polypodiaceous ferns were diversified much more recently than had been thought, and that the diversification occurred in the Cretaceous after angiosperms; that is the polypodiaceous ferns were likely to be diversified in the shadow of angiosperms. This hypothesis is consistent with the hypothesis by Kawai et al (2003). Therefore, it is very likely that the gain of *PHY3* gene in polypodiaceous ferns have enhanced their adaptation under low light conditions in a well-developed canopy.

17.5 Concluding Remarks

Molecular genetic analyses of photoreceptors using *A. thaliana* have provided a great deal of information on the function of photoreceptors and signal transduction in light-induced responses in seed plants. However, we had not until very recently known functions of photoreceptors in other plant species, particularly in lower plant species. Recent identification and characterization of phytochromes, cryptochromes, and phototropins from lower plants not only strengthen the conclusions from *Arabidopsis* research on photoreceptors, but also provide many unexpected results such as the identification of *phy3* gene in ferns, the modulation of auxin signaling by cryptochromes in *P. patens*, the involvement of a phototropin in sexual life cycle in *C. reinhardtii*, and so on. In addition to molecular genetic analyses using *A. thaliana*, further analyses of photoreceptors in lower plants will be necessary to understand the function and evolution of the photoreceptor genes in plants.

Acknowledgements

We thank Dr. Takatoshi Kagawa and Dr. Masahiro Kasahara for sharing their unpublished results and critical reading of the manuscript; Dr. Akeo Kadota and Dr. Hiroko Kawai for providing photographs; Dr. Tomoaki Nishiyama for the construction of the phylogenetic tree.

References

Babourina, O., I. Newman, and S. Shabala (2002) *Proc. Natl Acad. Sci. USA* **99**, 2433–2438.

Baum, G., J. C. Long, G. I. Jenkins, and A. J. Trewavas (1999) *Proc. Natl Acad. Sci. USA* **96**, 13554–13559.

Beck, C. F. and M. A. Haring (1996) *Int. Rev. Cytol.* **168**, 259–302.

Bögre, L., L. Ökrész, R. Henriques, and R. G. Anthony (2003) *Trends Plant Sci.* **8**, 424–431.

Briggs, W. R., J. M. Christie, and M. Salomon (2001) *Antiox. Redox Signaling* **3**, 775–788.

Briggs, W. R. and J. M. Christie (2002) *Trends Plant Sci.* **7**, 204–210.

Brücker, G., M. Zeidler, T. Kohchi, E. Hartmann, and T. Lamparter (2000) *Planta* **210**, 529–535.

Cashmore, A. R., J. A. Jarillo, Y. J. Wu, and D. Liu (1999) *Science* **284**, 760–765.

Christie, J. M., M. Salomon, K. Nozue, M. Wada, and W. R. Briggs (1999) *Proc. Natl Acad. Sci. USA* **96**, 8779–8783.

Christie, J. M., T. E. Swartz, R. A. Bogomolni, and W. R. Briggs (2002) *Plant J.* **32**, 205–219.

Crosson, S. and K. Moffat (2002) *Plant Cell* **14**, 1067–1075.

Dharmasiri, S. and M. Estelle (2002) *Plant Mol. Biol.* **49**, 401–409.

Ebrey, T. G. (2002) *Proc. Natl Acad. Sci. USA* **99**, 8463–8464.

Esch, H. and T. Lamparter (1998) *Photochem. Photobiol.* **67**, 450–455.

Esch, H., E. Hartmann, D. Cove, M. Wada, and T. Lamparter (1999) *Planta* **209**, 290–298.

Furuya, M., M. Kanno, H. Okamoto, S. Fukuda, and M. Wada (1997) *Plant Physiol.* **113**, 677–683.

Guo, H., H. Duong, N. Ma, and C. Lin (1999) *Plant J.* **19**, 279–287.

Hanelt, S., B. Braun, S. Marx, and H. A. W. Schneider-Poetsch (1992) *Photochem. Photobiol.* **56**, 751–758.

Haupt, W. and R. Scheuerlein (1990) *Plant Cell Environ.* **13**, 595–614.

Hayami, J., A. Kadota, and M. Wada (1986) *Plant Cell Physiol.* **27**, 1571–1577.

Hayami, J., A. Kadota and M. Wada (1992) *Photochem. Photobiol.* **56**, 661–666.

Hegemann, P., M. Fuhrmann, and S. Kateriya (2001) *J. Phycol.* **27**, 668–676.

Holzer, W., A. Penzkofer, M. Fuhrmann, and P. Hegemann (2002) *Photochem. Photobiol.* **75**, 479–487.

Huala, E., P. W. Oeller, E. Liscum, I.-S. Han, E. Larsen, and W. R. Briggs (1997) *Science* **278**, 2120–2123.

Huang, K. and C. F. Beck (2003) *Proc. Natl Acad. Sci. USA* **100**, 6269–6274.

Huang, K., T. Merkle, and C. F. Beck (2002) *Physiol. Plant.* **115**, 613–622.

Hughes, J. E., T. Lamparter, and F. Mittmann (1996) *Plant Physiol.* **112**, 446.

Imaizumi, T., T. Kanegae, and M. Wada (2000) *Plant Cell* **12**, 81–95.

Imaizumi, T., A. Kadota, M. Hasebe, and M. Wada (2002) *Plant Cell* **14**, 373–386.

Imaizumi, T., H. G. Tran, T. E. Swartz, W. R. Briggs, and S. A. Kay (2003) *Nature* **426**, 302–306.

Jarillo, J.A., H. Gabrys, J. Capel, J. M. Alonso, J. R. Ecker, and A. R. Cashmore (2001) *Nature* **410**, 952–954.

Kadota, A., M. Wada, and M. Furuya (1982) *Photochem. Photobiol.* **35**, 533–536.

Kadota, A. and M. Wada (1999) *Plant Cell Physiol.* **40**, 238–247.

Kadota, A., Y. Sato, and M. Wada (2000) *Planta* **210**, 932–937.

Kagawa, T. and M. Wada (1994) *J. Plant Res.* **107**, 389–398.

Kagawa, T., T. Sakai, N. Suetsugu, K. Oikawa, S. Ishiguro, T. Kato, S. Tabata, K. Okada, and M. Wada (2001) *Science* **291**, 2138–2141.

Kagawa, T., M. Kasahara, T. Abe, S. Yoshida, and M. Wada (2004) *Plant Cell Physiol.* **45**, 416–426.

Kanegae, T. and M. Wada (1998) *Mol. Gen. Genet.* **259**, 345–353.

Kasahara, M., T. E. Swarts, M. A. Olney, A. Onodera, N. Mochizuki, H. Fukuzawa, E. Asamizu, S. Tabata, H. Kanegae, M. Takano, J. M. Christie, A. Nagatani, and W. R. Briggs (2002) *Plant Physiol.* **129**, 762–773.

Kasahara, M., T. Kagawa, Y. Sato, T. Kiyosue, and M. Wada (2004) *Plant Physiol.* 135, 1388–1397.

Kawai, H., T. Kanegae, S. Christensen, T. Kiyosue, Y. Sato, T. Imaizumi, A. Kadota, and M. Wada (2003) *Nature* **421**, 287–290.

Kinoshita, T., M. Doi, N. Suetsugu, T. Kagawa, M. Wada, and K. Shimazaki (2001) *Nature* **414**, 656–660.

Kleiner, O., S. Kircher, K. Harter, and A. Batschauer (1999) *Plant J.* **19**, 289–296.

Kolukisaoglu, H. Ü., B. Braun, W. F. Martin, and H. A. W. Schneider-Poetsch (1993) *FEBS Lett.* **334**, 95–100.

Lagarias, D. M., S-H. Wu, and J. C. Lagarias (1995) *Plant Mol. Biol.* **29**, 1127–1142.

Lamparter, T., H. Esch, D. Cove, J. Hughes, and E. Hartmann (1996) *Plant Cell Environ.* **19**, 560–568.

Lamparter, T., H. Esch, D. Cove, and E. Hartmann (1997) *Plant Cell Physiol.* **38**, 51–58.

Lin, C. and D. Shalitin (2003) *Annu. Rev. Plant Biol.* **54**, 469–496.

Martínez-García, J. F., E. Huq, and P. H. Quail (2000) *Science* **288**, 859–863.

Matsushita, T., N. Mochizuki, and A. Nagatani (2003) *Nature* **424**, 571–574.

Nagy, F. and E. Schäfer (2002) *Annu. Rev. Plant Biol.* **53**, 329–355.

Nozue, K., S. Fukuda, T. Kanegae, and M. Wada (1998a) *Plant Physiol.* **118**, 711.

Nozue, K., T. Kanegae, T. Imaizumi, S. Fukuda, H. Okamoto, K-C. Yeh, J. C. Lagarias, and M. Wada (1998b) *Proc. Natl Acad. Sci. USA* **95**, 15826–15830.

Nozue, K., J. M. Christie, T, Kiyosue, W. R. Briggs, and M. Wada (2000) *Plant Physiol.* **122**, 1457.

Okamoto, H., Y. Hirano, H. Abe, K. Tomozawa, M. Furuya, and M. Wada (1993) *Plant Cell Physiol.* **34**, 1329–1334.

Okamoto, H., K. Sakamoto, K. Tomizawa, A. Nagatani, and M. Wada (1997) *Plant Physiol.* **115**, 79–85.

Pasentsis, K., N. Paulo, P. Algarra, P. Dittrich, and F. Thümmler (1998) *Plant J.* **13**, 51–61.

Peterson, J. L., D. W. Lang, and G. D. Small (1999) *Plant Mol. Biol.* **40**, 1063–1071.

Quail, P. H (2002) *Nat. Rev. Mol. Cell Biol.* **3**, 85–93.

Reisdorph, N. A. and G. D. Small (2004) *Plant Physiol.* **134**, 1546–1554.

Russell, A. J., D. J. Cove, A. J. Trewavas, and T. L. Wang (1998) *Planta* **206**, 278–283.

Sakai, T., T. Kagawa, M. Kasahara, T. E. Swartz, J. M. Christie, W. R. Briggs, M. Wada, and K. Okada (2001) *Proc. Natl Acad. Sci. USA* **98**, 6969–6974.

Sakamoto, K. and W. R. Briggs (2002) *Plant Cell* **14**, 1723–1735.

Salomon, M., J. M. Christie, E. Knieb, U. Lempert, and W. R. Briggs (2000) *Biochemistry* **39**, 9401–9410.

Sancar, A (2000) *Annu. Rev. Biochem.* **69**, 31–67.

Sato, Y., M. Wada, and A. Kadota (2001) *J. Cell Sci.* **114**, 269–279.

Sato, Y., M. Wada, and A. Kadota (2003) *Planta* **216**, 772–777.

Schaefer, D. G. (2002) *Annu. Rev. Plant Physiol. Plant Mol. Biol.* **53**, 477–501.

Schneider, H., E. Schuettpelz, K. M. Pryer, R. Cranfill, S. Magallon, and R. Lupia (2004) *Nature* **428**, 553–557.

Shalitin, D., H. Yang, T. C. Mockler, M. Maymon, H. Guo, G. C. Whitelam, and C. Lin (2002) *Nature* **417**, 763–767.

Shalitin, D., X. Yu, M. Maymon, T. C. Mockler, and C. Lin (2003) *Plant Cell* **15**, 2421–2429.

Sineshchekov, O. A. and E. G. Govorunova (1999) *Trends Plant Sci.* **4**, 58–63.

Small, G. D., B. Min, and P. A. Lefebvre (1995) *Plant Mol. Biol.* **28**, 443–454.

Stoelzle, S., T. Kagawa, M. Wada, R. Hedrich, and P. Dietrich (2003) *Proc. Natl Acad. Sci. USA* **100**, 1456–1461.

Suetsugu, N. and M. Wada (2003) *Curr. Opin. Plant Biol.* **6**, 91–96.

Suzuki, T., S. Takio, I. Yamamoto, and T. Satoh (2001) *Plant Cell Physiol.* **42**, 576–582.

Taylor, T. N. and E. L. Taylor (1993) in *The Evolution of Plants* (Prentice Hall, Englewood Cliffs), 346–722.

Thümmler, F., M. Dufner, P. Kreisl and P. Dittrich (1992) *Plant Mol. Biol.* **20**, 1003–1017.

Tidwell, W. D. and S. R. Ash (1994) *J. Plant Res.* **107**, 417–442.

Wada, M. and M. Furuya (1978) *Planta* **138**, 85–90.

Wada, M. and A. Kadota (1989) *Annu. Rev. Plant Physiol. Plant Mol. Biol.* **40**, 169–191.

Wada, M. and H. Sei (1994) *J. Plant Res.* **107**, 181–186.

Wada, M., T. Kanegae, K. Nozue, and S. Fukuda (1997) *Plant Cell Environ.* **20**, 685–690.

Wada, M., T. Kagawa, and Y. Sato (2003) *Annu. Rev. Plant Biol.* **54**, 455–468.

Wang, H., L. G. Ma, J. M. Li, H. Y. Zhao, and X. W. Deng (2001) *Science* **294**, 154–158.

Weissig, H. and C. F. Beck (1991) *Plant Physiol.* **97**, 118 – 121.

Willis, K. J. and J. C. McElwain (2002) in *The Evolution of Plants* (Oxford Univ. Press, Oxford), 130–155.

Winands, A. and G. Wagner (1996) *Plant Mol. Biol.* **32**, 589–597.

Wu, S-H. and J. C. Lagarias (1997) *Plant Cell Environ.* **20**, 691–699.

Yang, H-Q., Y-J. Wu, R-H. Tang, D. Liu, Y. Liu, and A. R. Cashmore (2000). *Cell* **103**, 815–827.

Yang, H-Q., R-H. Tang, and A. R. Cashmore (2001) *Plant Cell* **13**, 2573–2587.

Yatsuhashi, H., A. Kadota, and M. Wada (1985) *Planta* **165**, 43–50.

Yatsuhashi, H., M. Wada, and T. Hashimoto (1987) *Acta Physiol. Planta.* **9**, 163–173.

18
Neurospora Photoreceptors

Jay C. Dunlap and Jennifer J. Loros

18.1
Introduction and Overview

Filamentous fungi are well known to display a variety of responses to light, chiefly to blue light. Among these organisms, *Neurospora crassa* has emerged as a tractable and informative model system. The photobiology of Neurospora has been well reviewed, with more recent surveys tending to focus on the genetic and molecular studies aimed at identifying the photoreceptor and describing the spectrum of genes regulated by light (Ballario and Macino, 1997; Degli Innocenti and Russo, 1984a; Linden et al., 1997). The principal blue light photoreceptor was identified recently as WC-1 (Froehlich et al., 2002; He et al., 2002), and its identity highlighted the LOV protein domain already associated with blue light photoreception in plants (Christie et al., 1998; Huala et al., 1997). In addition, the recent availability of the Neurospora genomic sequence, followed by the genomes of other fungi, has yielded additional putative photoreceptors (Borkovich et al., 2004; Galagan et al., 2003). This chapter will begin with a brief overview of blue light photobiology in fungi and then go on to describe the nature of the known and suspected photoreceptors.

18.2
The Photobiology of Fungi in General and Neurospora in Particular

18.2.1
Photoresponses are Widespread

Although most research on photobiology in the fungi has been driven by Neurospora, a great deal of excellent research has also used *Phycomyces blakesleeanus* which also elaborates carotenoids in response to blue light as well as displaying well-described developmental and phototropic responses (Cerda-Olmedo, 2001). Research in this organism has been hampered by the lack of a reliable transformation system (Obraztsova et al., 2004). In addition to the many light responses in Neurospora and

Handbook of Photosensory Receptors. Edited by W. R. Briggs, J. L. Spudich

ISBN 3-527-31019-3

Phycomyces, asexual spore production is light-induced in *P. blakesleeanus*, *Trichoderma harzianum*, and *Aspergillus nidulans* as well as in *N. crassa*, and both cell-wall branching and fruiting-body production are induced by light in *Schizophyllum commune* as well as in *N. crassa* (reviewed in Lauter, 1996). Interestingly, while blue-light responses are known in all of these organisms, *Aspergillus* also exhibits a red (far-red reversible) light response in its requirement for light in the formation of asexual spores (Mooney and Yager, 1990); the photoreceptor for this response is not known.

18.2.2
Photobiology of Neurospora

Light signals influence many aspects of both the sexual and asexual (vegetative) stages of life (Figure 18.1), and a brief description of the life and times of Neurospora (Davis, 2000) will serve to set the stage. Neurospora comes in two mating types, **A** and **a**. Sexual spores are activated by fires in the wild, the heat allowing germination; Neurospora is classified as a Pyrenomycete. The organism spends most of its life

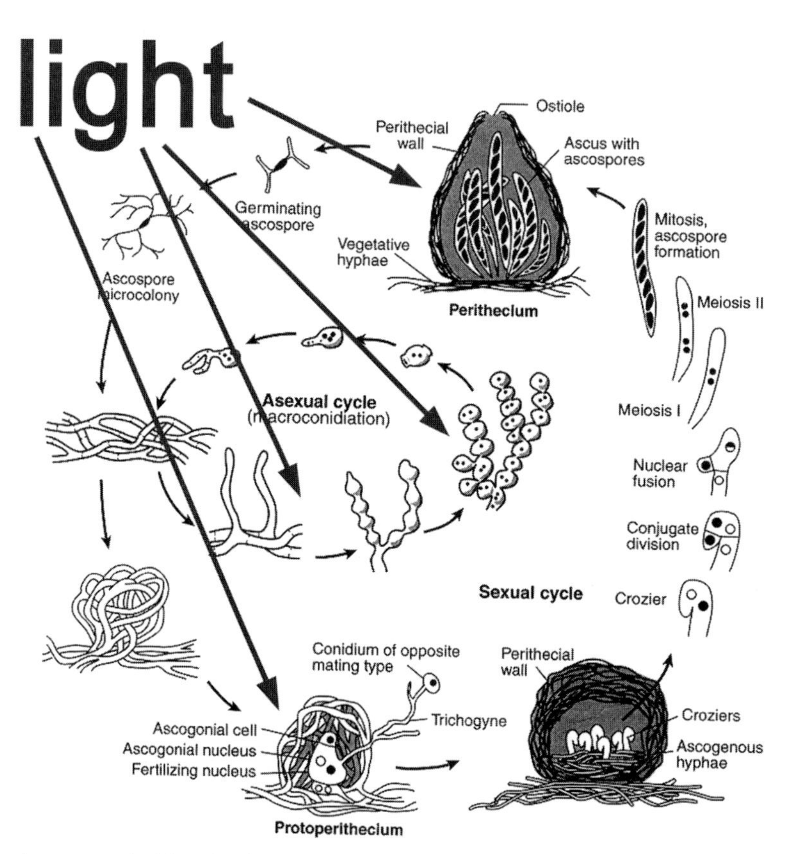

Figure 18.1 The life cycle of Neurospora. Light influences many aspects of both the sexual and asexual cycles. See text for detail. Adapted from Davis (2000) with permission.

growing vegetatively on the charred substrate as a syncytium with incomplete cell walls separating cellular compartments. As a vegetative culture, Neurospora can exist as surface mycelia or can elaborate aerial hyphae. The tips of aerial hyphae form morphological distinct structures called conidia that act as asexual spores and are easily dispersed by wind. When nutrients get scarce, vegetative Neurospora of either mating type induce sexuality by forming a fruiting body (a protoperithecium) which can be fertilized by a nucleus from a piece of mycelium or conidium of the opposite mating type. The mature reproductive fruiting body is the perithecium. Pairs of nuclei representing each parent replicate in tandem and eventually fuse to make a transient diploid that immediately undergoes meiosis to produce an 8-spored ascus, each ascus containing the products of meiosis from a single diploid nucleus.

The many effects of light on Neurospora are shown in Figure 18.2. During the asexual phase of the life cycle, light acts acutely to induce conidiation. More conidia are produced in light and they are produced faster (Klemm and Ninneman, 1978; Lauter, 1996). Carotenogenesis in mycelia is light-induced (Harding and Shropshire, 1980), and this response is quite rapid, being observable within the first 30 minutes after exposure to light. Pigmentation of the conidia is constitutive though, and in a strain defective for light perception or transduction of the light signal this results in the production of white mycelia underlying yellow/orange conidia, a phenotype that has

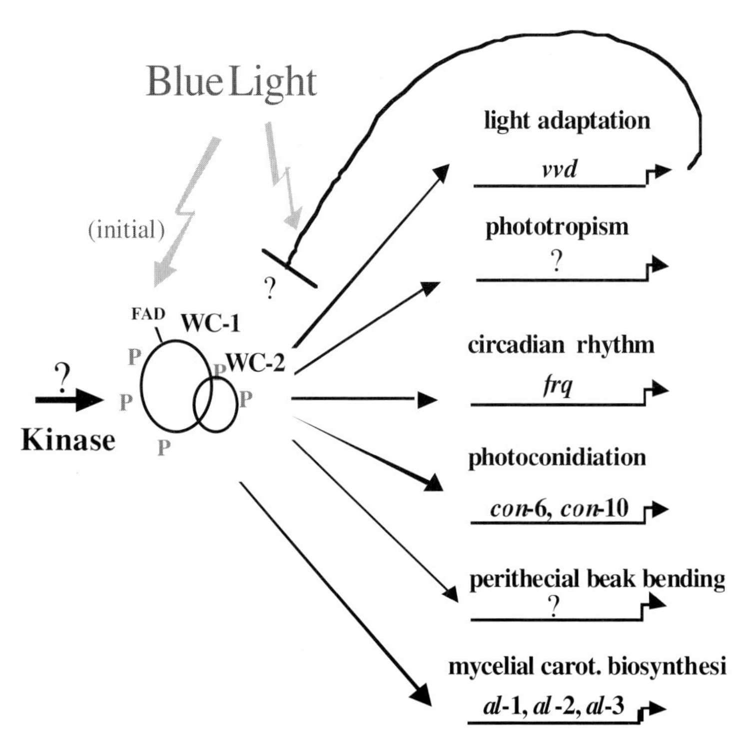

Figure 18.2 Summary of light responses in Neurospora. (Figure courtesy of C. Schwerdtfeger; all rights reserved.)

been of central importance in the genetic identification of the photoreceptors as described below. Although aerial hyphae are reported to display a phototropism (e.g. Siegel et al., 1968) in that they preferentially form on the lighted side of a dish, it is not clear to what extent this response is distinct from the overall light induction of conidiation. Reports also exist of changes in membrane conductivity (hyperpolarization and an increase in input resistance) in response to light (e.g. Potapova et al., 1984).

The most global effect of light during the asexual cycle (in the center of Figure 18.1) is to set the phase of the endogenous biological clock which acts to regulate a variety of aspects of the life cycle of the organism (reviewed in Loros and Dunlap, 2001; Sargent and Briggs, 1967). Light is used to set the phase of the clock that controls the daily timing of a developmental switch that leads to conidiation; a light-to-dark transfer is interpreted as dusk, and a dark to light transfer as dawn. Continued light also acts to suppress the expression of the clock. Conidiation involves a major morphological change that requires many novel gene products. Although the production of asexual spores is the best characterized light-phased circadian rhythm in *Neurospora*, other persisting rhythms at the physiological level have been described, which include the production of CO_2, lipid and diacylglycerol metabolism [e.g. (Lakin-Thomas and Brody, 2000; Ramsdale and Lakin-Thomas, 2000; Roeder et al., 1982)], a number of enzymatic activities [e.g. (Hochberg and Sargent, 1974; Martens and Sargent, 1974)], heat shock proteins (Rensing et al., 1987), and even growth rate (Sargent et al., 1966). These can be considered as secondary light responses.

The overall initiation of the sexual phase of the life cycle is enhanced by light (Degli Innocenti and Russo, 1983). During this time, carotenogenesis of perithecial walls is light induced (Perkins, 1988) and once mature perithecia are formed, ejection of spores from them is induced by light as is the direction in which the spores are shot: that is, the tips ('beaks') of the perithecia display a distinct phototropism (Harding and Melles, 1983). The number of spores shot from perithecia is also regulated by the light-phased circadian clock.

The involvement of light in the above-mentioned processes is acute and obvious, but Neurospora is also capable of more subtle responses to light characteristic of added regulatory sophistication: Neurospora responds to changes in the level of ambient light, from dark to dim or dim to bright in a process known as photoadaptation that is manifested in two ways. First, when the organism initially sees light, a response leading to transcription of light-induced genes is triggered that peaks within 15 to 30 minutes, but this response generally decays away within two hours or so, and if the organism is exposed to light within this two-hour period, no additional response is seen. Also, if the lights remain on, the response decays away, but after the two-hour latency period, the organism can respond again if the ambient level of light is increased.

Figure 18.2 not only summarizes these light responses but also indicates some of the genes associated with them. This overview also serves to emphasize the point that most of the well-characterized responses are tied directly to light-induced transcription.

18.3
Light Perception – the Nature of the Primary Blue Light Photoreceptor

18.3.1
Flavins as Chromophores

All known photoresponses in Neurospora are specific to light in the blue-green region of the spectrum. Action spectra developed in early efforts to identify the photoreceptor (Sargent, 1985) were consistent with the involvement of either carotenoids or flavins; however, since loss of the carotenogenesis genes *al-1, al-2,* and *al-3* eliminated nearly all of these pigments from cells but had no effect on the light responses (Russo, 1986), interest soon focused on flavins as chromophores. Consistent with this possibility, flavin biosynthesis mutants *rib-1* and *rib-2* showed reduced and flavin-dependent light responses including induction of carotenogenesis and phase shifting and photosuppression of circadian rhythmicity (Paietta and Sargent, 1981). Also, supplementation of the mutants with 1-deazariboflavin and roseoflavin yielded light responses having appropriately altered action spectra (Paietta and Sargent, 1983a). A number of flavoproteins have been suggested as possible blue light photoreceptors including nitrate reductase (Klemm and Ninneman, 1978) and cryptochromes. In the end, however, it was neither of these but was instead an entirely novel type of molecule, and genetic analysis of the phenomenon pointed the way to its identification.

18.3.2
Genetic Dissection of the Light Response

Screens for genes encoding the photoreceptor were based on the observation that blind strains are completely normal except for the fact that mycelia elaborate no carotenoids. Thus, colonies on agar are white if examined early (before conidia form) and slants kept in the light on the top of a lab bench will have white mycelia on the surface of the agar beneath a ring of yellow/orange conidia. Because of this appearance, when examined from the side, the collar of agar and mycelia at the top of the slant of such mutants will be white, hence the origin of the *white collar* genes. Such screens executed in the Russo and Macino laboratories identified a number of blind *white collar* strains all mapping to two loci, *white collar-1* and *white collar-2* (*wc-1* and *wc-2*). True loss-of-function *wc* mutants appear to be blind to all known light responses, including the light induction of carotenogenesis (Harding and Shropshire, 1980), light-induction of conidiation (Ninneman, 1991), induction of protoperithecia, and pointing of perithecial beaks (Harding and Melles, 1983; Degli Innocenti and Russo, 1984b). Further, photoadaptation is also lost because there is no primary light response to be adapted. A recent report of photoresponses in *wc* null strains (Dragovic et al., 2002) was traced to poor genetic craftsmanship in that the strains used were not true nulls; see (Lee et al., 2003) and (Cheng et al., 2003b) for more discussion. Genes encoding proteins that by sequence look like additional photoreceptors have been detected in the Neurospora genome, so novel assays for light responses may identify additional photoreceptor genes as described below.

Although *wc-1* and *wc-2* null mutants appear to be blind, a number of genes affecting photoresponses have been identified. The *poky* gene encodes a mitochondrial 19S RNA and its mutation appears to cause a b-type cytochrome deficiency that in some way reduces sensitivity for photosuppression of clock-regulated conidiation (Brain et al., 1977). Using a clever colony-based assay for the circadian rhythm, Paietta and Sargent identified several *light-insensitive* (*lis*) mutants that showed reduced photosuppression of the circadian rhythm (Paietta and Sargent, 1983b). Lastly, the *vivid* gene, now known to encode the photoreceptor required for photoadaptation (Schwerdtfeger and Linden, 2001), was described based upon a spontaneous mutation that resulted in a much more intense orange color (Hall et al., 1993).

18.3.3
New Insights into Photoreceptors from Genomics

Ongoing genomics efforts in Neurospora have turned up several additional putative photoreceptors (Figure 18.3) that cannot as yet be assigned to any photobiology. The first of these was NOP-1, a Neurospora opsin that appeared during an extensive EST sequencing project aimed at identifying genes regulated by the circadian clock (Zhu et al., 2001). Analysis of the protein following heterologous expression in *Pischia* showed that it bound a retinal cofactor and could undergo a photocycle, but the work failed to identify a solid light-regulatory function in Neurospora (Bieszke et al., 1999). The complete genomic sequence of *Neurospora* includes several bacteriophytochromes and a cryptochrome (Galagan et al., 2003) (Borkovich et al., 2004), genes for each of which are expressed. The CRY protein (Froehlich et al., 2004a) is found in both the cytoplasm and nucleus, is strongly light induced in a WC-1-mediated pathway, is circadianly regulated, and has been shown to bind FAD in vitro. Since CRYs are often characterized as proteins having sequences similar to DNA photolyases but lacking in photolyase function, it is of significance that complete deletion of CRY has no impact on photoreactivation repair (A. Froehlich, J. Loros, J. Dunlap, in preparation).

Also expressed are two genes encoding proteins having extensive similarities to bacteriophytochromes (Figure 18.3), members of the Cph1 family of phytochromes, with two PAS domains and a PHY domain followed by an ATPase and a response-regulator domain (Froehlich et al., 2004b). An N-terminal fragment of 515 amino acids of PHY-2 expressed in E. coli has been shown to bind either biliverdin or PCB and to undergo spectral shifts when exposed to red or far red light, and after red exposure there is a gradual decay back to the red absorbing form (Froehlich, A., Vierstra, R., Loros, J., and Dunlap, J. C., unpublished). Unfortunately, there are as yet no clock or light-related phenotypes that can be associated with loss of either PHY or CRY, and microarray analyses have so far failed to detect transcripts whose regulation is altered either by red or far-red light, or by loss of either PHY or CRY.

Sequence homologs to WC-1 have also appeared in other fungi. Although Saccharomyces has no apparent photobiology, the recent expansion of genomics efforts

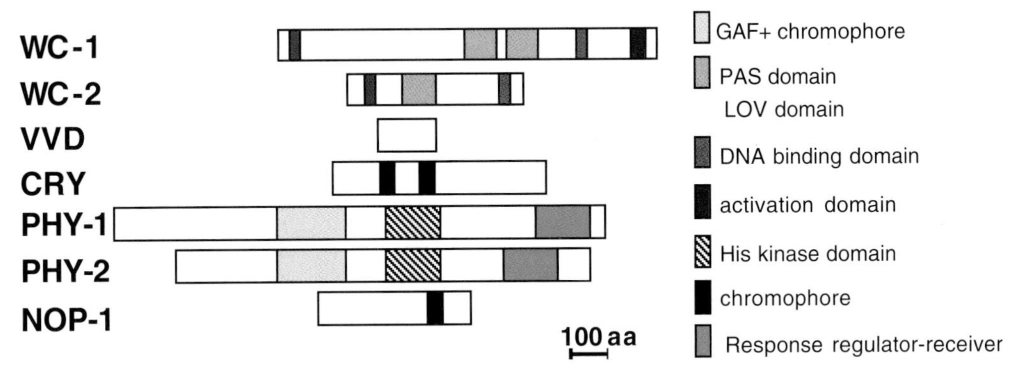

Figure 18.3 Real and putative photoreceptors in Neurospora. The identity and approximate location of various functional domains in Neurospora proteins are shown. In the list, only WC-1, VVD and CRY are known to bind chromophores and only the first two of these to be true photoresponse mediators, although WC-2 is required for the function of WC-1. The remaining genes have been identified as a result of various genomics-based efforts. See text for details.

from yeasts first to Neurospora and more recently to other complex filamentous fungi has identified a number of WC-1 homologs elsewhere. The common feature of these molecules is the LOV domains used as the photosensory module (see below). Based on strong sequence similarities, functional homologs of WC-1 probably exist in a number of fungi including *Podospora, Sordaria, Aspergillus, Cochliobolus, Magnaporthe, Coprinus* and *Fusarium*, although functional studies are still rare.

18.4 How do the Known Photoreceptors Work?

18.4.1 WC-1 and WC-2 contain PAS Domains and Act as a Complex

Molecular dissection of photoresponses in Neurospora began with the cloning of the *wc* genes by Macino and colleagues (Ballario et al., 1996; Linden and Macino, 1997). WC-1 is a 117 kD protein (Ballario et al., 1996) that contains a Zn-finger DNA-binding domain, a polyglutamine stretch in addition to two acidic domains of the types typical for transcriptional activators, and three PAS domains (Lee et al., 2000). In general, PAS domains are believed to mediate protein–protein interactions as has been shown to be the case with the WC proteins. The second and third PAS domains of WC-1 are of this type, but the first is a LOV domain, a subclass of PAS domains associated with proteins that sense light, oxygen, and voltage (Christie et al., 1999; Taylor and Zhulin, 1999). In a BLAST search, WC-1 shows strong similarity to a number of photoreceptors via the LOV domain and also to animal circadian clock proteins; subsequent work has shown that, consistent with its dark function in the circadian clock, WC-1 is a sequence and functional homolog of the mammalian clock protein BMAL1 with which it is 48% similar or identical over the entire length of BMAL1(Lee et al., 2000). WC-1 is a nuclear protein whose intracellular location is not regulated by light, whose expression is regulated post-transcriptionally (Lee et al., 2000), and whose activity and stability are controlled both through phosphorylation (Arpaia et al., 1999; Lee et al., 2000) and protein–protein interactions (Cheng et al., 2002; Denault et al., 2001; Talora et al., 1999). In contrast to WC-1, WC-2 is constitutively expressed, giving rise to a 57 kDa (530 amino acid) nuclear protein having a single activation domain, PAS domain, and Zn-finger that does not appear to be highly regulated (Denault et al., 2001; Schwerdtfeger and Linden, 2000). WC-1 and WC-2 interact in the nucleus via their PAS domains to form the White Collar Complex (WCC) (Talora et al., 1999; Schwerdtfeger and Linden, 2000; Ballario et al., 1998; Denault et al., 2001) and with FRQ (Denault et al., 2001; Cheng et al., 2001) in the feedback loop comprising the circadian oscillator (see below). WC-1 is the limiting factor in the complex, since WC-2 is always present in excess (Denault et al., 2001; Cheng et al., 2001), but WC-2 mediates the interactions between the regulated components of the WCC.

18.4.2
WC-1 is the Blue Light Photoreceptor

Based on the genetics, Harding and Shropshire and later Macino and colleagues proposed that the *white collar* genes encoded a photoreceptor (Ballario et al., 1998; Ballario et al., 1996; Harding and Shropshire, 1980; Linden and Macino, 1997), a sound guess that was later verified (Froehlich et al., 2002; He et al., 2002). The genetics were equally consistent with a model that posited either WC-1 and WC-2 as the photoreceptor or as essential elements in the signal transduction cascade leading from a photoreceptor to downstream responses. Interestingly, the identification in 1997 of an exclusively dark function for the WC proteins in the circadian clock (Crosthwaite et al., 1997) appeared to support the second model. In any case, to prove that a protein is a photoreceptor, of course, it is necessary not only to prove that mutants lacking it have no photoresponses, and to show that it binds a chromophore, but also that the protein itself undergoes photochemistry with a fluence and wavelength response consistent with the known biology. These were the tests used to establish WC-1 as the blue-light photoreceptor in Neurospora (Froehlich et al., 2002).

18.4.2.1 WC-1 utilizes FAD as a Chromophore

The sequence of the LOV domain of WC-1 contains the diagnostic NCRFLQ region that is conserved among all of the flavin-binding phototropins, and was expected to bind a flavin. It was thus satisfying when the WCC was epitope tagged, purified from Neurospora, and shown to contain a noncovalently bound flavin adenine dinucleotide (FAD) (He et al., 2002) (Figure 18.4). This result was consistent with the observation that FAD, but not FMN, is essential for function of the in vitro transcribed protein (Froehlich et al., 2002), and with the earlier biochemical genetic studies.

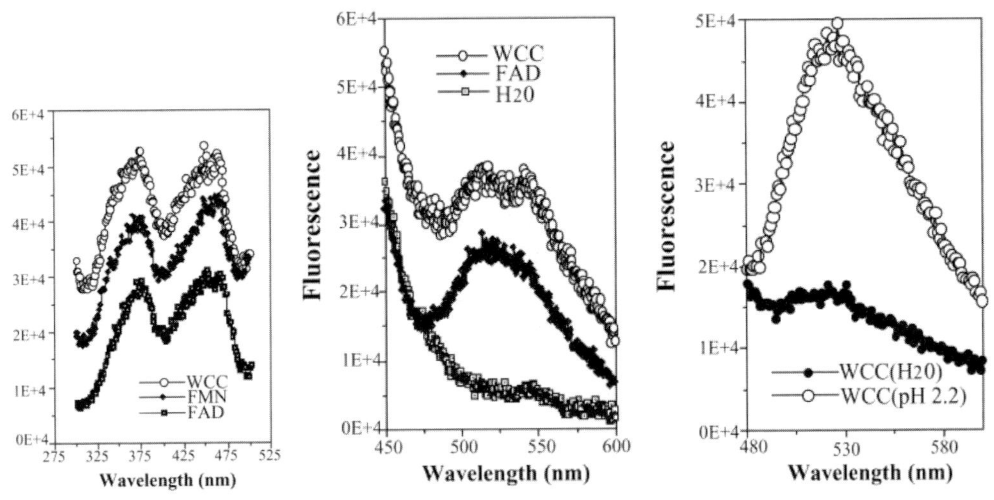

Figure 18.4 The affinity–purified white collar complex from *Neurospora* is associated with FAD based on fluorescence excitation peaks at 370 and 450 nm (left), and an emission peak at 520 (middle) that is pH sensitive (right) (from He at al., 2002).

18.4.2.2 **Proof of Photoreceptor Function**

An in vitro assay of photoreceptor function was used to establish unequivocally the role of WC-1 as a photoreceptor. Froehlich et al. (2002) first dissected the light-regulatory regions of a known highly light-inducible gene, the circadian clock gene *frq*, which turned out to have two separate light-regulatory elements (LREs). Light induction of *frq* was known to require both WC-1 and WC-2. Using nuclear protein extracts from both wild-type and definitive null strains, antisera that recognized both WC-1 and WC-2, and the LREs as targets in electrophoretic gel mobility shift assays (EMSAs), they showed that the WCC bound to each LRE. Mutants lacking either WC had no binding capacity, and FRQ was not found in the complex, although it is known to bind to the WCC in solution both in vivo and in vitro (Froehlich et al., 2002). The signal biochemical observation, however, was that the apparent size (mobility) of the complex in the shifted band was different when nuclear extracts were made in the light versus in the dark, and moreover that this difference could be recapitulated when extracts of dark grown cultures were extracted in the dark or under red light and later exposed to light – that is, the extracts themselves retained their ability to be photoresponsive (Figure 18.5). Based on these results, the biochemical nature of the photoreceptor could be determined, after controls were completed verifying that the in vitro reaction corresponded to the in vivo photobiology.

The important controls (Figures 18.6 and 18.7) established that the action spectrum for the conversion of dark complex to light complex had the same fluence and wavelength response (action spectrum) as had previously been shown for photosuppression of circadian rhythmicity (Sargent and Briggs, 1967) and also for blue light-induced phase shifting of the circadian clock (Crosthwaite et al., 1995). These data es-

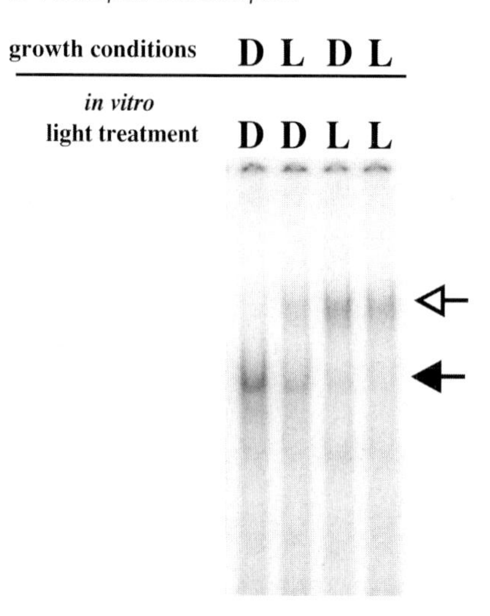

Figure 18.5 Electrophoretic mobility gel shift assays (EMSAs) using Neurospora nuclear extracts reveal an in vitro light-sensitive complex formation. Nuclear protein extracts were made from cultures grown the dark (D) or light (L) and used in EMSAs run either in D or L. Extracts from light cultures yield a slower mobility band (open arrow). Lane 3 shows that even after extraction, extracts retain the ability to respond to light and generate the lower mobility complex.

tablished the photoresponse in the test tube as indistinguishable from the biological responses. From this point it was logically straightforward to express the proteins in vitro in a coupled transcription/translation system, so that no other Neurospora proteins would be present, and to show that the light response (the change in mobility on the gel shift) could be achieved. In the absence of any added chromophore, both light and dark mobility complexes were seen, but the addition of FAD (but not FMN) promoted the formation of the dark complex in the dark and the light complex in the light.

Taken together, the data suggest the model shown in Figure 18.8 for the initial primary photoreception response. The complex of WC-1 and WC-2 binds to DNA at a specific consensus sequence, the light responsive element or LRE. Because the apparent mobility of the complex increases on exposure to light even when the complex is composed of proteins made in vitro in a test tube, the simplest interpretation is that light results in a multimerization of the components WC-1 and WC-2. This interpretation is consistent with the observation that self association of WC-1 in complexes is seen but self association of WC-2 is not seen (Cheng et al., 2003b). The proteins both contain GATA-type Zn fingers and the WCC binds to LREs having imperfect repeats of GATN (not GATA) (Froehlich et al., 2002), a sequence similar to one previously identified for the Neurospora *al-3* gene (Carattoli et al., 1994). Since the photoactive part of WC-1 is the LOV domain, everything supposed about the photochemistry of the events is based on better-studied LOV domains in the phototropins (Chapter 13, Christie and Briggs; Chapter 15, Crosson). Thus, it is expected that when the FAD chromophore absorbs blue light, the C(4a) position of the FAD undergoes a transient

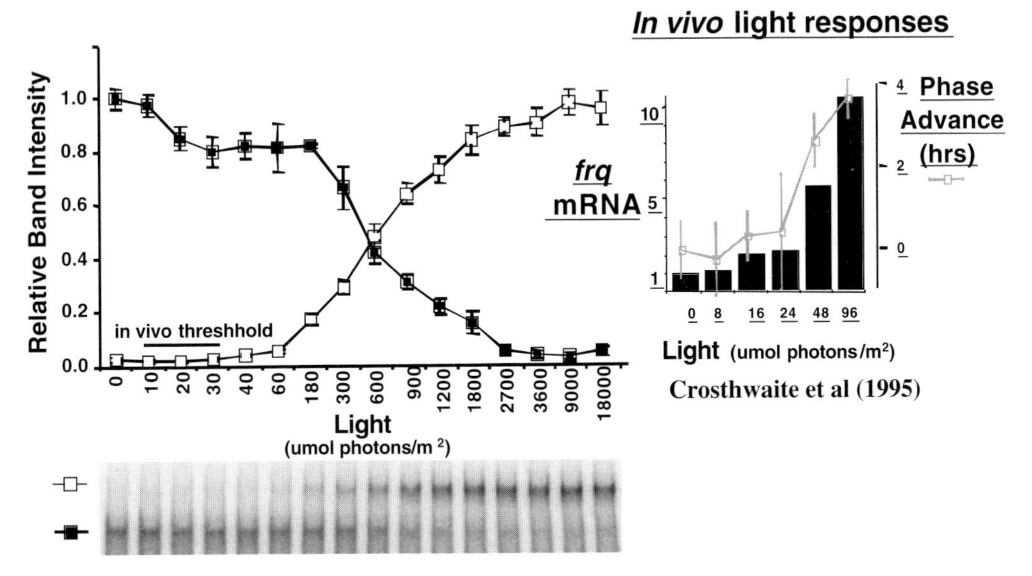

Figure 18.6 The in vitro light-induced WCC/LRE mobility change occurs at biologically relevant light intensities. The amount of material in the EMSA faster-mobility dark band (closed symbol) or slower mobility light band (open symbol) is plotted versus the amount of light (Crosthwaite et al, 1995). Figure adapted from Froehlich et al.(2002).

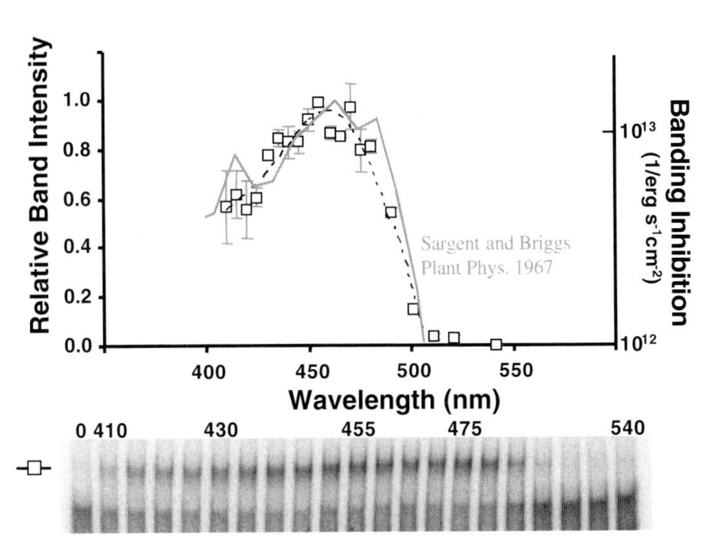

Figure 18.7 The in vitro light-induced WCC/LRE mobility change occurs at biologically relevant wavelengths. An action spectrum was derived by using the amount of light seen to give a half maximal response in Figure 18.6, but of different wavelengths. The in vitro reaction shows a peak in the blue that matches the action spectrum previously observed for photosuppression of circadian banding (Sargent and Briggs, 1967). Figure adapted from Froehlich et al. (2002).

covalent interaction with the cysteine in the NCRFLQ sequence of WC-1, thereby inducing a conformational change in the protein. Consistent with this expectation, point mutations changing this Cys to Ser or to Met yield a blind WC-1 (Cheng et al., 2003a). Since Froehlich et al. showed that the response occurs in a test tube containing only in vitro translated proteins and FAD, the WCC by itself must be sufficient with FAD to function as the photoreceptor. This conformational change in turn appears to bring about the quaternary interaction between WCCs, apparently resulting in formation of a multimer and in enhanced transcriptional activation of WCC-bound light responsive promoters.

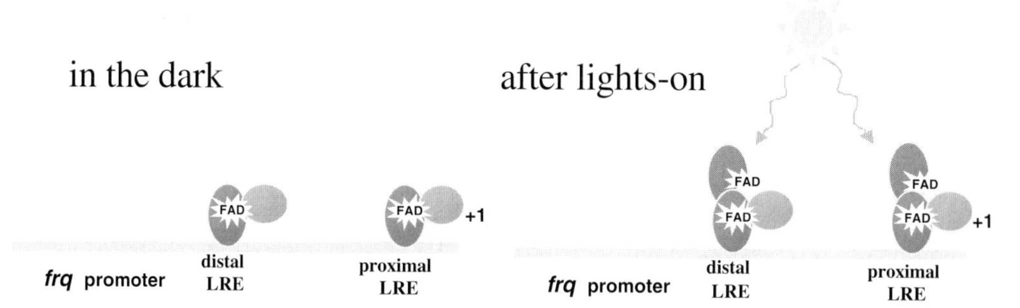

Figure 18.8 A model for blue light photoreceptor activation of transcription in Neurospora. Two light-regulatory sequences within the *frq* promoter in Neurospora bind to complexes of WC-1 and WC-2. The large oval represents WC-1 binding to FAD as the chromophore, and the small oval represents WC-2. See text for details. (Figure courtesy of A. Froehlich.)

18.4.3
Post-activation Regulation of WC-1

Following exposure to light WC-1 undergoes quantitative and qualitative changes that further modulate its activity. Macino and colleagues have advanced a model, based on inhibitor studies, in which WC-1 is phosphorylated and degraded soon after lights-on, followed by synthesis of non-phosphorylated WC-1 (Talora et al., 1999). The kinetics of light induction of most rapidly induced genes closely parallel this pattern, and the decrease in phosphorylated WC -1 is accompanied by a refractory period to light stimulation. In *vvd*^{null} strains (see below) total WC-1 decreases after light exposure but phosphorylated WC -1 levels remain high instead of decreasing, and the refractory period is missing; this is consistent with a model in which phosphorylated WC-1 is the active form of the protein (Heintzen et al., 2001; Schwerdtfeger and Linden, 2001; Talora et al., 1999).

18.4.4
A Non-photobiological Role for WC-1 and the WCC

As an aside, it should be noted that, exclusive of their role as the primary and initial blue-light photoreceptor complex in Neurospora, WC-1 and WC-2 serve as positive elements in the circadian feedback loop. Circadian rhythmicity is inextricably linked with light responses in most organisms: even though the clock will by design continue to run in the absence of light cues, it is light and dark transitions that most commonly provide the cues required to phase the internal clock oscillation correctly so that internal subjective day as experienced in the cell corresponds to external day as defined by the earth's rotation.

The WCC is the positive factor that drives expression of the negative element in the negative feedback loop comprising the core of the circadian oscillator (Crosthwaite et al., 1997). Briefly, in the transcriptional/translational feedback loops comprising the core of circadian oscillators of eukaryotes in the fungal/animal lineage, negative elements (like FRQ in *Neurospora*, PER/TIM in *Drosophila*, and PER1, PER2, CRY1, and CRY2 in mammals) block activation by heterodimeric PAS domain-containing positive elements (WC-1/WC-2 in *Neurospora*, dCLK/CYC in *Drosophila*, and CLOCK/BMAL1 in mammals). In turn, the positive elements activate expression of the negative elements, thereby giving rise to the negative feedback loop that lies at the core of the clock (Dunlap, 1998). Predictions that follow from this role of the WCC in the clock have been confirmed: (1) proteins such as WC-2 should play a role in temperature compensation (Collett et al., 2001); (2) defects in WC-2 should lengthen the circadian period length in the dark; (3) WC-2, FRQ, and WC-1 should physically interact in solution. Thus, in the *Neurospora* circadian clock, FRQ acts to depress the level of its own transcript at least in part by interfering with the formation of the WCC, thereby blocking activation of the *frq* gene by the WCC (Froehlich et al., 2003). The essential role of the WCC in the operation of the circadian clock, and of heterodimers of PAS domain proteins (such as CLOCK/BMAL in mammals or CLK/CYC in *Drosophila*) in circadian feedback loops was not known until the work of Crosthwaite in 1997 (Crosthwaite et al., 1997). That report provided the first precedent for the role of PAS:PAS heterodimers as activators in a circadian feedback loop, although it was followed just two weeks later by the cloning and description of the mammalian CLOCK gene (King et al., 1997) in which similar conclusions in the mammalian circadian system were independently reached.

In having an important and clearly distinct dark-only function, the WCC is probably unique among eukaryotic photoreceptors. To begin to gauge the importance of this regulation, Lewis et al (Lewis et al., 2002) expressed WC-1 under a regulatable promoter and examined gene regulation using microarrays. Although many genes were found to be regulated directly or indirectly in the dark by WC-1, most were not subsequently found to be light-induced. Conversely they found 22 light-induced genes that included 4 (like *frq*) that were also activated in the dark by WC-1. These data would be consistent with a role for WCC as a master regulator that acts to turn on downstream regulators that, in turn, more directly regulate target genes.

18.5
VIVID, a Second Photoreceptor that Modulates Light Responses

18.5.1
Types of Photoresponse Modulation

Although WC-1 is required for all known light responses, its actions are not sufficient to mediate all Neurospora photobiology. For there to be any light-dependent modulation of a light response, it follows that there must be additional photoreceptors. The VIVID protein (VVD) fulfills this role by modulating the organism's response to light signals subsequent to an initial exposure, a process known as photoadaptation (Schwerdtfeger and Linden, 2001; Schwerdtfeger and Linden, 2003; Shrode et al., 2001). VVD is, like WC-1 and WC-2, a member of the PAS protein superfamily (Heintzen et al., 2001) and its action describes an autoregulatory negative feedback loop that closes outside of the core circadian oscillator to affect both the clock and photoresponses. VVD as a photoreceptor binds a flavin (either FMN or FAD) as a chromophore (Schwerdtfeger and Linden, 2003), and as might be inferred from its place in the regulatory scheme, when *vvd* is lost all photoresponses in the organism are elevated.

When Neurospora sees light it exhibits an immediate early light response, for instance the rapid induction of *vvd* or al-*1* gene expression that peaks within 15–30 minutes and then decays away even when the light stays on (Figure 18.9). Interestingly, the photoresponse system remains refractory such that a second light exposure of the same fluence any time within two hours has no additional inductive effect, although exposure to brighter light can yield induction (Figure 18.10); this process requires de novo light-induced synthesis of VVD as well as phosphorylation of unspecified pro-

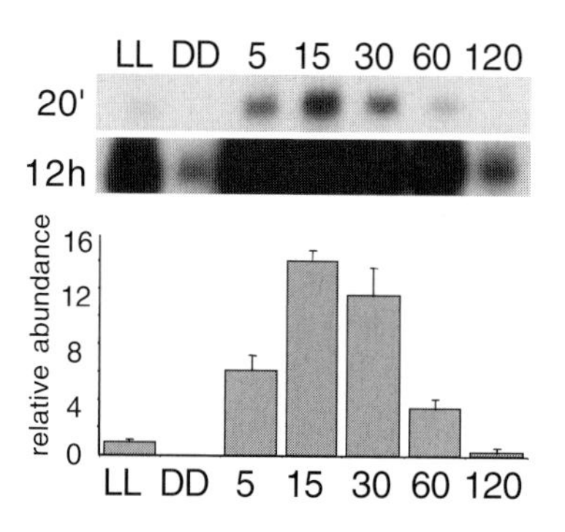

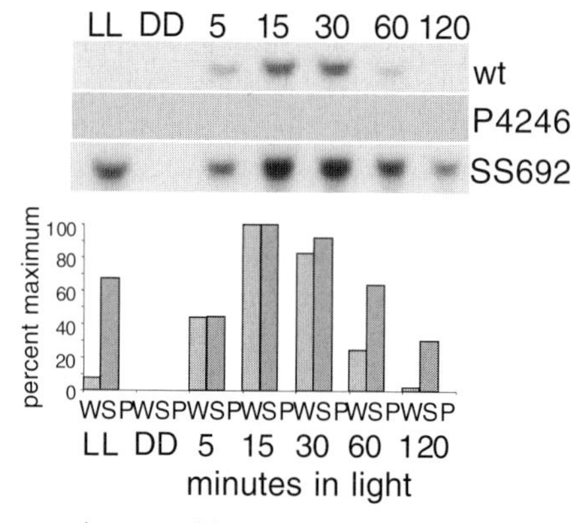

Figure 18.9 *vvd* is rapidly light induced, and loss of *vvd* slows the rate of decay of light-induced transcripts. On the right, P and S stand for two alleles, P which makes no transcript and S that makes transcript but no protein; W denotes wild type. Below the Northerns are shown the results of densitometric analysis relating the relative abundance of the *vvd* transcript seen above (From Heintzen et al., 2001).

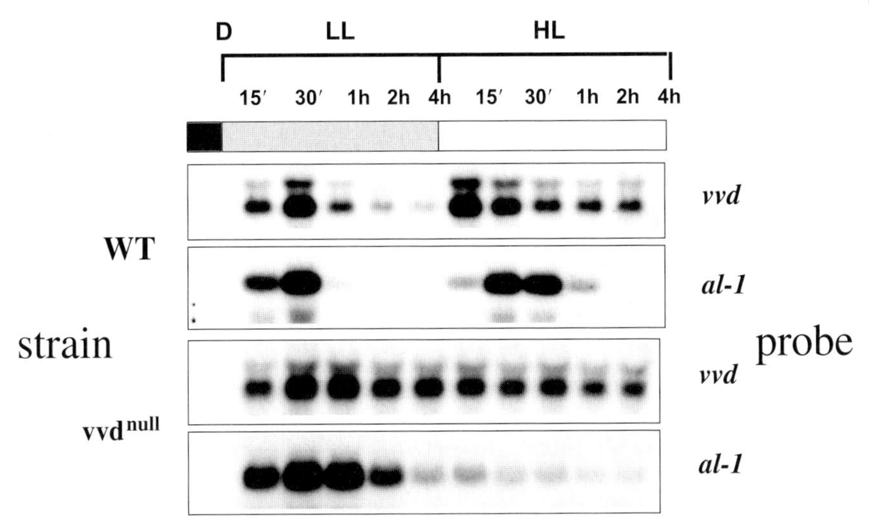

Figure 18.10 The ability to respond to increases in ambient light is lost in *vvd*- null strains. 15 μE m^{-2} s^{-1} of light was delivered at time 0, followed after 4 hrs by an increase to 55 μE m^{-2} s^{-1}. RNA was isolated from cultures and analyzed by Northern with *vvd* or *al-1* gene probes as shown. In WT the increased light elicited an increased transcriptional response which was not seen in a *vvd*-null strain. The *vvd*-null strain used is allele SS692 which makes transcript but no VVD protein. Adapted from Schwerdtfeger and Linden (2001 and 2003).

teins by protein kinase C. When VVD is lost, the rate of decay of light–induced gene expression is slowed, and more importantly perhaps, second exposures to light within two hours of a first exposure have additive effects (Schwerdtfeger and Linden, 2001; Schwerdtfeger and Linden, 2003; Shrode et al., 2001). As an aside it is worth noting that not all genes are subject to photoadaptation; for instance, the clock gene *frq* is simply induced by light (Crosthwaite et al., 1995) and its level does not decay over time to the same extent as other more typical genes.

Because VVD is strongly induced by light and in turn acts as a repressor of light responses, VVD affects circadian entrainment, the process by which the internal clock is set by the daily light/dark cycle. *vvd* is expressed at a significant level only during the first day in constant darkness, and VVD thus probably accounts for the finding that light signals have little if any clock-resetting effects when delivered on the first day after a light-to-dark transfer. Furthermore, *vvd* expression is controlled in part by the clock, and because of this control, the effect of a brief exposure to light on gene expression or on the clock will depend on the subjective time of day when the light is seen. This is an aspect of regulation known as circadian gating. VVD contributes to gating but loss of VVD does not result in complete loss of gating, suggesting that there are other factors involved (Heintzen et al., 2001). Clock regulation of the immediate and transient repressor VVD contributes to circadian entrainment by making dark-to-light transitions more discrete (Heintzen et al., 2001).

18.5.2
Proof of VVD Photoreceptor Function

VVD is a small protein, basically just a short N-terminal extension preceding a LOV domain (Figure 18.3). Its sequence thus suggested a photoreceptor function even before Schwerdtfeger and colleagues reported chromophore binding and a photocycle as seen in Figure 18.11 (Schwerdtfeger and Linden, 2003). VVD undergoes a classic photocycle; based on precedents set with the LOV domain of PHOT1, the expectation is that the C(4a) carbon of the flavin bound by VVD undergoes a transient covalent addition to the Cys in the NCRFLQ flavin binding site, resulting in structural changes in the protein that affect its activity and ability to interact with partners (Briggs and Christie, 2002; Harper et al., 2003). In support of this model, mutation of the Cys to Ala abrogates the ability of VVD to undergo the photocycle (Schwerdtfeger and Linden, 2003). A photoresponsive role for this LOV domain has also been supported by domain-swap experiments in which LOV domains from various proteins were swapped into WC-1 and the chimeras assayed in vivo for rescue of *frq* light induction and other related responses (Cheng et al., 2003a). A current model is that VVD will interact with the transcriptional activator WCC, down-regulating its activity and thereby influencing the expression of WCC-controlled genes, although this has not yet been verified biochemically.

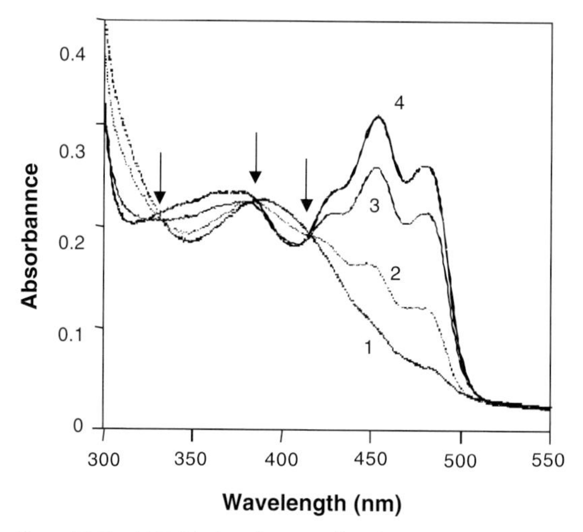

Figure 18.11 VVD binds a flavin and undergoes a photocycle. Repeated absorption spectra of VVD in solution taken (1) 10 s; (2) 10 min; (3) 2 h and (4) 5 h after initial exposure to 100 μmol photons $m^{-2} s^{-1}$ for 30 s. Three isosbestic points are shown by arrows. From Schwerdtfeger and Linden (2003).

18.6
Complexities in Light Regulatory Pathways

The straightforward expectation initially arising from experiments was that the WCC would bind to LREs in the promoters of light-regulated genes and activate them in concert, in most cases being modified by the action of VVD. However, the story does not appear to be this simple. Diverse induction kinetics are seen with light induced genes, and some genes are much more sensitive to the dose of WCC than are others. For instance, point mutations in the Zn-finger DNA-binding domain of WC-2 abolished light induction of some target genes but allowed induction of *frq* to proceed unaffected (Collett et al., 2002). Similarly, mutations in WC-1 having little effect on *frq* induction can abrogate light responses of other genes (Lee et al., 2003; Cheng et al., 2003a), and when expression of the WC proteins is placed under the control of inducible promoters, normal light-induced expression of *frq* is seen at basal levels of induction (Lee et al., 2003; Cheng et al., 2002).

An example from the literature illustrates errors that have arisen from this unexpected dosage dependence. In the MK1 allele of WC-1, a frame shift mutation places a STOP codon in the N-terminal region before the LOV domain; however, a very low level of WC-1 (less than 1% of the normal amount) is made due to reinitiation of translation and is sufficient to rescue low dosage functions such as light-induction of *frq* (Lee et al., 2003; Cheng et al., 2002). A similar *wc-1* allele, having point mutations in the N-terminal region of the gene that introduced STOP codons, was assumed to be a null and used to assess the importance of WC-1 to light-induction of *frq* and operation of the clock. Not appreciating the differential requirements for WCC levels in light-induction, or the importance of using true knockout strains, the authors mistakenly reported that "blind" mutants of Neurospora still showed robust light responses (Dragovic et al., 2002); actually WC-1 was still being made, as in MK1. More generally, it now appears that light-induced genes can be sorted into three groups based on their sensitivity to WC levels. The group most sensitive to WCC dose includes the genes encoding components of the carotenoid biosynthesis pathways (*al-1, al-2,*and *al-3*) and photoconidiation (e.g. *con-6* and *con-10)* pathways. Less sensitive is *vvd*. In a class by itself, *frq* requires log orders less WCC than the *albino* genes. The molecular basis of this sensitivity difference remains obscure.

18.7
Summary and Conclusion

Neurospora has for a long time been useful for dissecting the role of light in biological systems. Most photoresponses in Neurospora can be traced to light induction of transcription, and to date, responses only to blue light have been identified. Initial responses are due to the blue-light photoreceptor WC-1 that uses an FAD cofactor bound in a LOV domain; WC-1 partners with WC-2 to make a heterodimer which acts as a light-inducible transcription factor. Unique perhaps among photoreceptors, the WCC also displays a prominent function in the dark by playing the central role of key transcriptional activator in the negative feedback loop at the core of the circadian system. The other verified photoreceptor, VVD, uses a flavin bound in a LOV domain and acts independently as a photoreceptor to modulate the WCC response. Within the past few years, genomic studies in Neurospora have identified an opsin, a cryptochrome, and two phytochromes that may act in light responses, but to date no significant photobiological phenotypes have been associated with loss of these genes.

References

Arpaia, G., F. Cerri, S. Baima, and G. Macino (1999) *Mol Gen Genet* **262**, 314–322.

Ballario, P., and G. Macino (1997) *Trends Microbiol* **5**, 458–462.

Ballario, P., C. Talora, D. Galli, H. Linden, and G. Macino (1998) *Molec Microbiol* **29**, 719–729.

Ballario, P., P. Vittorioso, A. Magrelli, C. Talora, A. Cabibbo, and G. Macino (1996) *EMBO J* 15, 1650–1657.

Bieszke, J. A., E. L. Braun, L. E. Bean, S. Kang, D. O. Natvig, and K. A. Borkovich (1999) *Proc Natl Acad Sci USA* **96**, 8034–8039.

Borkovich, K., L. Alex, O. Yarden, M. Freitag, G. Turner, N. Read, S. Seiler, D. Bbell-Pedersen, J. Paietta, N. Plesofsky, *et al.* (2004) *Molec and Microb Rev* 68, 1–108.

Brain, R., D. Woodward, and W. Briggs (1977) Carnegie Inst Washington Yearbk **76**, 295–299.

Briggs, W. R., and J. M. Christie (2002) *Trends Plant Sci* **7**, 204–210.

Carattoli, A., C. Cogoni, G. Morelli, and G. Macino (1994) *Molec Microbiol* **13**, 787–795.

Cerda-Olmedo, E. (2001) FEMS *Microbiolog Rev* **25**, 503–512.

Cheng, P., Q. He, Y. Yang, L. Wang, and Y. Liu (2003a) *Proc Natl Acad Sci USA* 100, 5938–5943.

Cheng, P., Y. Yang, K. H. Gardner, and Y. Liu (2002) *Mol Cell Biol* **22**, 517–524.

Cheng, P., Y. Yang, and Y. Liu (2001) *Proc Natl Acad Sci USA* **98**, 7408–7413.

Cheng, P., Y. Yang, L. Wang, Q. He, and Y. Liu (2003b) *J Biol Chem* **278**, 3801–3808.

Christie, J. M., P. Reymond, G. K. Powell, P. Bernasconi, A. A. Raibekas, E. Liscum, and W. R. Briggs (1998) *Science* **282**, 1698–1701.

Christie, J. M., M. Salomon, K. Nozue, M. Wada, and W. R. Briggs (1999) *Proc Natl Acad Sci USA* **96**, 8779–8783.

Collett, M., J. C. Dunlap, and J. J. Loros (2001) *Molec Cell Biol* **21**, 2619–2628.

Collett, M. A., N. Garceau, J. C. Dunlap, and J. J. Loros (2002) *Genetics* **160**, 149–158.

Crosthwaite, S. C., J. C. Dunlap, and J. J. Loros (1997) *Science* **276**, 763–769.

Crosthwaite, S. C., J. J. Loros, and J. C. Dunlap (1995) *Cell* **81**, 1003–1012.

Davis, R. H. (2000). Neurospora, *Contributions of a Model Organism* (Oxford, UK, Oxford University Press).

Degli Innocenti, F., and V. E. A. Russo (1983) *Photochem Photobiol* **37**, 49–51.

Degli Innocenti, F., and V. E. A. Russo (1984a). Genetic analysis of blue light-induced responses in *Neurospora crassa*. In *Blue Light Effects in Biological Systems*, H. Senger, ed. (Berlin, Springer Verlag), pp. 213–219.

Degli Innocenti, F., and V. E. A. Russo (1984b) *J Bacteriol* **159**, 757–761.

Denault, D. L., J. J. Loros, and J. C. Dunlap (2001) *EMBO J* **20**, 109–117.

Dragovic, Z., Y. Tan, M. Gorl, T. Roenneberg, and M. Merrow (2002) *EMBO J* **21**, 3643–3651.

Dunlap, J. C. (1998) *Science* **280**, 1548–1549.

Froehlich, A., J. J. Loros, and J. C. Dunlap (2004a) in preparation.

Froehlich, A., J. J. Loros, and J. C. Dunlap (2004b) in preparation.

Froehlich, A. C., J. J. Loros, and J. C. Dunlap (2002) *Science* **297**, 815–819.

Froehlich, A. C., J. J. Loros, and J. C. Dunlap (2003) *Proc Natl Acad Sci USA* **100**, 5914–5919.

Galagan, J., S. Calvo, K. Borkovich, E. Selker, N. Read, W. FitzHugh, L.-J. Ma, SmirnovN., S. Purcell, B. Rehman, *et al.* (2003) *Nature* **422**, 859–868.

Hall, M. D., S. N. Bennett, and W. A. Krissinger (1993) *Georgia J Sci* **51**, 27 (Abstr.).

Harding, R., and S. Melles (1983) *Plant Physiol* **72**, 745–749.
Harding, R. W., and W. J. Shropshire (1980) *Ann Rev Plant Physiol* **31**, 217–238.
Harper, S. M., L. C. Neil, and K. H. Gardner (2003) *Science* **301**, 1541–1544.
He, Q., P. Cheng, Y. Yang, L. Wang, K. Gardner, and Y. Liu (2002) *Science* **297**, 840–842.
Heintzen, C., J. J. Loros, and J. C. Dunlap (2001) *Cell* **104**, 453–464.
Hochberg, M. L., and M. L. Sargent (1974) *J Bacteriol* **120**, 1164–1175.
Huala, E., P. W. Oeller, E. Liscum, I. S. Han, E. Larsen, and W. R. Briggs (1997) *Science* **278**, 2120–2123.
King, D., Y. Zhao, A. Sangoram, L. Wilsbacher, M. Tanaka, M. Antoch, T. Steeves, M. Vitaterna, J. Kornhauser, P. Lowrey, *et al.* (1997) *Cell* **89**, 641–653.
Klemm, E., and H. Ninneman (1978) *Photochem Photobiol* **28**, 227–230.
Lakin-Thomas, P. L., and S. Brody (2000) *Proc Natl Acad Sci USA* **97**, 256–261.
Lauter, F.-R. (1996) *J Genet* **75**, 375–386.
Lee, K., J. C. Dunlap, and J. J. Loros (2003) *Genetics* **163**, 103–114.
Lee, K., J. J. Loros, and J. C. Dunlap (2000) *Science* **289**, 107–110.
Lewis, Z. A., A. Correa, C. Schwerdtfeger, K. L. Link, X. Xie, R. H. Gomer, T. Thomas, D. J. Ebbole, and D. Bell-Pedersen (2002) *Mol Microbiol* **45**, 917–931.
Linden, H., P. Ballario, and G. Macino (1997) *Fungal Genet Biol* **22**, 141–150.
Linden, H., and G. Macino (1997) *EMBO J* **16**, 98–109.
Loros, J. J., and J. C. Dunlap (2001) *Annu Rev Physiol* **63**, 757–794.
Martens, C. L., and M. L. Sargent (1) *J Bacteriol* 974 **117**, 1210–1215.
Mooney, J. L., and L. N. Yager (1990) *Genes and Develop* **4**, 1473–1482.
Ninneman, H. (1991) *Photochem Photobiol* **9**, 189–199.
Obraztsova, I. N., N. Prados, K. Holzmann, J. Avalos, and E. Cerda-Olmedo (2004) *Fungal Genet Biol* **41**, 168–180.
Paietta, J., and M. Sargent (1981) *Proc Natl Acad Sci USA* **78**, 5573–5577.
Paietta, J., and M. Sargent (1983a) *Plant Physiol* **72**, 764–766.
Paietta, J., and M. L. Sargent (1983b) *Genetics* **104**, 11–20.
Perkins, D. D. (1988) *Fung Genet Newsl* **35**, 38–39.
Potapova, T., N. Levina, T. Belozerskaya, M. Kritsky, and L. Chailakhian (1984) *Arch. Microb* **137**, 262–265.
Ramsdale, M., and P. L. Lakin-Thomas (2000) *J Biol Chem* **275**, 27541–50.
Rensing, L., A. Bos, J. Kroeger, and G. Cornelius (1987) *Chronobiol Int* **4**, 543–549.
Roeder, P. E., M. L. Sargent, and S. Brody (1982) *Biochemistry* **21**, 4909–16.
Russo, V. (1986) Planta **168**, 56–60.
Sargent, M. (1985) Neurospora *Newsletter* **2**, 12–13.
Sargent, M. L., and W. R. Briggs (1967) *Plant Physiol* **42**, 1504–1510.
Sargent, M. L., W. R. Briggs, and D. O. Woodward (1966) *Plant Physiol* **41**, 1343–1349.
Schwerdtfeger, C., and H. Linden (2000) *Eur J Biochem* **267**, 414–422.
Schwerdtfeger, C., and H. Linden (2001) *Molec Microbiol* **39**, 1080–1086.
Schwerdtfeger, C., and H. Linden (2003) *EMBO J* **22**, 4846–55.
Shrode, L. B., Z. A. Lewis, L. D. White, D. Bell-Pedersen, and D. J. Ebbole (2001) *Fungal Genet Biol* **32**, 169–181.
Siegel, R. W., S. Matsuyama, and J. Urey (1968) *Experiencia* **24**, 1179–1181.
Talora, C., L. Franchi, H. Linden, P. Ballario, and G. Macino (1999) *EMBO J* **18**, 4961–4968.
Taylor, B. L., and I. B. Zhulin (1999) *Micro & Molec Biol Rev* **63**, 479–506.
Zhu, H., M. Nowrousian, D. Kupfer, H. Colot, G. Berrocal-Tito, H. Lai, D. Bell-Pedersen, B. Roe, J. J. Loros, and J. C. Dunlap (2001) *Genetics* 157, 1057–1065.

19
Photoactive Yellow Protein, *the* Xanthopsin

Michael A. van der Horst, Johnny Hendriks, Jocelyne Vreede, Sergei Yeremenko, Wim Crielaard and Klaas J. Hellingwerf

19.1
Introduction

19.1.1
Discovery of the Photoactive Yellow Protein

In 1985 T.E. Meyer reported the isolation of a series of colored proteins from the halophilic phototrophic bacterium *Ectothiorhodospira (Halorhodospira) halophila* (Meyer 1985). One of these proteins was yellow and was named 'Photoactive Yellow Protein' (PYP) in a subsequent study (McRee et al. 1986), because of its observed photoactivity. *H. halophila* is a unicellular phototrophic purple sulfur bacterium that deposits sulfur extracellularly. It was first isolated from the shores of Summer Lake, Lake County, Oregon (Raymond and Sistrom 1967; Raymond and Sistrom 1969), and subsequently also from the extremely saline lakes of the Wadi el Natrun in Egypt (Imhoff et al. 1978). Both locations are salt lakes and indeed *H. halophila* is a halophilic organism.

As a phototroph, *H. halophila* is able to exploit the free energy available from sunlight. However, like most organisms, *H. halophila* is not immune to the effects of UV-radiation, and by consequence protective mechanisms have evolved. Like most phototrophs *H. halophila* can perceive the quality and the quantity of the ambient radiation: It is attracted by red light that can be absorbed through its photosynthetic machinery, but repelled by high intensities of white and in particular blue light. This latter response has a wavelength dependence that fits the absorption spectrum of the photoactive yellow protein (Sprenger et al. 1993). This fit indicates that PYP may be the light-sensor in this blue-light response of *H. halophila*. Accordingly, the function of PYP may be similar to that of sensory rhodopsins, particularly to sensory rhodopsin II from *Halobacterium salinarum*, an *archae*bacterium that is also abundant in (solar) salt lakes. Rhodopsins (Spudich *et al.* 2000; Hoff *et al.*, 1997) form a large family with members in organisms in all kingdoms of life, from unicellular bacteria to *Homo sapiens sapiens*. It is the most extensively studied family of photorecep-

Handbook of Photosensory Receptors. Edited by W. R. Briggs, J. L. Spudich

ISBN 3-527-31019-3

tor proteins, especially via some of the visual rhodopsins, and bacteriorhodopsin, the latter being a light-activated proton pump found in the cytoplasmic membrane of *H. salinarum*. Sensory rhodopsins are structurally and photochemically closely related to bacteriorhodopsin and are present in the same organism. The similarity between PYP and the sensory rhodopsins was already noted after the first characterization of PYP (McRee et al. 1986). A major difference between PYP and sensory rhodopsins is that PYP is highly water soluble, whereas the rhodopsins, being membrane proteins, are not. The two proteins belong to structurally completely different families of photoreceptor proteins (which, however, may be similar in function; see above).

19.1.2
A Family of Photoactive Yellow Proteins: the Xanthopsins

After the initial biochemical characterization of PYP, reverse genetics led to the identification of the *pyp* operon that encodes this photoreceptor protein In *H. halophila* (Kort et al. 1996). Meanwhile, Southern hybridizations and PCR led to the discovery of a photoactive yellow protein in six other organisms – all purple bacteria – and a second one has been discovered in *H. halophila* itself (unpublished observation). This family of photoactive yellow proteins has been christened the Xanthopsin family (Kort et al. 1996). Presently, the eight known xanthopsins can be divided into three sub-groups, based on mutual sequence homology. The first sub-group is formed by the proteins found in *H. halophila [pyp(A)]* (Meyer 1985), *Rhodothalassium salexigens* (Meyer et al. 1990), and *Halochromatium salexigens* (Koh et al. 1996). The second is formed by proteins found in *Rhodobacter sphaeroides* (Kort et al. 1996), and *Rhodobacter capsulatus* (Jiang and Bauer 1998), and the third sub-group consists of proteins found in *Rhodospirillum centenum* (Jiang et al. 1999) and *Thermochromatium tepidum* (Cusanovich and Meyer 2003). In these latter two xanthopsins the homology to PYP is limited to their amino-terminal domain; they both have a larger bacteriophytochrome domain as well. Recently a second *pyp* gene (i.e. *pyp(B)*) was discovered in *H. halophila* (M.A. van der Horst et al., unpublished observation). Based on mutual homologies this xanthopsin can also be classified in sub-group III.

19.1.3
Differentiation of Function among the Xanthopsins

Though all these xanthopsins presumably absorb blue light, their role in the various organisms differs. In *H. halophila* PYP is reported to mediate intiation of a photophobic tactile response (Sprenger et al. 1993). It is relevant to note that this does not hold for the xanthopsin from *Rb. sphaeroides*, provided that no genetic redundancy exists in this organism (Kort et al. 2000). The results of the sequence comparison shown in Figure 19.1, revealing three sub-groups, are consistent with this conclusion. The xanthopsin from *Rs. centenum* transcriptionally regulates chalcone synthesis (Jiang et al. 1999). These different functions of the xanthopsin members coincide with the sub-group assignments. Though the function of only a few xanthopsins has been elucidated, it may well be that its members within the different sub-groups have

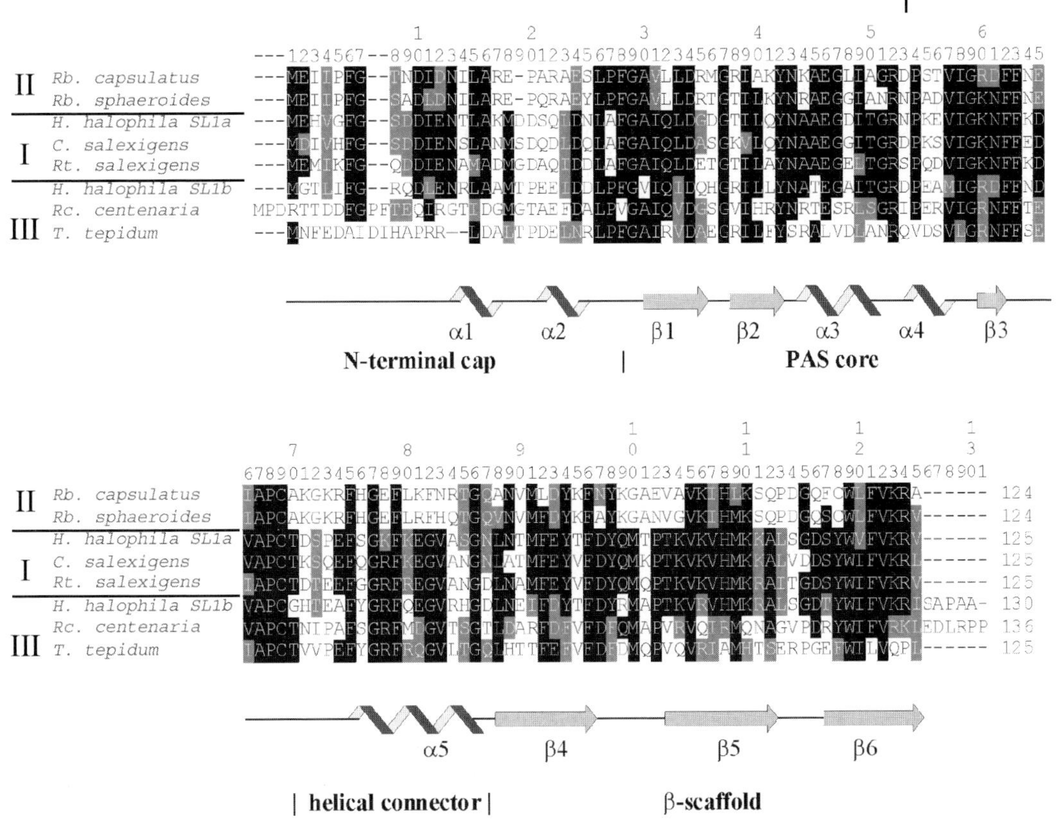

Figure 19.1 Multiple sequence alignment of all currently known xanthopsins. Alignments were made using the program ClustalW at http://www.ebi.ac.uk/clustalw/ (Thompson et al. 1994). Residues displayed with black background are conserved in more than 50% of the sequences; residues with a grey background are similar in more than 50% of the sequences. The Roman numbers indicate the subgroups to which the particular xanthopsin belongs. The location of the elements of secondary structure is shown via cartoons below the sequence alignment, together with the names of the sub-domains.

a similar biological function, while these functions may differ between the subgroups.

19.1.4
PYP: The Prototype PAS Domain

The xanthopsins thus far have only been identified in proteobacteria. Its members, however, also show a striking similarity with the much larger family of the PAS domains. These PAS domains have been identified in proteins from all three kingdoms of life, i.e. in the *Bacteria*, the *Archaea*, and the *Eucarya*. PAS is an acronym formed from the names of the proteins in which the PAS motive was first recognized (Nambu et al. 1991): the *Drosophila* period clock protein (**PER**), the vertebrate aryl hydrocarbon receptor nuclear translocator (ARNT), and the *Drosophila* single-minded protein (SIM).

Proteins containing PAS domains are predominantly involved in signal transduction. Over 2000 proteins have been identified that contain one or more PAS domains (Taylor and Zhulin 1999). Most of these proteins are receptors, signal transducers, or transcriptional regulators. PAS domains are usually present in proteins with a multi-domain architecture, such that a single protein can have up to six of these domains. Although most are present in the intracellular compartment, examples of periplasmic sensing domains with a PAS structure have also been described (Reinelt et al. 2003). The entire Photoactive Yellow Protein from *H. halophila* can be considered a single PAS domain with a ~30 amino acid N-terminal extension. As it is the first protein from the PAS domain family for which the 3-D structure has been elucidated, PYP has been proclaimed to be the structural prototype of the PAS domain fold (Pellequer et al. 1998).

19.2 Structure

The backbone of PYP is folded into an α/β-fold with a six-stranded anti-parallel β-sheet as a scaffold, flanked by several helices (Borgstahl et al. 1995). The loops containing helices $\alpha3$ and $\alpha4$, as well as the one containing helix $\alpha5$, fold on top of it to form the major hydrophobic core of PYP and a pocket in which the chromophore resides. Helices $\alpha1$ and $\alpha2$, which comprise the majority of the N-terminal domain fold behind it to form a second, smaller hydrophobic core. A spatial model of this arrangement of elements of secondary structure is shown in Figure 19.2.

In photoreceptor proteins the light-absorbing chromophore triggers functional activity. Its chemical structure typically correlates with the wavelength range in which the holo-protein is biologically active (Hellingwerf et al. 1996). Accordingly, the xanthopsins were shown to use an aromatic chromophore. After a long period of confusion this chromophore was identified in 1994 in *H. halophila* PYP as 4-hydroxycinnamic acid, covalently bound to the apo-protein via a thiol ester linkage with Cys69 (Baca et al. 1994; Hoff et al. 1994a).

19.2.1 Primary, Secondary, and Tertiary Structure

The amino acids that line the chromophore-binding pocket are (in the order in which they appear in the sequence): Ile31, Tyr42, Glu46, Thr50, Arg52, Phe62, Val66, Ala67, Cys69, Thr70, Phe96, Asp97, Tyr98, Met100, Val120 and Val122. When residing in this pocket the chromophore is completely buried in the major hydrophobic core of PYP and has no direct contact with solvent.

In the ground state of PYP, the chromophore is present as a phenolate anion (Baca et al. 1994; Kim et al. 1995). The resulting negative charge is delocalized over the chromophore via an extensive π-orbital system. It is additionally stabilized via a hy-

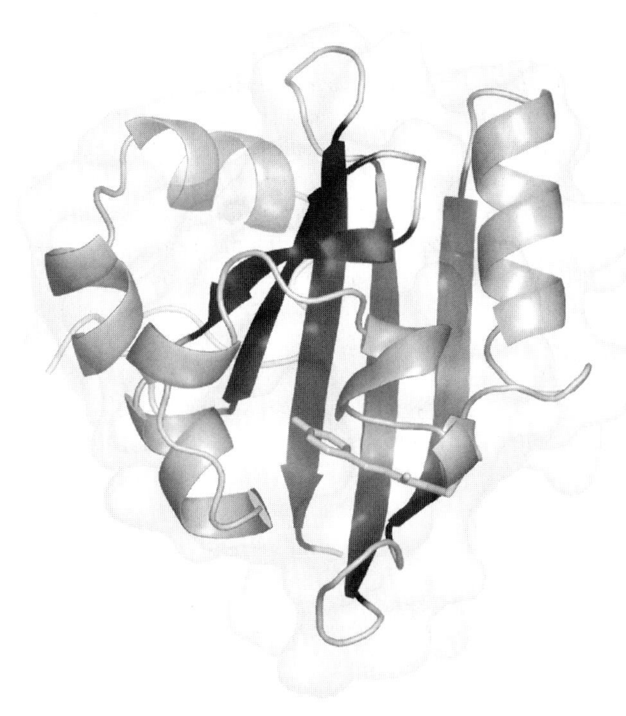

Figure 19.2 Tertiary structure of the Photoactive Yellow Protein. The backbone is represented as a ribbon with α helices, loops and β strands. The chromophore is shown in yellow. The transparent surface indicates the molecular surface of PYP. This picture was created using PyMOL (DeLano 2002).

drogen-bonding network and, presumably, by the positive charge of Arg52 (Borgstahl et al. 1995; Yoda et al. 2001; Groenhof et al. 2002b). The hydrogen-bonding network includes, besides the chromophore, residues Tyr42, Glu46, and Thr50, such that O_η from Tyr42 and $O_{\varepsilon,2}$ from Glu46 directly hydrogen-bond with the phenolate oxygen of the chromophore. $O_{\gamma,1}$ from Thr50 forms a hydrogen bond with O_η from Tyr42, yet does not line the chromophore-binding pocket.

19.2.2
Solution Structure vs. Crystal Structure

Up till now we have described the structure of PYP as determined by X-ray diffraction analyses, i.e. in the confines of a crystal lattice. In vivo, the protein is located in the cytoplasm, on the intracellular side of the cytoplasmic membrane (Hoff et al. 1994b), where it may have more freedom for dynamical alterations in its structure than in a crystal lattice. It is therefore relevant to know whether there are differences between its structure in a crystal and in aqueous solution [note that in the intact cell

PYP may also be confined or partly constrained in its movements by e.g. (a) transducer protein(s)].

The solution structure of PYP has been determined via multi-dimensional NMR spectroscopy (Dux et al. 1998) and is very similar to the structure determined with X-ray crystallography. Most of the elements of secondary structure are present in both structures, though they may start/end 1 to 2 residues earlier or later. However, helix α2 and the π-helix are not detectable in the solution structure. Additionally, there are three poorly defined regions in the solution structure comprising residues 1–5, 17–23, and 113–117. This is caused by lack of structural constraints in the NMR dataset, which might be caused by high side-chain and/or backbone mobility (Dux et al. 1998). In agreement with this suggestion, the crystal structure indicates that the corresponding regions have higher B-factors.

From an ensemble of structures, such as obtained with NMR and using Essential Dynamics, it is possible to determine eigenvectors that describe the path along which the atoms of the protein may move (van Aalten et al. 1998; Van Aalten et al. 2000). Strikingly, by using these eigenvectors, it is possible to transform the solution structure into the crystal structure and *visa versa*, indicating that the observed differences are within the confines of the intrinsic flexibility of the protein.

The solution structure, as determined with NMR, confirms the presence of the hydrogen-bonding network that stabilizes the anionic chromophore in the chromophore-binding pocket. However, there is one notable difference with the crystal structure: The side chain of Arg52 is present in two conformations, one in which Arg52 is clustered about 4 Å above the aromatic ring of the chromophore, and one with the guanidinium group of Arg52 positioned about 4 Å above the aromatic ring of Tyr98 (Dux et al. 1998). This latter position is in line with the observation that cations preferentially position themselves within 3.4 to 6 Å of the centroid of an aromatic ring [a phenomenon called π-stacking (Scrutton and Raine 1996)]. The conformation of Arg52 and Tyr98 as observed in the crystal is different from the two conformations for Arg52 and from the conformation of Tyr98 as detected in solution.

19.2.3 The Xanthopsins Compared

The xanthopsins can be divided into three sub-groups, based on their primary structure (see Section 19.1.2). Primary structure within the sub-groups is very similar, with identities around 75% (or: 87% similarity, i.e. including conserved substitutions) in pair-wise alignments. In a comparison of xanthopsins from sub-group I with xanthopsins from other groups, the similarity decreases with identities around 45% (67% similarity). Comparison of xanthopsins from sub-group II with those from sub-group III provides even poorer results.

Via comparison of the functional sub-domains of PYP in these sequence alignments, insight is obtained on which domains are most typical for a xanthopsin, and which domains are important for its function according to its sub-family classification. As alluded to above, PYP was proposed to be the prototype PAS domain. The PAS domain family is very large, spanning all three kingdoms of life. PAS domains are not so much defined by their primary structure, but rather by their secondary and tertiary structural elements. They can be divided into four sub-domains: the N-terminal cap, the PAS core, the helical connector, and the β-scaffold (Figure 19.1). In PYP from *H. halophila* these sub-domains comprise residues 1–28, 29–69, 70–87, and 88–125, respectively (Pellequer et al. 1998). With respect to secondary structure this implies that the N-terminal cap contains helices $\alpha 1$ and $\alpha 2$, the PAS core β-strands $\beta 1$ to $\beta 3$ and helices $\alpha 3$ and $\alpha 4$, the helical connector helix $\alpha 5$ and the β-scaffold β-strands $\beta 4$ to $\beta 6$. The residues that line the chromophore-binding pocket are all contained within the PAS core plus β-scaffold, which are sandwiched together.

Within the xanthopsin sub-families, no significant differences are detectable, with respect to the degree of homology, between the different PAS sub-domains. This indicates that mutations are spread evenly over the entire protein. However, in a comparison of all xanthopsins, a clear distinction can be made between the PAS sub-domains: the PAS-core and β-scaffold have mutual similarities of ~60 %, whereas this similarity amongst the N-terminal caps and the helical connectors is in the order of 30%. This difference suggests that the PAS-core and β-scaffold are the most typical domains of a xanthopsin and that the N-terminal cap and the helical connector determine the specific function of a particular xanthopsin (or: of one of its sub-families).

The N-terminal cap shows considerable similarity only between sub-families II and I. This similarity may be explainable by the fact that members of these sub-groups are complete proteins, whereas those from sub-group III are domains of a larger protein. This difference adds to the evidence that the N-terminal cap plays an important part in signal transduction (see further below).

19.3 Photoactivity of the Xanthopsins

Upon photoactivation, the xanthopsins enter a cyclic series of dark reactions called a photocycle. Chromophore *trans–cis* isomerization is the chemical trigger of this process. Conformational changes in the surrounding protein follow, which lead to formation of a signaling state. This self-regenerative cycle only requires the holo-protein to be sufficiently hydrated and does not require the presence of a membrane, additional proteins, or co-factors, *etc.* Most xanthopsin research is focused on characterization of the structural changes that underlie this photocycle, or a part of it. The best-studied xanthopsin by far is PYP from *H. halophila*.

19.3.1 The Basic Photocycle

Models of the photocycle of Photoactive Yellow Protein have become more and more complex over the years (see Figure 19.3 for a recent version). Key states in this cycle are: (i) The ground- or dark-adapted state, pG, with a deprotonated chromophore in *trans* configuration; (ii) pR states, spectrally red-shifted compared to pG; the first is formed in ~3 nanoseconds, with a chromophore that is still deprotonated but has the *cis* configuration; (iii) pB states, spectrally blue-shifted states with respect to the ground state pG, which are formed on a micro/millisecond time scale. The pB state is presumed to be the signaling state of the PYP photoreceptor protein. It is stable enough to allow for a signal to be transmitted to a signaling partner. In pB the chromophore is protonated while retaining the *cis* configuration.

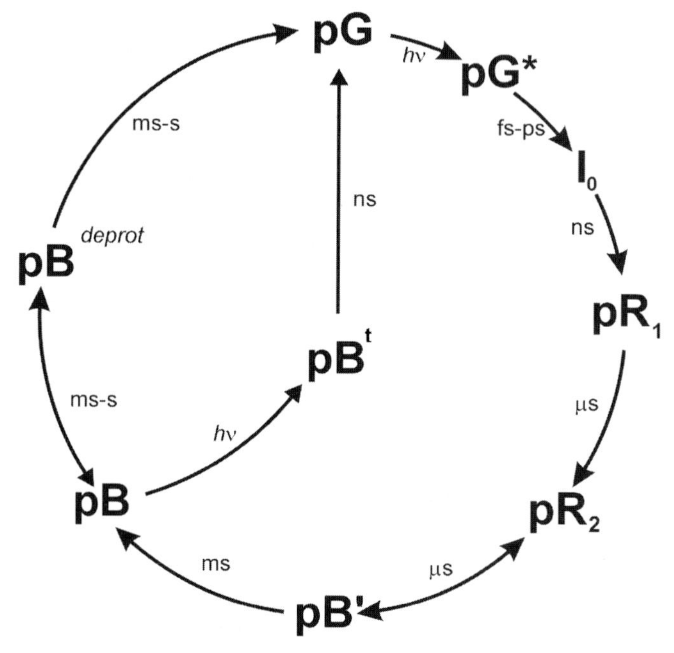

Figure 19.3 Up-to-date description of the photocycle of PYP at room temperature. pG and pG* refer to the ground and electronically excited state of PYP; states indicated with pR and pB have a red- and a blue-shifted visible absorption maximum, respectively, as compared to the pG state. Reactions elicited upon absorption of a photon are indicated with *hν*; for the others the approximate lifetime is indicated. The superscript t refers to light-driven *cis* to *trans* isomerization of the coumaryl chromophore of PYP.

These three major partial reactions, i.e. isomerization, protonation change, and recovery, can also be discerned in the sensory rhodopsins, which have a similar biological function as PYP, but are structurally very different. It is interesting to note that although the xanthopsins and sensory rhodopsins have evolved separately, the photochemistry they use to generate a signal from an absorbed photon is essentially the same.

19.3.2
Photocycle Nomenclature

Over the years, multiple nomenclatures for the photocycle intermediates of PYP have been introduced. They all make reference to the above-mentioned three basic photocycle intermediates and can therefore be compared using these three species as a reference. Initially, the ground-, red-shifted, and blue-shifted state were called P, I_1, and I_2 (Meyer et al. 1989). In 1994 Hoff et al. (Hoff et al. 1995) introduced the names pG, pR, and pB, thus stressing the wavelength shift in their UV/Vis absorption maximum. Yet another nomenclature (Imamoto et al. 1996) was introduced in 1996, in which the corresponding species are referred to as PYP, PYP_L, and PYP_M, i.e. a nomenclature based upon the similarities with bacterial rhodopsins. These nomenclatures have been made even more complicated by the use of sub- and/or superscripts that refer to a specific property of the species (e.g. the absorption maximum). As such properties may depend on the measurement conditions used, different subscripts are used for the same species.

More and more spectroscopic techniques are used to analyze the photocycle of PYP. This has led to the discovery of several new photocycle species that were undetectable previously. To provide these new intermediates with a new name that fits logically within one of the existing nomenclatures is close to impossible.

19.3.3
Experimental Observation: Context Dependence

It is of crucial importance to take the experimental context of the PYP protein into consideration when comparing different experiments. Four different experimental parameters are of basic importance: Temperature, nature and pH of the solvent, mesoscopic context (or phase), and illumination conditions. The first one, temperature, is obviously important when comparing kinetic experiments. But temperature may also allow one to trap certain photocycle species (Hoff et al. 1992; Imamoto et al. 1996) and prevent others from being formed. The second one, the nature of the solvent, is also very important, and is usually different between different experiments. Though pH is probably the most important solvent feature that has an effect on e.g. the kinetics of the photocycle (Genick et al. 1997b; Hendriks et al. 2003), other solvent features such as its viscosity, hydrophobicity (Meyer et al. 1989), type and concentration of solutes present (Meyer et al. 1987; Meyer et al. 1989), and the nature of the solvent itself (e.g. water vs. deuterium oxide (Hendriks et al. 2003)) can also have their effect on the performance of a dissolved photoreceptor protein. The third important parameter, the mesoscopic context or phase, dictates whether or not gross structural change is observed in the protein upon formation of the signaling state. This structural change is limited in a crystalline lattice, whereas when the protein in solution does show such a significant overall structural change (Xie et al. 2001). Recently, (single-crystal) absorption spectroscopy and X-ray diffraction experiments, as well as measurements in (partially) dehydrated protein films, have confirmed that the photocycle can proceed through a short-cut version under these conditions (Kort et al.

2003; Mano et al. 2003; Moffat et al. 2003; van der Horst et al. 2004). Also illumination conditions can have pronounced effects on the photocycle characteristics. Particularly light color and intensity are important. The choice of wavelength can already have an effect on which photocycle species are formed (Hoff et al. 1992; Imamoto et al. 1996). Also the duration of the excitation pulse can have an effect, because extended illumination allows photoactivation of photocycle species other than the ground state (Gensch et al. 1998). This may lead to hysteresis effects in ground state recovery (T. Gensch *et al.*, unpublished). However, the intensity of the probe beam can also influence the data, especially in protein variants with a high quantum yield of photochemistry or with a slowly recovering photocycle.

19.3.4
Mutants and Hybrids

PYP can be engineered genetically, via site-directed mutagenesis, and chemically, *e.g.* through the use of chromophore derivatives. The type of protein obtained in these two approaches is referred to as mutant and hybrid, respectively. Many of the residues that interact with the chromophore (e.g. Tyr42, Glu46, Thr50 and Cys69) have been mutated, resulting in (dramatically) altered photocycle characteristics. Various chromophore analogues, either with one or more ring substituents or "locked" in a specific configuration, have been used for holo-protein reconstitution.

19.3.5
Photo-activation in the Different Xanthopsins Compared

Besides the features of the photocycle of PYP from *H. halophila* described above, additional members of the xanthopsins have been characterized, albeit in much less detail. Both members of sub-group II (i.e. the PYP's from *Rb. sphaeroides* and *Rb. capsulatus*) show comparable spectroscopic features, which are significantly different from those of *H. halophila* PYP (Haker et al. 2000; Kyndt et al. 2003). Both show a second absorption maximum at 360 nm; this form is in a temperature- and pH-dependent equilibrium with the pG form. For PYP from *Rb. sphaeroides* it has been shown that, upon excitation, the species at 360 can also enter a photocycle, independent from the photocycle of the 446 nm species (Haker et al. 2003). Furthermore, in both proteins, the ground state is recovered after photoexcitation ~100-fold faster than in PYP from *H. halophila*.

As opposed to the faster photocycle recovery in these proteins, two xanthopsins have been found that show considerably slower ground state recovery after photoexcitation: Ppr from *Rhodospirillum centenum* (Jiang et al. 1999) and a recently discovered second PYP protein from *H. halophila* (PYP(B); Van der Horst *et al.*, unpublished results); in these, the recovery is ~300-fold and ~100-fold slower, respectively. Since the mechanism of the dark recovery of the ground state is poorly understood, an explanation for these differences cannot be given at this point.

19.4
The Photocycle of Photoactive Yellow Protein

19.4.1
Initial Events

The initial events in the photocycle of PYP are usually considered as transformations preceding the formation of the most stable red-shifted intermediate state (pR). Thus, they comprise the initiation and the first, (ultra) fast, phase of the photocycle. It is generally accepted that the first basic step of the photocycle is the *trans–cis* isomerization of the chromophore. However, the exact spatial description of this process still remains to be completed.

Femtosecond and picosecond transient absorption (Devanathan et al. 1999b; Baltuška et al. 1997), fluorescence (Mataga et al. 2003; Chosrowjan et al. 2004), and IR studies (Groot et al. 2003) have revealed that initiation of the photocycle occurs on the picosecond timescale. Femtosecond infrared spectroscopy – a method providing site- and bond-specific information – allowed the conclusion that breaking of the hydrogen bond of the chromophore's C=O group with the N–H group of Cys69 and the formation of a stable *cis* ground state occur in ~2 ps (Groot et al. 2003). This observation is consistent with the timescale of the main component in the fluorescence decay signal, which is usually associated with fluorescence quenching due to chromophore isomerization (Changenet et al. 1998; Mataga et al. 2003; Chosrowjan et al. 2004).

Several models for isomerization have been suggested, but it is not clear which (if any) of these models is correct. Some features have nevertheless been resolved: at least two different fluorescent states are formed after photo-excitation of PYP. In one of these states, which presents itself as an anisotropy difference with respect to the ground state, the chromophore configuration is clearly different from that in the ground state (Gensch et al. 2002). It has not been revealed whether these two fluorescent states are true photocycle intermediates (i.e. upon formation they return to the ground state via the regular photocycle intermediates) or whether they are photocycle dead ends. Femtosecond time-resolved fluorescence studies of wild type PYP plus several site-directed mutants and hybrids with chromophore analogues showed that the protein environment plays an important role in the kinetics and efficiency of photocycle initiation. Molecular-dynamics simulations and time-dependent density functional theory calculations also corroborate this conclusion. The simulations suggest that the protein regulates the isomerization of the chromophore via stabilization of the transition state in the isomerization pathway (Groenhof et al. 2002b; Groenhof et al. 2004).

Transient absorption measurements have revealed that, at room temperature, at least one photocycle intermediate is formed between the excited (fluorescent) state(s) and pR. This so called I_0 intermediate, is formed on a picosecond timescale and transforms at the nanosecond timescale into pR. In addition, the existence of yet another intermediate, between I_0 and pR, has been reported. This intermediate, referred to as $I_0^{\ddagger}$, has absorption characteristics very similar to those of I_0 (Ujj et al. 1998; Gensch et al. 2002). An alternative interpretation, however, holds that a long living ($\gg$2 ps) excited state causes these signals.

At cryogenic temperatures a branched photocycle pathway exists. Each of the two branches contains two intermediates. One branch proceeds via an intermediate that is similar to I_0 (PYP_B, converted into PYP_{BL}). The two branches merge upon formation of pR. Selective use of either of these two branches can be accomplished via tuning of the excitation wavelength. Shorter wavelengths make the branch with the I_0-like intermediate the preferred one (Hoff et al. 1992; Masciangioli et al. 2000). This might be a true PYP characteristic but it is also possible that PYP_B is photo-converted back to pG, which occurs with optimal efficiency at longer wavelengths. Similarly the PYP_H intermediate (converted in the dark into PYP_{HL}), at shorter wavelengths, can be photo-converted back to pG (Imamoto et al. 1996). This is one of the drawbacks when cryotrapping is used to characterize photocycle intermediates. Photoactivity, however, is not exclusive for the ground state; transient intermediates may be photoactive as well. For accumulation of a cryotrapped intermediate in the sample, extended illumination is usually needed. Consequently, the cryotrapped intermediate must also be significantly illuminated. As a result it can be photo-converted, e.g. to the ground state. It is also possible, however, that non-physiological intermediates, are formed as a result of these secondary photoreactions. A case in point may be the fluorescent species F_{430}, which may well be formed from the PYP_H intermediate (Hoff et al. 1992).

Of course it is important to identify spectroscopic intermediates correctly, but it is even more interesting to understand their spatial structure. It is evident that the protein itself displays only very modest structural changes during the first step of the photocycle. Time-resolved X-ray diffraction revealed that the aromatic ring of the chromophore of PYP stays approximately at the same position during its photoisiomerization. (Genick et al. 1998; Imamoto et al. 2001b; Ren et al. 2001). The only way to facilitate isomerization of the chromophore under these conditions is by rotation of the *thiol*-ester carbonyl. This carbonyl flip can be interpreted as a double isomerization around the $C_7{=}C_8$ double bond and the $C_9{-}S_\gamma$ single bond. The chromophore configuration accordingly changes from $C_7{=}C_8$-*trans*, $C_9{-}S_\gamma$-*cis* to $C_7{=}C_8$-*cis*, $C_9{-}S_\gamma$-*trans*. This model was first introduced on the basis of low temperature FTIR spectroscopy (Xie et al. 1996), and later confirmed (Brudler et al. 2001; Imamoto et al. 2001b; Xie et al. 2001).

The structure of two intermediates from the first steps in the photocycle, PYP_{BL} and pR, have also been determined with X-ray crystallography. In both intermediates, structural changes are largely limited to the immediate surroundings of the chromophore. The structure of the PYP_{BL} intermediate was obtained via cryotrapping and resolved down to a resolution of 0.85 Å. Based on the temperature and illumination conditions used, and the absorption spectrum of the sample, it was concluded that PYP_{BL} was the cryotrapped intermediate (Genick et al. 1998). It can not be excluded, however, that PYP_{HL} was also formed and thus that a mixture of intermediates was present. In the resolved structure the chromophore is in a distorted *cis* configuration; it has barely crossed the *trans* to *cis* transition point and by consequence isomerization is only completed in pR. An important conclusion is that the chromophore isomerizes in a way that minimizes movement of the chromophore within the chromophore pocket. Isomerization is achieved by rotating the carbonyl function 166.5°, with respect to the aromatic ring, breaking the hydrogen bond between the carbonyl function and the backbone amide group of Cys69. Recent low-temperature analyses, however, show that complete *trans* to *cis* isomerization of the chromophore is possible under these conditions (Kort et al. 2004). A key element in this latter study is the extensive correction for X-ray radiation damage.

The 1.9 Å resolution structure of the pR intermediate was obtained via the 1 ns time slice of a dataset encompassing time slices up to 1 ms, in a time-resolved X-ray crystallography experiment (at near room temperature) (Perman et al. 1998). The deposited pR structure has a completely *cis* isomerized chromophore for the C_7=C_8 double bond. In this structure the carbonyl function has rotated an additional 108.5° with respect to the cryo-trapped intermediate PYP_{BL}. This may facilitate the formation of a hydrogen bond between the carbonyl oxygen and the backbone amide of Tyr98.

The (free) energy content of PYP increases 120–160 kJ mol^{-1} upon formation of pR (van Brederode et al. 1995; Takeshita et al. 2000). Half of the energy of an absorbed photon is thus stored in the holo-protein at this point (a photon with a wavelength of 446 nm has an energy content of 268 kJ mol^{-1}). This energy should be (and is) enough to drive the remainder of the photocycle and make PYP return to the pG state. The other half of the energy of the absorbed photon is lost in e.g. thermal relaxations. FTIR analysis of the cryotrapped intermediates in the first step of the photocycle, suggests that there is very little structural difference between these intermediates (Imamoto et al. 2001b). Thus small movements induced by thermal relaxations may dictate the exact isomerization route of the chromophore.

19.4.2
Signaling State Formation and Ground State Recovery

Signaling of a blue photon in the cell occurs via interaction of the signaling state of PYP with a (putative) transducer protein. To allow protein–protein interactions the signaling state needs to have a relatively long lifetime. The PYP-transducer protein has not been identified yet; all conclusions so far are based on experiments with only purified PYP and therefore it is possible that certain characteristics of the signaling state will change when transducer protein is added. Such differences indeed can be observed in sensory rhodopsin I: Without the transducer protein sensory rhodopsin I acts as a proton pump. In the presence of the transducer protein this activity is lost and also the pH dependence of the photocycle kinetics differs (Spudich and Spudich 1993). Formation of the signaling state of PYP from the red-shifted photocycle intermediate pR is accomplished via two key events: protonation of the chromophore (neutralizing its negative charge) and a structural change of the protein. Ground state recovery, through reversal of these events, completes the photocycle. Insight in the formation of the signaling state has increased significantly over the past few years, but little still is known on the specifics of the recovery of the ground state.

Significant new insights in the mechanism of signaling-state formation have emerged recently. Early photocycle models considered only three states, the ground state pG, pR and the signaling state pB. Formation of pB (from pR) was initially described as a mono-exponential process but, when higher resolution was achievable, as a bi-exponential process. Because of this bi-exponential character of pB formation, a detailed photocycle model for PYP (Meyer et al. 1987) postulated an additional intermediate with similar spectral properties as pB. This model was abandoned in a subsequent paper (Meyer et al. 1989). Based on insights obtained with FTIR spectroscopy, recently, the intermediate pB' was introduced as an intermediate linking the pR and pB states (Xie et al. 2001). The role of this intermediate was subsequently confirmed in a detailed kinetic UV/Vis analysis of the photocycle. pB' was shown to be spectroscopically very similar to pB (Hendriks et al. 2003) and in equilibrium with pR, which explains the bi-exponential character of the pR to pB photocycle step (Meyer et al. 1987; Hoff et al. 1994c). Existence of the pB' intermediate was recently also confirmed with time-resolved resonance Raman measurements (Pan et al. 2004).

Besides the thermal recovery of the ground state (i.e. spontaneously in the dark), a light-induced recovery pathway also exists. This path is initiated after photon absorption by pB (and pB'). Via the light-induced *cis* to *trans* re-isomerization of the chromophore a thousand-fold increase in the rate of recovery, as compared to the thermally activated process, is observed (Hendriks et al. 1999b). For the thermal recovery it has been suggested that before isomerization can take place the chromophore must first be deprotonated. By consequence an intermediate must exist between pB and pG (Demchuk et al. 2000). This intermediate, pB^{deprot}, was recently detected in a study of the kinetic deuterium isotope effect on photocycle transitions in PYP (Hendriks et al. 2003).

19.4.3
Structural Relaxation of pR

Recently it was demonstrated that after pR has formed, significant relaxation in PYP structure takes place (Takeshita et al. 2002a; Takeshita et al. 2002b). Via transient grating and pulsed-laser photoacoustic methods, a µs dynamic component was shown to exist during the lifetime of pR. This indicates that after completion of structural changes in and immediately around the chromophore, additional structural changes take place further away from the chromophore. Accordingly, pR actually represents two intermediates, pR_1 and pR_2, and in hindsight evidence for the existence for these two pR states was already available in early UV/Vis data (Hoff et al. 1994c). In a recent UV/Vis study (Hendriks et al. 2003), kinetics were obtained that exactly fit the pR_1-to-pR_2 transition observed via transient grating. The spectra for both pR intermediates are very similar, pR_1 seems to have a slightly higher extinction coefficient than pR_2, but the λ_{max} values are indistinguishable.

19.4.4
Protonation Change upon pB' Formation

After formation of the pR intermediates, the next event is protonation of the chromophore. This results in the formation of pB'. The first experimental evidence for the existence of this intermediate was obtained in FTIR measurements (Xie et al. 2001). It was shown that deprotonation of Glu46 and protonation of the chromophore are simultaneous events that are followed by a structural change of the protein. Absorption changes, registered using UV/Vis spectroscopy, mainly display changes of the chromophore and its immediate surroundings. Since all the details of the structure of the surrounding protein cannot be probed with this technique, it is difficult to distinguish between pB' and pB with UV/Vis spectroscopy. Based on the FTIR evidence a pB' intermediate was, however, incorporated in a photocycle model, which was then used to analyze UV/Vis data in a study on the kinetic deuterium-isotope effect in the photocycle of PYP (Hendriks et al. 2003). From this approach it became clear that pR and pB' exist in an equilibrium that shifts towards pB' upon going to the extremes (both low and high) of pH. The observed kinetic deuterium isotope effect, for the whole pH range that was investigated (pH 5 to 11), is in line with proton transfer from Glu46 to the chromophore. For the reverse reaction the situation is more complex. As a function of pH, the mechanism of formation of pR from pB' may vary. The pH-dependence of the pR/pB' equilibrium can also explain the shift towards mono-exponential behavior of pB formation at very high or low pH.

Direct proton transfer from Glu46 to the chromophore is therefore considered as the most likely mechanism of the chromophore protonation upon the formation of the pB state. This conclusion also follows from the results of molecular dynamics simulations and semi-empirical calculations, which suggest that proton transfer from Glu46 to the phenolate oxygen of the chromophore is facilitated by the photo-induced isomerization of the chromophore (Sergi et al. 2001). It is relevant to note that, by necessity, the residue at position 46 cannot protonate the chromophore in the E46Q mutant.

19.4.5 Structural Change upon pB Formation

The non-linear Arrhenius kinetic behavior of the recovery reaction of pG (van Brederode et al. 1996) was the first evidence that the signaling state of PYP is at least partly unfolded. A mutant form of PYP, which lacks the first 25 N-terminal amino acids (van der Horst et al. 2001) displays close to normal Arrhenius behavior, indicating that its N-terminal region is largely responsible for the large structural change upon formation of the signaling state. Several additional methods, e.g. circular dichroism spectroscopy (CD) (Harigai et al. 2003), NMR (Rubinstenn et al. 1998; Craven et al. 2000; Derix et al. 2003), fluorescent-probe binding (Lee et al. 2001b; Hendriks et al. 2002), and FTIR spectroscopy (Hoff et al. 1999; Xie et al. 2001; Harigai et al. 2003) have confirmed this partial unfolding of PYP in its signaling state pB.

H/D exchange measurements have added to our current insight into the structural change that underlies signaling-state formation in PYP. An exposed hydrogen atom may be exchanged within seconds, whereas for a buried hydrogen atom this can take days. PYP contains 235 potentially exchangeable hydrogen atoms, 42 of which are from (de)protonatable groups. It was demonstrated, using electrospray ionization mass spectrometry (Hoff et al. 1999), that more hydrogen atoms were exchanged for deuterium atoms in the presence of light compared to PYP in the dark. The light-induced H/D exchange was independently confirmed by FTIR difference spectroscopy (Hoff et al. 1999). These experiments showed that there is a difference between the ground and signaling state of PYP with respect to H/D exchange protection. NMR spectroscopy (Craven et al. 2000) revealed some of the (exposed) sites involved: the rate of exchange of the backbone amide was recorded for 51 residues: 14 of which showed a significant change in exchange rate upon pB formation, only 2 less than the number predicted by mass spectrometry, a method which is not limited to the backbone amide hydrogen atoms. Most significant loss in protection was observed for residues Phe28, Glu46, and Thr70. The latter two are in the vicinity of the chromophore, Phe28 is close to Glu46.

A solution-NMR study of pB showed that, with respect to the ground state, this intermediate exhibits more structural and dynamic disorder (Rubinstenn et al. 1998). In a follow-up study on pB_{dark} formation (Craven et al. 2000) it was revealed that this intermediate is very similar to pB. These NMR data suggest that the pB intermediate is in fact a mixture of structurally perturbed forms and a form structurally similar to pB in crystalline PYP (Genick et al. 1997a).

The trigger for the large structural change upon formation of pB is the formation of a buried negative charge (i.e. on Glu46) upon proton transfer to the chromophore (Xie et al. 2001). While residing on the chromophore, this negative charge is stabilized by delocalization, by the hydrogen-bonding network, and possibly by the positive charge on Arg52. A comparable stabilization of this negative charge is not possible when it resides on Glu46. This situation then generates a free-energy stress within the protein, which can be relayed via several routes. One is return to the pR state, thus explaining the reversibility of pB' formation (see above). Other routes lead to pB formation; the extent of structural change upon formation of the pB intermediate then depends on the details of this route. One option is to relieve the buried negative charge via exposure to the solvent. This requires a large structural change of the protein. Another route is to protonate Glu46 via a donor different from the chromophore. Once Glu46 is protonated (large) structural changes are no longer a necessity.

As in these processes protonation changes play a key role, it is to be expected that they will display pronounced pH dependency. Also, the pH may influence the prevalence of one route over another. Indeed a pH dependence of the extent of structural changes has been observed to coincide with the protonation state of Glu46 (J. Hendriks and A. Xie et al., unpublished results). Significantly, both routes to pB imply that pB' is formed as an intermediate. An alternative possibility is that a direct route from pR to pB exists in which Glu46 stays protonated, i.e. it would not donate a proton to the chromophore in such a route. The chromophore then becomes protonated only after exposure to the solvent, or via another residue (e.g. Tyr42). A mechanism like this presumably would require less structural change, and might be preferred in a crystalline environment.

All of these routes may be possible and depending on the conditions one particular route may dominate. A key factor is what happens to the protonation state of both Glu46 and the chromophore. Consistently, in the Glu46Gln mutant of PYP the structural change upon formation of pB is significantly less than in wild type PYP (Xie et al. 2001); this has also been observed for the His108Phe mutant (Kandori et al. 2000).

The pH dependence of structural change has also been demonstrated via transient probe binding (Hendriks et al. 2002). Since in experiments monitoring the pH-dependent net proton uptake/release of PYP during pB formation (Hendriks et al. 1999a), and in FTIR spectra, a similar pH dependence was observed (Hendriks 2002), it is likely that the pK_a of Glu46 in pB is 5.5.

19.4.6
Recovery of the Ground State

For recovery of the ground state of PYP the chromophore has to re-isomerize, the protonation state of several functional groups needs to be restored and the protein needs to refold to its ground state conformation. Until recently, these partial reactions were thought to occur concertedly. The kinetic deuterium isotope effects (Hendriks et al. 2003), however, have shown that deprotonation of the chromophore occurs before re-isomerization. By consequence the intermediate pB^{deprot} is formed in the recovery of pG from pB. Deprotonation of the chromophore accelerates this re-isomerization (Sergi et al. 2001), although it remains a rate-controlling step, in which the exact protein conformation likely plays a crucial role. The pB^{deprot} intermediate therefore will have a deprotonated chromophore and a folding state that allows chromophore re-isomerization. By consequence, the absorption spectrum of this intermediate may be similar to that of pG, rather than to the pB intermediate with a deprotonated chromophore. In agreement with this possibility, its absorption maximum has been estimated to be around 430 nm (Hendriks et al. 1999a; Hendriks et al. 2003). Deprotonation of the chromophore upon formation of pB^{deprot} is most likely facilitated by a hydroxyl ion (Hendriks et al. 2003).

As discussed above, the recovery rate of the ground state increases a thousand-fold when re-isomerization of the chromophore is facilitated by blue light (Hendriks et al. 1999b). In the ensuing branching pathway an intermediate, pB^t (t for *trans*) is formed on a nanosecond time scale. The slight blue shift of pB^t, with respect to pB, can be explained by the altered chromophore configuration. UV/Vis spectroscopy did not reveal any additional intermediates between pB^t to pG. This lack indicates that re-folding and reformation of proper protonation states can proceed quickly once the chromophore is re-isomerized. The fact that this branching reaction exists can influence the interpretation of data collected in the presence of light. Activation of the pB intermediates, may allow recovery kinetics to appear faster than they really are in the absence of this (probe) light (Miller et al. 1993). Light-accelerated recovery can be exploited in the study of the very slow recovery variants, like M100A. It may have biological relevance in PYP variants with strongly decreased recovery rates, as in the second PYP discovered in *H. halophila* (i.e. PYP(B); M.A. van der Horst *et al.*, unpublished observations).

Several studies of the refolding of PYP have been performed. Often, these studies employ a denaturant to assist the initial unfolding of the protein. The exception is an NMR study (Rubinstenn et al. 1998), in which recovery of the ground state was measured on the basis of light-induced unfolding of the protein only. A dynamic process of refolding was observed in which the central β-sheet and parts of the α-helical structure refold slightly faster, after which the region around the chromophore returns to the ground state fold. The differences between the refolding rates are small, however, and it is unclear if these rates differ significantly. The observed trend is in line with the idea that for efficient re-isomerization of the chromophore, it not only needs to be deprotonated, but that also the protein needs to be in a relatively folded state.

In a study utilizing urea and guanidinium · HCl, refolding was studied with the unfolded ground and signaling state of PYP (Lee et al. 2001c). The major difference between these two denatured forms of the protein is the isomerization state of the chromophore. Refolding from the denatured ground state is a mono-exponential event, whereas refolding from the denatured signaling state is bi-exponential. The fast component in the latter process is identical to the refolding component from the denatured ground state. The slow exponent has a rate similar to the photocycle ground-state recovery rate under similar conditions. These observations indicate that after the signaling state renatures, it recovers to the ground state through regular photocycle states. Interestingly, extrapolation of the refolding kinetics to the absence of denaturant shows an approximate thousand fold faster rate for refolding with the chromophore in the *trans* state, compared to the *cis* state. This is similar to the differences observed between the thermal- and the photo-activated ground state recovery (Hendriks et al. 1999b). Similar results were obtained in experiments with the acid-denatured state of PYP (Lee et al. 2001a). Refolding from the acid denatured state with a *trans* chromophore, i.e. pB_{dark}, is 3 to 5 orders of magnitude faster than from the acid denatured state with the chromophore in the *cis* configuration.

For several PYP mutants, a severe decrease in the recovery rate has been reported (Glu46Asp (Devanathan et al. 1999a), Met100Ala (Devanathan et al. 1998), and Met100Leu (Sasaki et al. 2002)). These results indicate that Glu46 and Met100 are important for recovery. It was argued that the electron-donating character of the residue at position 100 influences the rate of recovery, presumably through interaction with Arg52 (Kumauchi et al. 2002).

19.5
Spectral Tuning of Photoactive Yellow Protein

Tuning of the UV/Vis absorption bands of PYP, away from the values of neutral and/or anionic coumaric acid to longer wavelengths, is of interest for several reasons. First, there is the tuning of the absorption band of the ground state structure. Here, the contribution of specific structural characteristics must be analyzed to explain the destabilization of the ground state and/or destabilization of the electronically excited state. However, during the photocycle of PYP, the maxima in the absorption spectrum change as well. These changes contain valuable information regarding the chromophore and its surroundings in the respective photocycle states. A proper understanding of the tuning of PYP will therefore aid to understand the events that occur during the photocycle.

19.5.1
Ground State Tuning

The interaction of the chromophore and apo-protein parts of PYP produces an absorption band with 446 nm as its UV/Vis absorption maximum. As the free chromophore, *trans*-4-hydroxycinamic acid, has an absorption maximum at 284 nm (in aqueous solvent around neutral pH) (Aulin-Erdtman and Sandén 1968), this interaction of the chromophore with the protein induces a large red shift. Several specific contributions to this large shift can be distinguished. For one, the thiol ester link of the chromophore with Cys69, causes a red shift of ~5713 cm^{-1} (from 284 to 339 nm). This conclusion follows from a comparison of the absorption maximum of 4-hydroxycinamic acid and the denatured form of PYP in aqueous solution at pH 7 (Kroon et al. 1996). An additional red-shift of 4310 cm^{-1} (from 339 to 397 nm) occurs when the chromophore becomes deprotonated, which follows from a comparison of the absorption maximum of the denatured form of PYP in aqueous solution at pH 7 and 11 (Kroon et al. 1996). This leaves a red shift of ~2767 cm^{-1} (from 397 to 446 nm) that is caused by interactions with the protein. Though this description is very illustrative, it does not provide specific insight regarding the nature of the interactions between the chromophore and the protein.

In a recent study, a closer look was taken at the mechanism(s) that lead(s) to the spectroscopic tuning of PYP in its ground state (Yoda et al. 2001). Two model compounds were used for comparison, i.e. a propyl ester and a propyl-thiol ester of 4-hydroxycinnamic acid. The thiol ester model compound was consistently red-shifted by ~1000 cm^{-1}, with respect to the ester model compound, irrespective of the protonation state of the chromophore. With regard to tuning in PYP, three tuning contributions were considered, i.e. a medium effect of the protein matrix (700 cm^{-1}), a counter ion effect (5300 cm^{-1}), and a hydrogen bonding effect (–1600 cm^{-1}). The medium effect of the protein matrix takes into account non-specific solvent effects. For the model compounds the absorption maximum shifted with the solvent that was used, e.g. for the thiol-ester model compound the absorption maximum ranged from 34800 cm^{-1} (287 nm) in pentane to 31500 cm^{-1} (317 nm) in pyridine. This difference is mainly caused by differences in the dielectric constant and refractive index between the solvents. To determine the medium effect in PYP the values for the dielectric constant and the refractive index were estimated for the protein. Absorption maxima of the thiol ester model compound in hexane and in a protein environment were calculated. Note, that only the solvent properties of the protein are considered here and not counter-ion and hydrogen-bonding effects. This leads to a contribution of ~700 cm^{-1} to the tuning of the chromophore in the protein.

The counter ion effect considers the difference in position of the counter ion of the thiol ester model compound in solution (sodium ion at 2.5 Å in a straight line from the phenolate oxygen bond) and in the protein (position of Arg52 in the crystal structure PDB ID: 2PHY (Borgstahl et al. 1995)). Here protein solvent conditions were used in the calculation of the absorption maxima. This leads to a contribution of ~5300 cm^{-1} to the tuning of the chromophore. The hydrogen-bonding effect was determined by placing methanol at the positions of the hydroxy groups of residues

Tyr42, Glu46, and Thr50, which are involved in the hydrogen-bonding network with the chromophore, in the calculations. By comparing the situation incorporating the medium effect of the protein matrix and the counter-ion effect, with the situation that also takes into account the hydrogen-bonding effect, the contribution of the latter was calculated as $-1600\ cm^{-1}$, i.e. a blue shift is caused by formation of a hydrogen bond to the phenolate oxygen of the chromophore! Furthermore, also with respect to the counter ion effect, an interesting observation was made. The position of Arg52 is such that it appears that the counter ion is infinitely far apart from the chromophore, i.e. it does not contribute to the tuning. Movement of the counter ion towards the thiol ester linkage would result in a red shift, whereas movement toward the phenolate oxygen would result in a blue shift. From NMR measurements (Dux et al. 1998) and molecular dynamics studies (Groenhof et al. 2002b) it is evident that Arg52, may have two distinct positions in the ground state. The two distinct positions obtained with the molecular dynamics studies would result in a difference of 20 nm between the absorption maxima of the two conformers (Groenhof et al. 2002b).

When the hydrogen-bonding network in PYP is weakened (by replacing the bridging hydrogen atoms with deuterium atoms) indeed a small red shift is observed, which would be expected when the hydrogen bonding effect contributes a smaller blue shift to the total tuning. Furthermore, the residues that are involved in the hydrogen-bonding network that stabilizes the anionic chromophore (i.e. Tyr42, Glu46, and Thr50) have been altered through mutagenesis. Mutants in these residues indeed result in a red shift of the UV/Vis absorbance maximum of the chromophore (e.g. Tyr42Ala, Tyr42Phe, Glu46Gln, Glu46Ala, Thr50Val, and Thr50Ala (Genick et al. 1997b; Mihara et al. 1997; Devanathan et al. 1999a; Brudler et al. 2000; Imamoto et al. 2001a)) reflecting the weaker hydrogen bonding effect in these mutants. In some mutants though, protein stability has been severely affected as indicated by a major, additional, blue-shifted absorption band (Tyr42Phe, Tyr42Ala, and Glu46Ala). Here, the weakened hydrogen-bonding network causes a shift in pK_a of the chromophore, leading to a protonated state (Meyer et al. 2003; El-Mashtoly et al. 2004).

Several mutants altered at the position of Arg52 have also been prepared. The Arg52Ala mutant is slightly red-shifted (Genick et al. 1997b), while the Arg52Gln mutant shows no shift of the absorption maximum (Mihara et al. 1997). As described above, removal of the counter ion would not lead to a change in absorption maximum, which is exactly what is observed in the Arg52Gln mutant. A small red shift that is observed in the Arg52Ala mutant may be explained by a more open structure of the chromophore-binding pocket, possibly allowing a solvent cation to act as a counter ion.

19.5.2 Spectral Tuning in Photocycle Intermediates

The negative charge on the chromophore is most effectively delocalized if the chromophore is planar. During isomerization planarity is lost, and the negative charge is not as efficiently delocalized, which results in a red shift of the absorption spectrum. This would explain why the intermediate I_0 is more red-shifted than pR, as in the former intermediate the chromophore may still be in a twisted form. In pR the chromophore is still not quite planar due to steric hindrance between the carboxylic oxygen and atoms of the phenolate ring (Groenhof et al. 2002a), which explains the persistent red shift. Additionally, in pR the chromophore has contracted ~0.5 Å (Groenhof et al. 2002a), while the structure of the protein is very similar to that of the ground state. Such a contraction could lead the counter ion Arg52 to become located closer to the phenolate oxygen of the chromophore, which would also lead to a red shift. The residues lining the binding pocket of the chromophore restrict its isomerization to the least volume-demanding mechanism (Imamoto et al. 2002).

Protonation of the chromophore leads to a large blue shift, which is exactly what is observed. In pB' the structure of the protein is still very similar to that of pR, and thus to the structure of the ground state. The transfer of the proton leads to a more relaxed conformation of the chromophore (Pan et al. 2004), but interactions with the protein likely persist (Genick et al. 1997a). When pB is formed in solution, the structure of the protein has dramatically changed, so that the interaction of the protein with the chromophore most likely diminishes, which is also indicated by the slight additional blue-shift of the pB intermediate with respect to pB'. However, even though the chromophore has become more exposed in pB, its absorption maximum is still slightly red-shifted compared to the situation in denatured protein (Lee et al. 2001c). The chromophore is therefore still tuned in pB, through interactions with the protein. The presence of these interactions is further demonstrated by the pK_a of the chromophore in pB which is ~10 and not 8.7 as is the case in fully denatured protein (Hendriks et al. 2003).

19.6 Summary and Future Perspective

After excitation of the ground state of Photoactive Yellow Protein, its chromophore will form a twisted excited state, which subsequently results in formation of the I_0 photocycle intermediate with the chromophore in *cis* configuration, be it still twisted. Further relaxation leads to formation of pR. Isomerization proceeds with a minimal amount of movement of the chromophore, i.e. through a concerted rotation around several bonds that can be described as a C_7=C_8-*trans* C_9-S_γ-*cis* to C_7=C_8-*cis* C_9-S_γ-*trans* multiple-bond isomerization, i.e. a rotary movement of the carbonyl oxygen. This isomerization increases the probability of proton transfer from Glu46 to the chromophore, which results in formation of pB'. As a result, the negative charge – which was stabilized via delocalization when it resided on the chromophore, via a hydrogen-bonding network, and via a counter ion – now resides on Glu46 where it is very localized. This is an energetically unfavorable situation that drives subsequent structural changes, both in the N-terminal domain of the protein and around the chromophore. Particularly the involvement of the independently folded N terminus in this partial unfolding is remarkable. Together with the sequence comparisons within the PAS domains, this suggests that indeed the N terminus is important for signaling. The extent of these structural changes depends on experimental conditions and may vary from very little structural change in crystals, to large structural changes in aqueous solution. This structural change will then have to elicit detectable structural change in a downstream signal transduction partner, be it in a domain of the same full-length protein like in Ppr, or via intermolecular interactions.

After formation of the signaling state, the ground state in the photocycle recovers, which requires chromophore re-isomerization and protein re-folding. Before re-isomerization can take place, the chromophore presumably needs to be deprotonated (leading to formation of the pB^{deprot} intermediate), and the protein needs to adopt a specific fold that allows for the re-isomerization to take place. Chromophore re-isomerization can also take place photochemically, which results in a rate of recovery that is three orders of magnitude faster than dark recovery. It is still unresolved whether this two-photon photocycle is biologically relevant.

During the past few years PYP has become popular as a model system. Three areas of research are relevant in this respect: (i) signaling state formation in PAS-domains, (ii) functional protein (un)folding and (iii) the primary photochemistry of photoreceptors. This popularity is based on favorable handling characteristics, availability of high-resolution structures, and the relatively simple structure of the chromophore of PYP. Significantly, several additional fields may benefit from the efforts to understand the details of signaling in PYP. An example is molecular-dynamics modeling, because this approach may benefit from the availability of transient reference structures from time-resolved X-ray diffraction. The field of signal transduction in photobiology could also benefit enormously from the detailed insight in the mechanism that leads to signal generation within PYP. However, for this to bear fruit the signal transduction component(s) that link PYP to the flagellar apparatus will have to be identified.

References

G. Aulin-Erdtman, R. Sandén. *Acta Chem. Scand.* **1968**, 22, 1187–1209.

M. Baca, G.E. Borgstahl, M. Boissinot, P.M. Burke, D.R. Williams, K.A. Slater, E.D. Getzoff. *Biochemistry* **1994**, 33, 14369–14377.

A. Baltuška, I.H.M. van Stokkum, A. Kroon, R. Monshouwer, K.J. Hellingwerf, R. van Grondelle. *Chem. Phys. Lett.* **1997**, 270, 263–266.

H.M. Berman, J. Westbrook, Z. Feng, G. Gilliland, T.N. Bhat, H. Weissig, I.N. Shindyalov, P.E. Bourne. *Nucleic Acids Res.* **2000**, 28, 235–242.

G.E. Borgstahl, D.R. Williams, E.D. Getzoff. *Biochemistry* **1995**, 34, 6278–6287.

R. Brudler, T.E. Meyer, U.K. Genick, S. Devanathan, T.T. Woo, D.P. Millar, K. Gerwert, M.A. Cusanovich, G. Tollin, E.D. Getzoff. *Biochemistry* **2000**, 39, 13478–13486.

R. Brudler, R. Rammelsberg, T.T. Woo, E.D. Getzoff, K. Gerwert. *Nat. Struct. Biol.* **2001**, 8, 265–270.

P. Changenet, H. Zhang, M.J. van der Meer, K.J. Hellingwerf, M. Glasbeek. *Chem. Phys. Lett.* **1998**, 282, 276–282.

H. Chosrowjan, S. Taniguchi, N. Mataga, M. Unno, S. Yamauchi, N. Hamada, M. Kumauchi, F. Tokunago. *J. Phys. Chem. B* **2004**, 108, 2686–2698.

C.J. Craven, N.M. Derix, J. Hendriks, R. Boelens, K.J. Hellingwerf, R. Kaptein. *Biochemistry* **2000**, 39, 14392–14399.

M.A. Cusanovich, T.E. Meyer. *Biochemistry* **2003**, 42, 4759–4770.

W.L. DeLano, *The PyMOL User's Manual*, DeLano Scientific

E. Demchuk, U.K. Genick, T.T. Woo, E.D. Getzoff, D. Bashford. *Biochemistry* **2000**, 39, 1100–1113.

N.M. Derix, R.W. Wechselberger, M.A. van der Horst, K.J. Hellingwerf, R. Boelens, R. Kaptein, N.A.J. van Nuland. *Biochemistry* **2003**, 42, 14501–14506.

S. Devanathan, R. Brudler, B. Hessling, T.T. Woo, K. Gerwert, E.D. Getzoff, M.A. Cusanovich, G. Tollin. *Biochemistry* **1999a**, 38, 13766–13772.

S. Devanathan, U.K. Genick, I.L. Canestrelli, T.E. Meyer, M.A. Cusanovich, E.D. Getzoff, G. Tollin. *Biochemistry* **1998**, 37, 11563–11568.

S. Devanathan, A. Pacheco, L. Ujj, M. Cusanovich, G. Tollin, S. Lin, N. Woodbury. *Biophys. J.* **1999b**, 77, 1017–1023.

P. Dux, G. Rubinstenn, G.W. Vuister, R. Boelens, F.A. Mulder, K. Hard, W.D. Hoff, A.R. Kroon, W. Crielaard, K.J. Hellingwerf, R. Kaptein. *Biochemistry* **1998**, 37, 12689–12699.

S.F. El-Mashtoly, M. Unno, M. Kumauchi, N. Hamada, K. Fujiwara, J. Sasaki, Y. Imamoto, M. Kataoka, F. Tokunaga, S. Yamauchi. *Biochemistry* **2004**, 43, 2279–2287.

U.K. Genick, G.E. Borgstahl, K. Ng, Z. Ren, C. Pradervand, P.M. Burke, V. Srajer, T.Y. Teng, W. Schildkamp, D.E. McRee, K. Moffat, E.D. Getzoff. *Science* **1997a**, 275, 1471–1475.

U.K. Genick, S. Devanathan, T.E. Meyer, I.L. Canestrelli, E. Williams, M.A. Cusanovich, G. Tollin, E.D. Getzoff. *Biochemistry* **1997b**, 36, 8–14.

U.K. Genick, S.M. Soltis, P. Kuhn, I.L. Canestrelli, E.D. Getzoff. *Nature* **1998**, 392, 206–209.

T. Gensch, C.C. Gradinaru, I.H.M. van Stokkum, J. Hendriks, K.J. Hellingwerf, R. van Grondelle. *Chem. Phys. Lett.* **2002**, 356, 347–354.

T. Gensch, K.J. Hellingwerf, S.E. Braslavsky, K. Schaffner. *J. Phys. Chem. A* **1998**, 102, 5398–5405.

G. Groenhof, M. Bouxin-Cademartory, B. Hess, S.P. De Visser, H.J. Berendsen, M. Olivucci, A.E. Mark, M.A. Robb. *J. Am. Chem. Soc.* **2004**, 126, 4228–4233.

G. Groenhof, M.F. Lensink, H.J. Berendsen, A.E. Mark. *Proteins* **2002a**, 48, 212–219.

G. Groenhof, M.F. Lensink, H.J. Berendsen, J.G. Snijders, A.E. Mark. *Proteins* **2002b**, 48, 202–211.

M.L. Groot, L. van Wilderen, D.S. Larsen, M.A. van der Horst, I.H.M. van Stokkum, K.J. Hellingwerf, R. van Grondelle. *Biochemistry* **2003**, 42, 10054–10059.

A. Haker, J. Hendriks, T. Gensch, K. Hellingwerf, W. Crielaard. *FEBS Lett.* **2000**, 486, 52–56.

A. Haker, J. Hendriks, I.H.M. van Stokkum, J. Heberle, K.J. Hellingwerf, W. Crielaard, T. Gensch. *J. Biol. Chem.* **2003**, 278, 8442–8451.

M. Harigai, Y. Imamoto, H. Kamikubo, Y. Yamazaki, M. Kataoka. *Biochemistry* **2003**, 42, 13893–13900.

K.J. Hellingwerf, W.D. Hoff, W. Crielaard. *Mol. Microbiol.* **1996**, 21, 683–693.

J. Hendriks **2002**, Shining light on the Photoactive yellow Protein from *Halorhodospira halophila*, University of Amsterdam

J. Hendriks, T. Gensch, L. Hviid, M.A. van Der Horst, K.J. Hellingwerf, J.J. van Thor. *Biophys. J.* **2002**, 82, 1632–1643.

J. Hendriks, W.D. Hoff, W. Crielaard, K.J. Hellingwerf. *J. Biol. Chem.* **1999a**, 274, 17655–17660.

J. Hendriks, I.H. van Stokkum, W. Crielaard, K.J. Hellingwerf. *FEBS Lett.* **1999b**, 458, 252–256.

J. Hendriks, I.H.M. van Stokkum, K.J. Hellingwerf. *Biophys. J.* **2003**, 84, 1180–1191.

W.D. Hoff, P. Dux, K. Hard, B. Devreese, I.M. Nugteren-Roodzant, W. Crielaard, R. Boelens, R. Kaptein, J. van Beeumen, K.J. Hellingwerf. *Biochemistry* **1994a**, 33, 13959–13962.

W.D. Hoff, S.L.S. Kwa, R. van Grondelle, K.J. Hellingwerf. *Photochem. Photobiol.* **1992**, 56, 529–539.

W.D. Hoff, H.C.P. Matthijs, H. Schubert, W. Crielaard, K.J. Hellingwerf. *Biophys. Chem.* **1995**, 56, 193–199.

W.D. Hoff, W.W. Sprenger, P.W. Postma, T.E. Meyer, M. Veenhuis, T. Leguijt, K.J. Hellingwerf. *J. Bacteriol.* **1994b**, 176, 3920–3927.

W.D. Hoff, I.H. van Stokkum, H.J. van Ramesdonk, M.E. van Brederode, A.M. Brouwer, J.C. Fitch, T.E. Meyer, R. van Grondelle, K.J. Hellingwerf. *Biophys. J.* **1994c**, 67, 1691–1705.

W.D. Hoff, K. H. Jung and J. L. Spudich. *Annu Rev Biophys Biomol Struct* **1997**, 26: 223–58

W.D. Hoff, A. Xie, I.H. Van Stokkum, X.J. Tang, J. Gural, A.R. Kroon, K.J. Hellingwerf. *Biochemistry* **1999**, 38, 1009–1017.

Y. Imamoto, M. Kataoka, R.S.H. Liu. *Photochem. Photobiol.* **2002**, 76, 584–589.

Y. Imamoto, M. Kataoka, F. Tokunaga. *Biochemistry* **1996**, 35, 14047–14053.
Y. Imamoto, H. Koshimizu, K. Mihara, O. Hisatomi, T. Mizukami, K. Tsujimoto, M. Kataoka, F. Tokunaga. *Biochemistry* **2001a**, 40, 4679–4685.
Y. Imamoto, Y. Shirahige, F. Tokunaga, T. Kinoshita, K. Yoshihara, M. Kataoka. *Biochemistry* **2001b**, 40, 8997–9004.
J.F. Imhoff, F. Hashwa, H.G. Trüper. *Arch. Hydrobiol.* **1978**, 84, 381–388.
Z. Jiang, E.C. Bauer. *unpublished* **1998**.
Z. Jiang, L.R. Swem, B.G. Rushing, S. Devanathan, G. Tollin, C.E. Bauer. *Science* **1999**, 285, 406–409.
H. Kandori, T. Iwata, J. Hendriks, A. Maeda, K.J. Hellingwerf. *Biochemistry* **2000**, 39, 7902–7909.
M. Kim, R.A. Mathies, W.D. Hoff, K.J. Hellingwerf. *Biochemistry* **1995**, 34, 12669–12672.
M. Koh, G. Van Driessche, B. Samyn, W.D. Hoff, T.E. Meyer, M.A. Cusanovich, J.J. Van Beeumen. *Biochemistry* **1996**, 35, 2526–2534.
R. Koradi, M. Billeter, K. Wuthrich. *J. Mol. Graph.* **1996**, 14, 51–55.
R. Kort, W. Crielaard, J.L. Spudich, K.J. Hellingwerf. *J. Bacteriol.* **2000**, 182, 3017–3021.
R. Kort, K.J. Hellingwerf, R.B.G. Ravelli. *J. Biol. Chem.* **2004**, in press.
R. Kort, W.D. Hoff, M. Van West, A.R. Kroon, S.M. Hoffer, K.H. Vlieg, W. Crielaand, J.J. Van Beeumen, K.J. Hellingwerf. *EMBO J.* **1996**, 15, 3209–3218.
R. Kort, R.B. Ravelli, F. Schotte, D. Bourgeois, W. Crielaard, K.J. Hellingwerf, M. Wulff. *Photochem. Photobiol.* **2003**, 78, 131–137.
A.R. Kroon, W.D. Hoff, H.P. Fennema, J. Gijzen, G.J. Koomen, J.W. Verhoeven, W. Crielaard, K.J. Hellingwerf. *J. Biol. Chem.* **1996**, 271, 31949–31956.
M. Kumauchi, N. Hamada, J. Sasaki, F. Tokunaga. *J. Biochem. (Tokyo)* **2002**, 132, 205–210.
J.A. Kyndt, F. Vanrobaeys, J.C. Fitch, B.V. Devreese, T.E. Meyer, M.A. Cusanovich, J.J. Van Beeumen. *Biochemistry* **2003**, 42, 965–970.
B.C. Lee, P.A. Croonquist, W.D. Hoff. *J. Biol. Chem.* **2001a**, 276, 44481–44487.
B.C. Lee, P.A. Croonquist, T.R. Sosnick, W.D. Hoff. *J. Biol. Chem.* **2001b**, 276, 20821–20823.
B.C. Lee, A. Pandit, P.A. Croonquist, W.D. Hoff. *Proc. Natl. Acad. Sci. USA* **2001c**, 98, 9062–9067.
E. Mano, H. Kamikubo, Y. Imamoto, M. Kataoka. *Spectr.-Int. J.* **2003**, 17, 345–353.
T. Masciangioli, S. Devanathan, M.A. Cusanovich, G. Tollin, M.A. el-Sayed. *Photochem. Photobiol.* **2000**, 72, 639–644.
N. Mataga, H. Chosrowjan, S. Taniguchi, N. Hamada, F. Tokunaga, Y. Imamoto, M. Kataoka. *Phys. Chem. Chem. Phys.* **2003**, 5, 2454–2460.
D.E. McRee, T.E. Meyer, M.A. Cusanovich, H.E. Parge, E.D. Getzoff. *J. Biol. Chem.* **1986**, 261, 13850–13851.
T.E. Meyer. *Biochim. Biophys. Acta* **1985**, 806, 175–183.
T.E. Meyer, S. Devanathan, T. Woo, E.D. Getzoff, G. Tollin, M.A. Cusanovich. *Biochemistry* **2003**, 42, 3319–3325.
T.E. Meyer, J.C. Fitch, R.G. Bartsch, G. Tollin, M.A. Cusanovich. *Biochim. Biophys. Acta* **1990**, 1016, 364–370.
T.E. Meyer, G. Tollin, J.H. Hazzard, M.A. Cusanovich. *Biophys. J.* **1989**, 56, 559–564.
T.E. Meyer, E. Yakali, M.A. Cusanovich, G. Tollin. *Biochemistry* **1987**, 26, 418–423.
K. Mihara, O. Hisatomi, Y. Imamoto, M. Kataoka, F. Tokunaga. *J. Biochem. (Tokyo)* **1997**, 121, 876–880.
A. Miller, H. Leigeber, W.D. Hoff, K.J. Hellingwerf. *Biochim. Biophys. Acta* **1993**, 1141, 190–196.
K. Moffat, S. Crosson, S. Anderson. *Abstr. Pap. Am. Chem. Soc.* **2003**, 226, U339–U339.
J.R. Nambu, J.O. Lewis, K.A. Wharton, Jr., S.T. Crews. *Cell* **1991**, 67, 1157–1167.
D. Pan, A. Philip, W.D. Hoff, R.A. Mathies. *Biophys. J.* **2004**, 86, 2374–2382.
J.L. Pellequer, K.A. Wager-Smith, S.A. Kay, E.D. Getzoff. *Proc. Natl. Acad. Sci. USA* **1998**, 95, 5884–5890.
B. Perman, V. Srajer, Z. Ren, T. Teng, C. Pradervand, T. Ursby, D. Bourgeois, F. Schotte, M. Wulff, R. Kort, K. Hellingwerf, K. Moffat. *Science* **1998**, 279, 1946–1950.
J.C. Raymond, W.R. Sistrom. *Arch. Mikrobiol.* **1967**, 59, 255–268.
J.C. Raymond, W.R. Sistrom. *Arch. Mikrobiol.* **1969**, 69, 121–126.
S. Reinelt, E. Hofmann, T. Gerharz, M. Bott, D.R. Madden. *J. Biol. Chem.* **2003**, 278, 39189–39196.
Z. Ren, B. Perman, V. Srajer, T.Y. Teng, C. Pradervand, D. Bourgeois, F. Schotte,

T. Ursby, R. Kort, M. Wulff, K. Moffat. *Biochemistry* **2001**, 40, 13788–13801.

G. Rubinstenn, G.W. Vuister, F.A. Mulder, P.E. Dux, R. Boelens, K.J. Hellingwerf, R. Kaptein. *Nat. Struct. Biol.* **1998**, 5, 568–570.

J. Sasaki, M. Kumauchi, N. Hamada, T. Oka, F. Tokunaga. *Biochemistry* **2002**, 41, 1915–1922.

N.S. Scrutton, A.R. Raine. *Biochem. J.* **1996**, 319, 1–8.

A. Sergi, M. Gruning, M. Ferrario, F. Buda. *J. Phys. Chem. B* **2001**, 105, 4386–4391.

W.W. Sprenger, W.D. Hoff, J.P. Armitage, K.J. Hellingwerf. *J. Bacteriol.* **1993**, 175, 3096–3104.

E.N. Spudich, J.L. Spudich. *J. Biol. Chem.* **1993**, 268, 16095–16097.

J.L. Spudich, C. S. Yang, K. H. Jung and E. N. Spudich. *Annu Rev Cell Dev Biol* **2000**, 16, 365–92

K. Takeshita, N. Hirota, Y. Imamoto, M. Kataoka, F. Tokunaga, M. Terazima. *J. Am. Chem. Soc.* **2000**, 122, 8524–8528.

K. Takeshita, Y. Imamoto, M. Kataoka, K. Mihara, F. Tokunaga, M. Terazima. *Biophys. J.* **2002a**, 83, 1567–1577.

K. Takeshita, Y. Imamoto, M. Kataoka, F. Tokunaga, M. Terazima. *Biochemistry* **2002b**, 41, 3037–3048.

B.L. Taylor, I.B. Zhulin. *Microbiol. Mol. Biol. Rev.* **1999**, 63, 479–506.

J.D. Thompson, D.G. Higgins, T.J. Gibson. *Nucleic Acids Res.* **1994**, 22, 4673–4680.

L. Ujj, S. Devanathan, T.E. Meyer, M.A. Cusanovich, G. Tollin, G.H. Atkinson. *Biophys. J.* **1998**, 75, 406–12.

D.M. van Aalten, W.D. Hoff, J.B. Findlay, W. Crielaard, K.J. Hellingwerf. *Protein Eng.* **1998**, 11, 873–879.

D.M.F. Van Aalten, W. Crielaard, K.J. Hellingwerf, L. Joshua-Tor. *Protein Sci.* **2000**, 9, 64–72.

M.E. van Brederode, T. Gensch, W.D. Hoff, K.J. Hellingwerf, S.E. Braslavsky. *Biophys. J.* **1995**, 68, 1101–1109.

M.E. van Brederode, W.D. Hoff, I.H. Van Stokkum, M.L. Groot, K.J. Hellingwerf. *Biophys. J.* **1996**, 71, 365–380.

M.A. van der Horst, I.H. van Stokkum, W. Crielaard, K.J. Hellingwerf. *FEBS Lett.* **2001**, 497, 26–30.

M.A. van der Horst, I.H. Van Stokkum, N. Dencher, K.J. Hellingwerf. **2004**, under review.

M.J. van der Meer **2000**, Femtosecond fluorescence studies of intramolecular reorientational motions, University of Amsterdam

A. Xie, W.D. Hoff, A.R. Kroon, K.J. Hellingwerf. *Biochemistry* **1996**, 35, 14671–14678.

A. Xie, L. Kelemen, J. Hendriks, B.J. White, K.J. Hellingwerf, W.D. Hoff. *Biochemistry* **2001**, 40, 1510–1517.

M. Yoda, H. Houjou, Y. Inoue, M. Sakurai. *J. Phys. Chem. B* **2001**, 105, 9887–9895.

20
Hypericin-like Photoreceptors

Pill-Soon Song

Abstract

The hypericin-derived pigments, stentorin and blepharismin, are the photosensory receptors for the photomotile responses of ciliates *Stentor coeruleus* and *Blepharisma japonicum*, respectively. Their photochemistry and its functional linkage to the photosensory signal transduction chain are not well established. In the light signal transduction of the giant single-cell ciliates, a light signal (intensity/wavelength) appears to be amplified by triggering Ca^{2+} ion influx into the cell. The photosensory signal-transduction pathways of the ciliates include the generation of membrane potentials. The heterotrimeric G-protein(s) may play a signal-transducing role. The photosensory signal transduction for the photophobic response of *Stentor coeruleus* is represented in the following scheme:

Stentorin → G-protein → phosphodiesterase →
cGMP/GMP-gated Ca^{2+}-channel → ciliary stroke reversal (photophobic response)

A similar scheme is likely for the closely related ciliate, *Blepharisma japonicum*.

20.1
Introduction

Photosensitive organisms exhibit varied behavioral responses to light stimuli such as intensity, propagation direction, polarization, wavelength, and light-dark rhythm. Both the ciliates *Stentor coeruleus* and *Blepharisma japonicum* respond to a sudden increase in light-stimulus intensity by displaying so-called step-up photophobic response. How do the colored ciliates reverse their cilia to stop and swim backwards momentarily when suddenly exposed to higher intensity light? Figure 20.1 shows the photophobic response of *Stentor coeruleus* upon encountering an illuminated area. Swimming movement of *Stentor coeruleus* is achieved by the clockwise beating of cilia. Figure 20.2 shows rows of cilia around the frontal pouch. The following scheme

Handbook of Photosensory Receptors. Edited by W. R. Briggs, J. L. Spudich

ISBN 3-527-31019-3

defines a sequence of the photosensory signal transduction pathway involved in the photophobic responses of the unicellular ciliates:

Light Signal

↓ *Signal Perception*

Photochemical Process

↓ *Signal Generation*

Biochemical Process

↓ *Signal Amplification*

Electrophysiological Event

↓ *Mechano-Transduction*

Mechanical Response (*Ciliary Stroke Reversal*)

Scheme 20.1

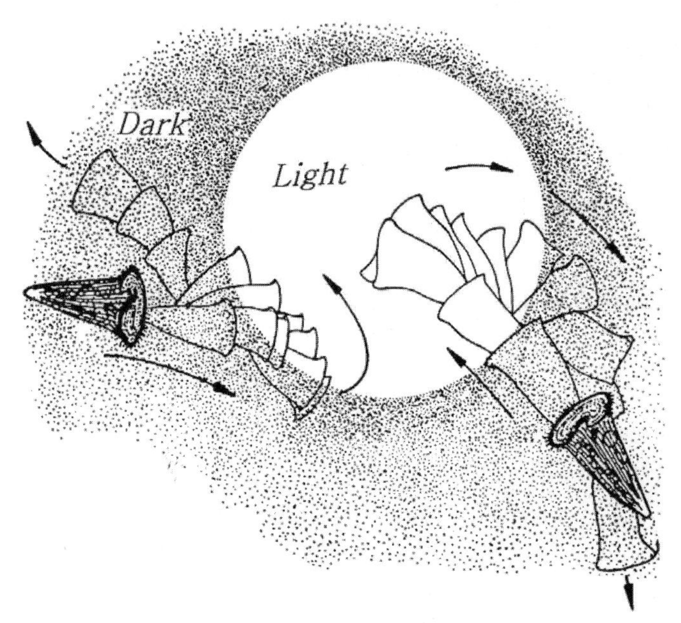

Figure 20.1 This figure shows 0.2-s sequential tracings from the microscopic video tape recordings for the two *Stentor coeruleus* cells which responded to a sudden increase in light intensity as they swim from the dark to the illuminated area of the medium. Each cell is seen to stop at the light-dark boundary, resulting from the reversal of cilia strokes when encountering the light (Song, 1983, 1999).

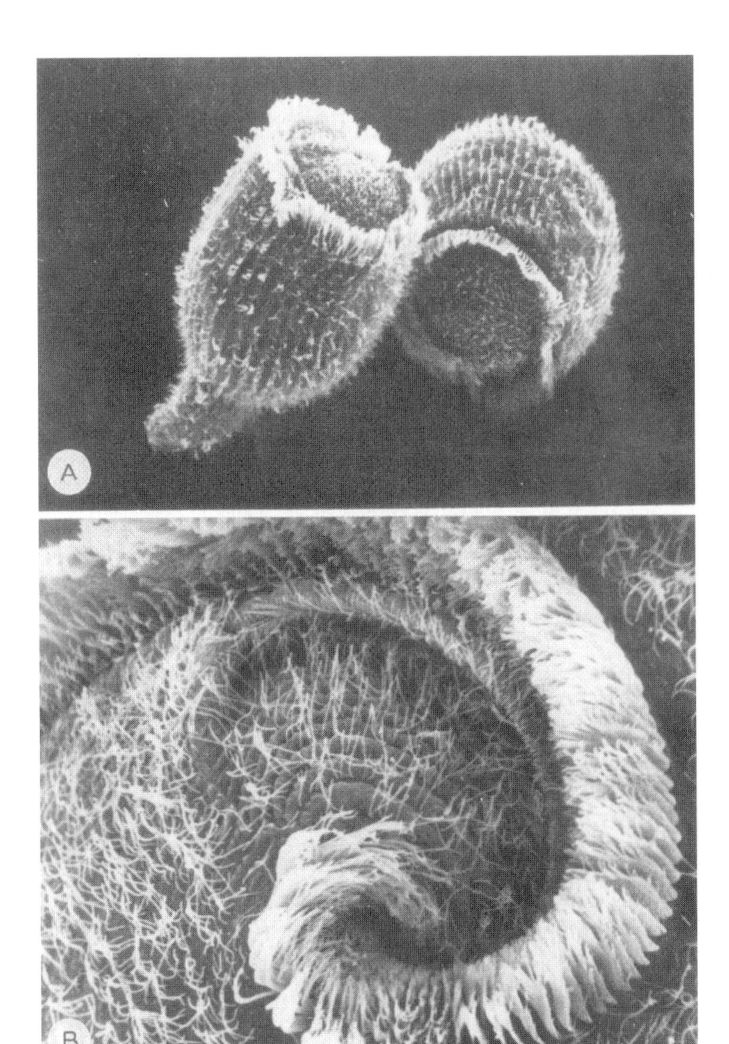

Figure 20.2 *Stentor coeruleus* (A) and cilia around the frontal pouch (B). SEM; top panel, magnification × 260; Ciliated oral fold (frontal pouch) and membranelle band, magnification × 1500. [Reproduced with copyright owner's permission]. *Blepharisma japonicum* possesses similar cilia along the cell body, but it does not have a ciliated oral fold. [Unpublished SEM micrographs by F. Verni and G. Rosati; private communication from F. Lenci].

Our understanding of the molecular mechanisms to describe the above scheme is still at its infancy. However, the nature of the chromophores in stentorin and blepharismin has recently been elucidated. Emphasis in this chapter is placed on the chemical structures of the photoreceptor pigments and their photochemical processes that may have functional implications. For relevant review of the photomotile responses of unicellular ciliates, several review articles are available in the literature

(Song, 1983, 1999; Kuhlman, 1998; Lenci et al., 2001). For the biology and photobiology of *Stentor* and *Blepharisma,* the two classic works by Tartar (1961) and Giese (1973), respectively, are recommended for the interested reader. Various photomovement modes such as phototaxis, photokinesis and photophobic responses in the ciliates have been described in a recent review (Marangoni et al., 2004).

20.2
Ciliate Photoreceptors

20.2.1
Action Spectra

Rhodopsin-like photoreceptors are among the most widely distributed chromoproteins in prokaryotic and eukaryotic organisms. However, the ciliates *Stentor coeruleus* and *Blepharisma japonicum* use the unique hypericin-like chromophore photoreceptors for their photophobic and phototactic responses. Action spectroscopy enables one to deduce the identity of the photoreceptors involved in various light-induced responses (Lipson, 1991). In both ciliates, a light stimulus triggers a millisecond change in membrane potential preceding an action potential arising from massive calcium influx. The nature of the "photoreceptor potential" preceding the latter is not well understood. However, it has been suggested that it arises from an interfacial electric potential across the pigment granular membrane induced by the light stimulus (see below). The action spectrum for the photophobic response in *Stentor coeruleus* resembles the absorption spectra of the whole cells (Figure 20.3) and the extracted pigment, identifying the photoreceptor as stentorin (Song et al., 1980a, 1980b; Kim et al., 1984). The fact that a colorless mutant as well as caffeine-bleached *Stentor coeruleus* (Tartar, 1972) either loses or reduces its photosensitivity further corroborate the identification of stentorin as the photoreceptor (Song, 1983). The action spectrum for the negative phototactic response is similar to that for the photophobic response (Song et al., 1980b). The action spectra for the light-induced receptor potential and ciliary stroke reversal are also consistent with the absorption spectra of the pigment stentorin (Song et al., 1980a, 1980b; Wood, 1976; Fabczak et al., 1993b). Both *Stentor coeruleus* and *Blepharisma japonicum* display action spectra for the membrane receptor potentials that spectrally correlated with the absorption spectra of stentorin and blepharismin, respectively (Fabczak et al., 1993b, 1993c).

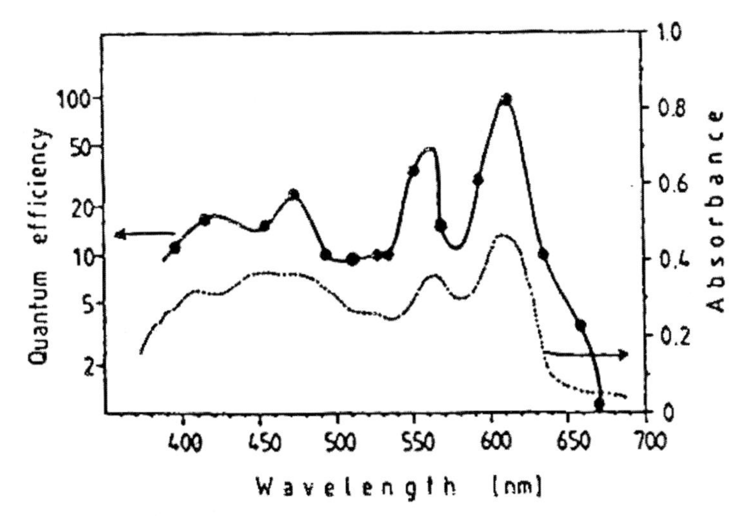

Figure 20.3 The action spectrum (arrow for the left ordinate) for the step-up photophobic response (Figure 20.1) of *Stentor coerulues*, overlaid above the absorption spectrum (arrow for the right ordinate) of the cell suspension (Song, 1983).

20.2.2
The Chromophores

The chromophore of stentorin is chemically derived from hypericin. Only relatively recently has the chemical structure of the stentorin chromophore been elucidated (Tao et al., 1993), as shown in Figure 20.4. The chromophore structures of blepharismin and its oxidized form, oxyblepharismin, have also been determined recently (Figures 20.4 and 20.5) (Checcucci et al., 1997; Maeda et al., 1997; Spitzner et al., 1998). There are several forms of blepharismin. The major form appears to be 2,4,5,7,2',4',5',7'-octahydroxy-6,6'-diisopropyl-1,11'-(hydroxybenzylidene)naphthodi-

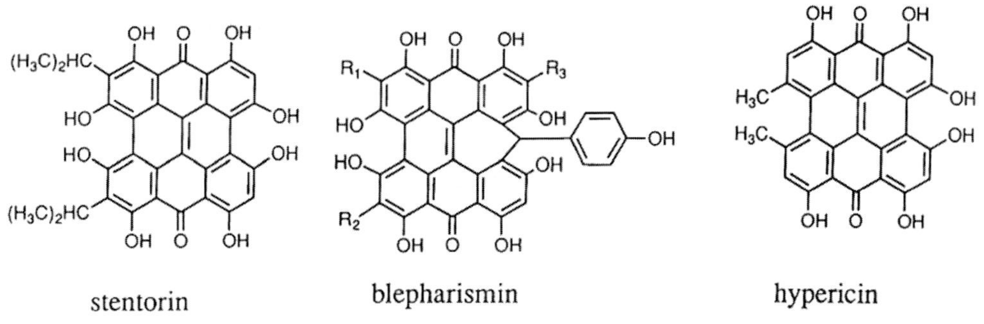

stentorin blepharismin hypericin

Figure 20.4 The chemical structures of stentorin, blepharismins and hypericin. For individual structures of blepharismins, see Figures 20.5 and 20.6.

anthrone, MW 698, shown in Figure 20.6 (Checcucci et al., 1997). There are four other forms having different combinations of substituents with MW 670, 684, 698, and 718 (Figure 20.5).

So far, these unique hypericin-like chromophores have not been found as the chromophore of photoreceptors in any other organisms. It is likely that hypericin-like chromophores are commonly present among the light sensitive ciliates. For example, *Fabrea salina* and *Ophryoglena flava* exhibit phototactic action spectra that qualitatively resemble the absorption spectra of hypericin, stentorin and blepharismin, al-

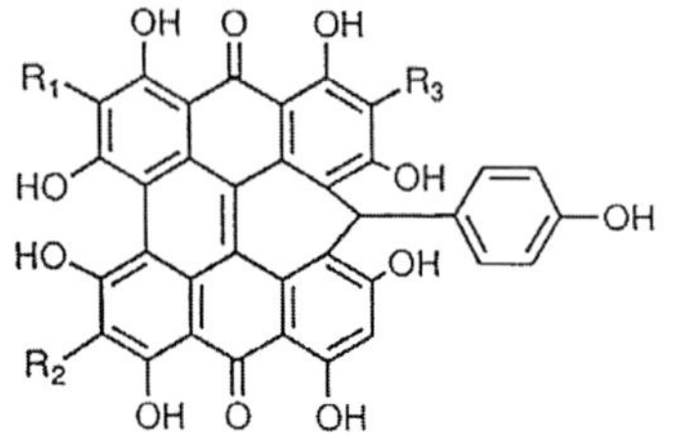

blepharismin

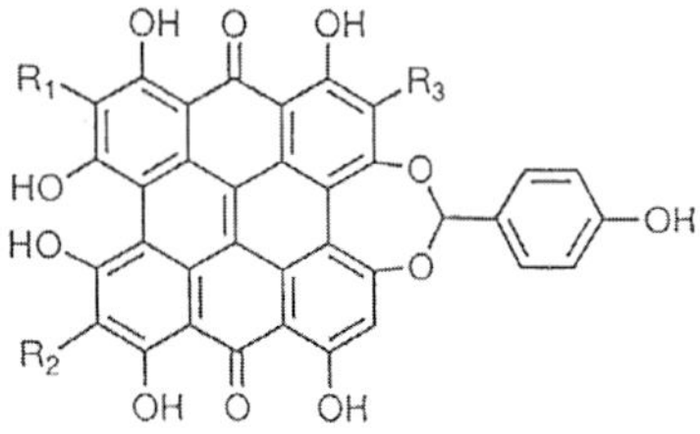

oxyblepharismin

	R1	R2	R3
A	Et	Et	H
B	Et	iPr	H
C	iPr	iPr	H
D	Et	iPr	Me
E	iPr	iPr	Me

Figure 20.5 The chemical structures of blepharimin and oxyblepharismin derivatives (Maeda et al., 1997). Substitutents: Et, ethyl; H, hydrogen; Me, methylethyl; iPr, isopropyl.

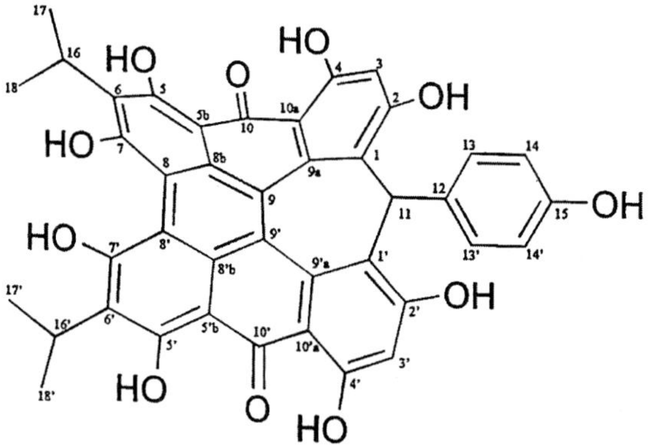

Figure 20.6 The chemical structure and numbering system of the major blepharismin chromophore (Checcucci et al., 1997).

though rhodopsin-like photoreceptors have also been suggested for the photomovement responses of these ciliates (Marangoni et al., 2004). A new genus of ciliate from Guam reefs, *Maristentor dinoferus,* contains a highly fluorescent hypericin-like pigment, apparently serving as a photoreceptor for positive phototaxis and/or step-down photophobic response (Private communication from Dr. Chris Lobban, www.uog.edu/dns/lobban.htm). It would be interesting to elucidate the nature of its photosensitivity.

20.2.3 Proteins and Localization

The pigment proteins are localized within a pigment granule, 0.3–0.7 μm in diameter. The pigment granules are distributed longitudinally along the cell body between the rows of cilia. The pigment granules in *Stentor coeruleus* are located in sub-pellicular ectoplasm and are distinct in nature from mitochondria which show a similar intracellular location (Weisz, 1950; Inaba et al., 1958).

The pigment granule may well be the "photoreceptor apparatus" of *Stentor coeruleus* (Kim et al., 1990). Figure 20.7 shows the pigment granules. Two forms of stentorin chromoproteins stentorin-1 and -2 can be chromatographically isolated from the cells. The red-wavelength absorbance maxima of stentorin-2 are red-shifted

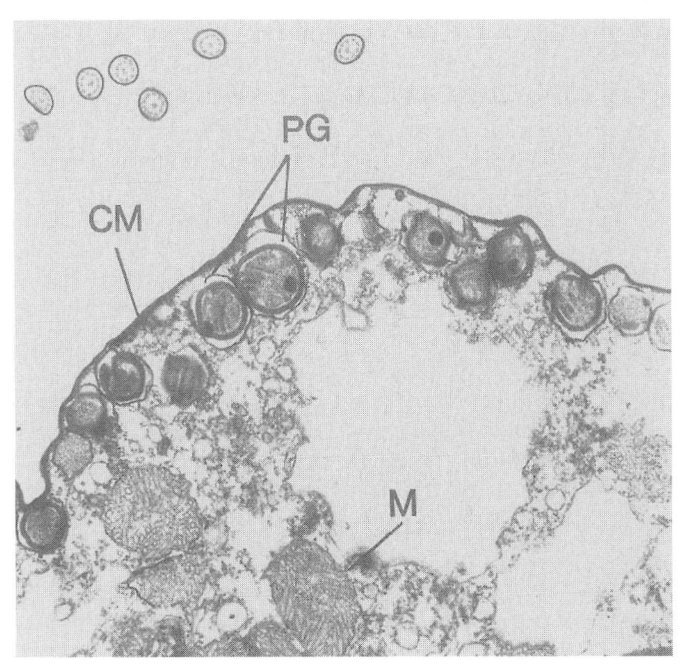

Figure 20.7 A scanning electron micrograph of *Stentor coeruleus'* transverse section showing cell membrane (CM), pigment granules (PG) and mictochondria (M) (Kim et al., 1990). Some internal structure is seen within the pigment granule (Kim et al., 1990). [Reproduced with copyright owner's permission].

by about 10 nm relative to those of stentorin-1. Stentorin-1 is strongly fluorescent while stentorin-2 is only weakly or non-fluorescent. Stentorin-1 is likely a chromophore–detergent complex. Detergent-solubilized stentorin-2 behaves like a large complex of molecular mass greater than 500 kDa. Stentorin-2 can be further resolved into stentorin-2A and -2B by hydrophobic interaction chromatography (Dai et al., 1995; Wynn et al., 1995). It contains the chromophore covalently bound to an approximately 50-kDa protein.

The apoprotein for stentorin is a 50-kDa protein (stentorin-2B). When isolated under native conditions, stentorin forms a large molecular assembly (stentorin-2, greater than 500 kDa in mass) comprised of the 50-kDa protein and additional non-chromophore-bearing protein subunit(s) (stentorin-2A). The chromophore structure of stentorin-2B appears to be identical with the stentorin structure shown in Figure 20.4.

Blepharismins are also localized within the pigment granules, ~0.5–0.7 μm in diameter. The granules surrounded by membrane are connected to the plasma membrane, and contain a honeycomb-like and stacked-lamella structure in rows parallel to the cilia. The TEM micrographs of freeze-fractured pigment granule confirmed the lamella structure (taken by T. Matsuoka, private communication via F. Lenci). *Fabrea salina*, a marine ciliate, also possesses membrane-enclosed pigment granules, apparently serving as the phototactic receptor organelle (Marangoni et al., 2004).

The blepharismin pigment has been reported to be either non-covalently bound to an apoprotein of molecular mass 38 000 dalton (Gioffre et al., 1993; Yamazaki et al., 1993b) or covalently bound to a molecular weight 200,000 protein (Matsuoka et al., 1993). Fabczak et al. (2001) reported that phosphorylation of two other proteins, 28- and 46-kDa proteins, of *B. japonicum* is down- and up- regulated by light, respectively. They suggest that the phosphorylated; 46-kDa protein may associate with or be a proteolytic digest product from the 200-kDa protein (Matsuoka et al., 2000). The large discrepancy between these two values of mol wt has not been resolved. Since both of these pigment-protein complexes are spectroscopically indistinguishable, and both are consistent with the action spectra of *Blepharisma japonicum*, other properties of the pigment proteins must be examined in order to determine their functional role. The reported mol mass value of 200,000 daltons for the blepharismin holoprotein seems anomalously large, in comparison to the 50-kDa stentorin-2B subunit. However, proteins with molecular masses comparable to the 200-kDa “blepharismin” are not uncommon. It remains to be seen which of the two proteins of extremely different molecular mass represent the functional blepharismin.

20.3 Photochemistry

20.3.1 Photosensitization?

Hypericin and free stentorin chromophore are efficient photosensitizers because of their ability to populate the triplet state and generate singlet molecular oxygen. However, native stentorin does not yield singlet oxygen in solution (Dai et al., 1992). The photochemical mechanism of stentorin function in *Stentor coeruleus* does not appear to be based on i) photodynamic action involving singlet oxygen or ii) gaseous singlet oxygen as a dissolved intracellular messenger in the light signal cascade. Hypericin, the chemical parent compound for stentorin and blepharismin, is an efficient photosensitizer, mediating various photodynamic actions *in vivo* and *in vitro* via singlet oxygen (Duran and Song, 1985; Yang et al., 1986). Singlet oxygen has been suggested to function as a signal messenger for the photomovements in microorganisms, such as *Anabaena variabilis, Physarum polycephalum,* and *Loxodes striatus* (Ghetti et al., 1992). However, at least in *Stentor coeruleus,* singlet oxygen does not seem to play a significant role in the photomovement transduction pathway, since stenorin-2 is a poor singlet-oxygen generator, unlike its free chromophore and unlike hypericin (Dai et al., 1992). A singlet oxygen quencher did not specifically inhibit the photophobic response of *Blepharisma japonicum,* also suggesting that the active oxygen is not a photosensory transducer component for this organism (Checcucci et al., 1991). What then is the primary photochemistry involved in triggering the photosensory transduction pathway in *Stentor* and *Blepharisma*? We can only speculate at the present time. In animals, the gaseous messenger molecule nitric oxide plays multiple roles in cellular signal transduction, but it is not known if NO is present and functional in *Stentor coeruleus* and *Blepharisma japonicum.*

20.3.2 Primary Photoprocesses

Photomotile responses in ciliates involve a chain of chemical and/or physical events triggered by the photoexcited photoreceptor molecule/apparatus. The nature and efficiency of the photomovement initiation trigger is provided by the primary photoprocess of the photoreceptor molecules. Stentorin-2 serves as the photosensor as implicated by the fact that stentorin-2 shows a very low fluorescence quantum yield with ultrafast decay in the picosecond time scale, whereas stentorin-1 is highly fluorescent (Song et al., 1990; Savikhin et al., 1993). The picosecond pump-probe spectroscopy study indicated that an initial photoprocess occurred within 3 ps in excited stentorin-2 but not in the free chromophore species, hypericin or stentorin-1 (Savikhin et al., 1993). Blepharismin in its protein bound form also exhibits a similar picosecond process (Yamazaki et al., 1993b).

The nature of the photoprocess remains to be elucidated. Unlike rhodopsin, photoactive yellow protein (PYP), phototropins, and phytochromes, both stentorin and blepharismin fail to exhibit a photoisomerization and/or photochemical transformation cycle that can be readily detected by spectrophotometry under a freeze-thaw cycling condition. Electron transfer-coupled proton transfer may be responsible for the initial photoprocesses in both stentorin and blepharismin.

Stentorin can mediate electron transfer processes in its excited state (Wells et al., 1997). The model compound, hypericin, undergoes two reversible one-electron reductions for hypericin at –0.87 V and –1.18 V vs. a normal hydrogen electrode as a reference electrode (Redpenning and Tao, 1993). Additionally, an oxidation is observed at ca. +0.90 V in DMSO. These formal potentials are consistent with the visible absorbance maximum of hypericin in DMSO at 599 nm (= 2.1 eV). Given this information, we estimate that the excited state potentials for hypericin (Hyp) are approximately –0.90 V for the formation of hypericin cation radical Hyp^+ (Because of uncertainty of these numbers, the oxidation potential of the excited state hypericin could be as high as +1.2 V (Song et al., 1990; Weller et al., 1997).

$$Hyp + h\nu \rightarrow Hyp^* \rightarrow Hyp^+ + e^-$$

and +0.90 V for the formation of hypericin anion radical Hyp^-,

$$Hyp + h\nu \rightarrow Hyp^* + e^- \rightarrow Hyp^-$$

where electron e^- is supplied by an electron donor including the ground state Hyp itself.

Thus, in the excited state, hypericin can be a good electron donor, as well as being a good electron acceptor, depending on the redox potentials of the donor/acceptor pairs present in solution. *p*-Benzoquinone, a classical electron acceptor, quenches the fluorescence of hypericin (Yamazaki et al., 1993a). The quenching is due to electron transfer from hypericin to *p*-benzoquinone at a diffusion-controlled rate (1.43×10^{10} M–1 s^{-1}). The electron donor property of hypericin in its excited state is consistent with the picosecond absorption difference measurements of hypericin. These measurements show that a new species having a transient absorption in the red/far-red region was formed in ~5 ps upon excitation. This species was not present in ground state absorption (Gai et al., 1993, 1994). The long-wavelength absorption is quenched by the electron-scavenging solvent, acetone that may have resulted from a solvated electron. The formation of cation and anion radicals can also be observed upon photolysis of hypericin (Weiner and Mazur, 1992).

Photolysis of stentorin-2 yields radical species, apparently producing stentorin cation radicals, in a manner similar to the scheme shown for the excited-state hypericin (Wells et al., 1997). Electron transfer probably occurs from the excited stentorin chromophore to a suitable acceptor/amino-acid residue. An efficient electron transfer process may account for the quenching of the fluorescence from the excited-state stentorin. The rate for this process can be estimated from the short fluorescence decay lifetime, ~1×10^{11} M–1 s^{-1}. This value is close to the diffusion-limited rate constant for electron transfer from the excited state hypericin to *p*-benzoquinone. This reaction is likely a reversible process in native stentorin.

Hypericin can also be photooxidized in the presence of dithiodiethanol ($HOCH_2CH_2$-S-S-CH_2CH_2OH) with a quantum yield of ca. 0.001. Photooxidation of hypericin with dithiodiethanol generates mercaptoethanol (HS-CH_2CH_2OH) and oxyhypericin (Dai, 1994). Similarly, photooxidation of stentorin (HO-ST-OH) would likely produce a new quinone form of oxy-stentorin (O=ST=O) according to the following scheme:

$$\text{HO-ST-OH} + h\nu \rightarrow \text{HO-ST-OH}^*$$
$$\text{HO-ST-OH}^* + \text{R-S-S-R} \rightarrow \text{HO-ST-OH}^{(+} + [\text{R-S-S-R}]^{\bullet -} \rightarrow \text{O=ST=O} + 2\text{H}^+ + 2\text{R-S}^-$$

In native stentorin, two electrons and two protons resulting from the similar reactions reduce the suitable amino acid residue(s) such as cystine. The potential for the oxidation of mercaptoethanol is +0.02 V at pH 8 (Zagal and Paez, 1989), while a potential of +0.6 V is necessary for the reduction of oxy-hypericin in DMSO (Redpenning and Tao, 1993). Thus, a spontaneous reduction of oxy-hypericin in the ground state is feasible. Similarly, an oxy-stentorin can be reduced to stentorin to complete the photo-cycle, with the possible vectorial release of protons from the stentorin-bound pigment granule to the cytoplasm. It remains to be seen if electron/proton transfer to and from the stentorin chromophore is accelerated by the apoprotein to initiate the functionally viable "photo-cycle" in *Stentor coeruleus.* The redox potential of the cysteine–cystine system is estimated to be −0.21 V at pH 7.0 (Cleland, 1964). This redox potential would enhance the photooxidation of stentorin when compared with the dithiodiethanol-mercaptoethanol system (−0.9 V at pH 10 for oxidation).

Electron transfer to and from the excited state of blepharismin is also likely to play a signal-initiating role in the photosensory transduction of *Blepharisma japonicum.* Because of the similarities in their molecular structures between blepharismin and hypericin (also stentorin), we expect blepharismin and its model compounds to exhibit similar electrochemical behaviors and efficient electron transfer in the excited state.

20.4 Photosensory Signal Transduction

How the primary photochemistry of the ciliate photoreceptors triggers the signal-transduction cascade for the cellular photoresponses is not understood. Photoexcitation of the photoreceptors generates membrane potentials that precede the influx of calcium ions and the reversal of ciliary beat stroke of *Stentor coeruleus* (Song, 1983; Matsuoka et al., 1992; Immitzer et al., 2000). However, little is known about the biochemical basis of the photosensory signal transduction in *Stentor, Blepharisma* and other ciliate cells.

20.4.1
Signal Generation

Light stimuli are transduced and eventually elicit mechanoresponses in photosensitive ciliates. The initial step of the signal transduction chain in *Stentor* and *Blepharisma* starts with generation of cellular signal(s) via the photoprocesses of the photosensor molecules. The photochemical mechanisms of stentorin and blepharismin are unknown at present. Steady-state photolysis of stentorin and blepharismin failed to demonstrate spectrophotometrically observable photochemical cycles. A pump-probe experiment showed that stentorin-2 undergoes an absorbance decrease (bleaching) in less than 3 ps; whereas free stentorin chromophore does not exhibit the ultrafast photoprocess (Savikhin et al., 1993). This process is faster than the fluorescence decay, occurring with a lifetime of about 10 ps. The ultrafast bleaching process, monitored at 565–630 nm, may be an intermolecular proton transfer to an appropriately situated amino acid residue in the photoreceptor molecule. Alternatively, the ultrafast process may involve intermolecular electron transfer. Stentorin and blepharismin are strong electron donors. The 3-ps absorption transient observed with stentorin-2 may result from electron transfer. Inter- and intra-molecular proton release/transfer may then follow the electron transfer.

Irradiation of Stentor cells leads to a lowering of intracellular pH. Fringelite D as a model compound for stentorin and blepharismin releases protons upon light excitation (Immitzer et al., 2000). Deprotonation of stentorin may serve as the initial photosensory transduction step. In a model system consisted of hypericin imbedded in phospholipid liposomes, a light-induced pH drop within the liposome has been reported (Fehr et al., 1995). An intracellular pH drop can also be triggered by the photoexcitation of hypericin incorporated in 3T3 mouse fibroblast cells (Chaloupka et al., 1998). These model studies suggest that the primary photoprocess can lead to an intracellular pH change. Such a transient pH change could serve as the initial signal for the subsequent signal transduction pathway involved in the photomovement of the ciliate cells.

Also, the photoresponses of *Stentor coeruleus* are dependent on pH. Thus, intracellular pH-modulating reagents (including ammonium chloride and the protonophores carbonylcyanide *m*-chlorophenyl-hydrazone (CCCP), and carbonylcyanide *p*-(trifluoromethoxy)-phenyl-hydrazone (FCCP)) inhibit the photo-motile responses of the ciliate cell, suggesting that some form of proton gradient is involved in the photosensory signal-transduction pathway (Dai et al., 1992). Perhaps the most convincing evidence for the generation of a transient pH change as an early intracellular transduction signal comes from the photoresponses of the ciliate cells in the presence of protonophores and exogenous ammonium chloride (Fabczak et al., 1993a). Ammonium chloride serves as a membrane-permeable weak acid that lowers the intracellular pH. An artificially lowered intracellular pH in the presence of NH_4Cl counteracts any light-induced pH drop as the early signal for the subsequent transduction cascade. Both *Stentor coeruleus* and *Blepharisma japonicum* show similar reductions in photoresponses to ammonium chloride (Fabczak et al., 1993a). Since blepharismin is structurally similar to stentorin, we expect that similar photochemistry occurs in blepharismin. An added proof is that the photophobic response in *Blepharisma japonicum* can be modulated by ammonium chloride, CCCP, and FCCP, as in *Stentor coeruleus*. An intracellular pH decrease upon light irradiation was also observed in *Blepharisma japonicum*.(Matsuoka et al., 1992).

20.4.2 Signal Amplification

A cellular signal generated by the photoprocess of the photosensor molecules must be amplified. Photons absorbed by stentorin in *Stentor coeruleus* results in an influx of Ca^{+} ions. The signal-amplification factor here appears to be 3 to 4 orders magnitude smaller than that in rhodopsin-triggered signal amplifications, the latter measured in terms of number of Na^{+} ions excluded (Na^{+}-channel closure) per photon absorbed. Consistent with the calcium influx as a key signal amplification, low extracellular Ca^{2+} ion concentrations and Ca^{2+} ion channel blockers, such as diltiazem and pimozide, all inhibit the photo-mechanoresponses of *Stentor coeruleus* and *Blepharisma japonicum*.

The biochemical reactions underscoring the signal generation and amplification are poorly understood at the present time. The photoexcitation of stentorin has been suggested to generate a transient intracellular pH change preceding the activation of a transducin-like G-protein. The G-protein is then coupled to the depolarizing action potential resulting from the influx of Ca^{2+} ions via a photoresponsive phosphodiesterase. A cGMP-dependent phosphodiesterase is a possible signal amplifier in the photosensory signal cascade in *Stentor coeruleus* and *Blepharisma japonicum*. A G-protein-coupled phosphodiesterase is likely to be a signal amplifier in these ciliate cells. In fact, the cytosolic cGMP level and its modulation play an important role in the photosensory transduction chain involved in both the photophobic and phototactic responses of *Stentor coeruleus* (Fabczak et al., 1993d; Walerczyk and Fabczak, 2001). L-cis-diltiazem, a cGMP-gated Ca^{2+} ion channel blocker, induced ciliary stroke reversal even in the dark and increased the photosensitivity of *Stentor coeruleus* cells (Walerczyk et al., 2003). For *Blepharisma japonicum*, an alternative signal-amplifer based on inositol-1,4,5-triphosphate as a second messenger has been proposed (Fabczak et al., 1996). Mastoparan is a drug known for its activity as a G-protein activator. This compound also raises the cytoplasmic inositol triphosphate level in animal and plant cells. Thus, it remains to be seen if a G-protein is coupled to the activation of phospholipase C and whether the two ciliates use different signal amplifying mechanisms based on cGMP and/or inositol triphosphate.

20.4.3 Signal Transduction

The motility of *Stentor coeruleus* is produced by ciliary strokes and its ultimate reaction to light is ciliary reversal, i.e. a photophobic response. Ciliary activity is controlled primarily through the regulation of electrical properties of the cell membrane (Eckert and Naitoh, 1972; Naitoh and Eckert, 1974), specifically through the internal ionic environment, especially the concentration of intracellular calcium ions (Eckert, 1972). The photophobic mechanoresponse of *Stentor coeruleus* is controlled by ion permeabilities of the cell membrane that determine resting, action, and mechanoreceptor potentials in the *Stentor* cell (Song et al., 1980a; Wood, 1982).

The final motile response in the photomovement of *Stentor coeruleus* consists of at least the following steps, namely, a stop reaction of its forward swimming, followed by brief backward swimming when light intensity is strong enough, and then resumption of forward swimming. When a light intensity higher than that required to elicit a receptor potential exceeding the threshold is applied, *Stentor coeruleus* cells reacted with a receptor potential that rapidly triggers an action potential, and after the decay of the action potential, a plateau of depolarization follows. With a light stimulus of similar intensity, *Stentor coeruleus* reacted by altering its swimming direction with a stop reaction and a period of backward swimming. The latency (delay time between the initial onset of the light stimulus and the appearance of action potential) of the stop reaction is shorter and the duration of backward swimming is longer with increasing light intensity. There is a temporal, and possibly functional correlation between the light-induced membrane potential changes and the phobic response (Fabczak et al., 1993b, 1993c).

20.5 Concluding Remarks

Further investigation of the ionic basis underlying the light-induced membrane potential changes is warranted to shed light on our understanding of the photosensory transduction pathway in *Stentor coeruleus* and *Blepharisma japonicum*. One direction of future work could be the application of patch-clamp techniques. Patch-clamp techniques allow one to measure the currents in single ionic channels. The patch-clamp technique has been successfully applied to single channel current recording in plasma membrane blisters of *Paramecium* (Saimi and Martinac, 1989). To confirm the involvement of cGMP in the photosensory transduction pathway (Walerczyk and Fabczak, 2001), the patch-clamp technique has been applied to identify cGMP-dependent ion channels in *Stentor* cells (Koprowski et al., 1997). Ion channels involved in rhodopsin-mediated visual excitation may be opened/closed by phosphodiesterase activated by a heterotrimeric G-protein. In some organisms, the channel opening/closure depends on PLC instead of PDE. *Blenpharisma japonicum* may well be such an organism (Fanczak et al., 1999; Fabczak and Fabczak, 2000). Eventually, molecular biological and genetic approaches to the mechanism of the ciliate photoresponses will open a new avenue of research in this field.

Acknowledgements

This work was in part supported by KISTEP/MOST through a National Research Laboratory grant and by KOSEF through an Environmental Biology Research Center/

GNSU grant. This is Kumho Life & Environmental Science Laboratory Publication No. 65.

References

Chaloupka, R., F. Sureau, E. Kocisova, and J. W. Petrich, *Photochem. Photobiol.* **1998**, 68, 44–50.

Checcucci, G., F. Lenci, F. Ghetti, and P.-S. Song, *J. Photochem. Photobiol. B: Biol.*, **1991**, *11*, 49–55.

Checcucci, G., R. K. Shoemaker, E. Bini, R. Cerny, N. Tao, J.-S. Hyon, D. Gioffre, F. Ghetti, F. Lenci, and P.-S. Song, *J. Am. Chem. Soc.* **1997**, *119*, 5762–5763.

Cleland, W. W. *Biochemistry* **1964**, *3*: 480–482.

Dai, R., *Ph. D. Dissertation*, University of Nebraska, Lincoln, NE, **1994**

Dai, R., R., P.-S. Song, J. L. Anderson, M. Selke, and C. S. Foote,, *IUBMB Conference on Biochemistry and Molecular Biology of Diseases Abstracts*, Nagoya, Japan, **1992**.

Dai, R., T. Yamazaki, I. Yamazaki, and P.-S. Song, *Biochim. Biophys. Acta* **1995**, *1231*, 58–68.

Duran, N., and P.-S. Song,. *Photochem. Photobiol.* **1985**, *43*, 677–680.

Eckert, R. *Science* **1972**, *176*, 473–481.

Eckert, R., and Y. Naitoh , *J. Protozool.* **1972**, *19*, 237–243.

Fabczak, H., and S. Fabczak, *Adv. Cell Biol.* **2000**, *27*, 425–440.

Fabczak, H., and S. Fabczak, P.-S. Song, G. Checcucci, F. Ghetti, F. Lenci, *J. Photochem. Photobiol. B: Biol.* **1993a**, *21*, 47–52

Fabczak, S, H, Fabczak, N., Tao, and P.-S. Song, *Photochem. Photobiol.* **1993b**, *57*, 696–701.

Fabczak, H., P. B. Park, S. Fabczak, and P.-S. Song. *Photochem. Photobiol.* **1993c**, *57*, 702–706.

Fabczak, H., N. Tao, S. Fabczak, and P.-S. Song, *Photochem. Photobiol.* **1993d,** *57*, 889–892.

Fabczak, H., B. Groszynska, and S. Fabczak, *Acta Protozool.* **2001**, *40*, 311 –315.

Fabczak, S., H. Fabczak, M. Walerczk, J. Sikora, B. Groszynska, and P.-S. Song, *Acta Protozool.* **1996**, *35*, 245–249.

Fabczak, H., M. Walerczyk, B. Groszynska, and S. Fabczak, *Photochem. Photobiol.* **1999**. *69*, 254–258.

Fehr, M. J.. M.A. McCloskey, and J. W. Petrich, *J. Am. Chem. Soc.* **1995,** *117*, 1833–1836.

Gai, F., M. J. Fehr, and J. W. Petrich, *J. Am. Chem. Soc.* **1993**, *115*, 3384–3385.

Gai, F., M. J. Fehr, and J. W. Petrich,. *J. Phys. Chem.* **1994**, *98*, 8352–8358.

Ghetti, F., G. Checcucci, and F. Lenci, *J. Photochem. Photobiol. Part. Biol.* **1992**, *15*, 185–198.

Giese, A. C. *Blepharisma: The Biology of a Light-Sensitive Protozoan*, Stanford University Press, Stanford, CA. **1973**.

Gioffre, D., F. Ghetti, F. Lenci, C. Paradiso, R. Dai, and P.-S. Song, *Photochem. Photobiol.* **1993**, *58*, 275–279.

Immitzer, B., C. Etzlstrofer, R. Obermueller, M. Sonnleitner, G. Schuetz, and H. Falk, *Mh. Chem.*, **2000**, *131*, 1039–1045.

Inaba, F., R. Nakamura, and S. Yamaguchi, *Cytologia (Tokyo)* **1958**, *23*, 72–79

Kim, I.-H., R.K. Prusti, P.-S. Song, D.P. Haeder, and M. Haeder,. *Biochim. Biophys. Acta* **1984**, *798*, 298–304.

Kim, I.-H., J.S. Rhee, J.W. Huh, S. Florell, B. Faure, K.W. Lee, M. Kahsai, P.-S. Song, N.Tamai, T. Yamazaki, and I. Yamazaki, *Biochim. Biophys. Acta* **1990**, *104*,: 43–57

Koprowski, P. M. Walerczyk, B. Groszynska, H. Fabczak, and A. Kubalski, *Acta Protozool.* **1997**, *36*, 121–124.

Kuhlmann, H.-W. Naturwissenschaften **1998**, *85*, 143–154.

Lenci, F., F. Ghetti, and P.-S. Song, In: ESP Review Series on Photobiology: Photomovements (D.-P. Haeder, M. Lebert, Eds.). pp.51. **2001**.

Lipson, E. D. In *Biophysics of Photoreceptors and Photomovements in Microorganisms* (F. Lenci, F. Ghetti, G. Colombetti, D.-P. Haeder, and P.-S. Song, eds), Plenum Press, New York. **1991**, 293–309.

Maeda, M., H. Naoki, T. Matsuoka, Y. Kato, H. Kotsuki, K. Utsumi, and T. Tanaka, *Tetrahedron Lett.* **1997**, 7411.

Marangoni, R., S. Lucia, and G. Colombetti, In: *CRC Handbook of Organic Photochemistry and Photobiology*, 2nd Ed. (Edited by W. Horspool and F. Lenci), CRC Press, Boca Raton, FL. **2004**, 122/1–122/18.

Matsuoka, T., Y. Murakami, T. Furukohri, M. Ishida, and K. Taneda, *Photochem. Photobiol.* **1992**, *56*, 399–402.

Matsuoka, T., Y. Murakami, and Y. Kato, *Photochem. Photobiol.* **1993,** *57*, 1042–1047.

Matsuoka, T., D. Tokumori, H. Kotsuki, M. Ishida, M. Matsushita, S. Kimura, T. Itoh, and G. Checcucci, *Photochem. Photobiol.* **2000**, *72*, 709–713

Naitoh, Y., and R. Eckert, In: *Cilia and Flagella* (M. A. Sleigh, ed.), Academic Press, London. **1974**, 305–352.

Redepenning, J., and N. Tao, *Photochem. Photobiol.* **1993**, *58*, 532–535.

Saimi, Y., and B. Martinac, *J. Membrane Biol.* **1989**, *112*, 79–89.

Savikhin, S., N. Tao, and P.-S. Song, W. Struve, *J. Phys. Chem.* **1993**, *9*, 12379–12386.

Song, P.-S. *Annu. Rev. Biophys. Bioengin.* **1983**, *12*, 35–68.

Song, P.-S. J. Photosci. **1999**, *6*, 37–45.

Song, P.-S., D.-P. Haeder, and K. L. Poff, *Arch. Microbiol.* **1980a**, *126*, 181–186.

Song, P.-S., D.-P. Haeder, and K.L. Poff, *Photochem. Photobiol.* **1980b**, *32*, 781–786.

Song, P.-S., I.-H. Kim, S. Florell, N. Tamai, T. Yamazaki, and I. Yamazaki, *Biochim. Biophys. Acta.* **1990**, *1040*, 58–65.

Spitzner, D., G. Hofle, I. Klein, S. Pohlan, D. Ammermann, and L. Jaenicke, Tetrahedron Lett. **1998**, *39*, 4003.

Tao, N., M. Orlando, J.-S. Hyon, M. Gross, and P.-S. Song,. *J. Am. Chem. Soc.* **1993**, *115*, 2526–2528

Tartar, V. *The Biology of Stentor*, Pergamon Press, New York. **1961.**

Tartar, V. *J. Exp. Zool.* **1972**, *181*, 245–252.

Walerczyk, M., and S. Fabczak, *Photochem. Photobiol.* **2001a**, *76*, 829–836.

Walerczyk, B. and S. Fabczak, *Acta Protozool.* **2001b**, *40*, 153–157.

Walerczyk, M., H. Fabczak, and S. Fabczak, *Photochem. Photobiol.* **2003**, *77*, 339–343.

Weiner, L., and Y. Mazur, *J. Chem. Soc. Perkin Trans.***1992,** *2*, 1439–1442.

Weisz, P. B., *J. Morphol.* **1950**, *86*, 177–184.

Wells, T. A., A. Losi, R. Dai, M. Anderson, J. Redepenning, P. Scott, S.-M. Park, J. Golbeck, and P.-S. Song,. *J. Phys. Chem.* **1997**, *101*, 366–372.

Wood, D. C. *Photochem. Photobiol.* **1976**, *24*, 261–266.

Wood, D. C. *J. Comp. Physiol.* **1982**, *146*, 537–550.

Wynn, J. L., I.-H. Kim, N. Tao, R. Dai, P.-S. Song, and T. M. Cotton,. *J. Phys. Chem.* **1995**, *99*, 2208–2213.

Yamazaki, T., N. Ohta, I. Yamazaki, and P.-S. Song, *J. Phys. Chem.* **1993a**, *97*, 7870–7875.

Yamazaki, T., I. Yamazaki, Y. Nishimura, R. Dai, and P..-S. Song, *Biochim. Biophys. Acta* **1993b**, *1143*, 319–326.

Yang, K. C., R. K. Prusti, P.-S. Song, M. Watanabe, and M. Furuya, *Photochem. Photobiol.* **1986**, *43*, 305–310.

Zagal, J. H., and C. Paez, *Electrochim. Acta* **1989**, *34*, 243–247.

21
The Antirepressor AppA uses the Novel Flavin-Binding BLUF Domain as a Blue-Light-Absorbing Photoreceptor to Control Photosystem Synthesis

Shinji Masuda and Carl E. Bauer

21.1
Overview

Visible light that is captured during photosynthesis has an important biological role as a cellular energy source. Shorter wavelength ultraviolet light also has detrimental affects on DNA. Because of the importance of these and other cellular processes, cells have evolved mechanisms of perceiving and responding to light to control various physiological responses. To date, three different types of flavin-binding domains have been described that function as photoreceptors by perceiving the intensity of blue-light. These photoreceptors are widely distributed among photosynthetic and non-photosynthetic prokaryotes and in eukaryotes. One class is the plant phototropins Phot1 and Phot2. These chromoproteins are photoreceptor kinases that contain two FMN-binding LOV domain as chromophore modules (Briggs and Christie 2002; Crosson et al., 2003). Phototropins control a number of events in plant cells including phototropism, chloroplast relocation, and stomatal opening. A second class of flavin-containing photoreceptors are FAD binding cryptochromes (Sancar, 2003; Lin 2000; Cashmore, 2003). Blue-light absorption by the flavin in plant cryptochromes mediates inhibition of hypocotyl cell elongation, promotion of cotyledon/leaf expansion, and regulation of flower development as well as phasing of circadian rhythms. Cryptochromes also mediate circadian rhythms in *Drosophila* and mice, and possibly humans, in response to blue-light (Cashmore, 2003).

A third type of flavin-containing photoreceptor has recently been described that controls blue-light regulation of photosystem synthesis in the purple bacterium *Rhodobacter sphaeroides* (Masuda and Bauer, 2002), as well as the blue-light avoidance response of the unicellular alga *Euglena gracilis* (Iseki et al., 2002). The photoactive flavin-binding domain of this class of photoreceptors is designated the "BLUF domain" for "sensor of **b**lue-**l**ight **u**sing **F**AD" (Gomelsky and Klug, 2002). Computer-aided similarity searches have indicated that the BLUF domain is present in proteins

Handbook of Photosensory Receptors. Edited by W. R. Briggs, J. L. Spudich

ISBN 3-527-31019-3

from various photosynthetic and nonphotosynthetic species. These proteins were shown to have a homology only at their FAD-binding domains, suggesting that this domain constitutes a new blue-light "input" domain used to control different cellular processes in response to blue-light.

In this review, we first summarize the discovery and biological activity of AppA, a BLUF-domain-containing photoreceptor from the purple bacterium *Rhodobacter sphaeroides*. We also discuss studies on the mechanism of the photocycle that is exhibited by the BLUF domain in AppA. Finally, we review studies recently undertaken on BLUF-domain-containing proteins from other species. A comprehensive description of the BLUF-domain-containing PAC protein from *E. gracilis* is also covered in Chapter 22 by Watanabe.

21.2
Oxygen and Light Intensity Control Synthesis of the Bacterial Photosystem

In most species of purple photosynthetic bacteria, photosynthesis is carried out only under anaerobic growth conditions. When oxygen is available, these organisms repress synthesis of their photosystem, and instead, grow using aerobic respiration as an energy source. In addition to oxygen regulation of photosystem synthesis, a classic study by Cohen-Bazire et al. (1956) demonstrated that light intensity has a regulatory function in which cells grown in the presence of high light synthesize less photosystem than do cells grown with low light. Interestingly, the process of photosynthesis, which generates reducing power, is itself not a signal for light regulation. Specifically, mutants that are defective in the synthesis of a functional photosystem still control the level of photopigments in response to alteration in light intensity (Yurkova and Beatty, 1996). In addition, Shimada and co-workers (Shimada et al., 1992) demonstrated that blue-light (around 450 nm) causes maximal repression of photosystem synthesis under semi-aerobic conditions in *R. sphaeroides*, implicating the existence of a blue light-absorbing photoreceptor controlling photosystem synthesis.

In *R. sphaeroides* there are two interacting proteins that appear to have a central role in coordinating both oxygen and light regulation of photosystem synthesis. The protein PpsR (also called CrtJ in other species) is a DNA-binding transcription factor that represses photosynthesis gene expression in response to the presence of molecular oxygen (Gomesky and Kaplan, 1995a; Penfold and Pemberton, 1991, 1994; Ponnampalam et al., 1995). The second protein is AppA that functions as both an oxygen and blue light-regulated antirepressor of PpsR (Braatsch et al., 2002; Gomesky and Kaplan, 1995b, 1997; Masuda and Bauer, 2002). The next two sections of this review cover the oxygen- and light-regulating activities of PpsR and AppA, respectively.

21.2.1
PpsR is a DNA-binding Transcription Factor that Coordinates both Oxygen and Light Regulation

Numerous in vitro studies have been undertaken with purified PpsR from *R. sphaeroides* and with the homologous CrtJ protein from *R. capsulatus*. These studies established that PpsR/CrtJ binds to the target DNA-sequence TGT-N_{12}-ACA positioned in the regulatory regions of several CrtJ/PpsR target genes (Gomesky and Kaplan, 1995a; Penfold and Pemberton, 1994; Ponnampalam et al., 1995). Gel filtration experiments also demonstrate that PpsR/CrtJ exists as stable tetramer in solution (dimer of dimer), and that PpsR/CrtJ exhibits cooperative binding to two target DNA-binding motifs (Elsen et al., 1998; Gomesky et al., 2000; Masuda and Bauer, 2002; Ponnampalam et al., 1998). Binding of PpsR/CrtJ is redox regulated with significantly higher binding affinity under oxidizing conditions (in the presence of oxidant such as ferricyanide or high levels of oxygen) than under reducing conditions (10 mM β-mercaptoethanol) (Masuda and Bauer 2002; Ponnampalam et al., 1998; Ponnampalam and Bauer 1997). Additional studies indicated that purified PpsR/CrtJ does not contain any cofactors but does have redox-active cysteines and that reversible oxidation and reduction of these cysteines are necessary for controlling its DNA-binding activity (Masuda and Bauer, 2002; Masuda et al., 2002). These results established that PpsR/CrtJ is a DNA-binding repressor of photosynthesis genes and that it has a redox-sensing capability. Surprisingly, deletion of PpsR from *R. sphaeroides* affects both oxygen and light control of photosystem synthesis suggesting that this protein also has a role in blue-light regulation of photosystem synthesis (Gomesky and Kaplan, 1997). This result was unexpected given that isolated PpsR/CrtJ has no identifiable light-absorbing chromophores.

21.2.2
Discovery of AppA, a Redox Responding, Blue Light Absorbing, Antirepressor of PpsR

Significant progress in understanding the molecular mechanism of oxygen and light regulation of photosystem synthesis was made upon discovery of AppA from *R. sphaeroides*. The *appA* gene was initially identified by Gomelsky and Kaplan (1995a) through a genetic screen for mutants that failed to activate the expression of the *puc* operon that encodes the light-harvesting antenna complex in *R. sphaeroides* (the gene designation of *appA* is based on: **a**ctivation of **p**hotopigment and ***p**uc* expression **A**). The *appA* gene encodes a 49-kDa protein with 450 amino acids that initially exhibited no obvious similarity to proteins in genetic databases. The phenotype of an *appA* null-mutant strain was a striking photosynthetically incompetent cell that failed to synthesize photopigments. In another study, the same authors identified several secondary suppressor mutations that allow the mutant strain to grow photosynthetically. These suppressors all mapped to the *ppsR* gene encoding the photosystem repressor PpsR (Gomesky and Kaplan, 1997). One important observation was that the *appA-ppsR* double mutant strain showed a phenotype identical to that of the *ppsR* deletion strain (Gomesky and Kaplan, 1997). This observation indicated that AppA is not an actual activator, but instead functioned as an antirepressor of PpsR.

Inspection of AppA's primarily structure indicates that it has at least two distinct domains, a C-terminal Cys-rich domain containing the motif Cys-X5-Cys-Cys-X4-Cys-X6-Cys-Cys, and an N-terminal domain that binds the flavin FAD (Gomesky and Kaplan, 1998; Masuda and Bauer, 2002). In vitro experiments showed that purified full-length reduced AppA had an ability to break a disulfide bond of PpsR and that reduction of the disulfide bond causes a 5-fold reduction in the DNA binding affinity of PpsR (Figure 21.1A) (Masuda and Bauer, 2002). Presumably, the Cys-rich C-terminal region of AppA is responsible for reduction of the disulfide bond in PpsR since it is this region that is necessary for redox-dependent antirepressor activity (Gomesky and Kaplan, 1995b). A recent in vivo study supports this model by showing that deletion of the N-terminal FAD-binding domain in AppA affects blue light-but not redox-dependent protein function (Braatsch et al., 2002; Gomesky and Kaplan, 1998).

In addition to disulfide-reducing activity, AppA has another interesting means of inhibiting the DNA-binding activity of PpsR. Gel filtration analysis indicated full length AppA is capable of forming a complex with PpsR where it is capable of converting $PpsR_4$ from an active tetramer into an inactive dimer (Figure 21.1B). The DNA-binding activity of PpsR is greatly reduced (>10-fold) when present as a $PpsR_2$AppA co-complex (Masuda and Bauer, 2002). The $PpsR_2$AppA co-complex can form irrespective of the redox state of the AppA and PpsR cysteine sulfhydryls. However, the formation of the $PpsR_2$AppA co-complex is inhibited by irradiation of AppA with blue-light. The mechanism of light inhibition of AppA antirepressor activity appears to involve a light-mediated shape change that occurs in AppA upon illumination of the flavin (Kraft et al., 2003; Masuda and Bauer, 2002). Thus, AppA is capable of regulating the DNA-binding activity of PpsR via two independent mechanisms (1) a reduction of Cys sulfhydryls in PpsR that reduces binding ~5-fold and (2) by the light regulated interaction of AppA to PpsR which causes a 10-fold reduction in DNA binding activity.

Even though redox regulation and light-mediated direct inhibition of PpsR activity by AppA appear to involve independent mechanisms, there is an interesting possibility that there may be convergence of redox- and light-mediated antirepressor activities of AppA. Specifically, blue light-dependent inhibition of photosynthesis gene expression is not observed *in vivo* when *R. sphaeroides* cells are grown under strictly anaerobic conditions (<3.5 μM dissolved oxygen) (Braatsch et al., 2002). This result indicates that some level of oxygen is needed for AppA to impart blue-light antirepressor activity. Even though the nature of the redox state of AppA is not known in anaerobically grown cells, it is possible that FAD bound to AppA is in a reduced state in anaerobically grown cells. If this were to occur, then AppA would have no intrinsic blue light-sensing capability given that reduced FAD ($FADH_2$) does not absorb light in the visible spectrum. Further experiments are needed to test this hypothesis.

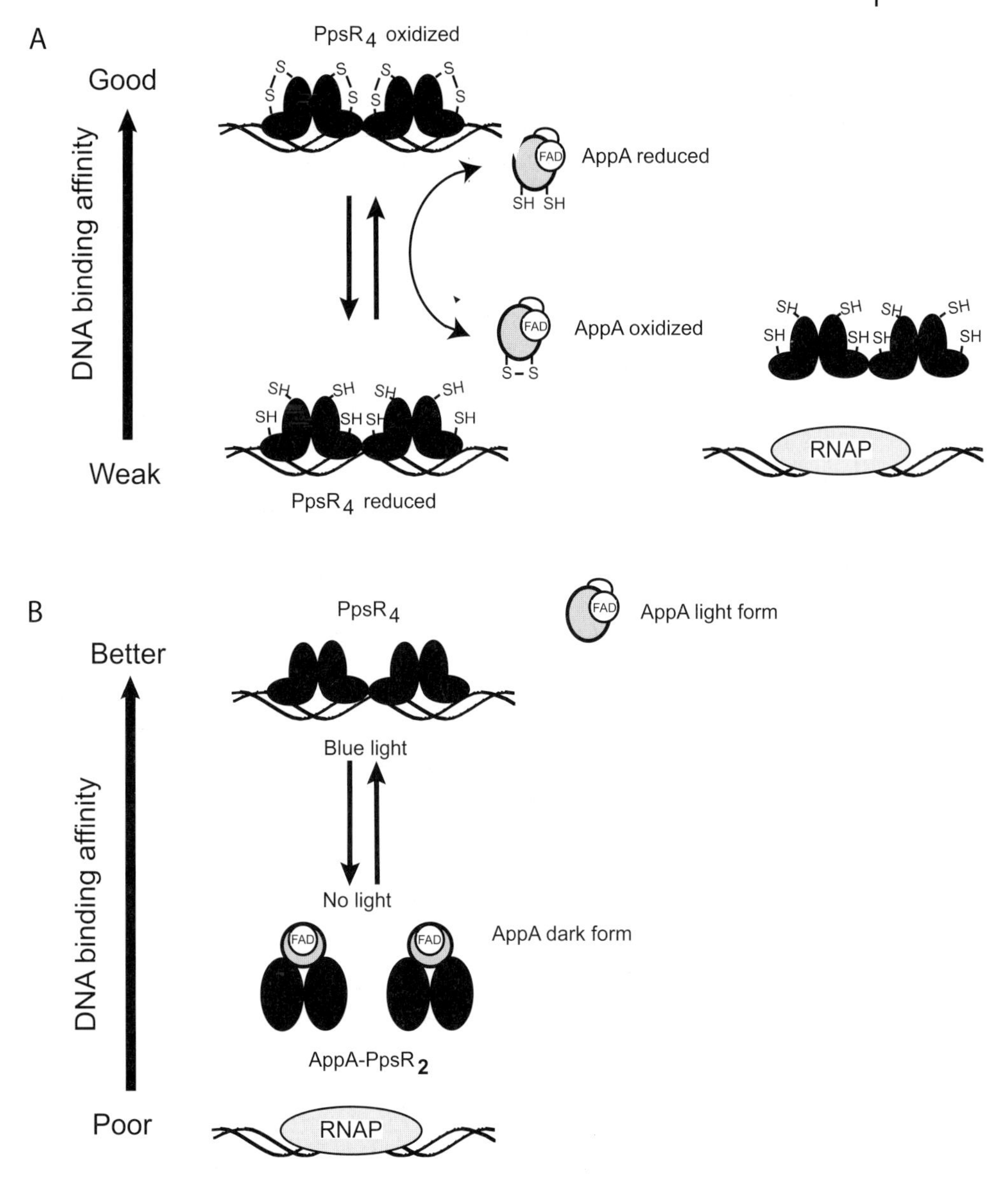

Figure 21.1 Current working model of the action of AppA and PpsR that controls photosynthesis gene expression in *R. sphaeroides* in response to oxygen- (redox) and blue light. AppA has two activities, one is to reduce a disulfide bond in $PpsR_4$ (A) and the other is to convert PpsR from a tetramer into an AppA-$PpsR_2$ complex that is incapable of binding DNA (B). The later reaction is inhibited by blue light excitation of the flavin FAD that is bound to AppA.

21.3
Mechanism of the BLUF Photocycle in AppA

In vitro data indicate that light-adapted AppA is functionally inactive as an antirepressor and that light-excited AppA needs more than 20 min to recover antirepressor activity when shifted to dark conditions (Masuda and Bauer 2002). Light excitation of the flavin in AppA results in a unique photocycle that is characterized by a very long-lived red-shift in the flavin absorption spectrum (Figure 21.2). The light-induced spectral shift decays back to the ground state over a remarkably long 50-min period with a calculated half time of approximately 10 min (Masuda and Bauer 2002) . A fluence-response curve generated from a series of light-minus-dark different spectra shows induction of the photocycle at a low-light intensity of 5 µmol m^{-2} s^{-1} with saturation of the cycle occurring at a moderate to high light intensity of 400 µmol m^{-2} s^{-1}. These values correlate with physiological light intensities needed for repression of photosystem synthesis (Masuda and Bauer 2002). This correlation suggests that the light-induced spectral red shift corresponds to a signaling state which converts AppA from a functional antirepressor in the dark form into an inactive state in the light form (Masuda and Bauer 2002).

Gel-filtration chromatographic analysis indicates that the BLUF domain of AppA undergoes a significant change in its elution profile upon light excitation. This result also indicates that there is a change in the protein's stable conformation and/or protein dynamics caused by light excitation of the bound chromophore (Kraft et al., 2003). Flash excitation experiments also demonstrate that light-induced formation of the stable red-shifted form involves a biphasic reaction with a fast phase having a rate constant of $>10^6$ s^{-1} followed by a slow phase at 690 s^{-1} (Kraft et al., 2003). As a result, it can be concluded that there are two phases during formation and of light-adapted

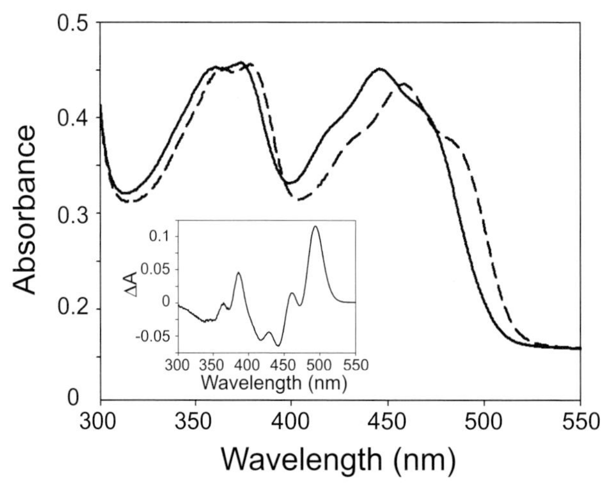

Figure 21.2 UV-visible absorption spectra of dark-adapted (solid line) and light-excited (dotted line) AppA. Inset: Light-minus-dark difference spectrum of AppA. The light-excited spectrum decays back to the dark state over a 30-min period.

AppA, which affect both the local environment of the FAD and the extended protein conformation (Kraft et al., 2003). It is likely that the first phase involves a local change in the orientation of the flavin, and the slow "second phase" is related to changes in the global protein moiety. The kinetics for recovery of the light-adapted AppA BLUF domain to its dark state is also bi-exponential having a fast phase which appears to be localized at the flavin, while the slow phase affects residues not directly interacting with the flavin (Kraft et al., 2003).

A major question being investigated is the mechanism of the AppA photocycle that leads to both flavin and protein conformational changes. In the flavin-based photoreceptor phototropin, a light induced structural change of the flavin in the LOV domain is accompanied by the formation of a transient covalent flavin-cysteine adduct (Crosson and Moffat, 2001; Iwata et al., 2002; Salomon et al., 2000; Salomon et al., 2001; Swartz et al., 2002). Cryptochromes also contain flavin as a chromophore, and have a light-induced electron transfer event during the photocycle (Lin, 2002, Cashmore, 2003; Giovani et al., 2003; Sancar, 2003). Little is known about how light induced alterations in these flavin-containing photoreceptors actually result in a change of protein tertiary structure.

Regarding the mechanism of the AppA photocycle, the observed light-induced red shift of the UV-vis spectrum is very different from spectral changes seen with either phototropin or cryptochrome. However, a similar red-shift in flavin-absorption has been observed upon binding of substrates to flavoprotein enzymes (Axley et al., 1997; Gopalan and Srivastava, 2002; Jung et al., 1999; Pellett et al., 2001; Zhou and Swenson, 1996). Several mechanisms have been proposed to account for substrate-induced spectral red-shifts including alterations in stacking interactions of an aromatic amino acid ring with the flavin ring (Bjornberg et al., 1997; Rowland et al., 1998), interactions of keto groups with the p-system of the flavin ring (Gopalan and Srivastava, 2002), and H-bonding interactions at the O(=C4) or N5 positions of the flavin ring (Axley et al., 1997).

The similarity of the spectral shift exhibited by the AppA BLUF domain led Kraft et al. (2003) to investigate whether the AppA photocycle may involve an interaction of the flavin with an aromatic amino acid. Site-directed replacement of Tyr21 in AppA (which is conserved in all BLUF domains) with Phe, indeed abolished the photocycle of AppA (Kraft et al., 2003). Circular dichroism (CD) spectra of wild-type and the Y21F mutant protein showed that only the Phe substitution resulted in a local change in the flavin microenvironment without any significant peptide secondary structural changes. Nuclear magnetic resonance (NMR) analysis also suggested Tyr21 forms π–π[MSOffice2] stacking interactions with the isoalloxazine ring of FAD, and that stacking interactions were lost when Tyr was replaced with Phe. Light excitation of the flavin also leads to quenching of FAD fluorescence, as well as perturbation of π–π stacking interactions. Collectively, these results suggest that proper π–π stacking interactions between Tyr21 and the flavin ring are necessary for the photocycle of AppA (Kraft et al., 2003). The decay of the AppA photocycle to the ground state in deuterium (D_2O) is also two times slower than what is observed in H_2O (S. Masuda and T. Ono, unpublished result), indicating that there are H-bond rearrangements and/or proton transfers in the BLUF domain upon photo-excitation. From these observations, it was suggested that strengthening of H-bonding interactions with the isoalloxazine ring of FAD might be responsible for the observed changes in the visible region of the spectrum upon photochemical excitation of AppA (Kraft et al., 2003). This mechanism can also account for differing degrees of light-induced FAD fluorescence quenching caused by a change in the relative orientation of the bound flavin (Kraft et al., 2003).

Recently, light-induced Fourier transform infrared (FTIR) difference spectroscopy of BLUF domain was performed (Laan et al., 2003; Masuda et al., 2004). The FTIR spectrum of the BLUF domain shows alteration of a dominant absorption band derived from C(4)=O stretching vibrations of the flavin isoalloxazine ring (Laan et al., 2003; Masuda et al., 2004). This band is down shifted to a lower wave-number upon photo-excitation, indicating a weakening of the C(4)=O bonding by light illumination. This result suggests that light absorption induces formation of an additional H-bond at O(=C4) of the flavin ring (Masuda et al., 2004). In addition, recent FTIR analysis of a cyanobacterial BLUF domain also gave a similar band shift upon light excitation (Masuda et al., 2004). In fact, ab initio molecular orbital calculation for isolumazine as a model for flavin suggests that formation of a hydrogen bond at O(=C4) leads to the observed visible flavin red-shift (Neiss et al., 2003). Together with these observations, it may be concluded that an additional hydrogen bond is formed between O(=C4) of FAD with a specific amino acid or peptide backbone upon photoexcitation causing the observed absorption red shift in the visible region spectrum (Masuda et al., 2004).

Laan et al. (2003) also described an alternative model suggesting that there is reversible intramolecular proton transfer occuring from N(3)-H of the flavin ring to the anionic Tyr21, during the initial photochemistry of the BLUF-domain photocycle in AppA. If one assumes this model, the FTIR spectral bands for the light state (positive bands in light-minus-dark spectrum) would be insensitive to deuteration since the isoalloxazine ring of FAD has no exchangeable hydrogen after deprotonation. However, this may not be the case since recent data shows that some of the corresponding bands in FT-IR spectrum of cyanobacterial BLUF domain are actually sensitive to deuteration (Masuda et al., 2004).

Several additional questions remain to be solved regarding the BLUF photocycle. For example, how does blue-light absorption result in perturbation of π–π stacking and hydrogen bond characteristics of the flavin ring? Why do small changes in the flavin microenvironment cause global changes of the BLUF-domain peptide conformation? Clearly additional structural and spectroscopic analyses of BLUF domains are necessary for understanding of the molecular mechanism of the photocycle in this class of proteins.

21.4
Other BLUF Containing Proteins

As discussed above, a purified 156 amino acid amino-terminal segment of AppA exhibits spectroscopic properties identical to that of full-length AppA. This result indicates that the amino terminal region of AppA contains the FAD-binding BLUF domain, which is itself photoactive (Gomelsky and Klug, 2002; Kraft et al., 2003). A search of genome-sequence databases for proteins that have homology to the BLUF domain of AppA has resulted in the identification of 28 additional proteins having one or more copies of the BLUF domain. These proteins are present in a variety of species, ranging from photosynthetic and non-photosynthetic bacteria to the single celled eukaryote *Euglena gracilis* (Iseki et al., 2002; Gomelsky and Klug, 2002 ; S. Masuda unpublished data] (Figure 21.3). Other than the presence of a BLUF domain, the identified proteins are of variable length and exhibit no identifiable sequence similarity outside that of the BLUF domain.

Many BLUF-containing proteins are of a small molecular weight (~20 kDa) and are composed of a single BLUF domain plus an additional 30–70 amino acids. The small size of these proteins suggests that they may be involved in protein-protein interactions, which provide blue-light dependent control of signaling pathways. Recently, the small BLUF-containing protein Slr1694, from the cyanobacterium *Synechocystis* sp. PCC 6803 has been characterized by a reverse genetic approach (Okajima et al., 2003). Slr1694 is a 17 kDa protein of 151 amino acids that contains a single BLUF domain. Gene-disruption analysis indicated that the Slr1694 protein was essential for positive phototaxis of this bacterium (Okajima et al., 2003). As is the case for AppA, the purified Slr1694 protein is also photoactive, showing a red-shift in absorption upon blue-light irradiation (Masuda et al., 2004). A homolog of *slr1694* is also present in the sequenced genome of the cyanobacterium, *Thermosynechococcus elongates* (tll0078, Figure 21.3) suggesting that the Slr1694 protein may function as a blue-light sensor for phototaxis in a number of cyanobacteria (Masuda et al., 2004). In addition to AppA, the genome of *R. sphaeroides* contains two additional small BLUF-related proteins. Interestingly, one of BLUF-related proteins (orf193) is located near putative taxis-related genes on the chromosome (data not shown). From the gene arrangement on the chromosome, the orf193 is speculated to be a photoreceptor for a blue-light motility response exhibited by this organism (Laan et al., 2003).

In addition to the small BLUF-containing proteins, there is a class of larger BLUF containing proteins such as AppA from *R. sphaeroides*, PAC from the alga *E. gracilis*, YcgF from *E. coli*, and an unannotated ORF from *Magnetococcus* species (Masuda and Bauer, unpublished analysis of BLUF domains in genome databases). In these larger proteins, the BLUF domain is located at variable positions relative to the amino terminus. This variability suggests that this class of proteins may have a common blue light input sensor (the BLUF domain) coupled to different output domains that are responsible for controlling different cellular processes in response to blue light. For example, the BLUF domain in *E. coli* YcgF and the BLUF domains from the *Magnetococcus* ORF are linked to GGDEF and/or EAL domains that are involved in functional synthesis and/or hydrolysis of the cyclic nucleotides, bis-(3'5')-cyclic diguany-

```
Rs_AppA        16  LVSCCYRSL----AAPDLTLRDLLDIVETSQAHNARAQLTGALFYSQG--VFFQWLEGRPAAVAEVMT-HIQRDRRHSNVEILAEEPIAKRRFA-GWHMQ
Rs_6138c193     4  LVSLTYRSR----VRLADPVADIVQIMRASRVRNLRLGITGILLYNGV--HFVQTIEGPRSACDELFR-LISADPRHQEILAFDLEPITARRFP-DWSMR
Rs_5263c188     5  LLTVSYRSR----VCLADPVADLLGIVQASRVRNLRLGITGILLYNGF--HFVQTIEGPQNACRELFA-RISEDPRHSDIVAFGVRKIERRFP-DWSMR
Th_tll0078      3  LHRLIYLSC----ATDGLSYPDIRDIMAKSEVNNLRDGITGMLCYGNG--MFLQTLEGDRQKVSETYA-RILKDPRHHSAEIVEFKAIEERTFI-NWSMR
Sy_slr1694      3  LYRLIYSSQ----GIPNLQPQDIKDILESSQRNNPANGITGLLCYSKP--AFLQVLEGECEQVNETYH-RIVQDERHHSPQIIECMPIRRRNFE-VWSMQ
Sc_ORF          3  LYRLIYVSQ----AIAELEYRDIIEILQKSEFNNRRAGITGMLAFGDL--MFLQILEGSRRAISETYN-RILLDSRHTNAELIEFSEIDQRDFG-IWSMK
Ec_YcgF         2  LTTLIYRSH----IRDDEPVKKIEEMVSIANRRNMQSDVTGILLFNGS--HFFQLLEGPEEQVKMIYR-AICQDPRHYNIVELLCDYAPARRFG-KAGME
Ec_CFT073      28  LTTLIYRSH----IRGDEPVKKIEEMVSIANRRNMRSDVTGILLFNGS--HFFQLLEGPEEQVQMIYR-AICQDPRHYNIVELLCDYAPARRFG-KAGME
Ce_AP005222    16  LKYLVYTSI----ASPDLTEESVQSMLTTAREHNASAEITGLLLFRTG--TFIQFLEGAAEEIDQLIS-RIRLDDRHQDVRVIIEEPVDHRRFP-DWRMG
So_ORF          2  LFELLYTSI----APAGLSDTELLSILEKARAKNQALDITGMLVYHN--REIMQILEGEELKVRGLFQ-TIFQDERHTSVEVFYQGNIDHRAFS-EWSMA
Cv_AE016922     3  LQHLIYVST----ARRELDNADIEALLESCVRHNLRNDISGVLLYGPG--NFIQVLEGEADAVAETYQ-RILDDPLHHNLIEILLENIERRSFA-DWRMG
Pi_BX294134     2  LTELVYCSV----ATQAMSVEEIDSILEQSRENNARLNVTGILLHSGKTREFMQLLEGEQECIEKLMQ-VISNDERHTCVDMMYHDRIESRNFE-DWSMA
Xa_AE011973     6  LHALAYVSE----VRRPWTLGEIDRLLVSSSALNAQNAITGVLLYDGQ--RFFQYIEGAPAPLRDIYD-RIKRSSRHDLVAELYNDQITERYFP-NWQMA
Xa_AE011847    16  LQALVYVSE----VRRPMSMGEVDRLLVSSSAHNSLAAITGVLLYDGD--RFFQYFEGPPAGVQHTYD-RIKRSSRHHLIAEVYNGSIVERYFP-DWQMA
Rp_3782        13  LFRVLYVSRNRIEGPLDRLSSEIAGILATSRINNARAGITGALIVNSG--FFAQVLEGPRSAVEKVFE-TIQQDHRHSVVQVLLASPITVRAFP-NWAMA
Rp_552          7  LYRLLYVSENRIEGDVAHVTREIEGILATSRRNNDALGLTGALIFNCG--FFAQVLEGPRETVERIFE-VIRHDPRHNKIQVMSAAEVKDREFP-NWTMA
Rp_1328         5  LYRCVYYSKNLIAGGPARIAEEIESILAAARRNNPTLQVTGALVFNRG--LFAQVLEGPCPSVESLFE-KISRDERHGDIQVLAFRAVPERLFS-NWSMA
Kp_c593         2  LTTIIYRSH----ICDNVSFKSIEAMVARANERNGQADVTGILLFNGT--HFFQLIEGPEEKVQDIYQ-HICQDPRHYNLVELLCDYAPSRRFG-KVGME
Kp_c674         2  LTTLIYRSQ----VHPDRPPVDIDALVHRASSKNLPLGITGILLFNGL--QFFQVLEGTEEALESLFS-EIQSDPRHRDVVELMRDYSAYRRFH-GTGMR
Mc_1535         1  MIQLVYVSS----SVRLFTKDELIGILEVSRTHNQSKGITGILLYANG--NFLQLLEGEKKDVDALYE-KITQDKRHRGCIKIFAKDATSRLFP-DWSMG
Mc_1552       983  LKFLIYISR----PKFEITEEGIQAILKQSREFNLKSGITGYLFYLKG--AFAQYLEGDEEPLSQLYQ-SIQKDNRHSDLKIIAEGDLPTRLFA-GWEMG
Mc_2761         5  IFHLIYVSK-----TDVLKPGDLGDILFAARKNNPPLNLTGVLIYNSD--FFFQLLEGRASNVEKMYK-IICDDPRHIACKTLHTYTDYHRKFT-QWSMG
Mm_648          2  LVRLLYASR----ARQPLDADYLDAILDTSLEQNLRHGITGVLCHSNT--VFLQALEGDREAVSRLFL-NIGRDDRHHHVVLLSYEEIAERAFA-GWSMG
Mm_3675         2  LRCIGYVSD---IVNINNAKNDIANILSATAFNDGHNITGMMAYYDG--CILQFIEGESQVIGELYE-RIGNDSRHRDLVKIYDAEIERRSFD-GWKMA
Ac_ADP1         3  YVRLLYVSK--TAKQHHEIKHDIMEILTTAIRFNSINKIKGVLYYGNG--YFVQCLEGEEKVIYDVFYNRIIKDSRHQNCEILSCKAVNELLFQ-KWSMK
Un_AE008919    22  LIHLAYSST----QTRDMKAADIISLLTEVRGLNESRNISGLLLHKN--QSFFQVIEGSRARVQETFN-NIMRDQRHEGVEVLFDEPLEAREFS-NWQMG
Un_AE008920     9  LFHLGYVSI----ETSDLGSDGMVSLLDEARRTNAERNLTGILLHRD--RSFYQVLEGEESEVRRTFE-GIGKDKRHKAIDVLFEGHVEQREFA-DWQMG
Eg_PACa1       55  LRRLMYLSA--STEPEKCNAEYLADMAHVATLRNKQIGVSGFLLYSSP--FFFQVIEGTDEDLDFLFA-KISADPRHERCIVLANGPCTGRMYG-EWHMK
Eg_PACb1       56  LRRLMYLSK--STNPEECNPQFLAEMARVATIRNREIGVSGFLMYSSP--FFFQVIEGTDEDLDFLFA-KISADPRHERCIVLANGPCTGRMYG-DWHMK
Eg_PACa2      467  LITLTYISQ----AAHPMSRLDLASIQRIAFARNESSNITGSLLYVSG--LFVQTLEGPKGAVVSLYL-KIRQDKRHKDVVAVFMAPIDERVYGSPLDMT
Eg_PACb2      471  LTTLTYISQ----ATRPMSRLDLSAIMRTATRRNAQQSITGTLLHVNG--LFVQTLEGPKDAVVNLYL-RIRQDPRHTDVTTVHMAPLQERVYPSEWTLT
```

Figure 21.3 Alignment of deduced amino-acid sequences of BLUF domain containing proteins from the following species: Rs, *R. sphaeroides*; Th, *Thermosynechococcus elongates*; Sy, *Synechocystis* sp. PCC6803; Sc, *Synechococcus* sp. PCC7002; Ec, *E. coli*; Ce, *Corynebacterium efficiens*; So, *Shewanella oneidensis*; Cv, *Chromobacterium violaceum* ATCC12472; Pi, *Pirellula* sp. strain 1; Xa, *Xanthomonas axonopodis*; Rp, *Rhodopseudomonas palustris*; Kp. *Klebsiella pneumoniae*; Mc, *Magnetococcus* MC-1; Mm, *Magnetospirillum magnetobacterium*; Ac, *Acinetobacter* sp. ADP1; Un, uncultured proteobacterium; and Eg, *E. gracilis*. Conserved and similar amino acids are indicated by black and gray shading, respectively. The alignment was performed using a previous-reported alignment of BLUF domains (accession number ALIGN_000426 in the data base of European Bioinformatics Institute) (Gomelsky and Klug 2002) with additional amino-acid sequences from following gene products: *E. coli* CTF073 ORF (accession number AE016759), *C. efficiens* ORF (accession number AP005222), *C. violaceum* ORF (accession number AE016922), *Pirellula* sp. ORF (accession number BX294134), *X. axonopodis* ORFs (accession number AE011973 and AE011847), and uncultured strain ORF (accession number AE008920).

late, c-di-GMP, and cyclic adenosine 3'5' monophosphate (Tal et al., 1998; Ausmees et al., 2001; Galperin et al., 2001; Pei and Grishin, 2001). These cyclic nucleotides are thought to function as regulatory signaling molecules in bacteria (Ross et al., 1987, 1991; Chang et al., 2001; Sasakura et al., 2002). Thus, the BLUF domain in these proteins may function as a regulatory domain that modulates the catalytic activity of cyclic-nucleotide enzymatic activity in response to blue light. A similar scenario occurs in *E. gracilis* which has two BLUF containing PAC proteins that control the blue-light photoavoidence response. The PAC proteins each contain two BLUF domains as well as an adenyl cyclase domain (Figure 21.3). The adenyl cyclase enzymatic activity of PAC proteins require blue-light irradiation demonstrating that the BLUF domains in PAC have a role in controlling enzymatic activity in response to blue-light (Iseki et al., 2002). Furthermore, the well-characterized blue light-induced avoidance response exhibited by *E. gracilis* is absent in cells that have reduced or absent levels of PAC, indicating a direct role of light absorption by BLUF domain(s) in triggering the blue light avoidance response.

21.5 Concluding Remarks

The redox- and blue light-antirepressor activity of AppA is the first protein identified that controls transcription in response to alterations in both redox state and blue-light intensity. Modulating the oxidation and reduction state of AppA cysteine thiol groups appear to be the mechanism of redox sensing, whereas, a shape change in AppA caused by blue light absorption by FAD appears to control the blue-light response. The mechanism of the BLUF photocycle appears to involve light-induced alterations of both π–π stacking interactions and of a hydrogen bonding network of the C(4)=O of the isoalloxazine ring. Additional structural studies with AppA will be required to address finer details of the molecular mechanism of the photocycle reaction of the BLUF domain, which is totally different from what has been described of other flavin containing photoreceptors. Such studies are warranted given the novel features of this photocycle.

Finally, even though the redox- and light-absorption activities of AppA are novel, the use of an antirepressor such as AppA to control the DNA binding activity of PpsR/CrtJ may be a common theme. Recent genomic-sequence data have revealed that both *Rhodopseudomonas palustris* and *Bradyrhizobium* strain ORS278 contain *ppsR* genes adjacent to genes encoding for proteins with homology to the chromophore (bilin) attachment domain of bacteriophytochromes (Giraud et al., 2002). Unlike most bacteriophytochromes, the carboxyl-terminal domain of these bacteriophytochrome-like proteins contains no histidine kinase domain. Instead, these bacteriophytochrome-like proteins contain PAS domains known to be involved in signal perception and/or mediating protein-protein interactions. A mutational loss of the bacteriophytochrome-like gene in *Bradyrhizobium* results in constitutive repression of photosystem synthesis in a manner that is very similar to what is observed upon loss of AppA in *R. sphaeroides* (Giraud et al., 2002). Furthermore, unlike blue light-dependent antirepression of the photosystem synthesis in *R. sphaeroides*, *Rhodopseudomonas* and *Bradyrhizobium* are known to undergo red-light dependent antirepression (740 nm) of photosystem synthesis (Giraud et al., 2002). Thus, *Rhodopseudomonas* and *Bradyrhizobium* may be utilizing a bacteriophytochrome-like protein to function as a red light-absorbing antirepressor of PpsR/CrtJ in the manner similar to that of AppA. The use of different photoreceptors to control PpsR/CrtJ activity in these different species presumably reflects adaptation to different environmental light conditions in nature.

Acknowledgement

S. M. thanks Koji Hasegawa and Taka-aki Ono for helpful discussions and Aaron Setterdahl for careful reading of the manuscript. AppA research for S. M. and C. B. are supported by Special Postdoctoral Researchers Program at RIKEN, and by NIH GM53940-6, respectively.

References

Ausmees, N., Mayer, R., Weinhouse, H., Volman, G., Amikam, D., Benziman, M., and M. Lindberg (2001) *FEMS Microbiol. Lett.* 204, 163–167.

Axley, M. J., Fairman, R., Yanchunas, J., Villafranca, J. J., and J. G. Robertson (1997) *Biochemistry* 36, 812–822.

Bjornberg, O., Rowland, P., Larsen, S., and K. F. Jensen (1997) *Biochemistry* 36, 16197–16205.

Braatsch, S., Gomelsky, M., Kuphal, S., and G. Klug (2002) *Mol. Microbiol.* 45, 827–836.

Briggs WR, J. M. Christie (2002) *Trends Plant Sci.* 7, 204–210.

Briggs, W. R., and E. Huala (1999) *Ann. Rev. Cell Dev. Biol.* 15, 33–62.

Cashmore, A. R. (2003) *Cell* 114, 537–543.

Chang, A., Tuckerman, J. R., Gonzalez, G., Mayer, R., Weinhouse, H., Volman, G., Amikam, D., Benziman, M., and M. Gilles-Gonzalez (2001) *Biochemistry* 40, 3420–3426.

Cohen-Bazire, G., Sistrom, W. R., and R. Y. Stanier (1956) *J. Cell. Comp. Physiol.* 49, 25–68.

Crosson, S., and K. Moffat (2001) *Proc. Natl. Acad. Sci. USA* 98, 2995–3000.

Crosson, S., Rajagopal, S., and K. Moffat (2003) *Biochemistry* 42, 2–10.

Elsen, S., Ponnampalam, S. N., and C. E. Bauer (1998) *J. Biol. Chem.* 273, 30762–30769.

Galperin, M. Y., Nikolskaya, A. N., and E. V. Koonin (2001) *FEMS Microbiol. Lett.* 203, 11–21.

Giovani, B., Byrdin, M., Ahmad, M., and K. Brettel (2003) *Nature Struct. Biol.* 10, 489–490.

Giraud, E., Fardoux, J., Fourrier N., Hannibal, L., Genty, B., Bouyer, P., Dreyfus, B., and A. Verméglio (2002) *Nature* 417, 202–205.

Gomelsky, M., and S. Kaplan (1995a) *J. Bacteriol.* 177, 1634–1637.

Gomelsky, M., and S. Kaplan (1995b) *J. Bacteriol.* 177, 4609–4618.

Gomelsky, M., and S. Kaplan (1997) *J. Bacteriol.* 179, 128–134.

Gomelsky, M., and S. Kaplan (1998) *J. Biol. Chem.* 273, 35319–35325.

Gomelsky, M., Horne, I. M., Lee, H.-J., Pemberton, J. M., McEwan, A. G., and S. Kaplan (2000) *J. Bacteriol.* 182, 2253–2261.

Gomelsky, M., and G. Klug (2002) *Trends Biochem. Sci.* 27, 497–500.

Gopalan, K. V., and D. K. Srivastava (2002) *Biochemistry* 41, 4638–4648.

Iseki, M., Matsunaga, S., Murakami, A., Ohno, K., Shiga, K., Yoshida, K., Sugai, M., Taka-

hashi, T., Hori, T., and M. Watanabe (2002) *Nature* 415, 1047–1051.
Iwata, T., Tokutomi, S., and H. Kandori (2002) *J. Am. Chem. Soc.* 124, 11840–11841.
Jung, Y.-S., Roberts, V. A., Stout, C. D., and B. K. Burgess (1999) *J. Biol. Chem.* 274, 2978–2987.
Kraft, B. J., Masuda, S., Kikuchi, J., Dragnea, V., Tollin, G., Zaleski, J. M., and C. E. Bauer (2003) *Biochemistry* 42, 6726–6734.
Laan, W., van der Horst, M. A., van Stokkum, I. H., and K. J. Hellingwerf (2003) *Photochem. Photobiol.* 78, 290–297.
Lin, C. (2000) *Trends Plant Sci.* 5, 337–342.
Masuda, S., and C. E. Bauer (2002) *Cell* 110, 613–623.
Masuda, S., Dong, C., Swem, D. L., Setterdahl, A. T., Knaff, D. B., and C. E. Bauer (2002) *Proc. Natl. Acad. Sci. USA* 99, 7078–7083.
Masuda, S., Hasegawa, K., Ishi, A., and T. Ono (2004) submitted.
Neiss C., Saalfrank P., Parac, M., and S. Grimme (2003) *J. Phys. Chem. A* 107, 140–147.
Okajima, K., Yoshihara, S., Geng, X., Katayama, M., and M. Ikeuchi (2003) *Plant Cell Physiol.* 44, supple, 162.
Pei, J., and N. V. Grishin (2001) *Proteins* 42, 210–216.
Pellett, J. D., Becker, D. F., Saenger, A. K., Fuchs, J. A., and M. T. Stankovich (2001) *Biochemistry* 40, 7720–7728.
Penfold, R. J., and J. M. Pemberton (1991) *Curr. Microbiol.* 23, 259–263.
Penfold, R. J., and J. M. Pemberton (1994) *J. Bacteriol.* 176, 2869–2876.
Ponnampalam, S. N., and C. E. Bauer (1997) *J. Biol. Chem.* 272, 18391–18396.
Ponnampalam, S. N., Buggy, J. J., and C. E. Bauer (1995) *J. Bacteriol.* 177, 2990–2997.
Ponnampalam, S. N., Elsen, S., and C. E. Bauer (1998) *J. Biol. Chem.* 273, 30757–30761.
Ross, P., Mayer, R., Benziman, M. (1991) *Microbiol. Rev.* 55, 35–58.
Ross, P., Weinhouse, H., Aloni, Y., Michaeli, D., Weinberger-Ohana, P., Mayer, R., Braun, S., de Vroom, E., van der Marel, G. A., van Boom, J. H., and M. Benziman (1987) *Nature* 325, 279–281.
Rowland, P., Bjornberg, O., Nielsen, F. S., Jensen, K. F., and S. Larsen (1998) *Protein Sci.* 7, 1269–1279.
Salomon M, Christie JM, Knieb E, Lempert U, and WR Briggs (2000) *Biochemistry* 39, 9401–9410.
Salomon, M., Eisenreich, W., Dürr, H., Schleicher, E., Knieb, E., Massey, V., Rüdiger, W., Müller, F., Bacher, A., and G. Richter (2001) *Proc. Natl. Acad. Sci. U.S.A.* 98, 12357–12361.
Sancar, A. (2003) *Chem. Rev.* 103, 2203–2237.
Sasakura, Y., Hirata, S., Sugiyama, S., Suzuki, S., Taguchi, S., Watanabe, M., Matsui, T., Sagami, I., and T. Shimizu (2002) *J. Biol. Chem.* 277, 23821–23827.
Shimada, H., Iba, K., and K. Takamiya (1992) *Plant Cell. Physiol.* 33, 471–475.Swartz, T. E., Wenzel, P. J., Corchnoy, S. B., Briggs, W. R., and R. A. Bogomolni (2002) *Biochemistry* 41, 7183–7189.
Tal, R., Wong, H. C., Calhoon, R., Gelfand, D., Fear, A. L., Volman, G., Mayer, R., Ross, P., Amikam, D., Weinhouse, H., Cohen, A., Sapir, S., Ohana, P., and M. Benziman (1998) *J. Bacteriol.* 180, 4416–4425.
Yurkova, N., and Beatty, J. T. (1996) *FEMS Microbiol. Lett.* 145, 221–225.
Zhou, Z., and R. P. Swenson (1996) *Biochemistry* 35, 15980–15988.

22
Discovery and Characterization of Photoactivated Adenylyl Cyclase (PAC), a Novel Blue-Light Receptor Flavoprotein, from *Euglena gracilis*

Masakatsu Watanabe and Mineo Iseki

22.1
Introduction

Photoactivated adenylyl cylase (PAC), a novel blue-light receptor flavoprotein with an intrinsic effector function, was first isolated from the photosensory organelle (paraflagellar body, PFB) of *Euglena gracilis,* a unicellular eukaryotic photosynthetic uniflagellate alga. This flavoprotein was originally shown to mediate one *Euglena* behavioral response namely photoavoidance (the step-up photophobic response) but not a second behavioral response, photoaccumulation (the step-down photophobic response) (Iseki et al., 2002). Here we summarize (1) the story of discovery and characterization of this unique photoreceptor molecule and (2) its additional characterization made thereafter, as well as (3) its future research prospects. All of the above research findings are the result of combined multidisciplinary efforts by investigators with independent minds, different capabilities, and mutual respect.

22.2
Action Spectroscopy

In the late 1980s, Sugai and Watanabe began a collaboration to extend the reported UVA/blue action spectra of *Euglena* photomovements (Checcucci et al., 1976) to the UV-B/C region of the spectrum in order to examine the possible flavoprotein nature of the putative UV-A/blue light receptor(s) in this organism. Individual-cell-level methods for assay of photomovement had been developed by the joint efforts of Watanabe, Aono, Kondo, Takahashi and Kubota (Kondo et al., 1988; Matsunaga et al., 1998). At the beginning of the study we made the embarrassing finding that replacement of an organic culture medium by a simple inorganic medium (Diehn's resting medium) (Diehn, 1969) before the photomovement assay caused the *Euglena* cells to

Handbook of Photosensory Receptors. Edited by W. R. Briggs, J. L. Spudich

ISBN 3-527-31019-3

respond only to step-up signals in a photophobic response. Matsunaga solved this problem by addition of small amount of Hyponex, a commercial plant fertilizer, which caused the cells to respond only to step-down signals in a photophobic response. Subsequent careful examination revealed that ammonium ion is the responsible factor that converts the *Euglena* cells from responding photophobically only to a step-up light signal to cells responding photophobically only to a step-down light signal (Figure 22.1; Matsunaga et al., 1999). On the basis of this methodological finding it became possible to determine reliable action spectra for both of these photophobic responses at the Okazaki Large Spectrograph (Figure 22.2; Matsunaga et al., 1998; Watanabe, 2004).

In the UV-A and visible regions of the spectrum, the shapes of the action spectra were of the so-called UV-A/blue type, originally named the cryptochrome type by Gressel (1979). In the newly studied UV-B/C region, an action peak was found at 270 nm for the step-down response and at 280 nm for the step-up response. The absorption spectrum of flavin adenine dinucleotite (FAD) appeared to fit the action

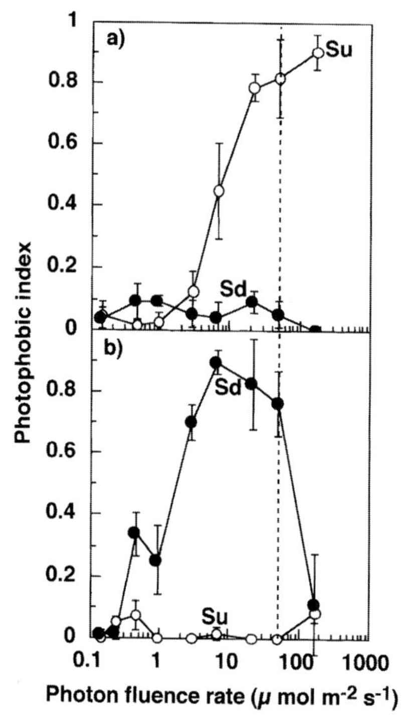

Figure 22.1 Fluence rate-response curves for step-up and step-down photophobic responses at 440 nm (Matsunaga et al. 1999). Step-up (Su, open circle) and step-down (Sd, closed circle) photophobic responses were measured for (a) the cells exponentially grown in Koren-Hutner's medium for 3 days, or, (b) the cells transferred on the 5th day to Diehn's resting medium with addition of 1 mM NH_4Cl and kept for 2 d until measurement of the responses. The technique for measuring these responses is described elsewhere (Matsunaga et al., 1998).

spectrum for the step-up response (Figure 22.2a), whereas the shape of the step-down action spectrum, which has a UV-A peak (at 370 nm) that is higher than the blue peak (at 450 nm), appeared to be mimicked by the absorption spectrum of a mixed solution of 6-biopterin and FAD (Figure 22.2b). These observations could also account for the fact that the UV-B/C peak wavelength at 270 nm of the action spectrum for the step-down response is shorter by 10 nm than the homologous peak in the action spectrum for the step-up response at 280 nm.

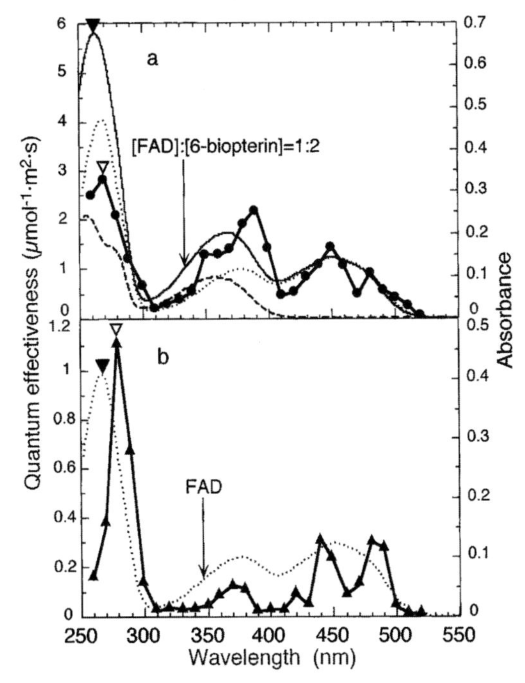

Figure 22.2 Action spectra for step-down and step-up photophobic responses in *Euglena* and absorption spectra of aqueous solutions of FAD, 6-biopterin, and their mixture (Matsunaga et al. 1998). (a) Action spectrum for the step-up photophobic response (closed circle), (b) action spectrum for the step-down photophobic response (closed triangle). Absorption spectra of aqueous solutions (in 50 mM potassium phosphate buffer, pH 8.0) of 1×10^{-5} M FAD [dotted line in (a) and (b)], of (2×10^{-5} M) 6-biopterin [broken line in (a)], mixed aqueous solution of FAD and 6-biopterin with molar ratio of 1:2 [solid line in (a)] are shown.

22.3
PAC Discovery and its Identification as the Blue-light Receptor for Photoavoidance

In searching for protocols to isolate the PFB, we compared two methods: (1) detachment of the flagellum together with PFB, by low temperature, high Ca^{2+}, or UVA treatment (e. g. Brodhun and Häder, 1990, Schmidt et al., 1990; Lebert, 2001) and (2) non-selective cell disruption by French press, Parr press, or sonication as suggested by Akio Murakami (personal communication). The cell disruption method worked well in liberating the PFB from the *Euglena* cells with a very short fragment of the flagellum attached, whereas the former seemed to retain the PFB within the cell body as indicated from examination by fluorescence microscopy (Matsunaga, pers. comm.). This simple but effective technique provided us with a sound basis for further steps of analysis.

The isolated PFBs (Figure 22.3a) were lysed by sonication in an appropriate buffer and a flavoprotein of about 400 kDa was purified chromatographically as follows: The crude extract (CE) was subjected to anion-exchange chromatography (AE) followed by gel filtration (GF). The fluorescence of the 400-kDa flavoprotein was faint but showed significant peaks at 370 nm and 450 nm in the excitation spectrum, consistent with the action spectra for photophobic responses of *Euglena* (Matsunaga et al. 1998). The fluorescence became greater after boiling, indicating that the protein noncovalently binds flavins (Figure 22.3b). The fluorescence intensity showed an obvious pH dependency that is characteristic of flavin adenine dinucleotide (FAD), suggesting that the chromophore of the protein is most probably FAD (Figure 22.3c).

From the results of SDS-polyacrylamide gel electrophoresis, the 400-kDa flavoprotein appeared to be a heterotetramer composed of two 105-kDa- and two 90-kDa subunits (Figure 22.3d). We determined N-terminal amino acid sequences of the subunits and amplified cDNAs encoding the subunits by PCR. Nucleotide sequences of the cDNAs were determined by direct sequencing of the PCR products and extended by 5'- and 3' rapid amplification of cDNA ends (RACE) methods. The deduced amino acid sequences of the 105-kDa and 90-kDa polypeptides indicated that they are composed of 1019 and 859 amino acids, respectively, and are homologous over the region where they overlap: the sequence identity over the 800 amino acids of the N-terminal portion is 75% (Figure 22.4a, b). In both sequences we found four characteristic regions: a pair of homologous regions (F1 and F2) that are separated by one (C1) of the other homologous pairs (C1 and C2) (Figure 22.4a). F1 and F2 show similarity to the flavin-binding domain of AppA (**a**ctivation of **p**hotopigment and ***puc*** expression) in *Rhodobacter sphaeroides* (Gomelsky and Kaplan, 1998; Masuda and Bauer, 2002) and several hypothetical proteins found in genome sequences of other prokaryotes (Gomelsky and Klug, 2002) (see Chapter 21, Masuda and Bauer). The apparent molar ratio of FAD to the 400-kDa protein of *Euglena*, estimated from flavin fluorescence and Bradford assay of the protein, varied from 2.5:1 (FAD:protein) to 7:1 in several experiments. Taking into account the sequence similarity and the limited accuracy of this estimation, we tentatively conclude that the 105-kDa and 90-kDa subunits each bind two FAD molecules per polypeptide, at F1 and F2.

The other characteristic regions, C1 and C2, quite unexpectedly showed similarity with class III adenylyl cyclase (AC) catalytic domains, especially with bacterial ones. Although the sequence identity over the full length of the AC catalytic domains is not so high, four consensus amino acid sequences in class III ACs (Danchin, 1993) are well conserved in C1 and C2 (Figure 22.4c). To examine whether the 400-kDa protein is actually an adenylyl cyclase, we determined the AC activity after each of its consecutive purification steps. Whereas the AC activity in each sample was low but significant in darkness, it was elevated up to 80-fold under blue light (Figure 22.5). The blue-

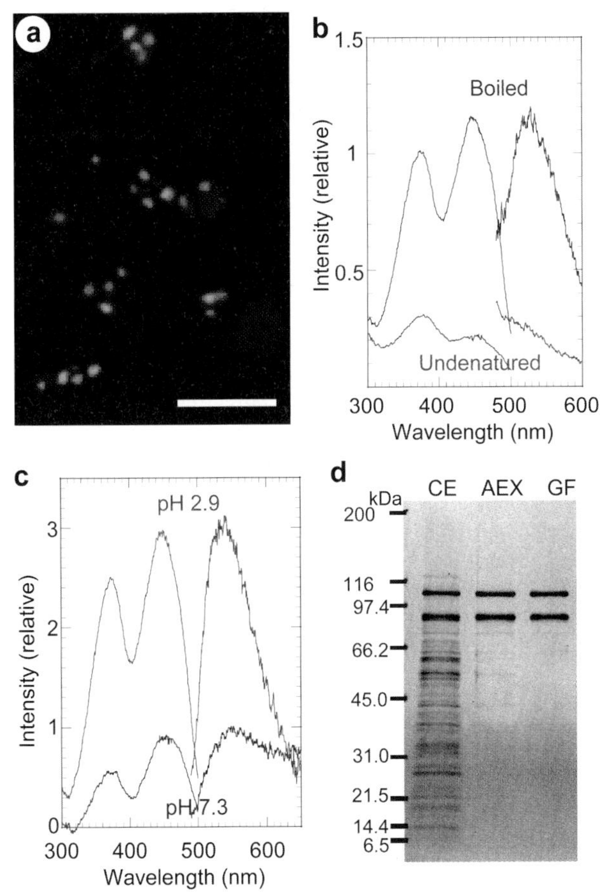

Figure 22.3 Purification of the flavoprotein from isolated paraflagellar bodies (Iseki et al., 2002) (a) Fluorescence micrograph of isolated PFBs (green fluorescent particles) observed under excitation by blue-violet light. Scale bar, 10 μm. (b) Fluorescence excitation (left, 530 nm emission) and emission (right, 370 nm excitation) spectra of the boiled and undenatured 400-kDa fractions (25 μg ml^{-1} protein). (c) Fluorescence excitation (left, 530 nm emission) and emission (right, 370 nm excitation) spectra of the boiled 400-kDa fraction at different pHs (pH 2.9, adjusted with citrate buffer; pH 7.3, the same pH as the gel-filtration buffer at room temperature.) (d) SDS-PAGE of PFB proteins at consecutive purification steps: CE, crude extract of PFB; AEX, the 530-nm fluorescence peak fraction of anion exchange chromatography; GF, the 400-kDa fraction from gel filtration. A 5–20% acrylamide gradient gel, stained by silver is shown in (d).

light-activated AC activity of the 400-kDa flavoprotein was saturated at 50 μM ATP, with a Vmax of 3500 pmol min^{-1} mg^{-1}, and an estimated Km value of 0.5 μM for the substrate ATP, values comparable to those of cyanobacterial AC (Kasahara et al., 1997). From these results, we concluded that the flavoprotein isolated from PFBs is an FAD-binding adenylyl cyclase with its activity regulated by blue light. We thus des-

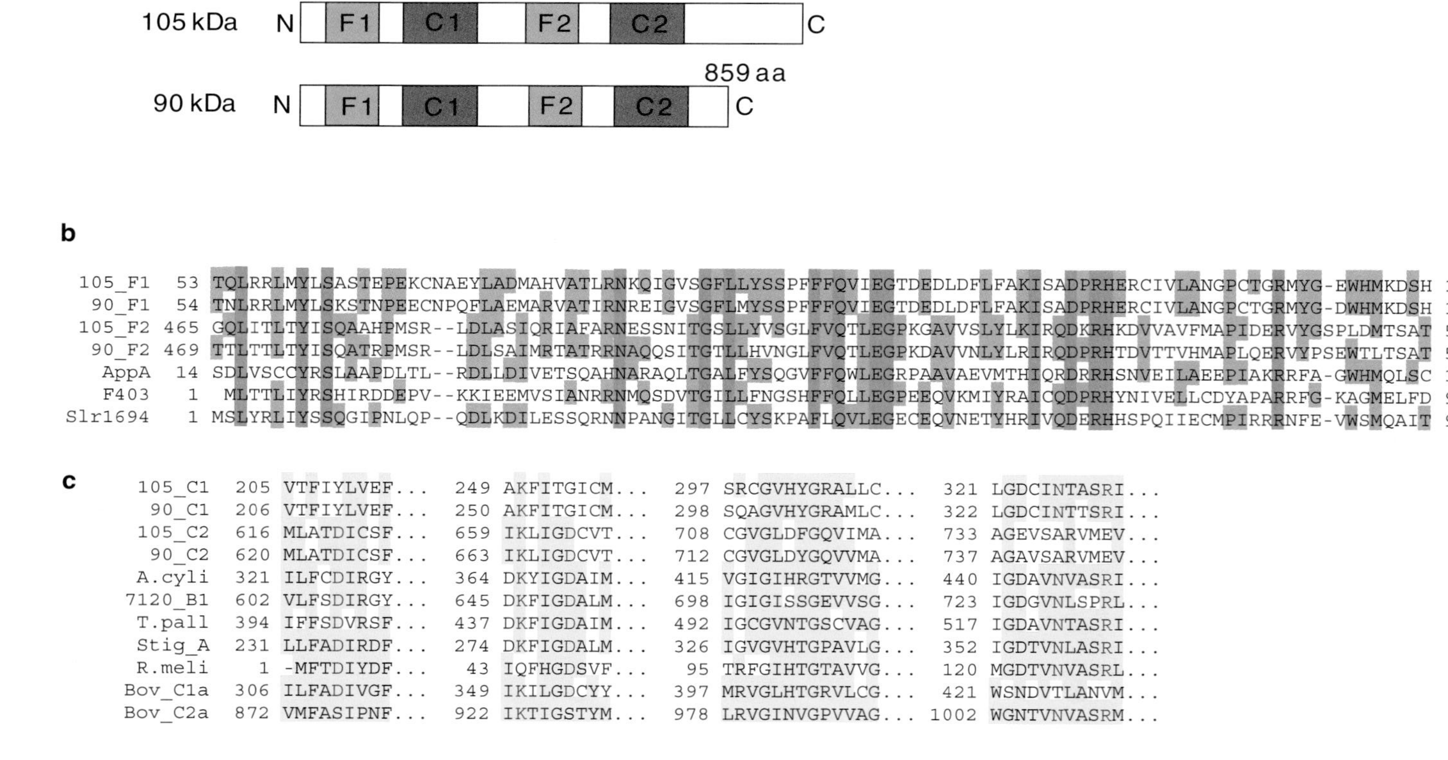
a
1,019 aa
105 kDa N F1 C1 F2 C2 C
859 aa
90 kDa N F1 C1 F2 C2 C
b
105_F1 53 TQLRRLMYLSASTEPEKCNAEYLADMAHVATLRNKQIGVSGFLLYSSPFFFQVIEGTDEDLDFLFAKISADPRHERCIVLANGPCTGRMYG-EWHMKDSH 151
90_F1 54 TNLRRLMYLSKSTNPEECNPQFLAEMARVATIRNREIGVSGFLMYSSPFFFQVIEGTDEDLDFLFAKISADPRHERCIVLANGPCTGRMYG-DWHMKDSH 152
105_F2 465 GQLITLTYISQAAHPMSR--LDLASIQRIAFARNESSNITGSLLYVSGLFVQTLEGPKGAVVSLYLKIRQDKRHKDVVAVFMAPIDERVYGSPLDMTSAT 562
90_F2 469 TTLTTLTYISQATRPMSR--LDLSAIMRTATRRNAQQSITGTLLHVNGLFVQTLEGPKDAVVNLYLRIRQDPRHTDVTTVHMAPLQERVYPSEWTLTSAT 566
AppA 14 SDLVSCCYRSLAAPDLTL--RDLLDIVETSQAHNARAQLTGALFYSQGVFFQWLEGRPAAVAEVMTHIQRDRRHSNVEILAEEPIAKRRFA-GWHMQLSC 110
F403 1 MLTTLIYRSHIRDDEPV--KKIEEMVSIANRRNMQSDVTGILLFNGSHFFQLLEGPEEQVKMIYRAICQDPRHYNIVELLCDYAPARRFG-KAGMELFD 96
Slr1694 1 MSLYRLIYSSQGIPNLQP--QDLKDILESSQRNNPANGITGLLCYSKPAFLQVLEGECEQVNETYHRIVQDERHHSPQIIECMPIRRRNFE-VWSMQAIT 97
c
105_C1 205 VTFIYLVEF... 249 AKFITGICM... 297 SRCGVHYGRALLC... 321 LGDCINTASRI...
90_C1 206 VTFIYLVEF... 250 AKFITGICM... 298 SQAGVHYGRAMLC... 322 LGDCINTTSRI...
105_C2 616 MLATDICSF... 659 IKLIGDCVT... 708 CGVGLDFGQVIMA... 733 AGEVSARVMEV...
90_C2 620 MLATDICSF... 663 IKLIGDCVT... 712 CGVGLDYGQVVMA... 737 AGAVSARVMEV...
A.cyli 321 ILFCDIRGY... 364 DKYIGDAIM... 415 VGIGIHRGTVVMG... 440 IGDAVNVASRI...
7120_B1 602 VLFSDIRGY... 645 DKFIGDALM... 698 IGIGISSGEVVSG... 723 IGDGVNLSPRL...
T.pall 394 IFFSDVRSF... 437 DKFIGDAIM... 492 IGCGVNTGSCVAG... 517 IGDAVNTASRI...
Stig_A 231 LLFADIRDF... 274 DKFIGDALM... 326 IGVGVHTGPAVLG... 352 IGDTVNLASRI...
R.meli 1 -MFTDIYDF... 43 IQFHGDSVF... 95 TRFGIHTGTAVVG... 120 MGDTVNVASRL...
Bov_C1a 306 ILFADIVGF... 349 IKILGDCYY... 397 MRVGLHTGRVLCG... 421 WSNDVTLANVM...
Bov_C2a 872 VMFASIPNF... 922 IKTIGSTYM... 978 LRVGINVGPVVAG... 1002 WGNTVNVASRM...

◀ **Figure 22.4** Sequences of the M_r 105K and 90K polypeptides (Iseki et al., 2002). (a) Structural features of the two polypeptides. Four characteristic regions are shown as blue (F1, F2) and red (C1, C2) boxes. The DDBJ/EMBL/GenBank accession numbers for the M_r 105K and 90K polypeptides are AB031225 and AB031226, respectively. (b) Sequence alignment of F1 and F2 with the AppA flavin (FAD) binding domain and its homologues. Regions of similarity (blue) and identity (green) are highlighted. Alignment and similarity identification were performed by the GCG software. AppA, a photoreceptor in *Rhodobacter sphaeroides* that serves as a redox regulator (L42555); F403, *E. coli* hypothetical protein (AE000215); Slr1694, *Synechocystis* sp. PCC6803 hypothetical protein (D90913). (c) Partial sequence alignment of C1 and C2 with class III adenylate cyclase catalytic domains. Amino acids consistent with the four consensus sequences of class III adenylyl cyclases (yellow) and those identical with important amino acids for catalysis in mammalian adenylyl cyclase (red) are highlighted. A. cyli, *Anabaena cylindrica* (D55650); 7120_B1, *Anabaena* sp. PCC7120 CyaB1 (D89623); T. pall, *Treponema pallidum* (AE001224); Stig_A, *Stigmatella aurantiaca* CyaA (AJ223796); R. meli, *Rhizobium meliloti* (M35096); Bov_C1a and Bov_C2a, bovine Type I (M25579).

ignated the flavoprotein as photoactivated adenylyl cyclase (PAC) and its subunits, 105-kDa and 90-kDa polypeptide, as PACα and PACβ, respectively (Iseki et al., 2002).

To examine whether PAC actually mediates the photophobic responses of *Euglena*, we suppressed expression of PACα and PACβ by RNA interference (RNAi) (Iseki et al., 2002): double stranded RNAs (dsRNAs) of PAC subunits were synthesized and electroporated into *Euglena* cells. As a consequence, endogenous PAC mRNAs were undetectable and PFBs were absent from the transformed cells (Figure 22.6a). Observation by a computerized video motion analyzer (Matsunaga et al. 1998) showed that the step-up photophobic response disappeared in the dsRNA-transformed cells even at a high enough intensity of light sufficient to cause saturation of the response in the control cells. However, the step-down response of the dsRNA-transformed cells was essentially the same as that of the control cells (Figure 22.6b). Thus the introduction of the dsRNAs impaired only the photosensing system for the step-up photophobic response without affecting the overall machinery that is needed for the cells

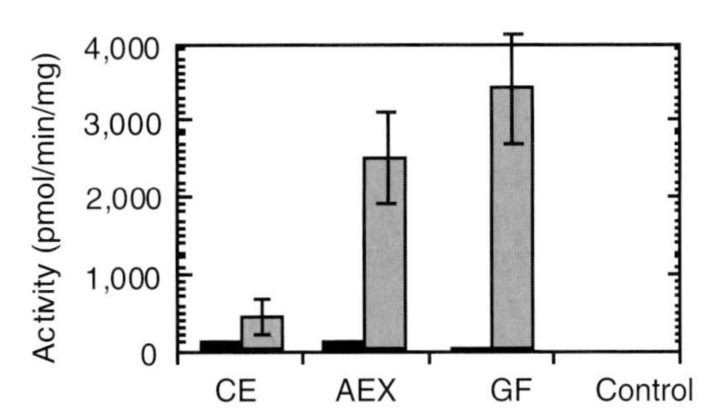

Figure 22.5 Adenylyl cyclase activity of the flavoprotein purified from PFBs (Iseki et al., 2002). Adenylyl cyclase activity of proteins at consecutive purification steps: CE, crude extract of PFB; AEX, the 530-nm fluorescence peak fraction of anion exchange chromatography; GF, the M_r 400 Kd fraction of gel filtration; Control, gel-filtration buffer as a control. Means and standard error of the mean from three measurements in darkness (black) or under blue light (gray) at 10 μmol m^{-2} s^{-1} are shown.

to exhibit phobic responses. From these results, we conclude that PAC is the major constituent of the PFB and that it acts as the photoreceptor for the step-up photophobic response in *Euglena*, whereas it is not involved in the step-down photophobic response.

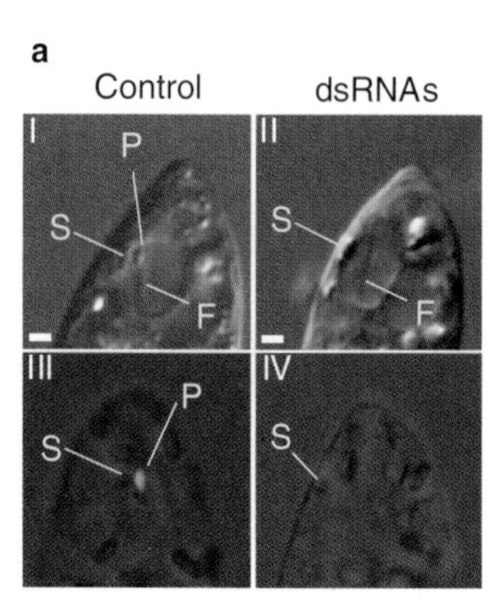

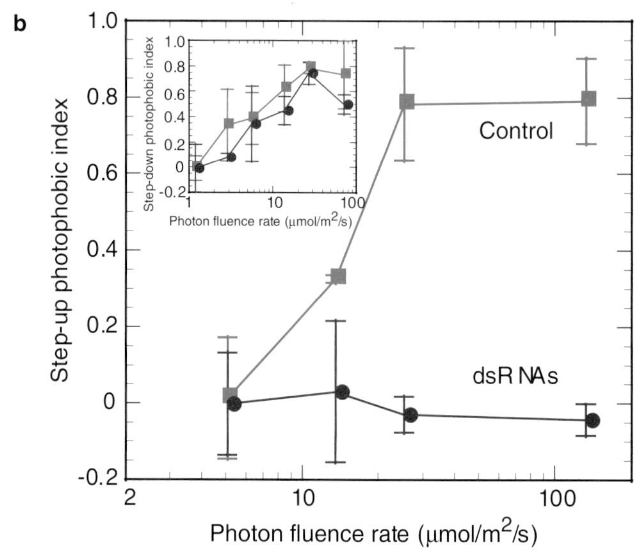

Figure 22.6 Suppression of gene expression of PAC by RNAi. (a) Photomicrographs of anterior portion of *Euglena* cells electroporated with buffer only (I, III) or dsRNAs of PACα and PACβ (II, IV). I, II: Nomarski optics; III, IV: Fluorescence images, under excitation by blue-violet light, superimposed with bright field images of the same cell. S, stigma; F, flagellum; P, PFB. Scale bar, 1 μm. (b) Fluence rate-response curves for the step-up photophobic response of the dsRNAs-introduced cells (circles) and control cells (squares) at 450 nm. Fluence rate-response curves for the step-down photophobic response are also shown (inset). Means and standard deviation from three measurements are shown.

Considering the fact that Ca^{2+} affects flagellar movements of *Euglena* (Doughty and Diehn, 1979), the process of step-up photophobic response might be explained simply as follows (Figure 22.7): blue light activates PAC to induce a local increase in the

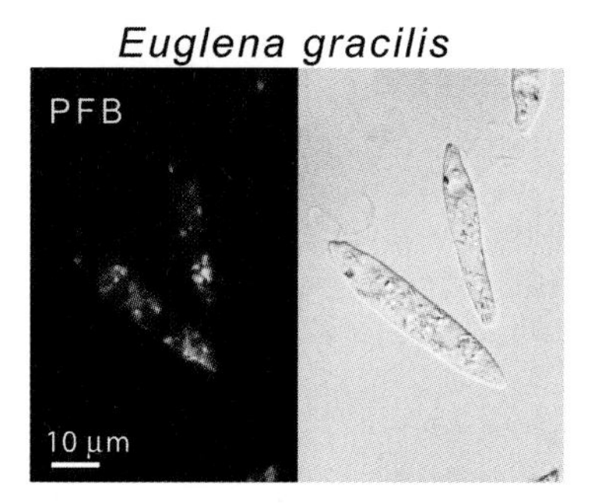

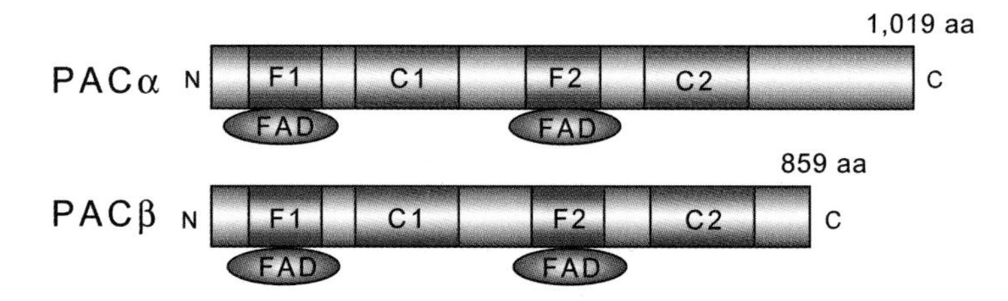

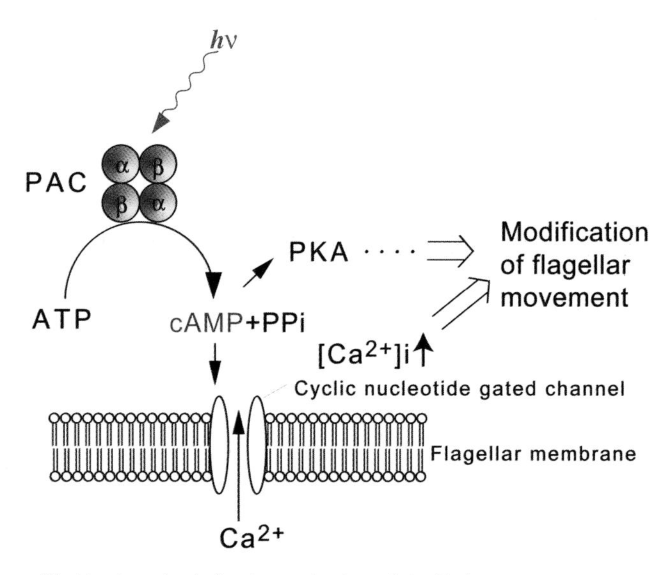

Figure 22.7 Working hypothesis for the mechanism of the *Euglena* step-up photophobic response mediated by PAC in PFB.

cAMP concentration around the PFB. The elevated cAMP level induces Ca^{2+} entry into the flagellar apparatus and the Ca^{2+} increase somehow modifies the direction of flagellar beating. Another possibility is that the elevated cAMP level causes phosphorylation of flagellar proteins by activating a cAMP-dependent protein kinase (PKA), which affects the flagellar motility as reported in other unicellular organisms (Hasegawa et al., 1987, Noguchi et al., 2000). Further work is needed to elucidate the exact mechanism.

22.4
PAC Involvement in Phototaxis

The PAC discovery stimulated a close collaboration between our laboratory and that of Donat Häder. This collaboration yielded discovery of the first PAC homologues (AlPAC alpha and beta) from *Astasia*, a close colorless relative of *Euglena*. It also showed by RNAi that PAC mediates the *Euglena* phototaxis response, an oriented, steering, behavioral response toward or away from the light source (Figure 22.8; Ntefidou et al., 2003).

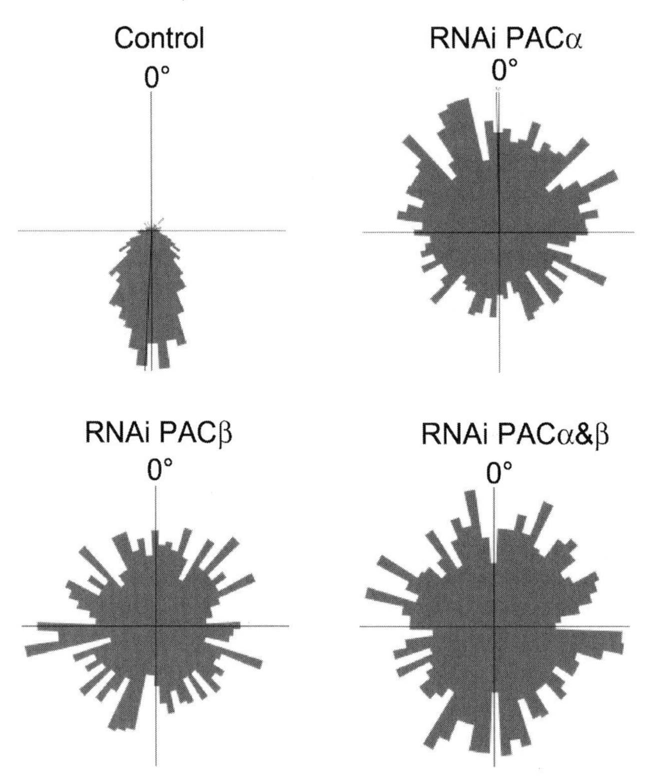

Figure 22.8 Tracks of *Euglena* cells at high irradiances summarized in circular histograms (Ntefidou et al. 2003). Control cells swim with high precision away from the light source (800 W m^{-2}) whereas RNAi-treated cells show random swimming.

22.5
PAC Origin

To gain an insight into the evolution of this unique photoreceptor protein, we searched for similar sequences in several euglenoids by RT-PCR using degenerate primers (Koumura et al., 2004). Two similar transcripts were thus detected in each of the four phototrophic euglenoids, *Euglena stellata, Colacium sideropus, Eutreptia viridis, Eutreptiella gymnastica,* and in an osmotrophic species (i. e., obtaining nutrients by absorption), *Khawkinea quartana,* but not in a phagotrophic species (i. e. obtaining nutrients by phagocytosis), *Petalomonas cantuscygni*. Each of the PAC-like sequences detected appeared orthologous to PACα and PACβ, respectively, and had the same domain structure as PAC subunits each of which is composed of two flavin binding domains, F1 and F2, each followed by an adenylyl cyclase catalytic domain, C1 and C2, respectively. These observations imply that the the encoded proteins constitute a functional photoactivated adenylyl cyclase similar to the *Euglena* PAC. Phylogenetic analysis of the adenylyl cyclase catalytic domains revealed that they belong to a bacterial cluster, not to a trypanosomal one (Figure 22.9). In addition, two trypanosome-type adenylyl cyclases were discovered in *E. gracilis*. In contrast to PAC, deduced amino acid sequences of the trypanosome-type adenylyl cyclases indicated that they are integral membrane proteins with a membrane-spanning region at their midpoint, followed by an adenylyl cyclase catalytic domain which seems cytoplasmic. Thus, we propose that PAC might have been transferred to euglenoids on the occasion of secondary endosymbiosis (Figure 22.10).

22.6
Future Prospects

The molecular identity of the putative photoreceptor for the step-down photophobic response still presents a challenging open question. This photoreceptor may be a sensor for a light-off signal whereas PAC is a sensor for a light-on signal. The elucidation of the mechanisms allowing both sensors to bring about the abrupt changes in flagellar motion resulting in the turn in the direction of cell movement will also require much interdisciplinary research, and will also provide a sound basis for biotechnological applications of these complimentary light sensors (see below).

PAC is a unique protein that can act both as a photoreceptor and an effector to catalyze cAMP synthesis in contrast with G-protein-coupled receptor systems in which three different proteins sequentially act to modulate the cyclic nucleotide level (Torre et al., 1995). This simple mechanism indicates that PAC may be a promising tool to photo-manipulate the intracellular cAMP level in various heterologous cell systems for cell biological and biotechnological purposes. In nerve and other cells, this would enable us to pinpoint control of cellular processes regulated by cAMP such as axon guidance, synaptic long term potentiation, synaptic long term depression, and cell differentiation. It would be even more fascinating if we succeeded in mutating a PAC into a PGC (photoactivated guanylyl cyclase, which produces cGMP), and replacing

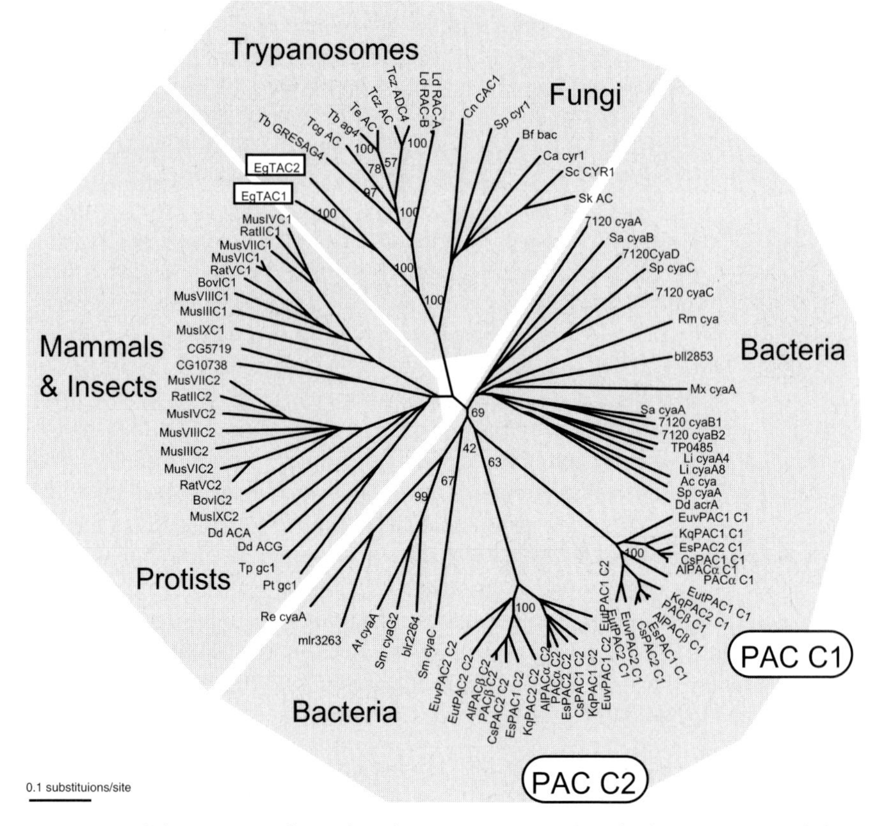

Figure 22.9 Phylogenetic tree for catalytic domains of class III adenylyl cyclases (Koumura et al., 2004). The tree was generated by the Neighbor-Joining method using Clustal X. Bootstrap values for the trypanosome-clade and the bacterial/PAC-clade are shown as percentage of 1000 replications.

the flavin chromophore with analogs that have different absorption peaks. This would enable us to use these combinations as wavelength-sensitive switches of biological activities (Lewis, 2002).

The above idea is actually encouraged by a recent success in functional heterologous expression of PAC subunits in the *Xenopus* oocyte, demonstrating that either of them alone is sufficient for photoactivated adenylyl cyclase activity (Schröder-Lang et al., 2004).

It is extremely interesting and surprising that recently an archaerhodopsin-type protein, Cop5, containing either an adenylyl cyclase or a guanylyl cyclase domain (AC/GC) was found in the *Chlamydomonas* genome (Kateriya et al., 2004). This discovery suggests further possibilities for finding examples of adenylyl cyclase- or guanylyl cyclase-containing photoreceptor proteins, whether membrane-bound (like Cop5), cytosolic, or paracrystaline (like PAC in PFB).

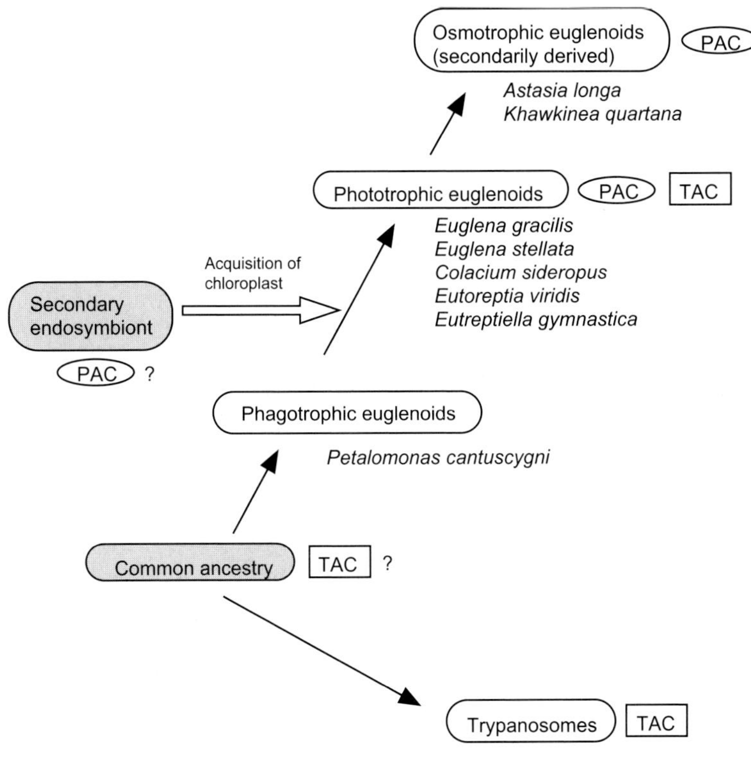

Figure 22.10 Working hypothesis on the origin of PAC (Koumura et al., 2004). Presence of orthologues of PAC subunits ('PAC' in an oval) and trypanosome-type adenylyl cyclases ('TAC' in a rectangle) are indicated on the well-accepted evolutionary history of euglenoids.

Acknowledgements

We acknowledge grant supports from Ministry of Education, Science, Cultue, and Sports of Japan and Japan Space Forum. M. I. Acknowledges postdoctoral fellowships from National Institute for Basic Biology (NIBB) and Bio-oriented Technology Research Advancement Institution during part of the studies summarized here. Part of the studies was done under the NIBB Cooperative Research Program to use the Okazaki Large Spectrograph.

References

Brodhun, B., and Häder, D.-P. (1990) *Photochem. Photobiol.* 52, 865–871.

Checcucci, A., Colombetti, G., Ferrara, R., and Lenci, F. (1976) *Photochem. Photobiol.* 23, 51–54.

Danchin, A. (1993) *Adv. Second Message Phosphoprotein Res.* 27, 109–162.

Diehn, B. (1969) *Exp. Cell Res.* 56, 375–381.

Doughty, M.J., and Diehn, B. (1979) *Biochem. Biophys. Acta* 588, 148–168.

Gomelsky, M., and Kaplan, S. (1998) *J. Biol. Chem.* 52, 35319–35325.

Gomelsky, M., and Klug, G. (2002) *Trends Biochem. Sci.* 27, 497–500.

Gressel, J. (1979) *Photochem. Photobiol.* 30: 749–754.

Hasegawa, E., Hayashi, H., Asakura, S., and Kamiya, R. (1987) *Cell Motil. Cytoskeleton* 8, 302–311.

Iseki, M., Matsunaga, S., Murakami, A., Ohno, K., Shiga, K., Yoshida, K., Sugai, M., Takahashi, T., Hori, T., and Watanabe, M. (2002) *Nature* 415, 1047–1051.

Kasahara, M., Yashiro, K., Sakamoto, T., and Ohmori, M. (1997) *Plant Cell Physiol.* 38, 828–836.

Kateriya, S., Nagel, G., Bamberg, E., and Hegemann, E. (2004) *News Physiol. Sci.* 19, 133–137.

Kondo, T., Kubota, M., Aono, Y., and Watanabe, M. (1988) *Protoplasma* Suppl. 1, 185–192.

Koumura, Y., Suzuki, T., Yoshikawa, S., Watanabe, M., and Iseki, M. (2004) *Photochem. Photobiol. Sci.* 3, 580–586.

Lebert, M. (2001) in Comprehensive Series in Photosciences 1: Photomovement (eds Häder, D.-P. and Lebert, M.) 297–341 (Elsevier, Amsterdam)

Lewis, R. A. (2002) *Biophotonics International* May 2002, 40–41.

Masuda, S., and Bauer, C.E. (2002) *Cell* 110, 613–623.

Matsunaga, S., Hori, T., Takahashi, T., Kubota, M., Watanabe, M., Okamoto, K., Masuda, K., and Sugai, M. (1998) *Protoplasma* 201, 45–52.

Matsunaga, S., Takahashi, T., Watanabe, M., Sugai, M., Hori, T. (1999) *Plant Cell Physiol.* 40, 213–221.

Noguchi, M., Ogawa, T., and Taneyama, T. (2000) *Cell Motil. Cytoskeleton* 45, 263–271.

Ntefidou, M., Iseki, M., Watanabe, M., Lebert, M., and Häder, D.-P. (2003) *Plant Physiol.* 133, 1517–1521.

Schmidt, W., Galland, P., Senger, H., and Furuya, M. (1990) *Planta* 182, 375–381.

Schröder-Lang, S., Ollig, D., Schiereis, T., Hegemann, P., and Nagel, G. (2004) *Abst. 11th Internatl. Conf. Retinal Proteins, Frauenchiemsee, Germany, June 20–24,* P60.

Torre, V., Ashmore, J.F., Lamb, T.D., and Menini, A. (1995) *J. Neurosci.* 15, 7757–7768.

Watanabe, M. (2004) in *CRC Handbook of Organic Photochemistry and Photobiology, 2nd ed.* (eds. Horspool, W., and Lenci, F.) 115-1~115–16 (CRC Press, Boca Raton)

Index

Handbook of Photosensory Receptors. Edited by W. R. Briggs, J. L. Spudich

ISBN 3-527-31019-3

t

u